T0261240

GALEN
METHOD OF MEDICINE
II

LCL 517

GALEN

METHOD OF MEDICINE

BOOKS 5–9

EDITED AND TRANSLATED BY

IAN JOHNSTON

AND

G. H. R. HORSLEY

HARVARD UNIVERSITY PRESS

CAMBRIDGE, MASSACHUSETTS

LONDON, ENGLAND

2011

First published 2011

LOEB CLASSICAL LIBRARY® is a registered trademark
of the President and Fellows of Harvard College

Library of Congress Control Number 2011921281
CIP data available from the Library of Congress

ISBN 978-0-674-99679-3

*Composed in ZephGreek and ZephText by
Technologies 'N Typography, Merrimac, Massachusetts.
Printed on acid-free paper and bound by
The Maple-Vail Book Manufacturing Group*

CONTENTS

MANUSCRIPTS   vii

ABBREVIATIONS   ix

SYNOPSIS OF CHAPTERS   xi

METHOD OF MEDICINE

    BOOK V   2

    BOOK VI   118

    BOOK VII   236

    BOOK VIII   344

    BOOK IX   450

MANUSCRIPTS

Kühn (= K) vol. X has provided the base text for ours printed here.

The Latin text printed in K (abbreviated by us as KLat in this volume) was produced three hundred years before K was printed, and so is not a translation, or a correction of K's Greek text. It clearly draws on other MSS than those used by K, reflecting different readings in the Greek. This may sometimes alert us to a problem in K; but generally we have not privileged the Latin translation over K's Greek text.

For a list of MSS of the *MM*, see Diels, pp. 91–92. The following manuscripts are referred to in our textual notes with the abbreviation listed.

B—British Library MS Add. 6898 (London; 12th cent.)[1]

P1—Parisinus Gr. 2160 (Paris; 14th cent.)

P2—Parisinus Gr. 2171 (Paris; 15th cent.)[2]

---

1 We are not attempting to provide here a full collation of B against K. B exhibits many other differences, e.g., in word order and in orthography; and it is interesting for other reasons as well. But we have very rarely noted these.

2 We have not directly consulted the two Paris MSS. We have made use of some of its readings for Books 1 and 2 as they are reported by R. J. Hankinson (1991), App. 1, pp. 235–37.

Boulogne (2009), pp. 11, 31, draws upon these three MSS, and also upon three others as well as two fragmentary MSS for his translation. We have included no references to these other five in our textual notes.

ABBREVIATIONS

Ce   Celsus. *De Medicina*. Translated by W. G. Spencer. Loeb Classical Library. 3 vols. Cambridge, MA: Harvard University Press, 1935–1938.

CMG   Corpus Medicorum Graecorum

Cu   Nicholas Culpepper. *The English Physician Enlarged* (*Culpepper's Herbal*). London: Folio Society, 2007 [1653].

D   Dioscorides. *The Greek Herbal of Dioscorides*, translated by John Goodyer [1653]. Edited by R. T. Gunther. New York: Hafner, 1968 [1933].

EANS   *The Encyclopedia of Ancient Natural Scientists*. Edited by P. T. Keyser and G. L. Irby-Massie. London: Routledge, 2008.

G   Galen. References to the *MM* are indicated by the Kühn page number; references to other Galenic works are indicated by the Kühn volume and page numbers. His three major pharmacological treatises are *De simplicium medicamentorum temperamentis et facultatibus*, XI.379–892K and XII.1–377K; *De compositione medicamentorum secundum locos*, XII.378–1007K and XIII.1–361K; *De compositione medicamentorum per genera*, XIII.362–1058K.

Gr M. D. Grmek. *Diseases in the Ancient Greek World*. Baltimore, MD: Johns Hopkins University Press, 1991.

L&S C. T. Lewis and C. Short. *A Latin Dictionary*. Oxford: Clarendon, 1993 [1879].

LCL Loeb Classical Library.

LSJ H. G. Liddell, R. Scott, and H. Stuart Jones. *A Greek-English Lexicon*. 9th ed. (1940), with revised suppl. by P. G. W. Glare. Oxford: Clarendon, 1996.

M C. C. Mettler. *The History of Medicine*. Philadelphia: Blakiston, 1947.

OCD *Oxford Classical Dictionary*. Edited by S. Hornblower and A. Spawforth. 3rd ed. Oxford: Clarendon, 1996.

OED *Oxford English Dictionary*. 12 vols. Oxford: Oxford University Press, 1978 [1933].

S *Stedman's Medical Dictionary*. 27th ed. Baltimore: Lippincott, Williams and Wilkins, 2000.

Si R. E. Siegel. *Galen on the Affected Parts*. Basel: S. Karger, 1976.

T Theophrastus. *Enquiry into Plants*. Translated by A. Hort. Loeb Classical Library. 2 vols. Cambridge, MA: Harvard University Press, 1916, 1926.

SYNOPSIS OF CHAPTERS

BOOK V

1. Galen details the principles of treating a wound, ulcer, or sore—that is, a dissolution of continuity—identifying the key components as conglutination, enfleshing, reduction of excess flesh, and cicatrization. Which is to be used depends on the exact circumstances and the part of the body involved. It is also important to deal with any general *dyskrasia* if it exists and any causative factors still operative. Drying agents are the most important medications.

2. Galen describes dissolution of continuity as affecting both *homoiomeres* and organic parts. Treatment depends, to some extent, on which of the two is involved—something lost on the followers of Erasistratus, who recognize organic parts only. He then deals with dissolution of continuity in blood vessels—both arteries and veins—beginning by identifying the various causative factors.

3. In treatment the immediate necessity is to stop the bleeding from the injured vessel. There are two options: close the opening and redirect the flow. The former is achieved by compression (manual, various tamponading devices, and bandaging), ligatures, eschar formation, positioning the wounded part, hemostatic (blood-stanching) medications, and cutting through the vessel completely.

The latter may occur naturally or may be helped by appropriate application of cupping glasses.

4. The next step is to enflesh the whole wound. No space should remain around the injured vessel which might allow the development of a (false) aneurysm. Various useful medications are detailed. Eschar formation and the risk of premature separation of the eschar are considered.

5. Deep hemorrhages are discussed. The problem of not being able to use ligatures, binding, or cauterization is examined. Treatment depends on revulsion, diversion, emplastic foods and drinks, and cooling and astringent medications. Instruments may be needed to deliver medications to deep sites. Various useful medications are listed.

6. Hemorrhages due to erosion are described. Medications are identified as the first line of treatment. Cooling, diversion and revulsion are also seen to have roles. The dangers of using cooling uncritically are considered.

7. The particular difficulties of treating wounds in arteries are examined and a case report is given. Galen claims arteries can heal primarily, particularly in women and children, and with small wounds. Medications to encourage healing and the methods of their delivery are listed.

8. Wounds in the lung with hemoptysis are considered. The conceptual difficulty associated with blood from the lung passing into the "rough arteries" (bronchial tree, trachea) is pondered. The treatment of a thoracic empyema and its bearing on this issue is described. Direct measures are not possible with bleeding in the lung. Other forms of treatment are offered. Treatment of wounds of the chest wall is outlined.

9. Wounds of the diaphragm are described. Conglutination is possible in the "fleshy" part but not in the "sinewy" part. The risks of inflammation and the particular medications that are useful are discussed.

10. Galen outlines the general principles of treating wounds, focusing on the importance of the part being treated in determining the type of treatment. The ways of the Methodics, Empirics, and Dogmatics are contrasted by means of a general discussion, a case report, and a reported dialogue between Galen and a Methodic—all, not surprisingly, unfavorable to the latter.

11. Galen examines how the means of delivery and the type, strength, and physical nature of the medications used in treatment depend on the part in which the wound exists.

12. Wounds and ulcers of the larynx, trachea, and bronchial tree are described. A case report of a plague victim is given. Galen emphasizes the particular value of milk, especially that used by the Tabiae. There is a digression on what features make milk good. Galen gives further consideration to the plague.

13. The treatment of hemorrhages with catarrh is described, illustrated by two case reports.

14. Galen focuses on what makes wounds of the lung difficult, or seemingly impossible, to cure, and how to deal with these problems. The importance of *kakochymia* and the difficulies involved in its correction are considered.

15. Management of hemoptysis is outlined. There are three key measures: phlebotomy, purging, and medications to strengthen the head. The inadequacies of the Erasistratean approach to treatment are highlighted. The possibility of treating wounds before inflammation sets in

is considered. Galen embarks on further criticism of Thessalus, this time for neglecting the importance of the affected part in determining treatment.

BOOK VI

1. Galen continues his account of dissolution of continuity or union, which can occur in all structures. The name for the class is self-explanatory. In soft tissues, it may be called a wound, ulcer, or sore. In other structures it may be given a specific name; for example, a fracture in bones. For treatment, there are general methods and there are methods specific to particular parts, a fact which apparently escaped the repeatedly pilloried Methodics.

2. Treatment of a simple puncture wound is discussed and, in particular, the risk of inflammation—how to recognize it in the individual patient and how to avoid it. Again the Methodics, and especially Thessalus, are criticized for their disregard of detailed technical knowledge. Nerve involvement in such wounds is considered. The basic principles of treatment are leaving the wound open for the outflow of ichors, cleansing the body of superfluities, and controlling pain in the wound. Various medications are described. Galen comes to the rescue in a patient treated incorrectly by a Thessaleian doctor. Details are given of one particular medication devised by Galen for the treatment of wounds of nerves. A further case report is given to emphasize the importance of the use of method and a knowledge of the theoretical basis of medicine, in particular here the elements.

3. Continuing with nerve injuries, Galen turns to division as opposed to puncture. The importance of the degree

of overlying skin damage and the exposure of the nerve is stressed. These factors determine the nature of the medications to be used. The primary aim is to dry in the least stinging way. Various preparations are described and a case report is given. Cooling substances must be avoided, due in part to the connection with the central nervous system and with muscles. The strength (capacity) of the patient's body must be assessed and medications chosen accordingly: strong for strong and weak for weak. There is a further attack on the Methodics for their idea of one treatment for all wounds. Contusion of nerves is also considered.

4. The treatments of ligament injuries and abdominal wounds are discussed. In the first case, there is the importance of distinguishing ligaments from nerves, and therefore of anatomical knowledge. In the second case, the technique of *gastrorraphy* (abdominal wall repair) is described, as are methods for reducing prolapsed bowel and omentum. Wounds of the alimentary tract itself are considered, taking into account the particular features of the individual parts: stomach, jejunum, small (thin) and large (thick) intestines. Finally, the treatment of prolapse and necrosis of the omentum is considered, providing a further opportunity to ridicule the Methodics.

5. The treatment of bone injuries (fractures) is discussed, beginning with a classification. The basic aim is, of course, union. A distinction is made between hard (as in adult males) and soft (as in children) bones in respect to the possibility of primary union. Conglutination is an alternative to union using natural materials built up at the fracture site. The main types of fractures are considered, beginning with the transverse. The first step, where possible,

is reduction; methods are described. The next step is immobilization, which includes bandaging, splint fixation, and inactivity. Bandaging does more than immobilize the fracture—by affecting the inflow and outflow of fluids, it contributes to healing and the prevention of inflammation. The position of the immobilized limb is also important. It should be natural and pain-free as far as possible. The importance of regimen in contributing to callus formation is stressed. Finally, a device for reduction and immobilization is described—the *glottocomon*.

6. Galen considers skull fractures, beginning with a classification based on the form and depth of the fracture. Several instruments for cleaning and preparing the bone for healing are described: knives, raspatories, and trephines. There is a detailed description of the treatment of depressed and elevated fractures. The problem of not being able to bind skull fractures is considered. The issue of adjunctive medications is briefly examined.

BOOK VII

1. Galen begins with some introductory remarks to Eugenianus, to whom the second, and rather delayed, component of the *Method of Medicine* is dedicated. He then reiterates his devotion to the pursuit of truth and his disdain for reputation.

2. Galen offers further general and introductory considerations, particularly on diseases of the *homoiomerous* parts. Differences between the three main schools—Empirics, Methodics, and Dogmatics—are identified. Issues of terminology and causation pertaining to inflammation are discussed. Galen praises Hippocrates as the "founding

father" of the use of method in treatment and criticizes Herophilus and Erasistratus for their failure to examine the nature of the primary bodies.

3. The claim is made that an understanding of *dyskrasia* requires a theory of the underlying structure of the body, and of matter in general. There is reference to the theory of elements described in Galen's *On the Elements according to Hippocrates*. The fundamental aim in preserving the health of the functioning parts of the body is to maintain their *krasis* in a proper balance of the four qualities. Galen provides a general statement of the method of cure of all diseases occurring in *homoiomeres*.

4. Galen argues that from a clear understanding of the underlying general principles it is possible to move theoretically to the treatment of individual cases. Specific details of the treatment of weakness (*atonia*) of the stomach are given, including a list of relevant medications. There is a case report in which Galen again saves the day.

5. Some differences in the treatment of the different *dyskrasias* relating to duration and risk are identified.

6. Varieties of abnormal dryness (dry *dyskrasias*) are described. The dangers of astringent medications, foods, and drinks are discussed. There is a further case report of a patient with a dry *dyskrasia* involving the stomach. A detailed consideration of the use of milk, both from asses and from women, is offered, the key factor being that it must retain its heat when given. The use of honey is described. There is a digression on the optimal care of the animals providing the milk. The case report continues. Aspects of a restorative diet, including wine, are listed. Some general aspects of a restorative regimen are discussed.

7. Cooling is seen as a sequel to chronic dry *dyskrasias*.

Once cooling is involved, treatment becomes more complicated. The treatment of a combined *dyskrasia* comprising dryness and coldness is discussed. Details of the various medications and preparations are given and the use of pitch is described.

8. A combined *dyskrasia* of dryness and heat is considered. The nature of the wine and honey to be used is described. Other useful foods are also considered. Galen provides a detailed case report with a bad outcome and acknowledges his early failings. A second case report follows. Galen, now much more experienced, achieves a better outcome.

9. Treatment of a hot *dyskrasia* combined, in the first instance, with a moist component and, in the second instance, with a dry component is discussed. The association with fever is identified. The use of cooling agents and their dangers are considered. A moist *dyskrasia* is seen as the easiest of the *dyskrasias* to cure. Cooling and astringent foods and drinks are the staples of treatment and these include both water and wine.

10. Galen provides a summarizing statement on the various *dyskrasias*, single and combined, involving the stomach.

11. *Dyskrasias* associated with excess moisture from an external source are described. The main treatment is purging by vomiting. There is further general consideration of *dyskrasias* due to inflowing material. The treatment options in general for a pathological flux are to stop the flux at its source and to prevent possible target structures from receiving the flux. Purging of the whole body is important. The treatment of *kakochymia* is discussed. Medications of value are listed.

12. Mixed conditions of the stomach involving both wall and lumen are described. The methodical approach to the sequencing of treatment is outlined. The importance of preserving the patient's capacity (strength) is emphasized. The management of *dyskrasias* of the stomach is taken as the model on which to base the management of *dyskrasias* in other structures.

13. The sequence of indications in treating *dyskrasias* is given. Five kinds of indication are listed. Some general observations on the treatment of *dyskrasias* are offered. Specific features of the functions of various organs that bear on the treatment of *dyskrasias* are identified. The importance of preserving the capacity is again stressed. Variations in indications according to the affected place and its sensory capabilities are considered.

BOOK VIII

1. Galen provides a recapitulation of basic theory concerning *dyskrasias* and then considers fever as a *dyskrasia* in which there is an excess of the hot quality. Fever is identified as one abnormality among several in a particular disease. The differentiae of fevers are set out. Two factors considered are whether a fever is "temporary" or "established" and whether the cause is still operative or not. The basis of treatment of a fever per se is cooling.

2. Blockage of the pores is described as a cause of fever. The main forms of treatment are listed, as are factors that are aggravating. The importance of baths in treatment is emphasized and the types of baths are described. A case report is given in which Galen, in collusion with the patient, outwits his bumbling colleagues and highlights the

danger of inappropriate fasting and bed rest. That fevers change in kind due to inappropriate treatment, particularly excessive fasting, is recognized. Reference is made to Hippocrates' recommendations on diet. Galen lists some of the specific components of the diet and notes the particular merits of ptisan. Other useful foods are listed. In the case of birds, fish, and animals, where they come from is important. The danger of bathing in astringent waters that cause blockage of the skin pores is recognized. Several clinical examples are given.

3. The importance of maintaining the patency of the skin pores in natures prone to fever is stressed. Issues of treatment, including timing, are considered. Details of the oils to be used as cooling agents are given. Different treatment strategies according to the cause of the fever are examined. Galen repeatedly points out that much of this material is covered in *On the Preservation of Health*. Diet, including the use of wine, in ephemeral fevers is described, as is the use of cold water as a drink. The basic rule is that opposites are the cures of opposites. The matter of the "three day period"—a target for Galen's scorn—is examined. Galen refers to his own background and his father's influence.

4. The importance of dealing with blockage of transpiration while it is still slight is stressed. When thick and viscid humors become impacted, the fever will not remain ephemeral. The role of *kakochymia* in determining the nature of the fever is identified. Signs to be sought in the pulse and urine are listed. The treatment of *kakochymia* is described. The importance of phlebotomy is noted. Attention must be given to the sequence of treatment—that is, deal with the *kakochymia* before attempting to deal with

the blockage. Recognition of blockage of the pores of the skin hindering transpiration is important. The severity of the fever is proportional to the magnitude of the blockage. Some cleansing agents for treating the blockage are listed.

5. Galen considers ephemeral fevers due to apepsia (failure of digestion) in certain natures (*krasias*). The differences in treatment, which depend on whether the stomach is retaining or expelling, are examined. Details of specific medications are given. Treatment if there appears to be inflammation of the stomach is described, including the use of suppositories and clysters. Some features of bradypepsia (slowing of digestion) and apepsia are described. Further references are made to *On the Preservation of Health*.

6. The relevance of ulcers and lymphadenopathy in ephemeral fevers is considered. There is further criticism of doctors who focus on the "three day period." Galen calls them "diatritarians," although he does not claim to have coined the term. A case report is provided.

7. Galen moves from ephemeral fevers in picrocholic natures to these fevers in people in whom hot and moist qualities prevail over cold and dry. A *dyskrasia* that is particularly prone to fevers is described. An important contributing factor is putrefaction of humors, to which such a *dyskrasia* predisposes if there are problems with transpiration. A hypothetical case is considered.

8. The relationship of different types of *dyskrasia* to fevers is discussed. Galen provides a ranking of the eight possible *dyskrasias* in terms of their predisposition to fever. Some issues of treatment are examined.

9. Observations are offered on the treatment of *dyskrasia* in fevers. In general, every fever requires a moist

and cold diet. The management of people who are healthy is discussed. Two possible objectives are to preserve the existing *eukrasia* and to improve it. Improving a *eukratic* state by diet is described. Galen discusses what to do when all the indications are similar and what to do when they are different. Management of a combination of diseases is considered. Diets for the different *dyskrasias* are given. Two hypothetical patients are compared. Galen examines the issue of weighing up a collection of indications in the individual case.

BOOK IX

1. Galen identifies the causes of ephemeral fevers, characterized by a single paroxysm and lasting one day, changing into another kind of fever. Other fevers described are those due to putrefaction of humors and the "hectic" fevers. Intermediate forms may exist. The consequences of mismanagement are considered. Fevers due to stoppage of the pores are described, as are the factors causing this. Issues of nomenclature are addressed.

2. Continuous (synochial or nonintermittent) fevers are described as having a single, continuous paroxysm extending over many days due to failure of "seething" humors to disperse. These fevers may be termed "polyhemeral."

3. Causes of persisting fever are identified. There are three kinds of persistence—continuing at the same magnitude, increasing in magnitude in a regular pattern, and increasing in an irregular pattern. The variations in increase and decrease are seen as being due to variable additions

and evacuations. Some determining factors of the kind of fever are listed. Diagnostic signs in the pulse and urine are described.

4. Two contrasting cases are recounted, one a free man and one a slave: the former has a continuous fever without putrefaction while the latter has such a fever with putrefaction. Galen offers an aside on what is required for teaching, and on his own background. The role of phlebotomy in continuous fevers is identified and its dependence on a strong capacity stressed.

5. Fevers due to stoppage of the pores caused by blockage are described. These are ephemeral if there is no putrefaction of humors and continuous (polyhemeral) if there is. The importance of phlebotomy is emphasized. The factors contributing to putrefaction and ways to prevent it are considered. Galen criticizes Erasistratus. The use of cold water to drink is discussed. The two main cures of continuous (polyhemeral) fevers are identified as phlebotomy and cold drinks/water.

6. The essential components of proper medical practice are outlined. The two "limbs" of theory and practice are discussed.

7. Galen examines the importance of the capacity (*dunamis*) in determining treatment. He considers errors of reasoning characteristic of "sophistical" doctors—presumably Methodics. The recurring issue of nomenclature is again addressed. The significance of the capacity is seen as varying in individual cases. According to Galen, all the indicators of treatment are subject to variation in the individual case.

8. Galen identifies himself as the one who is establish-

ing the proper method of treatment following the initial discovery of the right path by Hippocrates. He then develops the road metaphor in relation to Trajan.

9. The indicators of treatment are stated. The primary indicator/objective is recognized as being the eradication of the disease. Other indicators listed are bodily *krasis*, age, custom, place, season, climatic conditions, and the affected part. Of course, the patient's capacity is also important, as Galen reminds us.

10. Galen discusses capacities following Plato—capacity equates with soul. He gives Plato's threefold division. The brain is identified as the seat of the rational soul. The importance of timing and amount in treatment is considered, particularly in relation to preservation of the patient's capacity.

11. The purposes and indicators of phlebotomy are outlined. Galen stresses the importance of the strength of the patient's capacity in determing the place of phlebotomy in the individual case.

12. Galen states that each single indicator indicates one thing; for example, a simple *dyskrasia* indicates one thing while a compound *dyskrasia* indicates two.

13. The capacity, Galen claims, always has the one indication which is its own preservation. He offers some theoretical considerations on capacity and its fundamental importance for life. After the capacity, the indication comes from the condition being treated.

14. There is a recapitulation and summary of the therapeutic indicators, grouped under three general headings: the disease, the *krasis* of the body, and the ambient air. The distinction is made between *homoiomeres* and organic

parts. Further consideration is given to the role of the ambient air as either curative or causative.

15. Further aspects of classification of the therapeutic indicators are dealt with. The one general indicator of cures is identified as opposition. The issue of the "amount" of opposition is examined, as is the difficult issue of how to determine the *krasis* of the individual body—what it was in health and what it has become in disease.

16. The therapeutic role of cold drinks, presumably water, is defined. The importance of determining the existing *krasis* of the part or the person being treated with the cold drink is stressed, as is the importance of the customs of the patient with particular reference to cold drink. There is a digression on customs in general as indicators. The Empiric position is stated. How customs give insight into *krasis* is described.

17. Galen examines the effect of the substance of bodies, both animate and inanimate, and the influence of this on dispersal. Further observations on indicators are presented, particularly in the *dyskrasias*. The common indicator in dissolution of continuity is identified. Reference is made to Plato on the issue of division and classification. Galen offers some general considerations on the *dyskrasias*.

ΓΑΛΗΝΟΥ ΘΕΡΑΠΕΥΤΙΚΗΣ
ΜΕΘΟΔΟΥ

METHOD OF MEDICINE

ΒΙΒΛΙΟΝ Ε

1. Τὴν τῶν ἑλκῶν ἴασιν ὡς ἄν τις ἄριστα ποιοῖτο διὰ τῶν ἔμπροσθεν τοῦδε δυοῖν ὑπομνημάτων διερχόμενος, καὶ τοὺς ἄλλους μὲν ἅπαντας ἰατροὺς ὅσοι χωρὶς τοῦ ζητεῖσθαι τὰ στοιχεῖα τῶν ἁπλῶν ἐν ἡμῖν μορίων ἅπτονται τῆς τέχνης οὐ δυναμένους οὐδὲν ἰάσασθαι μεθόδῳ, πάντων δὲ μάλιστα τοὺς τὴν Θεσσαλοῦ πρεσβεύοντας αἵρεσιν, ἐπέδειξα. οἱ μὲν γὰρ ἄλλοι τά γ᾽ ἐν τοῖς διαφέρουσι μέρεσιν ἕλκη διαφόρως θεραπεύειν ἀξιοῦσιν ὑπὸ τῆς πείρας δεδιδαγμένοι· τοῖς δ᾽ ἀπὸ τοῦ
Θεσσαλοῦ διὰ τὸ περιττὸν τῆς σοφίας ἅπαν | ἕλκος ὁμοίας δοκεῖ δεῖσθαι θεραπείας, ἐν ᾧπερ[1] ἂν ᾖ μέρει τοῦ ζῴου γεγονός· εἰ μὲν γὰρ εἴη κοῖλον, ἀναπληρώσεως αὐτό φασι χρῄζειν· εἰ δ᾽ ὁμαλές, ἐπουλώσεως· εἰ δ᾽ ὑπερσαρκοῦν, καθαιρέσεως· εἰ δ᾽ ἔναιμόν τε καὶ πρόσφατον κολλήσεως· ὥσπερ τῷ ταῦτα γινώσκειν, ἀκόλουθον ἐξ ἀνάγκης τὸ θεραπεύειν ὀρθῶς, ἀλλ᾽ οὐχὶ κοινὸν μὲν τοῦτο καὶ πρὸς τοὺς ἰδιώτας ὑπάρχον, οὐδεὶς οὖν ἀγνοεῖ τῶν ῥηθέντων οὐδέν. οὐ μὴν οὔθ᾽ ὡς χρὴ σαρκῶσαι τὸ κοῖλον οὔθ᾽ ὡς ἐπουλῶσαι τὸ πεπληρωμένον, ἢ καταστεῖλαι τὸ ὑπεραυξανόμενον, ἢ συμφῦσαι τὸ καθαρόν τε καὶ μὴ κοῖλον ἐπίστανται·

2

BOOK V

1. I have shown how someone might best effect the cure of wounds, having gone through this in detail in the two books prior to this one. I have also shown that all other doctors who engage in the craft without seeking the elements of the simple parts in us are unable to cure anything by method, and that of all these, the most egregious are those representing the Thessaleian sect. Others, at least, think it right to treat wounds in the different parts differently, having been taught by experience. To the Thessaleians, in their excess of wisdom, every wound seems to require similar treatment in whatever part of the organism it might be. If it is hollow, they say it has need of filling; if flat, of cicatrization; if with exuberant flesh, of reduction; and if bleeding and recent, of conglutination. It is as if by knowing these things the correct treatment follows inevitably. But this [treatment] is not common [knowledge] shared by laymen, although no one is unaware of any of the things mentioned. What they don't know is how they should enflesh a hollow wound, or cicatrize one that has been filled, or repress the exuberant flesh, or unite what has been purified and is not hollow. Such things, I believe,

305K

306K

---

[1] B; ὅτῳπερ K

3

μόνον γὰρ, οἶμαι, τῶν ἰατρῶν τὰ τοιαῦτα τῶν ἔργων
ἐστὶν ἤτοι γ᾽ ἐκ λόγου τινός, ἢ ἐξ ἐμπειρίας, ἢ ἐκ
συναμφοτέρου τὴν εὕρεσιν ἔχοντα.

πάλιν οὖν ἀναλαβόντες ἐν κεφαλαίοις, ὑπὲρ αὐτῶν
διεξέλθωμεν ἕνεκα τοῦ συνάψαι τὴν ἀρχὴν τῶν μελ-
λόντων λεχθήσεσθαι τῇ τελευτῇ τῶν φθανόντων εἰρῆσθαι. πᾶν ἕλκος ἐδείχθη ξηραινόντων χρῄζειν φαρμά-
κων· ἀλλὰ τὸ μὲν κοῖλον ἧττόν τε τῶν ἄλλων καὶ πρὸς
307K τῷ μετρίως ξηραίνειν | τὸ ῥύπτειν ἐχόντων, τὸ δὲ τῆς
ἑνώσεως τῶν χειλῶν δεόμενον, οἷα τὰ καλούμενα πρὸς
τῶν ἰατρῶν ἐστιν ἔναιμα, τῶν τε μᾶλλον ἔτι ξηραινόντων καὶ χωρὶς τοῦ ῥύπτειν ἀτρέμα στυφόντων· ἔτι
δὲ μᾶλλον ὅσα τῶν ἑλκῶν εἰς οὐλὴν ἀχθῆναι δεῖται,
ξηραινόντων τε χρῄζει φαρμάκων καὶ στυφόντων οὐκ
ἀγεννῶς. εἰ δὲ ἡ σὰρξ αὐτῶν ὑπὲρ τὸ κατὰ φύσιν
ἀρθείη, δριμέων τε καὶ καθαιρετικῶν, ἅπερ ἐξ ἀνάγκης εἶναι θερμὰ καὶ ξηρά. συμπτώματος δέ τινος
ἑτέρου προσγινομένου τοῖς ἕλκεσιν ἀπὸ τῆς ἐκείνου
φύσεως ὁ σκοπὸς τῆς ἰάσεως ἐλαμβάνετο· κἀκ τούτου
πάλιν ἡ τῶν φαρμάκων δύναμις. εἰ μὲν γὰρ ῥύπος
ἐπιτραφείη, τῶν ἀφαιρούντων αὐτὸν εἶναι τὴν χρείαν·
ἅπερ ἅπανθ᾽ ὑπάρχει ῥυπτικὰ πολὺ δή τι πλέον ἢ
κατὰ τὰ σαρκοῦντα. πλέονος δ᾽ ὑγρότητος ἐν αὐτοῖς
φανείσης, ἔτι μᾶλλον ξηραίνειν χρῆναι τὸ φάρμακον,
οὐκ ἐξιστάμενον τῆς οἰκείας ἰδέας. ἀλλ᾽ εἰ μὲν κολλητικὸν εἴη, ξηραῖνόν τε καὶ στῦφον, εἰ δὲ σαρκωτικόν,
308K ξηραῖνόν τε καὶ ῥύπτον· ἐφ᾽ ἑκάστου τε τῶν | ἄλλων
ὥσπερ εἴρηται.

4

are only among the actions of doctors who have made the discovery either from some theory, or from experience, or from a combination of both.

Therefore, to summarize, let me go over these things again for the purpose of joining the beginning of what will be said to the end of what was said beforehand. Every wound was shown to need drying medications, but the hollow wound needs them less than others. Besides this, the medications should be moderately drying and cleansing. However, wounds that need union of the margins—those that doctors call "bleeding"—require medications that are more drying and are gently astringent without being abstersive. Those wounds that need to be brought to cicatrization require drying medications that are still more drying and are vigorously astringent. If their flesh is raised above the normal level, they require sharp and reducing medications, which, of necessity, are hot and dry. When some other symptom is added to the wounds, the indicator of the cure is taken from the nature of that, and from this in turn the potency of the medications is taken. If filth is built up, there is a need of those things that take this away, all of which are much more cleansing than enfleshing. If more moisture is apparent in these [wounds], there is a still greater need for a drying medication without going beyond the proper kind. But if the medication is conglutinating, [it needs to be] drying and astringent, whereas if it is enfleshing, [it needs to be] drying and cleansing; and in the case of each of the others, as was said.

307K

308K

5

δυσκράτου δὲ τῆς ὑποκειμένης γεγονυίας σαρκός,
ἐκείνης προτέρας ἰᾶσθαι τὴν δυσκρασίαν, τὴν μὲν
ξηροτέραν τοῦ κατὰ φύσιν ὑγραίνοντας, τὴν δ' ὑγρο-
τέραν ξηραίνοντας, καὶ τὴν μὲν θερμοτέραν ψύχοντας,
τὴν δὲ ψυχροτέραν θερμαίνοντας, καὶ εἰ κατὰ συζυγίαν δέ τινα δύσκρατος ὑπάρχοι, καὶ τὴν τοῦ φαρμάκου δύναμιν ἐξ ὑπεναντίου κατ' ἀμφοτέρας αἱρεῖσθαι τὰς συζυγίας. κοινὸν γὰρ εἶναι τοῦτό γε τῶν
παρὰ φύσιν ἀπάντων ὡς μηδὲν δύνασθαι πρὸς τὸ
κατὰ φύσιν ἐπανελθεῖν ἄνευ τῶν ἐναντίων ἑαυτῷ τῇ
δυνάμει. προσεπισκοπεῖσθαι δ' ἐν τῷδε καὶ τὰς αὐτῆς
τῆς δυσκρασίας αἰτίας εἴτε κοιναὶ τοῦ σώματος ἅπαντος ὑπάρχοιεν, εἴτε μορίων τινῶν ἐξαίρετοι, κατὰ συμπάθειαν ἀδικοῦσαι τὸ ἡλκωμένον.

ἐξιᾶσθαι δὲ προτέραν μὲν τὴν τῆς γινομένης δυσκρασίας αἰτίαν, ἐφεξῆς δὲ καὶ τῆς ἤδη γεγενημένης·
εἶναι γὰρ δεῖ καὶ τοῦτον τὸν σκοπὸν ἁπάντων κοινὸν
τῶν ὑπ' αἰτίου τινὸς ἐπιγιγνομένων. ἐδείξαμεν δὲ καὶ
τὰς ἀπὸ τῆς διαφορᾶς τῶν ἑλκῶν διαφόρους ἐνδείξεις
309K καὶ τὰς ἀπὸ τῆς | τῶν πεπονθότων σωμάτων κράσεως
ἑτέρως ἐχούσας ταῖς προειρημέναις. ἐκεῖναι μὲν γὰρ
ἀπὸ τῶν παρὰ φύσιν ὑπάρχουσαι τῶν ἐναντίων ἅπασι
χρῄζουσιν· αὗται δ' ἐκ τῶν κατὰ φύσιν ὁρμώμεναι τῶν
ὁμοίων· τὸ μὲν γὰρ ξηρότερον μόριον ξηραίνεσθαι δεῖ
μᾶλλον, τὸ δὲ ἧττον τοιοῦτον ἧττον ξηραίνεσθαι χρή.
κατὰ δὲ τὸν αὐτὸν τρόπον ἐπὶ τοῦ θερμαίνεσθαί τε καὶ
ψύχεσθαι· λαμβάνεσθαι δέ τινα κἀκ τοῦ κύριον ἢ μὴ

If the underlying flesh is *dyskratic*, its *dyskrasia* needs to be cured first; that is, what is drier than normal needs to be made moist, what is more moist needs to be dried, what is hotter needs to be cooled, what is colder needs to be heated, and if some conjunctive *dyskrasia* exists, the potency of the medication is chosen from the opposites of both conjunctive components. For this is a common feature of all things contrary to nature—that none [of them] can be restored to an accord with nature without things that are opposite to their own capacity. Also to be taken into consideration in this [situation] are the causes of the *dyskrasia* itself, whether they are common to the whole body or are peculiar to certain parts, and injure the wounded part by way of a sympathetic affection.

First, thoroughly cure the cause of the existing *dyskrasia*, and next, the *dyskrasia* that has already occurred, for this must be the common aim for all things that have arisen from some cause. I also showed that the different indications from the differentiae of the wounds and from the 309K *krasis* of the bodies which have been affected are different from the previously mentioned indications. For the indications that are from those things contrary to nature need their opposites in all cases, whereas the indications that are from those things in accord with nature need similars. The part that is drier needs to be dried more while the part that is less [dry] needs to be dried less. The same applies in relation to what is heated and cooled. However, it is necessary to take a different indication of cure depending on

τοιοῦτον εἶναι τὸ μόριον, εὐαίσθητόν τε καὶ δυσαίσθητον, διάφορον ἰάσεως ἔνδειξιν.

2. Ὅσον οὖν ὑπόλοιπον ἔτι τῆς τούτου τοῦ γένους θεραπείας ἐστὶν ἤδη λέγωμεν. ὀνομάζεται μὲν τὸ γένος τοῦτο πρὸς ἡμῶν ἕνεκα σαφοῦς διδασκαλίας ἑνώσεως λύσις, οὐδὲν διαφέρον εἰ καὶ συνεχείας εἴποιμεν. ἐγγίνεται δ᾽ οὐ μόνον τοῖς ὁμοιομερέσι τε καὶ ἁπλοῖς ὀνομαζομένοις μορίοις, ἀλλὰ καὶ τοῖς συνθέτοις τε καὶ ὀργανικοῖς. αἱ δὲ τῶν βοηθημάτων ἐνδείξεις ἕτεραι μὲν ἀπὸ τῶν ὁμοιομερῶν εἰσιν, ἕτεραι δὲ ἀπὸ τῶν ὀργανικῶν. ἀμφοτέρας μὲν οὖν οἱ τὴν Ἱπποκράτους μέθοδον ἀσπαζόμενοι γινώσκουσι, διότι
310K καὶ | τὴν ἑκατέρων τῶν μορίων φύσιν ἐπίστανται· τὴν δ᾽ ἑτέραν ἐξ αὐτῶν μόνην τὴν ἀπὸ τῶν ὀργανικῶν οἱ περὶ τὸν Ἐρασίστρατόν τε καὶ Ἡρόφιλον. ὥστε κἂν τοῖς ἐφεξῆς λόγοις ὅσα μὲν ἀπὸ τοῦ θερμοῦ καὶ ψυχροῦ καὶ ξηροῦ καὶ ὑγροῦ σώματος ἢ πάθους εἰς τὴν ἔνδειξιν λαμβάνεται, τούτων οὐδενὸς ἕξουσι μέθοδον οἱ περὶ τὸν Ἐρασίστρατόν τε καὶ Ἡρόφιλον· ὅσα δὲ ἀπὸ τῆς διαπλάσεως, ἢ θέσεως, ἢ κυριότητος, ἢ εὐαισθησίας, ἢ τῶν ἐναντίων, οὐκ ἀγνοήσουσι. φανείη δ᾽ ἄν σοι σαφέστερον ὃ λέγω τῶν σωμάτων αὐτῶν προχειρισθέντων.

ἐπεὶ τοίνυν ἔμπροσθεν ὑπὲρ τῶν ἐν σαρκώδεσι μέρεσι γινομένων ἑλκῶν ὁ πλεῖστός μοι λόγος ἐπεράνθη, νῦν δ᾽ ἤδη καιρὸς ὑπὲρ τῶν ἐν ἀρτηρίᾳ καὶ φλεβὶ καὶ νεύρῳ διελθεῖν, οὐδὲ περὶ τούτων ἁπλῶς, ἀλλὰ καὶ καθ᾽ ἕκαστον σπλάγχνον ἢ συλλήβδην

8

whether a part is a principal one or not, and the same with regard to normal and disturbed sensation.

2. Let me now speak about what still remains of the treatment of this class [of diseases]. I term this class, for the purpose of clear instruction, "dissolution of union," although it would make no difference if we were to speak of "[dissolution] of continuity." It happens not only in the *homoiomeres* and so-called simple parts, but also in the compound and organic parts. Some indications of remedies are from the *homoiomerous* parts and others from the organic [parts]. Those who adhere to the method of Hippocrates know both [kinds of indication] because they also 310K know the nature of each of the parts. Those who follow Erasistratus and Herophilus know only one of these indications—that from the organic parts. As a result, in the ensuing discussions, none of those who follow Erasistratus and Herophilus will possess a method for taking what is derived from the hot, cold, dry and moist body or affection toward the indication. However, they will not be unaware of [the indications] derived from conformation, position, importance, and normal sensation, or from their opposites. What I am talking about will appear clearer to you when the bodies themselves are introduced.

Since my major discussion about the wounds occurring in the fleshy parts was previously completed, it is now time to go over those [wounds] occurring in an artery, vein, or nerve, and not only these, but also those occurring in each internal organ, or to speak collectively, in each organic part

εἰπεῖν ὀργανικὸν τοῦ ζῴου μόριον. εἰ δέ τις ἀρτηρίαν ἢ
φλέβα τρωθείη μεγάλην, αἱμορραγία τε λαύρως ἐπι-
πίπτει τὸ παραυτίκα καὶ κολληθῆναι τῷ τοιούτῳ τραύ-
ματι χαλεπὸν μὲν κἂν τῇ φλεβί· κατὰ δὲ τὴν ἀρτηρίαν
311K οὐ | χαλεπὸν μόνον, ἀλλὰ ἴσως καὶ ἀδύνατον, ὥς τινες
ἀπεφήναντο τῶν ἰατρῶν. λεκτέον οὖν ὑπὲρ ἑκατέρων
ἐν μέρει· πρότερον μὲν τῶν αἱμορραγιῶν, δεύτερον δὲ
τῶν συμφύσεων. ἐπεὶ δὲ καὶ κατὰ ἀναστόμωσίν τε καὶ
τὴν καλουμένην διαπήδησιν αἱμορραγίαι γίνονται,
διὰ τὴν κοινωνίαν τῶν ἰαμάτων οὐδὲν χεῖρον ἐν τῷδε
μνημονεύειν αὐτῶν, εἰ καὶ δοκοῦσιν ἐξ ἑτέρου γένους
εἶναι νοσημάτων. ἐκχεῖται τοίνυν αἷμα φλεβὸς καὶ
ἀρτηρίας ἤτοι κατὰ τὸ πέρας ἀνεστομωμένων τῶν
ἀγγείων, ἢ τοῦ χιτῶνος αὐτῶν διαιρεθέντος, ἢ ὡς ἄν
τις εἴποι διηθούμενον, ἢ διϊδρούμενον. ὁ μὲν οὖν χιτὼν
διαιρεῖται τιτρωσκόμενός τε καὶ θλώμενος καὶ ῥηγνύ-
μενος καὶ διαβιβρωσκόμενος. ἀναστόμωσις δὲ γίνεται διά τε ἀτονίαν ἀγγείου καὶ πλῆθος αἵματος ἀθρόως ῥεύσαντος ἐπὶ τὸ στόμιον αὐτοῦ καί τινα ποιότητα
προσπίπτουσαν ἔξωθεν αὐτῷ δριμεῖαν. ἡ δὲ διαπήδησις ἀραιωθέντος μὲν τοῦ χιτῶνος, λεπτυνθέντος δὲ
τοῦ αἵματος ἀποτελεῖται· γένοιτο δ᾽ ἄν ποτε καὶ δι᾽
ἀναστόμωσιν ἀγγείων μικρῶν. λεκτέον οὖν ὑπὲρ ἑκάστης διαθέσεως ἐν μέρει, καὶ πρώτης γε τῆς διαιρέ-
312K σεως, ἣν | ἐξ ἀναβρώσεως καὶ τρώσεως καὶ θλάσεως
καὶ ῥήξεως ἔφαμεν γίγνεσθαι.

τὰ μὲν δὴ τιτρώσκοντα τῶν αἰτίων ὀξέα τ᾽ ἐστὶ καὶ
σύντομα, τὰ δὲ θλῶντα βαρέα τε καὶ σκληρά, τὰ δὲ

10

of the animal. If some large artery or vein is wounded and a sudden, fierce hemorrhage ensues, it is difficult for such a wound to be conglutinated, even in a vein. In an artery, however, it is not only difficult but perhaps also impossible, as some doctors declare. I must, therefore, speak about each of these in turn, dealing first with the hemorrhage and second with the closure [of the wound]. Since hemorrhages occur in relation to *anastomosis* and so-called *diapedesis*,[1] it would be no bad thing to mention these here, due to the common nature of the cures, even if they seem to be from a different class of diseases. Accordingly, blood flows out of an artery or vein either when the ends of the vessels are widely open, or when their wall is divided, or there is a filtering or transudation [through the vessel wall], as one might say. The [vessel] wall is divided when there is wounding, crushing, rupture or erosion. *Anastomosis* occurs due to a weakness of the vessel and a large amount of blood flowing all at once to its mouth, or some sharp quality falling on it from without. *Diapedesis* is brought about either when the vessel wall is made porous or when the blood is thinned. However, it might also occur at times due to *anastomosis* of the small vessels. I must, therefore, speak about each condition in turn, starting with division, which I said occurs from erosion, wounding, crushing, or rupture.

So then, the causes of wounding are acute and of short duration; of crushing, they are heavy and hard; of rupture,

311K

312K

[1] The Greek terms are retained here although their original meaning differs from current usage. Both terms are, in fact, defined in what immediately follows.

ῥηγνύντα διὰ μέσης μὲν ἅπαντα τῆς τάσεως ῥήγνυ-
σιν· ἔστι δὲ πλείω σφοδρότης ἐνεργείας καὶ πλῆθος
χυμῶν, οὐ τὸ πρὸς τὴν δύναμιν, ἀλλ' ὅταν ὑπὸ τῶν
περιεχόντων μὴ στέγηται, καὶ κατάπτωσις ἐξ ὑψηλοῦ
καί τι τῶν βαρέων καὶ σκληρῶν ἐμπεσόν. τὸ γὰρ
τοιοῦτον ἐπειδὰν μὲν ἤτοι κενοῖς, ἢ ὀλιγίστην οὐσίαν
περιέχουσιν ἀγγείοις ἐμπέσοι, θλάσιν ἐργάζεται,
μετὰ τοῦ καὶ τὸ ἀντιστηρίζον ἔχειν σκληρόν· ἐπειδὰν
δὲ πλῆρες ᾖ, φθάνει ῥηγνύειν πρὶν θλᾶσθαι, παρα-
πλησίου τοῦ συμβαίνοντος ὄντος ὥσπερ εἰ καὶ πλη-
ρώσας ἀσκὸν ἢ κύστιν ἐπιρρίψαις λίθον, ἢ εἰ κατὰ τοῦ
λίθου σφοδρῶς ἐπιρρίψαις τὸν ἀσκόν. ἔοικε δὲ μάλιστα τοῦτο τὸ γινόμενον ἐν ταῖς καταπτώσεσιν· ὃν γὰρ
ἔχει λόγον ὡς πρὸς τὸν λίθον ὁ ἀσκός, τοῦτον ὡς πρὸς
τοὔδαφος ὁ ἄνθρωπος. ὅσοι δὲ μέγιστον ἢ ὀξύτατον
313K βοήσαντες ἔρρηξαν ἀγγεῖον ἐν πνεύμονι, | διὰ συν-
τονίαν ἐνεργείας ἔπαθον οὕτως· ὡσαύτως δ' αὐτοῖς καὶ
ὅσοι βαρὺν ὄγκον σώματος ἢ τοῖς ὤμοις ἀναθέσθαι
προὐθυμήθησαν, ἢ ἄλλως πως ἐξᾶραι διὰ ταῖν χεροῖν.
ἔτι δὲ μᾶλλον ὅσοι διὰ δρόμον ὠκύν, ἢ πηδήσαντες
μέγιστα, ἤ πως ἄλλως ἰσχυρῶς τείναντές τι μόριον·
ἔστι γὰρ καὶ τοῦτο παραπλήσιον ὡς εἰ καὶ σχοινίον, ἢ
ἱμάντα διατείναις ἐπὶ πλεῖστον. ὅτι δὲ καὶ διὰ τὸ μὴ
στέγειν τὸ περιεχόμενον ἐν αὐτοῖς αἷμα, καὶ μάλισθ'
ὅταν ᾖ πνευματικόν, οὐκ ὀλίγα ῥήγνυνται, δηλοῦσιν οἵ
τε πίθοι πρὸς τοῦ γλεύκους ἀναρρηγνύμενοι καὶ ἄλλα
πολλὰ τῶν ἰσχυροτάτων σωμάτων.

3. Εἰ μὲν οὖν ἐκ τρώσεως, ἢ βοῆς, ἢ καταπτώσεως

12

every breakage is due to stretching. [And the same applies] when there is a greater degree of vigorous function or an abundance of humors, not in respect of the capacity but whenever they are not contained by the surrounding [vessel walls] due to a fall from a height, or something heavy and hard falling on the vessel. For whenever such a thing befalls vessels that are either empty or contain very little substance, this brings about a crushing because the hard object encounters resistance. However, when the vessel is full, it ruptures before it is crushed, this being similar to what happens if you fill a wineskin or bladder and throw a stone at it, or violently cast the wineskin against the stone. This particularly seems to occur in falls, where the wineskin has the same causal relation to the stone as the person has to the ground. Those who shout very loudly or sharply, if they rupture a vessel in the lung, suffer thus due to the 313K intensity of function similar to those who exert themselves in placing the heavy mass of a body on their shoulders or otherwise in lifting it up somehow with their hands. Even more does this happen due to running swiftly, or jumping very vigorously, or when people stretch some part severely in some other way. The same applies if you stretch a cord or thong to its greatest extent. Wine jars, when they are broken open by sweet wine, and many other very strong bodies, show that quite a number [of vessels] are ruptured due to not retaining the blood contained within them, and especially when distended with *pneuma*.

3. If rupture has occurred due to a wound, shout or fall,

ἡ ῥῆξις ἐγένετο, πέπαυται τὸ αἴτιον. εἰ δ᾽ ὑπὸ πλήθους, ἐγχωρεῖ καὶ νῦν ἔτι τὸ ἀγγεῖον ἐπὶ πλέον ἀναρ-
ρήγνυσθαι, μενούσης ἔτι τῆς ποιούσης αἰτίας. ἐπὶ μὲν
οὖν τῆσδε τῆς διαθέσεως ἐκκενωτέον ὅτι τάχιστα τὸ
πλῆθος· εἶθ᾽ οὕτως ἐπισχετέον τὸ αἷμα· κἄπειτα τὸ
ἕλκος θεραπευτέον. ἐφ᾽ ὧν δὲ οὐκ ἔστι τὸ αἴτιον,
314K   ἐπισχεῖν μὲν χρὴ πρῶτον τὸ | αἷμα, τὸ δ᾽ ἕλκος ἑξῆς
ἰᾶσθαι. πῶς οὖν ἐπισχῶμεν τὸ αἷμα; στεγνώσαντες
μὲν τὸ ἐρρωγός, ἐκτρέψαντες δὲ καὶ ἀποστρέψαντες
ἑτέρωσε τὸ δι᾽ αὐτοῦ φερόμενον, ὡς εἴ γε καὶ τοῦτ᾽
ἐπιρρέοι, καθάπερ ἐξ ἀρχῆς ὥρμησε, καὶ τὸ στόμιον
ἴσον ἑαυτῷ διαμένοι, θᾶττον ἂν ἀπόλοιτο τὸ οὕτως
αἱμορραγοῦν ζῷον ἢ τὸ αἷμα παύσαιτο κενούμενον.
ἀλλὰ καὶ τό γ᾽ ἕλκος, ἤτοι τῶν διεστώτων χειλῶν ἐς
ταὐτὸν ἀφικομένων, ἢ τοῦ στομίου φραχθέντος στε-
γνωθήσεται. συναχθήσεται μὲν οὖν ἐς ταὐτὸν ἀλλήλοις τὰ χείλη διά τε τῶν ἡμετέρων χειρῶν, ὅταν οὕτως
πρόχειρον ᾖ τὸ τραῦμα, καὶ δι᾽ ἐπιδέσεως, ὑπό τε² τῶν
ψυχόντων καὶ στυφόντων φαρμάκων. οὐ γὰρ δὴ ῥάπτειν γε δυνάμεθα τὸ τῆς ἀρτηρίας ἢ φλεβὸς τραῦμα,
καθάπερ ἂν εἴποιεν οἱ μηδεμίαν ἐκ τῆς τῶν τετρωμένων μερῶν οὐσίας τε καὶ φύσεως ἔνδειξιν γίνεσθαι
φάσκοντες. ἐμφραγήσεται δὲ τὸ στόμιον ὑπό τε θρόμβου καὶ τῶν ἔξωθεν ἐπιβαλλομένων αὐτῷ· δύναται δὲ
ἐπιβάλλεσθαι τά τε περικείμενα σαρκώδη καὶ τὸ δέρμα κατ᾽ ἐνίας τρώσεις ὅσα θ᾽ ἡμεῖς ἐπιτεχνώμεθα.
315K   ταῦτα δ᾽ ἐστὶν οἵ τε | καλούμενοι μοτοὶ καὶ τῶν
φαρμάκων ὅσα τ᾽ ἐμπλάσσει γλίσχρα καὶ παχέα ταῖς

14

the cause has ceased. If, however, it has occurred due to
abundance, it is now possible for the vessel to be ruptured
further since the effecting cause still remains. Thus, in this
particular condition, the abundance must be emptied (by
venesection) as quickly as possible, or the blood must be
stanched, and then the wound treated. However, in cases
where the cause no longer still exists, it is necessary first to
stanch the blood and next to cure the wound. How, then,   314K
do we stanch the blood? By closing off what is ruptured,
and by diverting and redirecting elsewhere what is being
carried through the wound because, if this keeps on flowing, set in motion as it was at the start, and the opening remains as it was, the animal hemorrhaging in this way may
die quicker than the blood emptying out stops. But in fact
the wound will be closed up when either the separated
margins are brought together or the hole is closed off. The
[wound] margins will be coapted with each other by our
hands whenever the wound facilitates this, or by a bandage, or by cooling and astringent medications. For in
reality, we are not able to suture the wound of an artery
or vein, just as those people might say who claim that no
indication arises from the substance and nature of the
wounded parts. The opening will, however, be blocked up
by thrombus and by those things applied to it externally. In
some wounds it is possible to apply to the surrounding
flesh and skin those things we might contrive for the purpose.

These are the so-called tampons and, among medica-   315K
tions, those that plug up, being viscous and thick in their

---

2 B; (cf. per KLat); ὁπότε K

15

οὐσίαις ὄντα, καὶ ὅσα τὴν καλουμένην ἐσχάραν ἐργά-
ζεται· καὶ γὰρ καὶ ταύτην οἷον φράγματι τοῖς τοιούτοις ἕλκεσι ἐξεῦρον οἱ πρόσθεν· ἐργάζονται δ' αὐτὴν
διά τε πυρὸς αὐτοῦ καὶ φαρμάκων ὁμοίαν πυρὶ τὴν
δύναμιν ἐχόντων. ἐκ τοιούτων μὲν δὴ στεγνοῦται τὸ
στόμιον, ἀποτρέπεται δὲ καὶ ἀποστρέφεται τὸ αἷμα
πρὸς ἕτερα μόρια παροχετεύσει τε καὶ ἀντισπάσει·
καὶ γὰρ καὶ ταῦθ' Ἱπποκράτους εὑρήματα κοινὰ
πάσης ἀμέτρου κενώσεως. παροχετεύεται μὲν οὖν εἰς
τοὺς πλησίον τόπους· ἀντισπᾶται δὲ ἐπὶ τοὺς ἀντι-
κειμένους· οἷον τῷ δι' ὑπερῴας κενουμένῳ διὰ ῥινῶν
μὲν ἡ παροχέτευσις, κάτω δ' ἡ ἀντίσπασις, ὥσπερ γε
τῷ δι' ἕδρας διὰ μήτρας μὲν ἡ παροχέτευσις, ἄνω δὲ ἡ
ἀντίσπασις.

οὕτως γοῦν καὶ ἡ φύσις αὐτὴ δρᾶν πέφυκε. γυναικὶ
μέν, φησίν, αἷμα ἐμεούσῃ τῶν καταμηνίων ῥαγέντων
λύσις. διὰ τοῦτ' ἄρα καὶ καταμηνίων ἀθρόων ἐκκενου-
μένων ἢ ὁπωσοῦν ἄλλως αἱμορραγούσης μήτρας ἀν-
316K τισπάσεις | ἄνω σικύαν μεγίστην ὑπὸ τοὺς τιτθοὺς
προσβαλών· ἔστι γὰρ καὶ τοῦτο Ἱπποκράτους εὕρη-
μα. διὰ τοῦτο δὲ καὶ τὰς ἐκ ῥινῶν αἱμορραγίας ἐπ-
έχουσιν αἱ κατὰ τῶν ὑποχονδρίων ἐπιβαλλόμεναι
μέγισται σικύαι. χρὴ δ' ὅταν ἐκ δεξιοῦ ῥέῃ μυκτῆρος,
ἐφ' ἥπατος ἐρείδειν, ὅταν δ' ἐξ ἀριστεροῦ, κατὰ σπλη-
νός, ὅταν δ' ἐξ ἀμφοτέρων, ἀμφοτέροις τοῖς σπλάγ-
χνοις ἐπιφέρειν τὰς σικύας. εἰ δὲ μηδέπω καταλελυ-
μένος ὁ κάμνων εἴη, καὶ φλέβα τέμνειν ἐξ ἀγκῶνος

16

substances, and those which create what is called an eschar. Furthermore, our predecessors discovered this eschar for such wounds by practice, as it were. They created
it by actual fire and by medications having a capacity like
fire. So then, by means of such things, the opening [in the
vessel] is closed off and the blood turned aside and redirected to other parts by diversion and revulsion. For actually these discoveries of Hippocrates were generally applicable to every excessive emptying out [of blood]. Thus
[the blood] is diverted to places nearby and is held back in
places lying opposite—for example, diversion is through
the nose for what is emptying out via the palate, while revulsion is downward, just as in fact for an emptying out via
the anus, diversion is through the uterus whereas revulsion is upward.

At any rate, Nature itself also does the same thing naturally. [Hippocrates] said: "For a woman vomiting blood,
the resolution is menstrual bleeding."[2] Because of this,
when there is excessive menstrual flow or uterine hemorrhage in any other way, you will draw this in the opposite
direction (i.e. effect revulsion) by applying a large cupping
glass above, under the breasts. This was also a discovery of
Hippocrates. On the same basis too, very large cupping
glasses, when applied to the hypochondrium, stop hemorrhages from the nose. However, when [blood] flows from
the right nostril, it is necessary to place [the cupping glass]
over the liver, while when it flows from the left nostril, it is
necessary to place it over the spleen, and when it flows
from both nostrils, over both viscera. If the patient is still
not relieved, it is necessary to cut a vein in the antecubital
fossa on the side of the hemorrhage, drawing off a little

316

[2] Hippocrates, *Aphorisms*, V.32.

τοῦ κατ᾽ εὐθύ, κενώσαντα δ᾽ ὀλίγον, εἶτα διαλιπόντα
μίαν ὥραν αὖθις ἐκκενοῦν· εἶτ᾽ αὖθις καὶ αὖθις ὡς
πρὸς τὴν δύναμιν τοῦ κάμνοντος.

οὕτω δὲ καὶ τἄλλα ῥεύματα σύμπαντα, κοινὸς γὰρ
ὁ λόγος, ἀντισπάσεις τε καὶ παροχετεύσεις, τὰ μὲν
διὰ γαστρὸς ἤτοι δι᾽ οὔρων ἢ μήτρας, τὰ δὲ δι᾽ οὔρων
ἤτοι διὰ μήτρας ἢ δι᾽ ἕδρας· ὡσαύτως δὲ καὶ τὰ διὰ
μήτρας ἤτοι δι᾽ οὔρων ἢ διὰ γαστρός· ἐπὶ δὲ τῶν κατ᾽
ὀφθαλμοὺς καὶ ὦτα καὶ ὑπερώαν, διὰ ῥινῶν ἡ προ-
χέτευσις. ἡ δ᾽ ἀντίσπασις ἄνω μὲν ἐπὶ τοῖς κάτω
πᾶσι, κάτω δ᾽ ἐπὶ τοῖς ἄνω· καὶ μὲν δὴ κἀκ τῶν δεξιῶν |
317K ἐπὶ θάτερα, κἀξ ἐκείνων ἐπὶ ταῦτα, κἀκ τῶν εἴσω πρὸς
τὰ ἔξω, κἀκ τούτων αὖ πάλιν πρὸς ἐκεῖνα. τρίψεις οὖν
τῶν ἀντικειμένων μερῶν καὶ μάλιστα διὰ φαρμάκων
θερμαινόντων, ἔτι τε δεσμὰ σφοδρότερα τῶν ἀντισπα-
στικῶν ἐστι βοηθημάτων· ὥσπερ γε καὶ οἱ ἀντικείμε-
νοι τῶν φυσικῶν πόρων ἀναστομωθέντες· εἴρηται δ᾽ ἐν
τοῖς περὶ φαρμάκων ὑπομνήμασιν ἡ τῶν τοῦτο δυναμένων ἐργάζεσθαι φαρμάκων ὕλη. περὶ μὲν οὖν ἁπάντων ῥευμάτων εἰπεῖν ἀναγκαῖον ἔσται κἀν τοῖς ἑξῆς
ὑπομνήμασι·

νυνὶ δὲ ἐπὶ τὰς αἱμορραγίας ἰτέον. ἐκ γάρ τοι τοῦ
γένους τῆς στεγνώσεώς ἐστί πως καὶ ὁ βρόχος ὁ
περιτιθέμενος αὐτοῖς τοῖς αἱμορραγοῦσιν ἀγγείοις, οἵ
θ᾽ ἡμέτεροι δάκτυλοι συνάγοντές τε καὶ σφίγγοντες
αὐτά. τούτου δ᾽ ἐστὶ τοῦ γένους καὶ ἡ ἐπίδεσις· καίτοί
γ᾽ οὐ κατὰ κύκλον τὸ ἀγγεῖον περιλαμβάνουσα,
καθάπερ ὁ βρόχος, ἀλλὰ τῷ συνάγειν τέ πως ἐκ

blood, and then, after an interval of one hour, drawing it again, then repeatedly, according to the capacity of the patient.

In the same way too in all other flows, for the argument is a general one, you will redirect and hold back—those from the stomach either via the urine or the vagina, and those from the urine either via the vagina or the anus. And similarly, you will redirect those from the vagina either via the urine or the stomach. In the case of those [flows] from the eyes, ears and palate, redirection is via the nostrils. However, revulsion is upward for all those below and downward for all those above, and further, revulsion is to the left from those on the right and vice versa, and toward the outside from those within, and vice versa again. Therefore, you will rub the oppositely lying parts. Also, heating medications especially, and the firmer bandages are among the revulsive remedies, as are the oppositely placed natural channels when they are opened up. The material of the medications capable of bringing this about is discussed in the treatises on medications.[3] It will be necessary to speak about all the fluxes in the books that follow.

317K

Now, however, we must go on with the hemorrhages. For surely the ligature which is placed around the hemorrhaging vessels is, in a way, from the class of closing off, as are our fingers when they bring vessels together and compress them. The bandage too is from this class. And indeed, among the remedies are things that do not surround the vessel in a circle, like a ligature, but in some way par-

[3] Galen's three major pharmacological treatises, listed in Book 3, note 14.

μέρους τὰ διεστῶτα χείλη τοῦ τετρωμένου μορίου καὶ
προστέλλειν τὰ ἐπικείμενα τῶν στεγνωτικῶν ἐστι
318K βοηθημάτων. | ἔξωθεν δὲ τῶν εἰρημένων ἁπάντων αἱ-
μορραγίας βοήθημ᾿ ἐστὶ τὸ ἐπιτήδειον σχῆμα τοῦ
τετρωμένου μορίου. ἐπιτήδειον δὲ γίνεται δυοῖν τούτοιν ἐχόμενον σκοποῖν, ἀνωδυνίας τε καὶ ἀναρροπίας·
εἰ δ᾿ ἤτοι κατάρροπον ἢ ὀδυνῶδες γίγνοιτο, καὶ τὰς
οὐκ οὔσας αἱμορραγίας ἐργάσεται, μή τοί γε δὴ τὰς
οὔσας οὐ παύσει. ταῦτ᾿ οὖν ἐπιστάμενός τις, ἢν ἐπι-
στῇ ποθ᾿ αἱμορραγοῦντι μορίῳ διὰ τρῶσιν, ἐπὶ τούτων
γὰρ ὁ λόγος μοι γιγνέσθω πρῶτον, αὐτίκα μὲν ἐπι-
βαλλέτω τὸν δάκτυλον ἐπὶ τὸ στόμιον τοῦ κατὰ τὸ
ἀγγεῖον ἕλκους, ἐρείδων πραέως καὶ πιέζων ἀνωδύ-
νως, ἅμα τε γὰρ ἐφέξει τὸ αἷμα καὶ θρόμβον ἐπιπήξει
τῇ τρώσει. καὶ μέντοι κἂν εἰ διὰ συχνοῦ βάθους εἴη τὸ
αἱμορραγοῦν ἀγγεῖον, ἀκριβέστερον ἂν καταμάθοι
τήν τε θέσιν αὐτοῦ καὶ τὸ μέγεθος· καὶ πότερα φλὲψ ἢ
ἀρτηρία ἐστί· μετὰ δὲ ταῦτα διαπείρας ἀγκίστρῳ
ἀνατεινέτω τε καὶ περιστρεφέτω μετρίως. μὴ ἐπι-
σχεθέντος δ᾿ ἐν τῷδε τοῦ αἵματος, εἰ μὲν φλὲψ εἴη,
πειράσθω χωρὶς βρόχου στέλλειν τὸ αἷμα τῶν ἰσχαί-
319K μων τινὶ φαρμάκων. | ἄριστα δ᾿ αὐτῶν ἐστι τὰ ἐμ-
πλαστικά, συντιθέμενα διά τε τῆς φρυκτῆς ῥητίνης
καὶ ἀλεύρου πυρίνου χνοῦ καὶ γύψου καὶ ὅσα τοιαῦτα.

εἰ δὲ ἀρτηρία ἐστί, δυοῖν θάτερον, ἢ βρόχον περι-
θείς, ἢ ὅλον διακόψας τὸ ἀγγεῖον, ἐφέξεις τὸ αἷμα.
βρόχον δ᾿ ἀναγκαζόμεθά ποτε καὶ ταῖς μεγάλαις
περιτιθέναι φλεψίν, ὥσπερ γε καὶ διατέμνειν πότ᾿

20

tially bring together the separated margins of the wounded part, and those of the occluding agents which are laid on to cover them. Apart from all the aforementioned reme- 318K dies for hemorrhage, there is the favorable form of the wounded part. Being favorable arises through having these two indicators: being painless and being inclined upward. If, however, the part is either inclined downward or painful, not only will it bring about hemorrhages not previously in existence, but also will not stop already existing hemorrhages. Therefore, if someone who knows these things should, on some occasion, come upon a part hemorrhaging due to wounding (for let my discussion start with such cases), he must immediately place a finger on the opening of the wound in the vessel, supporting it gently and compressing it painlessly, for this will, at one and the same time, hold back the blood and congeal the clot in the wound. Of course, if the hemorrhaging vessel is at a considerable depth, he should examine its position and size more accurately, and whether it is a vein or an artery. After this, let him lift it up, attempting this with a hook, and surround it moderately. If the blood is not stanched by this, should it be a vein, let him attempt to check [the flow] with one of the hemostatic medications apart from a ligature. The best of these are the emplastics compounded from 319K roasted pine resin, heated barley, wool, chalk and other such things.

If, however, it is an artery, there are two other [options]: stopping the blood either by placing a ligature around the vessel or by cutting through the whole vessel. Sometimes, we are also forced to place a ligature around [one of] the great veins, just as we are sometimes forced to cut

21

αὐτὰς ὅλας, ἐγκαρσίας δηλονότι. κατασταίη δ' ἄν τις
εἰς ἀνάγκην τοῦδε κατὰ τὰς ἐκ πολλοῦ βάθους ὀρθίας
ἀναφερομένας, καὶ μάλιστα διὰ στενοχωρίας τινὸς ἢ
μερῶν κυρίων. ἀνασπᾶται γὰρ οὕτως ἑκάτερον τὸ
μέρος ἑκατέρωθεν, καὶ κρύπτεται καὶ σκέπεται πρὸς
τῶν ἐπικειμένων σωμάτων ἡ τρῶσις. ἀσφαλέστερον δ'
ἄμφω ποιεῖν, βρόχον μὲν τῇ ῥίζῃ περιτιθέναι τοῦ
ἀγγείου, τέμνειν δὲ τοὐντεῦθεν. ῥίζαν δ' ἀγγείου καλῶ
τὸ πρότερον αὐτοῦ μέρος, ἤτοι τῷ ἥπατι συνάπτον ἢ
τῇ καρδίᾳ. τοῦτο δ' ἐν τραχήλῳ μέν ἐστι τὸ κάτωθεν,
ἐν χερσὶ δὲ καὶ σκέλεσι τὸ ἄνωθεν· ἐν ἑκάστῳ τε τῶν
ἄλλων μορίων, ὡς ἐξ ἀνατομῆς ἔνεστι μαθεῖν· ἣν οὐδ'
αὐτὴν οἱ ἀμέθοδοι προσίενται Θεσσάλειοι. |

320K   4. Ταῦτα δὲ πράξαντα σαρκοῦν ὅτι τάχιστα τὸ
τραῦμα, πρὶν ἀπορρυῆναι τοῦ ἀγγείου τὸν βρόχον. εἰ
μὴ γὰρ φθάσειεν ἡ ἐπιτραφεῖσα σὰρξ στεγνῶσαι τὸ
πέριξ χωρίον τῆς ἀποτετμημένης ἀρτηρίας, ἀλλ' εὑρε-
θείη που χώρα κενή, τὸ καλούμενον ἀνεύρυσμα γίνεται. διὰ τοῦτο καὶ τοῖς ἐμπλάττουσιν ἰσχαίμοις χρῆσθαι κελεύω μᾶλλον τῶν ἐσχαρούντων, ὅτι θᾶττόν τε
καὶ ἀκινδυνότερον ἐπ' αὐτοῖς σαρκοῦται τὸ τραῦμα·
κίνδυνος γὰρ ἐπὶ τῶν ἄλλων ἀποπιπτούσης τῆς ἐσχάρας αὖθις αἱμορραγῆσαι τὸ ἀγγεῖον. ἄριστον οὖν
ἁπάντων ὧν οἶδα φαρμάκων, ᾧ καὶ πρὸς τὰς ἐκ
μηνίγγων αἱμορραγίας ἀσφαλέστερον χρῆσθαι, τὸ
λεχθησόμενόν ἐστι· λιβανωτοῦ μέρος ἓν ἀλόης ἡμίσει
μεμίχθω μέρει· κἄπειτα τῆς χρείας ἐπιστάσης ῷοῦ τῷ
λευκῷ φυράσθω τοσούτῳ τὸ πλῆθος ὡς μελιτώδη

through the whole vein, obviously transversely. This necessity arises in a vein that passes straight up from a very deep position, and particularly if there is a confined space or there are important parts. For each part [of the vein] retracts on both sides in such a way that the wound is concealed and covered by the overlying bodies. However, it is safer to do both—that is, to place a ligature around the root of the vessel and to cut it at that point. I call the root of a vessel the proximal part of it, or the part connected to either the liver or heart. In the neck this is the part below while in the arms and the legs it is the part above. In each of the other parts it is that which it is possible to learn from anatomy, which is the very thing the amethodical Thessaleians don't believe in.

4. Having done these things, enflesh the wound as   320K
quickly as possible before the ligature on the vessel disintegrates because, if the regenerated flesh has not already closed off the space surrounding the cut artery but somewhere an empty space is to be found, what is called an aneurysm arises. And because of this I prescribe the use of the emplastic hemostatics rather than those that create an eschar in that the wound is enfleshed quicker and less dangerously with these. With the other [medications], there is a danger of the vessel hemorrhaging again if the eschar falls off. Therefore, the best of all the medications I know, and the one which is safer to use for hemorrhages from the meninges, is the one I shall speak of. Mix one part of frankincense with half a part of aloes, and then, when it comes to the time for use, mix in the white of an egg to such a degree that the whole has a honeylike consistency. Next, let this be

σύστασιν ἔχειν· εἶτ' ἀναλαμβανέσθω τοῦτο λαγωαῖς
θριξὶ ταῖς μαλακωτάταις· κἄπειτα τῷ ἀγγείῳ καὶ τῷ
ἕλκει παντὶ πλεῖστον ἐπιτιθέσθω. καταδείσθω δ' ἔξωθεν ἐξ ὀθόνης ἐν ὑποδεσμίδι· τὰς μὲν πρώτας ἐπιβολὰς τέτταρας ἢ πέντε κατ' αὐτοῦ αἱμορραγοῦντος

321K ἀγγείου ποιουμένων | ἡμῶν, ἐντεῦθεν δὲ ἐπὶ τὴν ῥίζαν
αὐτοῦ νεμομένων, ἐφ' ὧν ἐγχωρεῖ μερῶν ἐπὶ τὴν ῥίζαν
νέμεσθαι· σχεδὸν δ' ἐπὶ πάντων ἐγχωρεῖ πλὴν μηνίγγων. εἶτα λύσαντα διὰ τρίτης, εἰ μὲν ἀσφαλῶς ἔτι
προσέχοιτο τῷ ἕλκει τὸ φάρμακον, αὖθις ἕτερον ἐν
κύκλῳ περιχέοντα καὶ ὥσπερ ἐπιτέγγοντα τὸν ἐκ τῶν
τριχῶν μοτὸν ἐπιδεῖν, ὡς ἐξ ἀρχῆς ἐπέδησας· εἰ δ'
αὐτομάτως ὁ πρότερος ἀποπίπτοι μοτός, ἀτρέμα πι-
έζοντα τῷ δακτύλῳ τὴν ῥίζαν τοῦ ἀγγείου πρὸς τὸ
μηδὲν ἐπιρρυῆναι, τοῦτον μὲν ἀφελεῖν ἠρέμα, προσθεῖναι δ' ἕτερον. οὕτω σε θεραπεύειν δεῖ μέχρι περ ἂν
σαρκωθῇ τὸ ἀγγεῖον, ἀνάρροπον ἀπ' ἀρχῆς ἄχρι
τέλους φυλάττοντα τὸ μόριον. ἔστω δὲ μηδὲ τοῦτο
ἄμετρον τὸ σχῆμα· κίνδυνος γὰρ ὀδύνην γενέσθαι καὶ
αὖθις αἱμορραγῆσαι τὸ ἀγγεῖον. οὐδὲν γὰρ ὀδύνης
μᾶλλον αἱμορραγίας τε κινεῖ καὶ φλεγμονὰς αὐξάνει.

    τούτῳ τῷ φαρμάκῳ πολυειδέστατα χρῶμαι· ποτὲ
μέν, ὡς εἴρηται, μιγνύων ἀλόῃ τὸ διπλάσιον τοῦ λιβανωτοῦ, ποτὲ δ' ἐξίσης ἀμφοτέροις χρώμενος, ἢ
βραχεῖ πλείονι τῷ λιβανωτῷ τῆς ἀλόης, ἢ πλείονι
μέν, ἀλλ' οὔπω τῷ διπλασίῳ· καὶ μάννην δ' ἀντὶ

322K λιβανωτοῦ | ποτ' ἔβαλλον. ἔστι δὲ στυπτικώτερον μὲν
φάρμακον ἡ μάννη τοῦ λιβανωτοῦ· ὁ λιβανωτὸς δ'

24

taken up by the softest hairs of a hare, and then let it be applied in abundance to the vessel and to the whole wound. Bind externally with a linen cloth in an underbandage, making the first four or five turns on the hemorrhaging vessel itself, and from that point, distribute it to the root of 321K the vessel in those parts where it is possible to distribute to the root, which is almost all parts except for the meninges. Then, when you release it on the third day, if the medication is still adhering safely to the wound, apply another encircling bandage, moistening the tampon from the hairs, as you bound it initially. If the first tampon should fall off spontaneously, gently compress the root of the vessel with your finger until there is no further flow, then remove your finger carefully and apply another tampon. You need to treat in this way until the vessel is enfleshed, keeping the part tilted upward from start to finish. But don't take this position to the extreme because there is a danger of pain and of the vessel bleeding again. Nothing stirs up hemorrhage and increases inflammation more than pain does.

I use this medication in very many forms. Sometimes, as I said, I mix one part of aloes to two of frankincense, sometimes I mix an equal amount of both, or a little more frankincense than aloes, or a lot more, although not yet twice as much, and sometimes I put in manna instead of 322K frankincense. Manna is a more astringent medication than frankincense whereas frankincense is more emplastic than

ἐμπλαστικώτερος τῆς μάννης. εὔδηλον δ᾽ ὅτι τοῖς μὲν
σκληροῖς σώμασι τὴν ἀλόην χρὴ προσφέρεσθαι
πλείονα, τοῖς δὲ μαλακοῖς τὸν λιβανωτόν· ἔσται δὲ τὸ
μὲν ἕτερον αὐτῶν στυπτικώτερον, τὸ δ᾽ ἕτερον ἐμπλαστικώτερον. διὸ καὶ τὸν γλισχρότερον καὶ ὡς ἂν
εἴποι τις ῥητινωδέστερον αἱρεῖσθαι χρὴ λιβανωτόν,
ὅταν ἐμπλαστικώτερον ἐργάσασθαι τὸ φάρμακον ἐθέ-
λῃς· ἔστι δ᾽ οὗτος ἁπαλώτερος ἅμα καὶ λευκότερος καὶ
μασσώμενος οὐ θρύπτεται δίκην ἀλόης καὶ μάννης,
ἀλλὰ συνεχὴς ἑαυτῷ μένει, καθάπερ ἡ μαστίχη. ταυτὶ
μὲν οὖν ἤδη τῆς περὶ συνθέσεως φαρμάκων πραγματείας ἐστίν, ἧς ἀδύνατον μηδαμῶς ἅπτεσθαι κατὰ τὴν
ἐνεστῶτα λόγον. ἢ γὰρ οὐδόλως ἐχρῆν ἡμᾶς μεμνῆσθαι τῶν ἐν μέρει παραδειγμάτων, ἀλλ᾽ αὐταῖς μόναις
ἀρκεῖσθαι ταῖς καθόλου μεθόδοις, ἢ καὶ τῶν ἐν μέρει
προσαπτομένους ἅπτεσθαί ποτε καὶ τῆς σκευασίας
αὐτῶν. ἀλλ᾽ ὡς καὶ πρόσθεν εἴρηται, παράδειγμα τοῦ
γένους τῶν φαρμάκων τῶν κατὰ τὴν μέθοδον εὑρισκο-
323K μένων, ἐν ᾗ δύο ἐνταυθοῖ γράφοντας | ἀρκεῖσθαι
προσήκει. τούτου μὲν οὖν ἀεὶ χρὴ μεμνῆσθαι·

πάλιν δ᾽ ἐπανέλθωμεν ἐφ᾽ ὅπερ λέγοντες ἀπελίπο-
μεν. ἐμνήσθην γὰρ ἐπὶ πλέον τοῦ προειρημένου φαρμάκου, διότι πάντων αὐτὸ τῶν ἄλλων ἄμεινον εἶναι
πέπεισμαι· καὶ θαυμάζοιμ᾽ ἂν εἴ τις εὑρεῖν δύναται
βέλτιον. καὶ διὰ τοῦτ᾽ αὐτῷ χρῶμαι κατά τε τῶν
μηνίγγων ἀεὶ καὶ τῶν ἐν τῷ τραχήλῳ τρώσεων, ἄχρι

manna. Obviously, we must apply aloes more to hard bodies and frankincense more to soft bodies in that the former will be the more astringent of these and the latter the more emplastic. On which account, it is necessary to choose the more viscous and, as one might say, resinous frankincense, whenever you wish the medication to bring about greater adhesion. This is softer and, at the same time, whiter, and does not break up when kneaded as aloes and manna do, but retains its continuity like mastich. These particular matters are already part of the treatise on the composition of medications,[4] but it is impossible not to touch on them in some way in the present discussion. It is entirely unnecessary for me to make mention of examples individually but only so far as will suffice for the methods themselves alone in general or, if I do focus on them individually, it is to touch from time to time on their preparation. But as I said earlier, it is appropriate here to be content with writing about one or two examples of the class of medications discovered by method. It is always necessary to be mindful of this.

323K

However, let me return again to the point where I left off the discussion. I made mention of the medication previously spoken of because I am persuaded that it is better than all the others, and I would be amazed if someone were able to discover a better one. For this reason, I always use it on the meninges and on wounds in the neck, even as

[4] In relation to hemorrhagic wounds in particular, for aloes see *De compositione medicamentorum secundum locos*, XIII.316K; for frankincense, *De simplicium medicamentorum temperamentis et facultatibus*, XI.735K; and for manna, *De compositione medicamentorum per genera*, XII.845K.

καὶ τῶν σφαγιτίδων αὐτῶν· καὶ γὰρ καὶ τούτων ἐπέχει
τὸ αἷμα χωρὶς βρόχου. χρὴ δὲ μὴ σπεύδειν ἐν τῷ
ἔργῳ καθάπερ ἔνιοι τῶν ἐμπλήκτων χειρουργῶν, ἀλλὰ
τῇ μὲν ἑτέρᾳ χειρὶ τὸ κάτω μέρος τοῦ ἀγγείου προστέλλειν ἢ περιλαβόντα κατέχειν, τῇ δ' ἑτέρᾳ τὸ φάρμακον ἐπιθέντα τῇ τρώσει προστέλλειν ἀτρέμα, ἄχρι
περ ἂν προσπαγῇ, κἄπειθ' οὕτως ἐπιδεῖν ἄνωθεν
κάτω, μὴ καθάπερ ἐν τοῖς κώλοις κάτωθεν ἄνω· πρὸς
γὰρ τὰς ῥίζας τῶν ἀγγείων ἀφικνεῖσθαι χρὴ τὴν
ἐπίδεσιν ἀναστέλλουσαν τὸ ἐπιρρέον.

ἔστι δὲ καὶ ἄλλα πλείω φάρμακα τῶν ἐμπλαττόντων ἀλύπως· ἀλλ' οὐδὲν οὕτως σαρκοῖ. χρεία δ'
324K ἐστὶ μάλιστα κατὰ τὰ τοιαῦτα συμπτώματα | τοῦ
περισαρκοῦσθαι τὸ ἀγγεῖον, ἀποπίπτοντος τοῦ προτέ-
ρου φαρμάκου. τὰ δέ γε τὰς ἐσχάρας ἐπιπηγνύντα
γυμνότερον ἐργάζεται τὸ μέρος ἢ κατὰ φύσιν εἶχεν
ἀποπιπτούσης αὐτῆς. ἔστι γὰρ ἡ γένεσις τῆς ἐσχάρας ἐκ τῶν ὑποκειμένων καὶ περικειμένων σωμάτων,
ὡς ἂν εἴποι τις ἡμικαύτων γενομένων, ἵν' ὁποῖόν τι
χρῆμα τουτουσὶ τοὺς ἐσβεσμένους ἄνθρακας εἰς τὸν
χειμῶνα παρασκευάζονται, τοιοῦτο ὑπόλειμμα τῆς
διαπύρου σαρκὸς ἡ ἐσχάρα γένηται. ὥστ' ἀπόλλυται
τοῦ μορίου τοσοῦτον τῆς κατὰ φύσιν σαρκός, ὅσον
ἐσχαρούμενον ἐκαύθη. τοῦτ' οὖν ἅπαν λείπει τῷ μορίῳ
τῆς ἐσχάρας ἐκπιπτούσης, καὶ διὰ τοῦτο γυμνὸν καὶ
ἄσαρκον φαίνεται, καὶ πολλοῖς αἱμορραγία δυσεπί-
σχετος ἐπηκολούθησεν ἐπὶ ταῖς τῶν ἐσχαρῶν ἀπο-
πτώσεσιν. ὥστε κἂν τούτοις ὅστις ἂν ἐθέλῃ μεδόθῳ

far as those of the throat, for it also holds back the blood from these without a ligature. However, we must not be hasty in this action, as some foolish surgeons are, but with one hand to protect the part of the vessel below, or having encircled it, we must hold back [the blood] while applying the medication to the wound with the other hand, protecting it gently to the extent that it may be fixed, and then, in such a way, to bandage it from above downward and not, as in the limbs, from below upward, for it is necessary for the bandage which checks the flow to come toward the roots of the vessels.

There are also many other emplastic medications that are painless, but none that is enfleshing in the same way. There is, in such symptoms, a particular use in surrounding the vessel with flesh when the prior medication falls off. In fact, those medications that coagulate eschars leave the part more exposed than normal when the eschar falls off. For the genesis of the eschar is from the underlying and surrounding bodies being half burned, one might say, so that the the eschar may be such a remnant of the burned flesh—the kind of thing people prepare from quenched coals in wintertime. Consequently, as much of the natural flesh is destroyed as is burned in forming the eschar. This all leaves when the eschar falls off the part, and because of this, it appears bare and fleshless, and in many instances a hemorrhage which is hard to control ensues due to eschars falling off. So that even in these instances, let anyone who

324K

πάντα πράττειν, ἐσκέφθω τε πρὸ πολλοῦ τοὺς τρόπους ἅπαντας οἷς ἐφέξει τὸ αἷμα, τόν τ᾽ ἀκινδυνότερον αἱρεῖσθαι πειράσθω χρώμενος καὶ τοῖς ἄλλοις ἅπασιν, ὅταν ἀνάγκη βιάζηται. μεγίστην δ᾽ ἀνάγκην οἶδα τοῦ χρῆσθαι φαρμάκοις ἐσχαρωτικοῖς ἢ καυτηρίοις

325K διαπύροις, | ὅταν ἐξ ἀναβρώσεως σηπεδονώδους ἡ αἱμορραγία γίγνηται. καὶ μέντοι κἀπειδὰν ἐν ταῖς τοιαύταις διαθέσεσιν ἅπαν ἐκκόψωμεν τὸ σηπόμενον, ἀσφαλέστερον ἤτοι καίειν αὐτοῦ τὴν οἷον ῥίζαν ἢ φαρμάκοις ἐσχαρωτικοῖς χρῆσθαι. καὶ κατὰ τοῦτο ἐπ᾽ αἰδοίων καὶ ἕδρας εἰς τὴν τοιαύτην ἀνάγκην ἀφικνούμεθα πολλάκις, ὅτι ῥᾳδίως σήπεται τὰ μόρια διά τε τὴν σύμφυτον ὑγρότητα καὶ ὅτι περιττωμάτων εἰσὶν ὀχετοί.

σκοπὸς δ᾽ ἔστω σοι τῶν ἐσχαρούντων φαρμάκων οὐχ ἁπλῶς ἡ θερμότης, ἀλλ᾽ ἄμεινον ὅταν ἅμα τῷ στύφειν ὑπάρχῃ, καθάπερ ἐν χαλκίτιδι καὶ μίσυϊ καὶ χαλκάνθῳ καυθεῖσί τε καὶ ἀκαύστοις εἰς χρείαν ἀγομένοις. τὰ δὲ διὰ τῆς ἀσβέστου τιτάνου σφοδρότερα μὲν τούτων, ἀλλ᾽ οὐ μέτεστι τῇ τιτάνῳ στυπτικῆς δυνάμεως. ἀποπίπτουσι τοιγαροῦν θᾶττον αἱ τοιαῦται τῶν ἐσχαρῶν, ἔχονται δὲ τῶν σωμάτων ἐπὶ πλέον αἱ ἐκ τῶν στυφόντων· ἄμεινον δὲ τοῦτο μακρῷ· φθάνει γὰρ ὑποσαρκοῦσθαι τὰ κατὰ τὴν βάσιν αὐτῶν καὶ γίνεσθαι καθάπερ τι πῶμα τοῖς ἀγγείοις. ὅθεν οὐδ᾽

326K ἡμᾶς αὐτοὺς | χρὴ σπεύδειν, ὥσπερ ἔνιοι, τὰς ἐσχάρας ἐκβάλλειν, ἔνθα κίνδυνος αἱμορραγίας, ἀλλ᾽ ἐπ᾽ ἐκείνων μόνων τῶν διαθέσεων, ἐφ᾽ ὧν διὰ σηπεδόνα τὸ

30

wishes to do everything by method give prior consideration to all the ways by which he will hold back the blood,
and let him try to choose the less dangerous, also using all
the other [ways] whenever necessity dictates. I know there
is a great need to use eschar-producing medications or
burning cauteries when a hemorrhage arises from the ero-   325K
sion of putrefaction. Nevertheless, whenever we eradicate
the putrefaction entirely in such conditions, it is safer either to burn the root of it, as it were, or to use eschar-producing medications. And in respect of this (i.e. putrefaction), we often come to such a necessity in the case of the
external genitalia and the anus, in that the parts readily putrefy due to their natural moistness and because they are
conduits for excretory materials.

Make your objective with the eschar-producing medications not just heating. Rather, it is better whenever there
is astringency in conjunction with this, as with chalcitis,
misu and copperas water, used either burned or unburned.[5] The medications made from unslaked lime are
stronger than these but do not, with the lime, partake of an
astringent capacity. Accordingly, such eschars fall off quite
quickly whereas those made from medications partaking
of astringency adhere longer, and this is better by far
because the formation of flesh will be from the base of
these and will be like a cover for the vessels. Whence it is
not necessary for us to hurry ourselves to cast aside the   326K
eschars as some do when there is a danger of hemorrhage,
but only in those conditions in which we are compelled to

5 For these three agents, see Dioscorides, V.115, 117, and 114
respectively.

καυτήριον ἠναγκάσθημεν ἐπιφέρειν ἀνθρώπου σώματι. καλεῖται δ' οὐκ οἶδ' ὅπως ἅπασιν ἤδη τοῖς ἰατροῖς ἡ τοιαύτη διάθεσις νομή, διότι νέμεσθαι συμβέβηκεν αὐτὴν ἀπὸ τῶν πεπονθότων μορίων ἐπὶ τὰ κατὰ φύσιν ἔχοντα καὶ τούτων ἀεί τι προσεπιλαμβάνειν, ὡς ἀπὸ τοῦ συμβεβηκότος, οὐκ ἀπὸ τῆς οὐσίας τοῦ δηλουμένου πράγματος, ἔθεντο τὴν προσηγορίαν. ἡ δ' εὐπορία τῆς ὕλης ἁπάντων τῶν τοιαύτην ἐχόντων δύναμιν φαρμάκων ἐν ταῖς οἰκείαις εἰρήσεται πραγματείαις. οἰκείαις δὲ δηλονότι λέγω τήν τε περὶ τῆς τῶν ἁπλῶν φαρμάκων δυνάμεως καὶ τὴν περὶ συνθέσεως αὐτῶν.

5. Ἀναλαβόντες οὖν αὖθις τὰ κεφάλαια τῆσδε τῆς μεθόδου, καθ' ἣν ἰᾶσθαι χρὴ τὰς προχείρους αἱμορραγίας, ἑξῆς ἐπὶ τὰς διὰ βάθους γιγνομένας ἀφικώμεθα τῷ λόγῳ. τὸ τοίνυν ἐκχεόμενον αἷμα τῶν ἀγγείων ἢ τῷ μηκέτι ἐπιρρεῖν αὐτὸ παύσαιτ' ἄν, ἢ τῷ
327K στεγνωθῆναι τὴν διαίρεσιν, | ἢ ἅμα ἀμφοτέρων γιγνομένων, ὅπερ, οἶμαι, βέλτιστόν ἐστιν. ἐπιρρεῖν μὲν οὖν κωλύεται διά τε λειποθυμίαν καὶ ἀντίσπασιν καὶ παροχέτευσιν καὶ ψύξιν ὅλου τε τοῦ σώματος, καὶ πολὺ δὴ μᾶλλον αὐτοῦ τοῦ τετρωμένου μορίου. κατὰ τοῦτον γοῦν τὸν λόγον καὶ τὸ ψυχρὸν ὕδωρ ποθέν, ἐπέσχεν αἱμορραγίας πολλάκις, ἔξωθέν τε καταντλούμενον αὐτό τε τὸ ὕδωρ ψυχρόν, ὀξύκρατόν τε καὶ οἶνος στρυφνός, ὅσα τ' ἄλλα στύφειν ἢ ἁπλῶς ψύχειν πέφυκε. στεγνοῦται δὲ ἡ διαίρεσις ἢ μύσαντος, ἢ

apply the cautery to a person's body because of putrefaction. Such a condition is now called by all doctors "noma"[6] (I don't know how) because this happens to spread from the affected parts to those that are normal, and always attacks one of these besides, as if they applied the term by chance and not from the essence of the indicated matter. The means of provision of the material of all the medications having such a potency will be spoken of in the relevant treatises. And obviously in the appropriate treatises I speak about the means of provision in relation to both the potency of the simple medications and what is compounded from these.[7]

5. Therefore, having taken up again the chief points of this method by which we should cure the readily accessible hemorrhages, let us come next in our discussion to those occurring in the depths. [In those], the blood flowing out of the vessels may stop, either by virtue of the blood itself no longer flowing, or by the division being closed over, or by both of these things occurring at the same time, which I think is the best. Flow is prevented by fainting, revulsion, diversion, cooling of the whole body, and more particularly of course, of the wounded part itself. On this point, cold from any source often prevents bleeding, whether it be cold water poured on externally, as well as oxykratos, astringent wine and other such things that are astringent or simply cooling by nature. The division is closed when what

327K

---

[6] The term "noma" seems here to have a wider meaning than currently, when it is restricted to such spreading infections involving the mouth, lips, and cheeks (cancrum oris) or the labia majora; see, for example, S, p. 1125.   [7] A further reference to the three treatises listed in Book 3, note 14.

φραχθέντος τοῦ διῃρημένου. μύει μὲν οὖν στυφόμενόν
τε καὶ ψυχόμενον, ἐπιδέσει τε συναγόμενον καὶ βρόχῳ
διαλαμβανόμενον. ἐμφράττεται δὲ ἢ ἔνδοθεν, ἢ ἔξω-
θεν· ἔνδοθεν μὲν ὑπὸ θρόμβου, ἔξωθεν δὲ ὑπό τε
τούτου καὶ μοτῶν καὶ σπόγγων, ἐσχάρας τε καὶ τῶν
ἐμπλαστικῶν φαρμάκων, ἔτι τε τῶν περικειμένων σωμάτων προσταλέντων· ὅπως δ᾽ ἐργάσῃ τῶν εἰρημένων
ἕκαστον εἴρηται. τὰ δ᾽ ἐκ τοῦ βάθους αἱμορραγοῦντα
βρόχοις μὲν οὐκ ἄν τις οὐδ᾽ ἐπιδέσει θεραπεύσειεν,
328K   ὥσπερ οὐδὲ καυστηρίοις, | οὐδ᾽ ἁπλῶς εἰπεῖν ὅσα διὰ
τοῦ ψαύειν αὐτοῦ τοῦ διῃρημένου σώματος ἢ μέρους
ἐπιτεχνώμεθα· δι᾽ ἀντισπάσεως δὲ καὶ παροχετεύσεως
ἐδεσμάτων τε καὶ πομάτων ἐμπλαττόντων τε καὶ ψυχόντων καὶ στυφόντων φαρμάκων· ἄφθονον δὲ καὶ τὴν
τῶν τοιούτων ὕλην ἐν ταῖς οἰκείαις ἔχεις πραγματείαις.

ἡ δ᾽ ἀπὸ τῶν μορίων ἔνδειξις τοῖς κοινοῖς τῆς
θεραπείας σκοποῖς τοῖς εἰρημένοις ἐξ ἐπιμέτρου καθ᾽
ἕκαστον πάθημα πρόσεισιν· ὀργάνοις γοῦν ἐνίοτε
χρώμεθα διὰ τὴν ἰδιότητα τοῦ μέρους, ἄλλοις μὲν ἐπὶ
μήτρας, ἄλλοις δὲ ἐπὶ κύστεως, ἄλλοις δὲ ἐπὶ τῶν
παχέων ἁπάντων ἐντέρων. εἰς ταῦτα μὲν γὰρ διὰ
κλυστῆρος, εἰς μήτραν δὲ διὰ μητρεγχυτῶν τῶν ἐπιτηδείων τι φαρμάκων ἐνίεμεν· ὥσπερ γε καὶ εἰς κύστιν
διὰ τῶν εὐθυτρήτων καθετήρων. σπανιώτεραι μὲν οὖν
αἱ ἐκ τούτων αἱμορραγίαι· γίγνονται δ᾽ οὖν ποτε κἂν εἰ
μὴ τῷ λάβρῳ[3] κινδυνώδεις, ἀλλὰ τῷ γε χρονίζειν οὐκ

[3] Β; τῷ λαύρῳ Κ

is divided is either closed up or contracted. Thus, it closes off with astringency and cooling; it is drawn together with a bandage; and it is cut off by a ligature. It is blocked up either internally or externally: internally by a clot and externally by this, by tampons and sponges, by an eschar, and by emplastic medications, and further, by the surrounding bodies being closed up. I have stated how you may do each of the aforementioned things. You will not treat hemorrhages from the depths with ligatures, nor with bandaging, just as you will not with cauterization, nor, to speak simply, with those things which we devise to affect the divided body or part itself. Rather, you will treat them by revulsion, diversion, emplastic foods and drinks, and by cooling and astringent medications. You have material in abundance on such medications in the specific treatises.[8]

328K

The indication [taken] from the parts is to be added to the common indicators of treatment mentioned additionally in relation to each affection. At any rate, we sometimes use instruments because of the particular property of the part, using different instruments for the uterus, bladder and all the large intestines. We insert one of the suitable medications into these [intestines] via a clyster, into the uterus via a *metrenchytos*,[9] just as [we do] into the bladder via a straight catheter. Hemorrhages from these organs are relatively rare. They do, however, occur sometimes, and even if there are no dangers due to their violence, they are not without danger due to their chronicity. Anyway, I know

[8] See note 3 above.

[9] A type of syringe for injection into the uterus; see Galen, XIII.316K, and Soranus, 2.41.

ἀκίνδυνοι. τεττάρων γοῦν ἡμερῶν οἶδά ποτε διὰ μήτρας φερόμενον αἷμα, πρὸς οὐδὲν τῶν βοηθημάτων
329K εἶξαν ἄχρι τῆς τετάρτης, | χυλῷ δὲ χρησαμένων ἡμῶν
ἀρνογλώσσου παντάπασιν ἐπαύσατο. κάλλιστον μὲν
οὖν τοῦτο τὸ φάρμακόν ἐστι καὶ πρὸς τὰς ἐξ ἀναβρώ-
σεως αἱμορραγίας· εἴωθα δ' αὐτῷ μιγνύειν τι τηνικαῦτα καὶ τῶν ἰσχυροτέρων φαρμάκων ἄλλοτε ἄλλο,
πρὸς ὅλην ἀποβλέπων τὴν διάθεσιν, ὅπερ ἀεὶ χρὴ
πράττειν, οὐ μικρὸν ἔχοντας εἰς ἅπαντα παράγγελμα.
κατὰ γοῦν αὐτὰ ταῦτα τὰ ἐκ μήτρας, ἢ ἐκ κύστεως, ἢ
ἐντέρων αἱμορραγοῦντα, σκέπτεσθαι μὲν χρὴ καὶ τὸ
τῆς αἱμορραγίας ποσόν· ἵν' ἤδη τὸν μὲν πρῶτον
ἔχωμεν ἐν τῇ θεραπείᾳ σκοπὸν τοῦτον, ἢ τὸν δεύτερον·
οὐ μὴν οὐδὲ τῆς ὅλης διαθέσεως ἀσκέπτως ἔχειν.

εἰ μὲν γὰρ ἔρρωγεν ἀγγεῖόν τι τῶν μεγάλων, ἢ
ἰσχυρῶς ἀνεστόμωται, τῶν αὐστηρῶν φαρμάκων ἐστὶ
χρεία, ὡς βαλαυστίου καὶ ὑποκυστίδος καὶ ῥοῦ καὶ
ὀμφακίου καὶ ἀκακίας καὶ κηκίδος ὀμφακίτιδος καὶ
τῶν τῆς ῥοιᾶς λεμμάτων· εἰ δ' ἤτοι μικρόν ἐστι τὸ
ἐρρωγὸς ἀγγεῖον, ἢ ἀνεστόμωται μετρίως, ὡς ὀλίγον
εἶναι τὸ φερόμενον αἷμα, καὶ ἀλόη καὶ μάννα καὶ ὁ
τῆς πίτυος φλοιός, ἥ τε σφραγὶς ἡ Λημνία καὶ ὁ τῆς
330K Αἰγυπτίας ἀκάνθης καρπός, ἔτι τε κρόκος καὶ λίθος | ὁ
αἱματίτης ὀνομαζόμενος, ὅσα τ' ἄλλα τοιαῦτα δι'
οἴνου μεγάλως αὐστηροῦ χρηστὰ φάρμακα. μὴ παρόντος δὲ μήτ' οἴνου τοιούτου μήτ' ἀρνογλώσσου μήτε
στρύχνου, καὶ γὰρ καὶ ταῦτα ἐπιτήδεια, ἀφεψεῖν ἐν
ὕδατι βλαστοὺς βάτου καὶ κυνοσβάτου καὶ μυρσίνης

36

of blood being passed from the uterus for four days which responded to none of the remedies until the fourth day, when I used the juice of plantain and it stopped altogether.   329K This medication is also best for hemorrhages due to erosion. Nevertheless, I am accustomed to mix one of the stronger medications with it, different ones at different times, paying attention under these circumstances to the whole condition, which we must always do, having quite a few examples of all of them. In these same hemorrhages from the uterus, bladder and intestines, we must also give consideration to the amount of bleeding, so that we might already have this as the first indicator in the treatment, or the second. Nor should we be unmindful of the whole condition.

If one of the major vessels is ruptured or severely opened up, there is a use for the astringent medications, such as the flower of the wild pomegranate, hypocystis, sumac, omphacium, acacia, oak gall and the leaves of the corn poppy. If either the ruptured vessel is small, or it is only moderately opened up so the blood it carries is slight in amount, aloes, manna, pine bark, a troche of Lemnian earth and fruit of the Egyptian thorn, and even saffron and the stone which is called bloodstone (hematite) and other   330K such medications made from wine that is very harsh are useful. If no such wine is available, nor plantain, nor nightshade (for these things are also suitable), decoct berries of bramble, white rose, myrtle, mastich, ivy or, in a word, any-

καὶ σχίνου καὶ κισσοῦ καὶ ἁπλῶς εἰπεῖν ἁπάντων τῶν
στυφόντων, εἴτε ῥίζα τις, εἴτε καρπός, εἴτε φλοιός, εἴτε
βλάστημα τύχοι· διὰ τοῦτο καὶ τῶν στυφόντων μήλων
καὶ μάλιστα τῶν κυδωνίων, ἔτι τε μύρτων καὶ μεσπί-
λων ἐπιτήδειόν ἐστιν εἰς τὰ παρόντα τὸ ἀφέψημα.

6. Εἰ δ' ἐξ ἀναβρώσεώς τις αἱμορραγία γίγνοιτο,
τὰ πολλὰ μὲν οὐδὲ λαῦρος αὐτοῖς ἐστιν, ἀλλ' ὀλίγη
καὶ κατὰ βραχύ· καὶ διὰ τοῦτο τῷ Πασίωνος, ἢ Ἄνδρωνος, ἢ Πολυείδου τροχίσκῳ χρηστέον, ἢ ὡς εἴρηται τῷ ἡμετέρῳ τὴν αὐτὴν μὲν κατὰ γένος ἔχοντι
δύναμιν, ἰσχυροτέρῳ δὲ ὑπάρχοντι. παύει γὰρ ταῦτα
τὴν ἀνάβρωσιν ἅμα τῷ δηλονότι καὶ τοῦ παντὸς
σώματος προνοεῖσθαι, ὡς ἔμπροσθεν εἴρηται. λαύρου
δὲ ἐμπεσούσης αἱμορραγίας, τοῖς αὐστηροτάτοις
χρηστέον φαρμάκοις, ἄχρι περ ἂν ἐπισχῶμεν αὐτῆς
331K τὸ | σφοδρόν· εἶθ' οὕτως ἅμ' ἐκείνοις μιγνύναι τοὺς
τροχίσκους· εἶθ' ἑξῆς ἐπ' αὐτοὺς μόνους μετέρχεσθαι
μετά τινος τῶν εἰρημένων χυλῶν ἢ ἀφεψημάτων. ὅσα
δὲ ἔξωθεν ἐπιτίθεται τοῖς αἱμορραγοῦσι στύφοντά τε
καὶ χωρὶς στύψεως ψύχοντα, ταῦτ' οὐχ ἁπλῶς ἐπαινῶ
καθάπερ οἱ πολλοὶ τῶν ἰατρῶν, ἀλλά μοι δοκεῖ τοὐναντίον ἐνίοτε πᾶν οὗ δεόμεθα διαπράττεσθαι, συν-
ελαύνειν εἴσω τὸ αἷμα καὶ πληροῦν τὰς ἐν τῷ βάθει
φλέβας. οἶδα γοῦν τινας ἐκ πνεύμονος ἀναβήττοντας
αἷμα προφανῶς βλαβέντας ἐπὶ ταῖς τοῦ θώρακος
καταψύξεσιν· οὕτω δ' ἐνίους καὶ τῶν ἐμούντων αἷμα,
τῆς γαστρὸς ἔξωθεν ψυγείσης· ὡσαύτως δὲ καὶ τῶν ἐκ

thing that is astringent, whether it happens to be roots, fruit, bark or buds. This is why the decoctions of astringent apples, and particularly of the Cydonian quince, and further, of myrtle and medlar are suitable for the matters under present consideration.

6. If hemorrhages occur due to erosion, the majority are not fierce in these cases but slight and gradual and, because of this, you must use a troche of Pasion, Andron or Polyeides,[10] or, as I said, that of my own devising which has the same [type of] potency in terms of class, although it is stronger. For these [medications] bring the erosion to an end along with obviously providing for the whole body, as was said before. When fierce hemorrhage does occur, we must use the astringent medications up to the point where we control its violence. Then similarly mix the troches with those, and next in turn, proceed to the troches alone along with one of the aforementioned juices or decoctions. I absolutely do not recommend, as many doctors do, those things that are applied externally to parts that are bleeding, and are cooling without being astringent. Rather, it seems to me, we sometimes need to do the exact opposite; that is, to drive the blood inward and fill the veins in the depths. In fact, I know some people who, when they brought up blood from the lung, were clearly harmed by cooling of the chest—some to such an extent that they also vomited blood since the stomach was cooled externally. Similarly, I

331K

---

[10] For the composition of these troches or pastilles, see the list of medications in the Introduction (section 10).

ῥινὸς αἱμορραγούντων ἔνιοι φανερῶς ἐβλάβησαν ἐπὶ ταῖς τῆς κεφαλῆς ψύξεσιν.

οὔκουν ἁπλῶς, οὐδ' ἀδιορίστως, οὐδ' ἐν παντὶ καιρῷ συμβουλεύω ψύχειν τὰ κύκλῳ τῶν αἱμορραγούντων, ἀλλ' ἀποτρέψαντας εἰς ἕτερον πρότερον, οἷον ἐπὶ τῆς ῥινὸς ἤτοι φλεβοτομίᾳ χρησάμενον, ὡς εἴπομεν, ἢ τρίψει καὶ δεσμοῖς τῶν κώλων, ἢ σικύαις ἐπὶ τῶν

332K ὑποχονδρίων. οὐ δεῖ | οὖν οὐδ' οὕτω χρῆσθαι τοῖς ψύχουσιν εὐθέως ἐπὶ μετώπου καὶ κεφαλῆς[4], ἀλλὰ σικύᾳ πρότερον ἀντισπᾶν ἐρείδοντα κατ' ἰνίον. διττὴ γὰρ ἡ ἀντίσπασις ἐπὶ ταῖς ἐκ μυκτήρων αἱμορραγίαις, ἥ τε ἐπὶ τὰ κάτω μόρια τοῦ σώματος ὅλου καὶ ἡ ἐπὶ τὰ τῆς κεφαλῆς ὀπίσω· διότι καὶ αὐτὸ τὸ μόριον ἡ ῥὶς ἄνω τ' ἐστὶ καὶ πρόσω, ἀντίκειται δὲ τὸ μὲν ἄνω τῷ κάτω, τὸ δὲ ὀπίσω τῷ πρόσω· ἀλλὰ περὶ μὲν αἱμορραγίας ἱκανὰ καὶ ταῦτα.

δῆλον γὰρ ὅτι καὶ ἡ διαπήδησις ὑπὸ τῶν ψυχόντων καὶ στυφόντων θεραπευθήσεται· καὶ εἰ διὰ λεπτότητα γίγνοιτό ποτε τοῦ αἵματος, ὑπὸ τῆς παχυνούσης διαίτης· ὁποία δ' ἐστὶν αὕτη λεχθήσεται μέν που κἀν τοῖς ἑξῆς, εἴρηται δὲ δυνάμει κἀν τῷ Περὶ λεπτυνούσης διαίτης· ἐπὶ δὲ τὴν θεραπείαν ἰτέον ἤδη τῶν εἰρημένων διαθέσεων. τὸ μὲν οὖν τῆς φλεβὸς ἕλκος ἐπειδὰν πρόχειρον ᾖ, παραπλησίας δεῖται θεραπείας τοῖς ἐν σαρκὶ γιγνομένοις, ὑπὲρ ὧν ἐν δυοῖν ὑπομνήμασι τοῖς πρὸ τοῦδε διελέχθην. εἴτε γὰρ ἐκ τρώσεως ὑπογυίου

---

[4] B; μετώπῳ καὶ κεφαλῇ K

have seen some of those bleeding from the nose clearly harmed by cooling of the head.

I do not, therefore, simply or uncritically, or on every occasion, recommend cooling those parts which surround the sites of bleeding, but after diverting [the blood] to another part first. For example, in the case of the nose, I recommend the use of phlebotomy, as I said, or rubbing and bindings of the limbs, or cupping glasses [applied] to the hypochondrium. You must not use cooling agents in this way immediately in the case of the face and head, but first use a cupping glass affixed to the occiput to drive [the blood] back. For revulsion in bleeding from the nostrils is twofold: downward to the parts of the whole body and backward to the posterior parts of the head. This is because the nose itself is a part which is superior and anterior, and what is superior stands opposite to what is inferior and what is anterior stands opposite to what is posterior. But this is enough about hemorrhages.

332K

It is clear that *diapedesis* too will be treated by things that are cooling and astringent, and if it sometimes occurs due to a thinness of the blood, by a thickening diet. What sort of diet this is will be spoken of at some point in what follows, and has been spoken of, in one respect, in the work *On the Thinning Diet*.[11] We must now proceed to the treatment of the aforementioned conditions. The wound of the vein, whenever it is readily accessible, requires treatment similar to those for wounds occurring in the flesh, which I discussed in the two books prior to this one. If it has oc-

[11] *De victu attenuante*. This work is not in Kühn. There is an edition by K. Kalbfleisch (1898) and a recent English translation by P. N. Singer (1997).

333K τυγχάνει γεγονός, ἐγχωρεῖ συμφύειν αὐτὸ τοῖς | ἐναί-
μοις καλουμένοις φαρμάκοις· εἴτε κατ᾽ ἀνάβρωσιν,
ὅσα περὶ τῶν κακοηθῶν ἑλκῶν εἴρηται, ταῦτα τῷ λόγῳ
διορισάμενον, ἑξῆς ἔργῳ πράττειν αὐτὰ πειρᾶσθαι.
καὶ μέντοι καὶ ὁπότε βρόχῳ διαληφθείσης, ἢ ἰσχαίμοις φαρμάκοις, ἢ καυτηρίοις χρησαμένων, ὁ σκοπὸς
τῆς θεραπείας ἐστὶ περισαρκῶσαι τὰ χείλη τοῦ ἕλκους, τοῖς αὐτοῖς χρησόμεθα πάντως φαρμάκοις οἷς ἡ
μέθοδος ἡμᾶς ἐδίδαξεν ἐπὶ τῶν κοίλων ἑλκῶν.

7. Ἐπ᾽ ἀρτηρίας μέντοι τρωθείσης ἐρρήθη καὶ
πρόσθεν ὡς ἐνίοις τῶν ἰατρῶν ἀδύνατος ἡ σύμφυσις
εἶναι δοκεῖ. καὶ λέγουσι τοῦτό τινες μὲν ἀρκεῖσθαι
φάσκοντες μόνῃ τῇ πείρᾳ, τινὲς δὲ καὶ λόγῳ χρώμενοι
τοιῷδε, σκληρὸν καὶ χονδρῶδη φασὶν εἶναι τὸν ἕτερον
τῶν χιτώνων τῆς ἀρτηρίας· οὐδὲν δὲ τῶν τοιούτων
συμφῦναι δύνασθαι, μόνων γὰρ τῶν μαλακῶν σωμάτων ἰδίαν ὑπάρχειν τὴν σύμφυσιν, ὡς ἐπί τε τῶν ἐκτὸς
ἔνεστι θεάσασθαι, μήτε λίθου πρὸς λίθον ἑνωθέντος
ποτὲ μήτε ὀστράκου πρὸς ὄστρακον· ἐπί τε τῶν ἐν
ἡμῖν, μήτε χόνδρου χόνδρῳ συμφύντος οὔτ᾽ ὀστοῦ |
334K πρὸς ὀστοῦν ἑνωθέντος. οὐδὲ γὰρ τὰ κατάγματα διὰ
συμφύσεως, ἀλλὰ διὰ πώρου τὴν κόλλησιν ἴσχειν.

ἀρξώμεθα οὖν καὶ ἡμεῖς ἀπὸ προτέρας τῆς ἡμετέρας πείρας, εἰς τὸ κοινὸν κατατιθέντες ὅσα τυγχά-
νομεν ἑωρακότες. ἐθεασάμεθα γὰρ ἐπὶ μὲν γυναικῶν

curred from a recent wound, it is possible to bring it together with the so-called hemostatic medications. If it has   333K
occurred from an erosion, use those medications mentioned in regard to the *kakoethical* ulcers.[12] Having first
distinguished these things in theory, the next task is to attempt to carry them out in practice. And indeed, whenever
the bleeding has been controlled by a ligature, or if we
are using hemostatic medications or cauteries, the aim of
treatment is to cover the margins of the wound with flesh,
and we shall in general use the same medications which
the method taught us in the case of hollow wounds and
ulcers.

7. However, in the case of an artery that has been
wounded it was also stated earlier that to some doctors
union seems to be impossible. And some say they are content to make this claim on the basis of experience alone,
while others also use the following argument: they say that
one of the tunics of an artery is hard and cartilaginous, and
that nothing of this kind can unite because union is specific
to soft bodies alone, as it is possible to see in the case of
things outside [the body]—stone can never be united with
stone nor shell with shell. And in our own case, cartilage
does not unite with cartilage nor bone with bone, for frac-   334K
tures do not maintain union by growing together but
through a callus.

Therefore, let me also begin from my own prior experience, adding to the common pool those things I happen to
have seen. For I have seen, in the case of women and chil-

---

[12] On the use of this transliterated term, see the introductory
section on terminology (section 6). We have chosen to avoid "malignant" in view of the current, more specific meaning in relation
to an ulcer.

καὶ παιδίων ἀρτηρίας κολληθείσας τε καὶ περισαρκω-
θείσας ἐν μετώπῳ καὶ σφυρῷ καὶ καρπῷ. νεανίσκῳ δ᾽
ἀγροίκῳ ποτὲ καὶ τοιοῦτόν τι συνέπεσε, φλέβα τμη-
θῆναι βουληθέντι κατὰ τὴν ἐαρινὴν ὥραν, ἐν ἔθει δὲ
μάλιστα τοῦτό ἐστι τοῖς παρ᾽ ἡμῖν ἀνθρώποις, δήσαν-
τος τὸν βραχίονα τοῦ μέλλοντος αὐτὸν φλεβοτομεῖν
ἰατροῦ, συνέβη κυρτωθῆναι τὴν ἀρτηρίαν, ὥστε καὶ
διεῖλεν αὐτὴν ὁ ἰατρὸς ἀντὶ φλεβός. βραχεῖα μὲν οὖν
ἡ διαίρεσις ἐγένετο, ξανθὸν δὲ καὶ λεπτὸν καὶ θερμὸν
εὐθέως ἐξηκοντίζετο τὸ αἷμα σφυγμωδῶς. ὁ μὲν οὖν
ἰατρός, ἦν δὲ καὶ πάνυ νέος καὶ ἄπειρος τῆς τέχνης
τῶν ἔργων, ᾤετο διῃρηκέναι τὴν φλέβα. θεασάμενος
δ᾽ ἐγὼ τὸ γεγονὸς σὺν ἄλλῳ τινὶ τῶν παρόντων ἰατρῶν
πρεσβύτῃ, τῶν ἐναίμων τι φαρμάκων τῶν ἐμπλαστῶν
παρασκευάσας ἐπιμελῶς τε πάνυ συναγαγὼν τὴν
335K   διαίρεσιν, ἐπέθηκά τε τὸ | φάρμακον αὐτίκα καὶ σπόγ-
γον μαλακώτατον ἔξωθεν ἐπέδησα.

θαυμάζοντος δὲ τοῦ τέμνοντος τὴν ἀρτηρίαν τὸν
λόγον τῆς ἐπελθούσης ἡμῖν περὶ τὰ τοιαῦτα προνοίας,
ἐδηλώσαμεν αὐτῷ τὸ γεγονός, ἐξελθόντες τῆς κατ-
αγωγῆς τοῦ τμηθέντος· ἐκελεύσαμέν τε μήτε λῦσαι
χωρὶς ἡμῶν μήτε θᾶττον πρᾶξαι τοῦτο τῆς τετάρτης
ἡμέρας, ἀλλ᾽ ὡς εἶχεν ἡ ἐπίδεσις φυλάξαι ἐπιτέγγοντα μόνον τὴν σπογγιάν· ἐπεὶ δὲ λύσαντες ἐν τῇ
τετάρτῃ κεκολλημένην ἀκριβῶς εὕρομεν τὴν διαίρεσιν, ἐπιθεῖναί τε τὸ αὐτὸ φάρμακον αὖθις ἐκελεύσαμεν, ἐπιδήσαντά τε ὁμοίως ἡμερῶν πλειόνων μὴ λύειν.
οὕτω μὲν ἐθεραπεύθη τὴν διαίρεσιν τῆς ἀρτηρίας ὁ

dren, arteries being conglutinated and surrounded by flesh in the face, ankle and wrist. Such a thing happened on one occasion in a young country lad who wanted a vein to be cut in the spring, for this particularly is a custom among our countrymen. When the doctor bound the arm which he was about to phlebotomize, it happened that the artery was made to bulge so that the doctor cut this open instead of the vein. The division was small and the blood that immediately spurted forth in a pulsatile manner was yellow, thin and hot. The doctor, who was very young and inexperienced in the practice of the craft, thought he had cut through the vein. When I, along with one of the other, older doctors present, saw what had happened, I prepared one of the hemostatic medications of the emplastic type, then carefully brought the division together, immediately placed the medication on it, and bound on a very soft sponge externally.

335K

The doctor who had cut the artery was amazed at the rationale, encountering my prior knowledge in such matters. I indicated to him what had happened and went over the restoration of what had been cut. I directed him not to release [the bandage] without my being present, and not to do so sooner than the fourth day, but to keep the bandage as it was, only moistening the sponge. When, having released it on the fourth day, I found that the division had conglutinated completely, I ordered this same medication to be applied again, binding it in the same manner without releasing for many days. That man, whom I saw with an incision in an artery at the elbow, was the only one in whom a

ἄνθρωπος ἐκεῖνος μόνος ὃν εἶδον ἐν ἀγκῶνι τμηθέντα
ἀρτηρίαν· ἅπασι δὲ τοῖς ἄλλοις ἀνεύρυσμα τοῖς μὲν
μεῖζον ἐπεγένετο, τοῖς δ' ἧττον. ὅπως μὲν οὖν χρὴ
θεραπεύειν ἀνευρύσματα προϊόντος τοῦ λόγου κατὰ
τὸν οἰκεῖον εἰρήσεται καιρόν, ὅταν καὶ τῶν ἄλλων
ἁπάντων ὄγκων τῶν παρὰ φύσιν ἡ θεραπεία γράψη-
ται· νυνὶ δὲ ἐπὶ τὸ συνεχὲς ἴωμεν τοῦ λόγου.

336K   ἡ γάρ τοι φύσις ἡ | τῆς ἀρτηρίας ὄντως ἐνδείκνυται
τὸ δυσσύμφυτον, οὐ τὸ παντάπασιν ἀσύμφυτον τοῦ
σκληροῦ τῶν ἐν αὐτῇ χιτώνων· οὐ γὰρ οὕτως ἐστὶ
ξηρὸς καὶ σκληρὸς ὡς ὀστοῦν ἢ χόνδρος, ἀλλὰ πολὺ
μαλακώτερος τούτων καὶ σαρκωδέστερος. οὔκουν εὐλόγως ἄν τις ἀπογινώσκοι συμφῦναι τὴν διαίρεσιν,
ὅταν αὐτή τε γένηται μικρὰ καὶ τὸ σῶμα τἀνθρώπου
μαλακὸν ὑπάρχει φύσει. ἔοικε δὲ καὶ ἡ πεῖρα τῷ λόγῳ
μαρτυρεῖν· ἐπί τε γὰρ παίδων καὶ γυναικῶν ἐθεασάμην συμφύσασαν αὐτὴν διὰ τὴν ὑγρότητα καὶ
μαλακότητα τῶν σωμάτων· ἐφ' ἑνός τε νεανίσκου
βραχεῖαν ἴσχοντος διαίρεσιν, ὡς εἴρηται. δυσιατο-
τέρα μὲν οὖν ἐστιν ἡ ἀρτηρία τῆς φλεβός, οὐ μὴν
ἐξήλλακταί γε πολλῷ τῶν φαρμάκων ἡ χρῆσις ἐφ'
ἑκατέρου τῶν ἀγγείων, ἀλλ' ἔστιν ἡ αὐτὴ κατ' εἶδος,
ἐν τῷ μᾶλλόν τε καὶ ἧττον διαφέρουσα· τοσούτῳ γὰρ
δεῖται ξηραντικωτέρων φαρμάκων ἡ ἀρτηρία τῆς φλεβός, ὅσῳπερ καὶ φύσει ξηροτέρα τὴν κρᾶσίν ἐστιν. εἰ
δὲ περισαρκοῦσθαι δέοι, τῶν αὐτῶν ἀμφότεραι χρή-
ζουσι· σαρκὸς γάρ ἐστι γένεσις ἐν τῇ περισαρκώσει

46

division of an artery was treated in this way. In all other instances, an aneurysm arose, larger in some, smaller in others. How one should treat aneurysms will be spoken of when the discussion advances to the proper point, and when the treatment of all other swellings contrary to nature is written about.[13] For the present, however, let me proceed to the next stage of the discussion.

Surely the nature of an artery actually indicates difficulty of union, although not complete failure of union of the hard wall in it. For it is not dry and hard in the way that bone or cartilage is, but is much softer and fleshier than these. It is unreasonable, therefore, to despair of the division growing together whenever it is small and the body of the person is soft in nature. And it seems that experience bears out theory, for in children and women I have seen the division grow together due to the moistness and softness of their bodies, and the same in the case of the one young man having a slight division, as I said. An artery is, then, more difficult to heal than a vein, but the use of the medications is not in fact changed by much in each of the [two types of] vessel but is the same in terms of kind, differing only in terms of more or less. For an artery needs more drying medications than a vein to the extent that it is more dry in *krasis*. If the vessel needs to be surrounded by flesh, both require the same thing, because in the surrounding process the generation of flesh occurs in the

336K

13 The subject is addressed in the final two books of this treatise and in the specific work *De tumoribus praeter naturam*, VII.705–32K.

47

337K κατὰ τὸν αὐτὸν τρόπον γεννωμένης, | ὡς κἂν τοῖς κοίλοις ἕλκεσιν ἐδείχθη γιγνομένη.

αἱ δὲ κατὰ τὴν μήτραν, ἢ κύστιν, ἢ ἔντερα φλέβες καὶ ἀρτηρίαι, τῶν αὐτῶν δεόμεναι κατὰ γένος φαρμάκων ὅταν ἑλκωθῶσιν, ὀργάνων χρήζουσι τῶν εἴσω παραπεμψόντων αὐτά, μητρεγχυτῶν δηλονότι καὶ καθετῆρος εὐθυτρήτου καὶ κλυστῆρος. εἰς δὲ τὰ κατὰ τὸ ἀπευθυσμένον ἕλκη καὶ διὰ κύστεως ἐγχωρεῖ ἐγχεῖν φάρμακον τηκτὸν χλιαρόν, αὐλίσκον ἐχούσης κατὰ τὸ πέρας αὐτῆς εὐθύτρητον. ἐξελεγχθήσεται τοιγαροῦν ἅμα τῇ τῶν ὀργάνων ἰδέᾳ καὶ ἡ τῶν φαρμάκων σύστασις· οὐ γὰρ οἷόν τε παχέα φάρμακα τοῖς τοιούτοις ἐγχεῖν, ἀλλ᾽ ὑγροτέρων δηλονότι δέονται καὶ διὰ τοῦτο συμμέτρως θερμῶν ὡς τὰ πολλά· διὰ τοῦτο δὲ καὶ τῶν τηκτῶν καλουμένων φαρμάκων ἐπιτηδειότερα τὰ ξηρά· μίγνυται γὰρ ἑτοίμως, εἴτ᾽ ἀρνογλώσσου τις, εἴθ᾽ ἑτέρου τοιούτου βούλοιτο χυλῷ. τὰ δὲ τοιαῦτα φάρμακα κρόκος τ᾽ ἐστὶ καὶ πομφόλυξ καὶ ἀλόη καὶ τὰ κεφαλικὰ καλούμενα· κατὰ δὲ τὸν πρῶτον καιρὸν τῆς σαρκώσεως, τὸν συνάπτοντα τῇ ἐπισχέσει τοῦ αἵματος, καὶ ἡ Λημνία σφαγὶς ἀγαθὸν φάρμακον. |

338K 8. Τὰ δ᾽ ἐν τῷ πνεύμονι συνιστάμενα τῶν ἑλκῶν χαλεπωτέραν ἔχει τὴν ἴασιν. ἐνίοις δ᾽ οὐ χαλεπὴ μόνον, ἀλλὰ καὶ ἀδύνατος εἶναι δοκεῖ τῷ τε λόγῳ τεκμαιρομένοις καὶ τῇ πείρᾳ· τῷ λόγῳ μέν, ἐπειδὴ διὰ τὴν ἀναπνοὴν ἀεικίνητόν ἐστι σπλάγχνον ὁ πνεύμων, ἡσυχίας δὲ δεῖ τοῖς μέλλουσιν ἰαθήσεσθαι· τῇ πείρᾳ δ᾽, ὅτι μηδὲ πώποτε μηδένα τῶν τοῦτο παθόντων

same way as it was shown to occur in hollow wounds and 337K
ulcers.

The veins and arteries in the uterus, bladder and intestines, being in need of the same medications in terms of class whenever they are wounded, require instruments for conveying these medications inward: the *metrenchytos*, obviously, the straight catheter and the clyster. For wounds and ulcers in the rectum, it is also permissible to pour in a medication that is dissolved and lukewarm through the bladder by having a catheter passed straight through to the end of this. Therefore, both the form of the instrument and the consistency of the medication will have a bearing on this, for it is not possible to deliver thick medications with such instruments. Obviously, they require medications that are more liquid, and because of this, those that are moderately warm for the most part. Because of this too, dry medications are more suitable than the so-called dissolved medications, for they are more readily mixed with the juice of plantain, or some other such thing, should someone wish to do so. Medications of this sort are saffron, pompholyx, aloes and the so-called cephalics. However, on the first occasion of flesh production, which is in combination with the stoppage of the blood, Lemnian earth is also a good medication.

8. Wounds in the lung are more difficult to cure. How- 338K
ever to some who form a judgment based on reason and experience, they seem to be not only difficult but even impossible—by reason, because the lung is an organ in perpetual motion due to respiration, whereas it needs rest in those who are going to be cured, and by experience because they have never seen anyone of those so affected

ἐθεάσαντο θεραπευθέντα. τὸ μὲν δὴ τῆς πείρας, ἐντεῦθεν γὰρ ἄρξασθαι δίκαιον, ἀμφισβήτησιν ἴσως
ἕξει κατὰ τὴν διάγνωσιν. ἡμεῖς γοῦν ἐθεασάμεθα τὸν
μέν τινα τῷ βοῆσαι μέγα, τὸν δὲ τῷ καταπεσεῖν ἀφ᾽
ὑψηλοῦ, τὸν δ᾽ ἐν παλαίστρᾳ πληγέντα, παραχρῆμά
τε βήξαντα σφοδρότατα καὶ σὺν τῇ βηχὶ τοὺς μὲν
μίαν ἢ δύο κοτύλας, ἐνίους δὲ καὶ πλείους πτύσαντας.
αὐτῶν δὲ τῶν ταῦτα παθόντων ἔνιοι μὲν ἀνώδυνοι
παντάπασιν ἦσαν, ἔνιοι δ᾽ ὠδυνῶντο κατὰ θώρακα.
καὶ τοίνυν καὶ τὸ αἷμα τοῖς μὲν ὀδυνηθεῖσιν οὔτ᾽
ἀθρόον ἦν οὔτε πολὺ καὶ ἧττον ἐρυθρὸν καὶ ἧττον
θερμόν, ὡς ἂν πόρρωθεν ἧκον· τοῖς δ᾽ ἀνωδύνοις οὖσιν |

339K ἀθρόον τε καὶ πλεῖστον, ἐρυθρόν τε καὶ θερμὸν ἀν-
εβήττετο, σαφῶς ἐνδεικνύμενον ὅτι μὴ πόρρωθεν ἥκει.

  ἔχει μὲν οὖν ἀπορίαν καὶ ἄλλην οὐ σμικρὰν ἡ ἐκ
τοῦ θώρακος εἰς τὰς τραχείας ἀρτηρίας μετάληψις τοῦ
αἵματος, ἐζητημένην τοῖς ἰατροῖς ὅσοι ἀδύνατον εἶναι
νομίζουσι τὴν διὰ τοῦ περιέχοντος ὑμένος τὸν πνεύμονα φορὰν αὐτοῦ. καὶ τάχ᾽ ἄν, εἰ μὴ πολλοῖς τῶν
οὕτω παθόντων ὅ τε παραχρῆμα πόνος ἥ τ᾽ ἐξ ὑστέρου
τισὶν αὐτῶν ἐπιγενομένη φλεγμονὴ σὺν ἀποστάσει
σαφῆ τὴν ἔνδειξιν ἔφερε τοῦ κατὰ τὸν θώρακα
πάθους, ἀπεφήναντο μηδέποτ᾽ ἐκ θώρακος αἷμα διὰ
τῆς φάρυγγος ἀναβήττεσθαι. νυνὶ δ᾽ ὑπὸ τούτων δυσ-
ωπούμενοι συγχωροῦσι μὲν ἀπὸ τοῦ θώρακος φέρε-

who was treated [successfully]. Certainly, in terms of experience, for it seems right to begin here, perhaps there will be some dispute about the diagnosis. At all events, I have seen those who, due to shouting loudly, or falling from a height, or suffering a blow in a wrestling school, coughed very violently and immediately spat up blood with the coughing, some producing one or two cupfuls. Some of those so affected were altogether free of pain whereas others suffered pain in the chest. Moreover, the blood in those who suffered pain was not produced all at once or in great amount, and was not very red and hot, as if it were coming from afar. On the other hand, in those who were without pain, the blood was coughed up all at once and copiously, and was red and hot, clearly indicating that it did not come from afar.

339K

The transfer of blood from the chest to the rough arteries[14] presents a puzzle of some moment. It has been investigated by doctors who think that the passage [of blood] through the membrane surrounding the lung is impossible. And perhaps, if the distress is not immediate for many of those affected in this way, or in some of them when inflammation with an abcess later supervenes, which clearly bears the indication of an affection in the chest, they claimed that blood was never coughed up from the chest through the pharynx. For the present, however, those perplexed by these indications agree that [the blood]

---

[14] There is some doubt about terminology here. We have taken τὰς τραχείας ἀρτηρίας in the plural as "bronchial tree and trachea." When the singular is used, we have translated this as "trachea." Linacre here has the singular for the Greek plural, while Peter English translates as "sharp arterie" (p. 96).

σθαι, ζητοῦντες δ᾽ ἑτέραν ὁδὸν τῆς διὰ τοῦ περιέχοντος ὑμένος τὸν πνεύμονα, πολλὰ καὶ ἄτοπα λέγειν
ἀναγκάζονται.

καίτοι τό γε πῦον ἐπὶ τῶν ἐμπύων ἐν τῇ μεταξὺ
χώρᾳ θώρακός τε καὶ πνεύμονος ὁμολογοῦντες περιέχεσθαι βλέπουσιν, οἶμαι, σαφῶς ἀναβηττόμενον ἐκ
πνεύμονος. ἡμεῖς δὲ κἀπὶ τῶν ἀπόστημα τοιοῦτον ἐν
340K θώρακι ἐχόντων, ὡς σφακελίσαι τι καὶ τῶν | ὀστῶν,
ἐπεδείξαμεν αὐτοῖς ἐναργῶς τὸ ἐνιέμενον τῷ θώρακι
μελίκρατον ἀναβηττόμενον ἐκ πνεύμονος. ἴσμεν γὰρ
δήπου τὰς τοιαύτας διαθέσεις ἐν θώρακι συνισταμένας οὐκ ὀλιγάκις ἐν Ῥώμῃ μάλιστα τρόπῳ ῥευ-
ματικῷ, ὡς ἀναγκασθῆναι τὸ πεπονθὸς ἐκκόπτειν
ὀστοῦν. ἐπὶ τῶν πλείστων δ᾽ εὐθὺς ἅμα τῇ διεφθαρ-
μένῃ πλευρᾷ καὶ ὁ ὑπεζωκὼς αὐτὴν ὑμὴν ἔνδοθεν
εὑρίσκεται διασεσηπώς. εἰώθαμεν οὖν ἐν τῇ θεραπείᾳ
μελίκρατον ἐκχέοντες διὰ τοῦ ἕλκους, ἐνίοτε μὲν ἐγκε-
λεύεσθαι βήττειν αὐτοῖς ἐπικεκλιμένοις τῷ πεπονθότι
μέρει, πολλάκις δὲ κατασείειν ἠρέμα, καί ποτε καὶ
πυουλκῷ κομίζεσθαι τὸ ὑπολειπόμενον ἔνδον τοῦ
μελικράτου. ἐπειδὰν δὲ τοῦτο ποιήσαντες ἐκκεκλύσθαι
τότε πῦον ἅπαν καὶ τοὺς ἰχῶρας τοῦ ἕλκους ἐλπίσω-
μεν, ἐντίθεμεν οὕτως τὰ φάρμακα. κατὰ τὰς τοιαύτας
οὖν διαθέσεις, εἴ τις ἐάσειεν ἐν τῇ μεταξὺ πνεύμονος
καὶ θώρακος χώρᾳ τὸ ὑγρόν, αὐτίκα ἀναβηττόμενον
θεάσεται. θαυμάσαι δ᾽ ἐστὶν ἐπὶ τῇ δια[5] τοῦ πνεύμο-

[5] B; διά om. K

is carried from the chest, but when they seek a path other than that through the membrane surrounding the lung, they are forced to utter many absurdities.

Moreover, since they agree that in the case of the empyemas, they see the pus to be contained in the space between the chest wall and the lung, clearly, I think, what is being expectorated is from the lung. And even in the case of those who have an abscess in the chest wall of such a kind as to cause necrosis of the bones as well, I clearly demonstrated to them that melikraton put into the chest wall is expectorated from the lung. For we know, of course, that such conditions exist in the chest wall not infrequently in Rome in a manner particularly conducive to discharge, so that excision of the affected bone is obligatory. In the majority of these cases, along with the corrupted pleura, the membrane directly underlying it internally is found to be putrefied. Therefore, it is my custom in the treatment, when I pour melikraton in through the wound, sometimes to urge patients to cough while lying inclined toward the affected part, and often to shake [them] gently, and sometimes also to introduce the rest of the melikraton with a *pyulcus*.[15] When I have done this, being hopeful that I have washed away all the pus and ichors of the wound at that time, I apply the medications in this way. In such conditions then, if you leave some fluid in the space between the chest wall and the lung, you will see this immediately expectorated. If this is astonishing to those who are per-

340K

---

15 Described as an instrument for drawing off pus and treating fistulae; see Galen, *Ad Glauconum de methodo medendi*, XI.125–26K.

νος ὁδῷ τῶν ἀπορούντων, πῶς οὐ μᾶλλον ἀποροῦσιν
ἐπὶ τοῦ προχεομένου παχέος αἵματος ἐν ταῖς τῶν
341K καταιμάτων πωρώσεσιν· | αὐτό τε γὰρ τὸ προχεόμενον
οὐ σμικρῷ παχύτερόν ἐστι τοῦ κατὰ φύσιν, ἥ τε τοῦ
δέρματος οὐσία παμπόλλῳ δή τινι παχυτέρα τοῦ
παρὰ τῷ πνεύμονι χιτῶνος.

ὅπερ οὖν ἐλέγομεν ὅταν ἐκ καταπτώσεως, ἢ μεγάλης ἅμα καὶ ὀξείας φωνῆς, ἀγγείου ῥαγέντος ἐν πνεύμονι χωρὶς ὀδύνης ἀθρόον καὶ θερμόν, εὐανθές τε καὶ
πολὺ μετὰ βηχὸς ἀναφέρηται τὸ αἷμα, πεπεῖσθαι μὲν
χρὴ τὸ τραῦμα κατὰ τὸ σπλάγχνον ἐνυπάρχειν, ἐπιχειρεῖν δὲ τῇ θεραπείᾳ καθάπερ ἡμεῖς πολλάκις ἐπι-
χειρήσαντες ὀλιγάκις ἀπετύχομεν. αὐτῷ μὲν οὖν τῷ
κάμνοντι προστάξαι χρὴ μήτ᾽ ἀναπνεῖν μέγα καὶ
σιωπᾶν ἀεί· τέμνειν δ᾽ αὐτίκα φλέβα κατ᾽ ἀγκῶνα τὴν
ἔνδον· ἐπαφαιρεῖν δὲ δίς που καὶ τρίς, ἀντισπάσεως
ἕνεκα· τρίβειν τε καὶ διαλαμβάνειν ὅλα τὰ κῶλα
δεσμοῖς, ὥσπερ εἰώθαμεν. ἐπειδὰν δὲ ταῦτα πραχθῇ,
πρῶτον μὲν ὀξύκρατον ὑδαρές τε καὶ χλιαρὸν διδόναι
πίνειν, ὅπως εἴ τις εἴη θρόμβος ἐν τῷ σπλάγχνῳ
διαλυθεὶς ἐκβηχθείη· καὶ τοῦτ᾽ οὐδὲν κωλύει καὶ δὶς
καὶ τρὶς ἐν ὥραις τρισὶν ἐργάσασθαι. μετὰ δὲ ταῦτα
διδόναι τι τῶν ἐμπλαττόντων τε ἅμα καὶ στυφόντων
342K φαρμάκων, | ἤτοι δι᾽ ὑδαροῦς ὀξυκράτου τὴν πρώτην ἢ
δι᾽ ἀφεψήματος μήλων, ἢ μύρτων, ἤ τινος ἄλλου τῶν
στυπτικῶν. εἰς ἑσπέραν τε πάλιν ὁμοίως διδόναι τὸ
φάρμακον, εἴργοντας ἁπάσης τροφῆς, ἐὰν ἰσχυρὸς ᾖ.
εἰ δὲ μή, ῥοφήματος αὔταρκες διδόναι. κάλλιστον δὲ

54

plexed at the path through the lung, how are they not more perplexed in the case of the thick blood that is poured forth in the calluses of fractures, for what pours forth is to no small extent thicker than what accords with nature, and the substance of the skin is to a great degree thicker than the membrane adjacent to the lung. 341K

What I was saying, therefore, is that whenever, due to a collapse or a loud and sharp use of the voice, a vessel is ruptured in the lung without pain, and blood which is dense, warm, fresh and copious in amount is brought up with a cough, it is necessary to be convinced that the wound exists in the organ, and then it is necessary to attempt the treatment, often succeeding but occasionally failing. We must direct the patient himself not to breathe deeply and always to remain silent, and we must immediately cut a vein in the antecubital fossa and withdraw blood two or perhaps three times for the purpose of revulsion. And we must rub and bandage all the limbs, as we are accustomed to do. When this is done, first give oxykraton that is dilute and lukewarm to drink so that, if there is some blood clot in the organ, having been dissolved, it is coughed up. There is nothing to prevent this being done two or three times in a three-hour period. After this, give one of the medications that is emplastic and at the same time astringent, either through 342K the dilute oxykraton in the first place, or through a decoction of apples, or myrtle, or another of the astringents. Again, toward evening, give the medication in the same way, preventing all nourishment if [the patient] is strong. If he is not, give an adequate amount of thick gruel. It is also

καὶ κατὰ τὴν ὑστεραίαν ἐπαφελεῖν αὖθις ὀλίγον αἵματος ἀπὸ τῆς τετμημένης φλεβός, ἐὰν ἰσχυρὸς ᾖ· καὶ
τροφαῖς καὶ φαρμάκοις ὡσαύτως χρῆσθαι, μέχρι τῆς
τετάρτης ἡμέρας, ἐπιβρέχοντας ἐν κύκλῳ τὴν θώρακα, θέρους μὲν μηλίνῳ ἢ ῥοδίνῳ, χειμῶνος δὲ μύρῳ
ναρδίνῳ.

εἰ δὲ καὶ τῶν ἐμπλαστῶν τινι χρήσασθαι βούλοιο
φαρμάκων, κάλλιστον ἂν ἔχοις τὸ ἡμέτερον, ᾧ καὶ
πρὸς τὰ ἄλλα τραύματα θαρρῶν χρῶ· σύγκειται δὲ ἐξ
ἀσφάλτου καὶ ὄξους ὅσα τ' ἄλλα μίγνυται τοῖσδε
κατὰ τὰς καλουμένας ὑπὸ τῶν ἰατρῶν ἐναίμους βαρ-
βάρους. εἰ δὲ γυναῖκα θεραπεύοις, ἢ παῖδα, ἢ ὅλως
μαλακόσαρκόν τινα, τὸ διὰ τῆς χαλκίτεως ἱκανὸν
φάρμακον· οὗ κατὰ τὸ πρῶτον ἐν τοῖς περὶ φαρμάκων
343K συνθέσεως ἐδήλωσα τὴν δύναμιν. οὕτως ἡμεῖς | ἐθε-
ραπεύσαμεν οὐκ ὀλίγους, ἐπιφανέντες αὐτίκα τῷ πεπονθότι. τὸ γάρ τοι μέγιστον αὐτὸ τοῦτ' ἔστιν, ᾧ πάνυ
προσέχειν σε χρὴ τὸν νοῦν, εἰ παραχρῆμα τῆς ῥήξεως
τῶν ἀγγείων γεγενημένης, ὑπάρξαις τῆς θεραπείας,
ὥστ' ἔναιμον ὂν κολληθῆναι τὸ τραῦμα πρὶν ἄρξασθαι φλεγμαίνειν. εἰ δέ γε φλεγμήνειεν, ὀλίγη μὲν ἔτι
τοῦ κολληθῆναι τὸ τοιοῦτον ἡ ἐλπίς, εἰς χρόνον δ'
ἐκπίπτει πλείονα. καὶ τὸ χαλεπὸν ἢ ἀδύνατον τῶν ἐν
τῷ τοιῷδε τῆς ἰάσεώς ἐστι· ἐκπλύνειν γὰρ δηλονότι
χρὴ τό τε πῦον καὶ τοὺς ἰχῶρας τοῦ ἕλκους, λυομένης

---

16 On the compound medication, see also *Ad Glauconem de*

best, on the following day, to again withdraw a small amount of blood from the previously cut vein, if the patient is strong, and to use nutriments and medications in the same way until the fourth day, moistening the chest all around with honey or rosewater in summer and oil of spikenard in winter.

If you also wish to use one of the emplastic medicaments, it is best if you have mine which I also use confidently in other wounds. It is compounded from asphalt and vinegar and those other things called by doctors "foreign blood-stanchers." If you are treating a woman or child, or speaking generally, someone who is soft-fleshed, one of the copper-containing medications is sufficient, the potency of which I first demonstrated in the works on the composition of medications.[16] I treated quite a number [of patients] in this way, coming onto the scene as soon as they were shown to have been affected. Without doubt this is the most important thing, and the one to which you must particularly direct your attention—that immediately the rupture of vessels occurs, you begin the treatment, so that the wound which is bleeding is conglutinated before it begins to be inflamed. If, however, it is inflamed, there is little hope of such a wound still being conglutinated, as it degenerates into chronicity. And there is difficulty or impossibility of cure of wounds in this state, for it is clearly necessary to wash out the pus and ichors of the [chronic,

343K

methodo medendi, XI.126K, and In Hippocratis librum De fracturis commentarii, XVIIIB.537K. On the potencies of chalcitis, see De simplicium medicamentorum temperamentis et facultatibus, XII.241K, and De compositione medicamentorum per genera, XIII.568K, and also Dioscorides, V.115.

τῆς φλεγμονῆς. ἐκ μὲν δὴ τῶν κατὰ μήτραν καὶ κύστιν
ἐκκρίνεται μὲν καὶ αὐτόματα τῷ κατάντει τῆς φορᾶς,
ἐκκλύζεσθαι δὲ δύναται καὶ πρὸς ἡμῶν· ἐπὶ πνεύμονος
δὲ γίνεσθαι ἑκάτερον ἀδύνατον. ἀπολείπεται οὖν ἔτι
μόνη τοῖς ἐνταυθοῖ πᾶσιν ἕλκεσιν ἡ μετὰ τοῦ βήττειν
κάθαρσις. ἀλλ᾽ εἴπερ ὀρθῶς τήν τ᾽ ἄλλην ἡσυχίαν
αὐτοῖς συνεβουλεύομεν, ἀναπνεῖν τε καὶ καταβραχὺ
καὶ μηδ᾽ ὅλως φθέγγεσθαι προσετάττομεν ὑπὲρ τοῦ
κολληθῆναι τὸ τραῦμα, τίς ἂν εἴη τοῖς βήττουσι τῆς
344K   ἰάσεως ἡ ἐλπίς; οὔκουν ὅτι διὰ τὴν ἀναπνοὴν | ἀεικί-
νητόν ἐστι τὸ σπλάγχνον, ὡς οἱ πρὸ ἡμῶν φασιν
ἰατροί, ἄπορος ἡ ἴασις, ἀλλὰ διὰ τὴν τῶν ἰχώρων τε
καὶ τοῦ πύου κένωσιν. καὶ διὰ τοῦτο παραχρῆμα μὲν
εἰς σύμφυσιν ἀφικνεῖται καθ᾽ ὃν εἴρηται τρόπον
ἰαθέντα· φλεγμῆναι δὲ φθάσαντα χαλεπὴν καὶ ἄπο-
ρον ἴσχει τὴν ἐπανόρθωσιν· ὅ τε γὰρ ἰχὼρ καὶ τὸ πῦον
οὐκ ἀκριβῶς ἐκκενοῦται τῶν κατὰ τὴν πνεύμονα χωρίων, αἵ τε βῆχες ἱκανῶς σπαράττουσι τὰ πεπονθότα.

τὰ δ᾽ ἐκ θώρακος τριχῇ τῶν κατὰ τὸ σπλάγχνον
τοῦτο πλεονεκτεῖ· τά τε γὰρ ἀγγεῖα πολὺ σμικρότερα
καὶ ἡ ἐκροὴ τῶν ἰχώρων εἰς τὴν εἴσω γίνεται χώραν,
ὅλως τε σαρκωδέστερός ἐστι τοῦ πνεύμονος. οὐδὲ γὰρ
οὐδ᾽ ἡ τῶν τραχειῶν ἀρτηριῶν οὐσία κατ᾽ ἄλλο τι τοῦ
ζῴου μόριόν ἐστιν, ἱκανῶς ἀσάρκων καὶ ξηρῶν ὑπαρ-
χουσῶν· ὧν καὶ αὐτῶν δήπου ῥῆξις ἐν ταῖς τοιαύταις
πτώσεσι γίγνεται· ὡς εἴ γε τῶν ἄλλων μέν τι ῥαγείη,
μένοιεν δ᾽ ἀπαθεῖς αἱ τραχεῖαι ἀρτηρίαι, κατὰ τὰς

infected] wound if the inflammation is to be resolved. So then, from wounds in the uterus and bladder, not only is there spontaneous expulsion downward of what is brought forth, but we are also able to wash [these organs] out. Each [of these strategies] is, however, impossible in the case of the lung. The only thing left for all the [infected] wounds here is cleaning out by means of coughing. But if we are correct in advising the patients to be otherwise quiet, to take small breaths, and not to speak at all, as we recommend for the conglutination of the wound, what hope of cure would there be for those who cough? It is not, therefore, because the organ is in perpetual motion due to respiration, as the doctors before us said, that the cure is difficult, but because of the purging of the ichors and the pus. For this reason, those [wounds] that come to union immediately are cured according to the manner spoken of, whereas those that are already inflamed are difficult and uncertain in terms of restoration because the ichor and the pus are not completely evacuated from the spaces in the lung, and coughing excessively tears what is affected.

344K

Those things [brought forth] from the chest [wall] have a threefold advantage compared with the things evacuated from [within] this organ (i.e. the lung): the vessels are very much smaller, the outflow of the ichors occurs to the space externally, and it is altogether more fleshy than the lung. The substance of the rough arteries is unlike that of any other part of the organism, being excessively fleshless and dry. Presumably, a rupture of these particular [structures] occurs in such expectorations, as if with rupture the rough arteries remain unaffected compared to other structures, the transference [of blood] occurring only through the

συναναστομώσεις μόνας ἡ μετάληψις γίνεται. ὥστε
οὔτε θερμὸν οὔτ᾽ ἐρυθρὸν οὔτε πολὺ τοῖς τοιούτοις
ἀναφέρεται τὸ αἷμα. καὶ δοκεῖ μὲν εἶναι μετριώτερα
345K τὰ τοιαῦτα | παθήματα τήν γε πρώτην, ὡς ἂν μὴ
καταπλήττοντα τῷ πλήθει τῆς κενώσεως· ἔστι δὲ χεί-
ρω δι᾽ αὐτὸ τοῦτο· κωλύεται γὰρ ἡ σύμφυσις ὑπὸ τοῦ
περιθρομβουμένου τοῖς ἐρρωγόσιν ἀγγείοις αἵματος,
οὐκ ἔχοντος εὐπετῆ τὴν εἰς τὰς τραχείας ἀρτηρίας
διέξοδον.

9. Ὡσαύτως δὲ καὶ τὰ τοῦ διαφράγματος τραύματα, τά γε μὴ διασχόντα πρὸς τοὐκτός, ἐν μὲν τοῖς
σαρκώδεσι μέρεσι πολλάκις ἐκολλήθη, μηδὲν ὑπὸ
τῆς κινήσεως αὐτοῦ παρεμποδισθέντα· τὰ δ᾽ ἐν τοῖς
νευρώδεσιν ἀνίατα. χαλεπὴ δὲ καὶ ἡ ἐν τοῖς σαρ-
κώδεσιν ἴασις, εἰ φλεγμῆναι φθάσειεν, οὐκ αὐτῷ τῷ
διαφράγματι μόνῳ, ἀλλὰ καὶ τοῖς ἔνδον τοῦ περιτοναίου πᾶσιν· εἰς ταῦτα γὰρ ἀπορρέουσιν οἱ ἰχῶρες.
ἀλλὰ χρὴ ξηραίνειν παντὶ τρόπῳ πειρᾶσθαι, διά τε
τῶν ἔξωθεν ἐπιτιθεμένων φαρμάκων καὶ τῶν πινομένων δι᾽ ὕδατος ἢ οἴνου λεπτοῦ. ἔστι δ᾽ αὐτῶν ἐπιτη-
δειότατα τά τε διὰ σπερμάτων λεγόμενα καὶ τὸ σύνηθες ἡμῖν, ᾧ καὶ πρὸς τὰς τοῦ θώρακος συντρήσεις
ἀεὶ χρώμεθα, τῷ διὰ τῆς κασσίας. ἀλλὰ καὶ ταῦτα καὶ
346K τινες | ἐπιβροχαὶ τοῖς τοιούτοις τραύμασιν ἐπιτήδειοι
τῆς περὶ φαρμάκων εἰσὶ πραγματείας· ἐνταυθοῖ δὲ τὰς
μεθόδους λέγομεν μόνας ἐνδεικνυμένας τῶν φαρμάκων τὸ γένος· ὥστ᾽ εἴποτε λέγεταί τινα τῶν κατὰ

arteriovenous connections.[17] As a result, the blood that is brought up is neither hot, nor red, nor much in amount in such structures. Affections of this sort appear to be more moderate, at least initially, which would not be surprising   345K due to the amount of the evacuation. However, it is worse for this very reason in that the union is prevented by the blood undergoing clotting in the ruptured vessels and not having a favorable outflow path to the rough arteries.

9. Similarly also, wounds of the diaphragm, or at least those that do not penetrate to the outside, are often conglutinated in the fleshy parts, not being prevented by its movement. However, the wounds in the sinewy parts are incurable. The cure in the fleshy parts is also difficult if there is already inflammation, not only in the diaphragm alone, but also in all those [structures] within the peritoneum, for the ichors run off toward these. Yet it is necessary to attempt to dry in every way, both through the medications that are applied externally and those that are drunk, either with water or thin wine. The most suitable of these are the medications said to be made from seeds (*diaspermaton*)[18] and that which is usual for me, and which I also always use for perforating wounds of the chest, and is made from cassia. But both these and certain bathings for   346K such wounds are proper [subjects] for the treatise on medications. Here I speak of the methods alone as demonstrating the class of medications, so that, if ever some of the

17 On these presumed connections, see Galen, *De usu partium*, Book 6, chapter 17 (III.492K ff, M.T. May, 1968, vol. 1, p. 322ff).   18 This is presumably the medication described in *De compositione medicamentorum per genera*, XIII.978K. It is mentioned again in the present book (V.13, 372K).

μέρος, ὡς παραδείγματα σαφηνείας ἕνεκα γεγράφθαι
χρὴ νομίζειν αὐτά.

10. Αὖθις οὖν ἐπὶ τὰς μεθόδους ἀνέλθωμεν, ἐπιδεικνύντες ὅσον ἡ κοινὴ πάντων ἑλκῶν ἴασις ἐξ-
αλλάττεται κατ᾽ εἶδος ἐν ἑκάστῳ τοῦ ζῴου μορίῳ.
μαθησόμεθα γὰρ ἐκ τοῦδε μάλιστα μὲν ὃ δι᾽ ὅλης
πρόκειται τῆς πραγματείας, ἀναμαρτήτους εἶναι κατὰ
τὰς ἰάσεις, ἐν παρέργῳ δὲ καὶ τὴν τῶν ἑαυτοὺς ὀνο-
μασάντων Μεθοδικοὺς τόλμαν. εἰ γὰρ ἐμοὶ χρή τι
πιστεῦσαι, μηδὲν μήτε πρὸς χάριν εἰωθότι μήτε πρὸς
ἀπέχθειαν λέγειν, ἁπάντων τῶν ἰατρῶν ὄντες ἀμεθ-
οδώτατοι κατεγνώκασιν, ὥς γε δὴ γράφουσιν, οὐ μόνον τῶν ἄλλων παλαιῶν, ἀλλὰ καὶ τοῦ πασῶν τῶν
μεθόδων ἡμῖν ἡγεμόνος Ἱπποκράτους.

ὁ μὲν οὖν Ἐμπειρικὸς ἰατρὸς ἐκ πείρας φησὶν
347K ἐγνῶσθαι πάνθ᾽ ἅπερ ἡμεῖς | δι᾽ ἐνδείξεως εὑρίσκε-
σθαι δείκνυμεν. τρίτου δὲ οὐδενὸς ὄντος οὐδεμιᾶς
εὑρέσεως ὀργάνου παρά τε τὴν ἔνδειξιν καὶ τὴν πεῖραν, οὐδετέρῳ χρώμενοι Μεθοδικοὶ καλεῖσθαι δικαιοῦσιν. ἀκολουθησάτωσαν οὖν ἡμῖν καὶ νῦν γοῦν εἰς
τὴν τῶν καθ᾽ ἕκαστον μέρος ἑλκῶν θεραπείαν, ἐπι-
δειξάτωσάν τε μίαν ἁπάντων ἴασιν. ἀκούσωμεν αὐτῶν
ὅπως θεραπεύουσιν ἕλκος, ἤ τινας ἐνδείξεις ἀπ᾽ αὐτοῦ
λαμβάνουσιν· ἆρ᾽ ἄλλας τινὰς ἢ τὸ μὲν ὁμαλὲς ἐπου-
λῶσαι κελεύουσι, τὸ δὲ κοῖλον σαρκῶσαι, τὸ δ᾽ ἔναιμον συμφῦσαι; πῶς οὖν εὕρω τά τε ἐπουλωτικὰ καὶ τὰ
κολλητικὰ καὶ τὰ σαρκωτικὰ καὶ τὰ συμφυτικά; Μὴ
κάμνε, φασίν· εὕρηται γάρ, ἀλλ᾽ οὐκ οἶδ᾽ εἰ καλῶς ἢ εἰ

medications are spoken of individually, we must regard them as having been written about as examples for the sake of clarity.

10. Therefore, let me return once more to the methods to show how much the general cure of all wounds is changed in kind in each part of the organism. For we shall learn from this particularly what is set forth right through the whole treatise, which is that the methods are without fault when it comes to cures, while as a by-product, we shall also learn of the audacity of those who call themselves Methodics. If I am to be believed—I who am not accustomed to say anything, either for the sake of favor or out of enmity—they are to be condemned as being the most amethodical of all doctors, not only on the basis of what our own predecessors write, but also of what Hippocrates, the authority for all our methods, writes.

Thus the Empiric doctor says he knows everything from experience which we show is discovered by indication. Although there is no third instrument of discovery besides indication and experience, the Methodics, who use neither, think it right to so name themselves. Let them then follow us, and now at least demonstrate in the treatment of wounds in each part that there is one cure of all. Let us hear from them how they treat a wound or take certain indications from it. Do they urge other indications than to cicatrize the flat, enflesh the hollow, and bring about union in what is bleeding? How then shall I discover those things that are cicatrizing, conglutinating,[19] enfleshing and uniting? "Don't worry," they say, "they have been

347K

19 Linacre (p. 257) has a marginal note to the effect that this is an addition.

πάντα· δύναται γὰρ ἤτοι γε εὑρῆσθαι μέν τινα τῶν
φαυλοτέρων, οὐχ εὑρῆσθαι δὲ τὰ βελτίω· ἢ καὶ τὰ
δοκοῦντα εὑρῆσθαι, καὶ αὐτὰ εἶναι μοχθηρά· καὶ διὰ
τοῦτο τὰ μὲν μηδόλως θεραπεύεσθαι τῶν ἑλκῶν, τὰ δὲ
ἐν χρόνῳ πλέονι καὶ σὺν ἀλγηδόσι καὶ λιμαγχονίαις
οὐκ ἀναγκαίαις. οὐδὲ γὰρ οὐδὲ τολμᾷ τις ἐπιθέσθαι
348K νεωτέραν | πεῖραν, πρὶν ἢ ἑαυτὸν πεῖσαι τῆς ἔμ-
προσθεν ἀγωγῆς ὀρθῶς κατεγνωκέναι· οἷον αὐτίκα
περὶ τῶν ἐν πνεύμονι συνισταμένων ἑλκῶν, ἃ μηδὲ
διαγνῶναι τὴν ἀρχὴν οἷόν τ’ ἐστὶν ἄνευ τῆς ἀνατομῆς
τε καὶ τῆς τῶν ἐνεργειῶν γνώσεως, ἅπερ ἀμφότερα
φεύγουσιν.

ἀλλὰ συγκεχωρήσθω κατά γε τὸ παρὸν ἐγνῶσθαι
τοῖς περὶ τὸν Θεσσαλὸν ἕλκος ἐν πνεύμονι γεγονός.
πότερον οὖν ὥσπερ τοῦτο συνεχωρήσαμεν αὐτοῖς,
οὕτω καὶ ὅτι ῥυπαρόν, ἢ καθαρόν, ἢ ὅτι κοῖλον, ἢ ὁμαλές, ἢ ἰχώρων, ἢ πύου μεστόν, ἐπίστασθαι συγχωρήσομεν, ἤτοι γ’ ἐξ ἐπιπνοίας ἐνθέου,[6] ἢ ὄναρ ἰδοῦσιν;
ἢ κἂν τούτῳ γοῦν ἐρωτήσομεν αὐτοὺς τὰς διαγνώσεις,
ἢ χωρὶς τοῦ διαγνῶναι διδόναι συγχωρήσομεν ὅ τι
βούλοιντο[7] φάρμακον εἰδέναι; ἐγὼ μὲν γὰρ ἡγοῦμαι
καθῆραι μὲν χρῆναι τῶν ῥυπαρῶν πρότερον τὴν ῥύ-
πον· σαρκῶσαι δὲ τὸ καθαρὸν ἅμα καὶ κοῖλον· ἀπο-
καθῆραι δὲ τοὺς ἰχώράς τε καὶ τὸ πῦον, ᾧ περικέχυται
ταῦτα· κἄπειθ’ οὕτως ἐπουλοῦν. οἱ δὲ οὐκ οἶδ’ ὅπως
ἰάσονται καὶ τὰ τοιαῦτα τῶν ἑλκῶν· τὴν ἀρχὴν γὰρ

discovered." But I do not know if [they have been discovered] properly, or if all [of them have been discovered]. It is, in fact, possible for those that are more trivial to be discovered, while those that are better are not discovered, or for those that seem to be discovered to be those that are bad. Because of this, some wounds are not treated at all, while some are treated over a long time, and with pain and weakness due to a reduced diet that is unnecessary. Someone should not dare to apply a newer experience before 348K persuading himself that he has rightly condemned the previous method of treatment. An example concerns wounds existing in the lung which, to begin with, it is impossible to recognize without anatomy and the knowledge of function, both of which they shrink from.

But let it be conceded, at least for the present purposes, that the followers of Thessalus recognize that a wound has occurred in the lung. Shall we, then, just as we allow them this, in like manner also concede that they know whether it is filthy or clean, hollow or flat, or filled with ichors or pus, and that they see this either by divine inspiration or in a dream? And even in this, shall we question them anyway on the diagnosis, or apart from the diagnosis, allow that they know to give whatever medication they might wish? For I think there is a need to cleanse the filth from wounds that are filthy as a first measure, and to enflesh those that are clean and at the same time hollow, and to clear away the ichors and pus by which these wounds are surrounded, and then, in this way, to cicatrize them. I also do not know how they will cure such [infected] wounds, for to begin

---

6 nos (*cf.* ex divino afflatu KLat; ἐπιπνοίας θεῶν B; ἐπινοίας ἐνθέου K   7 B; βούλοιτο K

349K οὐδ' ἔγραψέ τι | Θεσσαλὸς ὑπὲρ αὐτῶν, ἵν' ἤτοι τὴν
ἄγνοιαν ἢ τὴν ἀσυμφωνίαν αὐτοῦ πρὸς ἑαυτὸν ⟨μὴ⟩
ἐπιδείξῃ.[8] ἤτοι γὰρ ἄλογα καὶ ψευδῆ περὶ αὐτῶν
ἐροῦσιν, ἢ εἴπερ ἀληθῆ τίς φησι, τὴν ἔνδειξιν αὐτῶν
ἀναγκαῖον ἐκ τῆς οὐσίας γενέσθαι τοῦ μορίου καὶ
προσέτι θέσεώς τε καὶ διαπλάσεως. ὑποκείσθω γὰρ
αὐτοὺς ἐπίστασθαι διαγινώσκειν ἕλκος ἐν πνεύμονι
ῥυπαρὸν καὶ καθαρόν, ἀφλέγμαντόν τε καὶ φλεγμαῖνον, ὁμαλές τε καὶ κοῖλον· ἔτι δὲ καὶ τοῦτ' αὐτοῖς
συγχωρείσθω γιγνώσκειν· οὐ μὴν οὐδ' αὐτοῦ καίτοι
γε μικροτάτου γε ὄντος ἔχουσιν εἰπεῖν ἰδίαν εὕρεσιν,
ὅτι τὸ τοιοῦτον ἕλκος ὑπὸ τοῦ χλωροῦ φαρμάκου
γίγνεται καθαρόν, εἰ βούλει τοῦ Μαχαιρίωνος ἢ τῆς
Ἴσιδος, οὐδὲν γὰρ διαφέρει.

τί ποτε οὖν ποιήσουσιν ἐπὶ τοῦ κατὰ τὸ σπλάγχνον
ἕλκους, ἀποκρινέσθωσαν ἡμῖν. ἆρά γε καταπιεῖν δώσουσι τοῦ χλωροῦ φαρμάκου; γελοῖον μὲν ὅλως τὸ
πρᾶγμα. ἐπειδὴ γὰρ ἐκ τῆς ἀνατομῆς ἔγνωσται ἡ εἰς
τὸν πνεύμονα τῶν φαρμάκων δίοδος, οἱ δ' ἀπὸ τοῦ
Θεσσαλοῦ τὴν ἀνατομὴν διαπτύουσι, πόθεν γνώσον-
ται ὅπως εἰς τὸν πνεύμονα τὸ φάρμακον πορεύεται;
350K λεγέτωσαν οὖν ἡμῖν, | πόθεν ἴσασιν ἐνεχθησόμενον
εἰς τὸν πνεύμονα τὸ τοιοῦτον φάρμακον; εἰ δ' ἄρα καὶ
τοῦτ' ἴσασιν, ἀλλ' ὅτι γε φυλάξει τὴν αὐτὴν δύναμιν,
ἣν ἐπὶ τῶν ἑλκῶν τῶν ἐκτὸς εἶχεν, οὐχ οἷόν τε γιγνώ-

---

[8] πρὸς ἑαυτὸν ἐπιδείξῃ K; ⟨μὴ⟩ nos (cf. ne KLat); sed πρὸς
ἑαυτὸν ἐπιδείξω B, recte fort.

with, Thessalus did not write anything about them, so that   349K
he did not reveal either his own ignorance or his inconsistency. The Thessaleians will say irrational and false things
about these, or if someone does say something true, it is
that the indication necessarily arises from the substance of
the part, and over and above this, from the position and
conformation. Let us assume they know how to recognize a
wound in the lung that is filthy or clean, noninflamed or
inflamed, or level or hollow. And let us further concede to
them that they know these things but do not have to state a
specific discovery, and indeed this is a very small matter,
that such a wound becomes clean by means of the green
medication, or if you prefer, the medication of Makhairion
or Isis,[20] for it makes no difference.

Let them tell me, then, what they will do in the case of a
wound in the organ. Will they give [the patient] some of
the green medicine to drink? This is absolutely ridiculous
because, since the pathway of medications into the lung is
known from anatomy whereas the followers of Thessalus
spit on anatomy, from what source will they know how the
medication is carried to the lung? Let them tell me from   350K
what source they know that such a medication will be
borne to the lung? If, however, they do also know this, it
will nevertheless be impossible for them to know that the
medication will preserve the same potency which it has in

[20] It is not clear what exactly these medications were; see
EANS, p. 446. On the green medication, see *De compositione
medicamentorum per genera*, XIII.470K.

GALEN

σκεῖν αὐτοῖς. ἀλλὰ δὴ καὶ γινωσκέτωσαν ὅτι τε φυλάξει τὴν αὐτὴν δύναμιν ἣν ἐπὶ τῶν ἐκτὸς ἑλκῶν εἶχε
καὶ ὅτι καθαριεῖ τὸν ῥύπον· οὐ μὴν ὅτι γε βῆχα
κινήσει δυνατὸν αὐτοῖς ἐπίστασθαι· χωρὶς δὲ τοῦ διὰ
βηχὸς ἐκκαθαρθῆναι τὸν ῥύπον οὐδὲ τοῦ ῥύπτοντός
ἐστι χρεία φαρμάκου. συγχωρείσθω δὲ οὖν καὶ τοῦτ'
αὐτοῖς. ἀλλὰ τό γε διάφορον ἐφ' ἕλκους πεποιῆσθαι
θεραπείαν ἐν μηρῷ καὶ πνεύμονι γεγονότος οὐ δύνανται φυγεῖν, εἴ γε τὸ μὲν ὕδατι περιπλύνουσι, τὸ δὲ ταῖς
βηξὶν ἐκκαθαίρουσιν.

ὑποκείσθω δὲ πάλιν ἐν τῷ πνεύμονι περικεχύσθαι
τῷ ἕλκει παχὺ πῦον· ἆρά γε καὶ νῦν τὸ χλωρὸν
δώσουσι φάρμακον, ἢ μέλιτος ἐκλείχειν κελεύσουσιν;
ἀλλὰ καὶ τοῦτ' αὐτὸ πόθεν εὑρήκασι λεγέτωσαν. οὐ
γὰρ δὴ τοῦτό γε φήσουσιν, ὅτι λεπτυντικήν τινα καὶ
τμητικὴν ἔχει δύναμιν, αὐτοί γε ἑκόντες ἀποστάντες
τοῦ ζητεῖν τὰς τοιαύτας δυνάμεις. οὐ μὴν οὐδ' ὅτι τοῖς
351K Ἐμπειρικοῖς | εὕρηται τὸ μέλι κατὰ τοιάνδε συνδρομὴν
ἐπιτήδειον, ἔνεστιν αὐτοῖς ὁμοίως ἐκείνοις χρῆσθαι·
πρῶτον μὲν ὅτι τῆς ἐμπειρίας καταφρονοῦσιν· ἔπειτα
δὲ ὅτι κατὰ τὰς τοιαύτας συνδρομὰς ὁ Ἐμπειρικὸς
ἥτις μέν ἐστιν ἡ ἐν τῷ πνεύμονι διάθεσις ἀγνοεῖν
φησί, τετηρῆσθαι δ' ἐκ πείρας ἑαυτῷ τὰ συμφέροντα.
Θεσσαλῷ δὲ οὐκ ἀρκεῖ θεραπεύειν ὃ μηδ' ὅλως οἶδεν,
ἀλλ' ἀπὸ τῆς τῶν παθῶν ἐνδείξεως ὁρμᾶται. εἰ δὲ δὴ
καὶ πάντ' αὐτῷ συγχωρήσαιμεν, ὥσπερ καὶ πρόσθεν,
ὁμοίως ἡμῖν ἐπίστασθαι, τό γ' ἐπὶ τοῖς διαφέρουσι
μέρεσιν ἐξαλλάττεσθαι τὴν θεραπείαν κατ' εἶδος οὐκ

68

the case of wounds that are external. But let it be the case that they do know it will preserve the same potency it has in external wounds and it will clear away the filth. It is certainly not possible for them to know that it will cause coughing, although without the filth being cleared away by coughing, it is of no use as a cleansing medication. Therefore, let us also concede this to them. But at least they cannot escape the fact that a different treatment is employed for a wound that has occurred in the thigh and one that has occurred in the lung, since they wash the one with water and cleanse the other with coughing.

Let us assume again that thick pus has been disseminated by the wound in the lung. Will they now give the green medication or will they direct [the patient] to lick honey? But let them also say from what source they discovered this. They will certainly not say this—that it has a thinning and cutting potency—these men who deliberately distance themselves from the search for such potencies. It is not because honey was found by the Empirics to be suitable in this sort of syndrome that it is possible for them (i.e. the Methodics) to use it like those men do—first because they despise experience and second because, in such syndromes, the Empiric says he is ignorant as to what the condition in the lung is, although those things that are useful have been observed by him through experience. It is not enough for Thessalus to treat what he does not know at all, but he proceeds from the indication of the affections. Doubtless, if we were to concede to him, as we did before, that he knows everything as we do, he would never shrink from changing the kind of treatment in the different parts.

351K

ἄν ποτε ἐκφύγοι. οὐ γὰρ δήπου ταὐτόν ἐστιν ἢ μελίκρατον εἰς μήτραν ἐγχέαι δι᾽ ἕλκος ῥυπαρόν, ἢ μέλιτος ἐσθίειν, ἢ καταπλύνειν σπόγγῳ τὸ ἕλκος· ἀλλὰ
ταῦτα μὲν ἔτι σμικρά, μέγιστα δὲ ἐκεῖνα.

χρόνιον ἀφλέγμαντον ἕλκος ἐν πολλοῖς ὑποκείσθω
μέρεσιν, ὀφθαλμῷ καὶ ἀκοῇ καὶ μυκτῆρι καὶ στόματι
καὶ μηρῷ καὶ γαστρὶ καὶ μήτρᾳ καὶ ἕδρᾳ καὶ αἰδοίῳ·
προσυποκείσθω δέ, εἰ βούλει, τὸ ἕλκος ἢ ὁμαλὲς ἀκριβῶς, ἢ ὀλίγου δεῖν ὑπάρχειν ὁμαλές· ἀποκρινάσθω-
352K σαν | ἡμῖν οἱ ἀπὸ Θεσσαλοῦ, τοῦ μηδὲν ὅλως περὶ τῶν
τοιούτων διορισαμένου, πῶς ἐπουλώσομεν αὐτὸ τῷ διὰ
τῆς καδμείας, νὴ Δία· τοῦτο γὰρ ἐπουλώσει καλῶς τὸ
κατὰ τὸν μηρὸν ἕλκος· ἆρ᾽ οὖν καὶ τὸ κατὰ τὸν
ἀκουστικὸν πόρον; ἄπιστον μὲν ἴσως ἐρῶ, ἀλλ᾽ ἴσασιν οἱ θεοί, χρόνιον ἕλκος εὑρόν ποτε τῶν σοφωτάτων
τινὰ Θεσσαλείων οὕτω θεραπεύοντα. θᾶττον δ᾽ ἂν
ἐσάπη τὸ οὖς τἀνθρώπῳ καὶ σκώληκας ἔσχεν, ἢ τῷ
διὰ τῆς καδμείας ἐπουλώθη φαρμάκῳ· συγχωρήσαν-
τες δ᾽ ὅμως αὐτῷ πλείοσι χρήσασθαι ἡμέραις, ἐπειδὴ
καθ᾽ ἑκάστην δυσωδέστερόν τε τὸ οὖς ἐγένετο καὶ
ἰχώρων μεστόν, ἀπιστότερον ἔτι τοῦ πρόσθεν ἐθεασάμεθα τολμηθέν. οἰηθεὶς γὰρ φλεγμαίνειν ἐν τῷ
βάθει τὸν πόρον, ἐπὶ τὴν τετραφάρμακον ἧκεν, ὃ πολὺ
δὴ μᾶλλον ἔμελλε σήψειν τὸ μόριον· οὐδὲ γὰρ ξη-

---

[21] See Dioscorides, V.84 (pp. 623–24) on the source of this
medication and the efficacy of cadmia in the treatment of chronic
wounds and ulcers.

Obviously, to pour melikraton into the uterus because of a filthy wound, to eat some honey, and to wash the wound with a sponge are not the same. But these latter points are minor; those former issues are, however, major.

Let us assume there is a chronic wound or ulcer without inflammation in many parts—eye, ear, nose, mouth, thigh, stomach, uterus, anus and external genitalia. And let us assume in addition, if you wish, that the wound or ulcer is either completely or almost completely flat. Let the followers of Thessalus reply to us, since he made absolutely no distinction among such places, how, by Zeus, we shall cicatrize this with the medication made from cadmia,[21] for this will effectively cicatrize the wound in the thigh. Will it, then, also cicatrize the wound in the external auditory canal? Perhaps I shall say this is incredible, but the gods know that, on occasion, I found one of the most sapient Thessaleians treating a chronic wound here in this way. The ear in the person would more quickly putrefy and have grubs in it than it would be cicatrized by the medication made from cadmia. If, however, we allow him to use a greater number of days, since each day the ear becomes more foul-smelling and full of ichors, we see that what was undertaken was even more incredible than before. For thinking that the channel is inflamed in the depths, he proceeds to the "tetrapharmaceutical potency"[22] which is undoubtedly going to putrefy the part much more because

352K

[22] On the composition of this medication, see Galen, *De simplicium medicamentorum temperamentis et facultatibus*, XII.328K, and *De compositione medicamentorum secundum locos*, XII.601K.

ραίνειν ὅλως ἕλκη δυνατόν ἐστιν, ἀλλὰ πέττειν τὰ
φλεγμαίνοντα. ἅτε δ᾽ ἐναντιωτάτῳ τῆς διαθέσεως αὐτοῦ χρησαμένου φαρμάκῳ, μετὰ μίαν ἡμέραν ἢ δύο
353K ἐπολυπλασιάσθη | μὲν αὐτίκα τὸ τῶν ἰχώρων πλῆθος,
ἀφόρητος δ᾽ ἦν ἡ δυσωδία.

οὔκουν ἔτι συνεχώρουν οἱ οἰκεῖοι τοῦ ὠτὸς ἅπτεσθαι τῷ Θεσσαλείῳ· ὁ δ᾽ ὑπ᾽ ἀναισχυντίας τε ἅμα καὶ
ἀναισθησίας ἠξίου μὴ μόνον ἐντιθέναι τι τῆς τετρα-
φαρμάκου δυνάμεως, ἀλλὰ καὶ καταπλάττειν ἔξωθεν
αὐτῷ χαλαστικῷ καταπλάσματι. τῶν δ᾽ οἰκείων ἀπ-
ελαυνόντων τε καὶ δεδιότων ἐν μεγάλῳ κακῷ τὸν
πάσχοντα γενέσθαι, παρεκαλέσαμεν ἡμεῖς ἔτι μίαν
ἡμέραν ἐπιτρέψαι τῷ Θεσσαλείῳ τὴν θεραπείαν αὐτοῦ. ἔμελλον δὲ δήπου κατὰ τὴν ὑστεραίαν ὅ τε ἰχὼρ
ἔσεσθαι πολλαπλάσιος ἥ τε ὀδμὴ δυσωδεστάτη. καθ᾽
ἣν ἐπειράθην εἰ οἷόν τ᾽ εἴη μεταπεῖσαι τὸν Θεσσάλειον ὄνον, ὅπως μὴ πάντας ἐπιτρίβῃ τοὺς πάσχον-
τας, ἀλλά τινας ἤδη ποτὲ κἂν ὀλίγους δυνηθείη σῶ-
σαι, τῆς ἀμεθοδωτάτου αἱρέσεως ἀποστάς. ἠρξάμην
μὲν οὖν λόγου τοιοῦδε πρὸς αὐτόν.

Ἆρα οὐχὶ φλεγμονή σοι δοκεῖ κατὰ βάθος εἶναι
τοῦ πόρου καὶ διὰ τοῦτο χαλαστικοῖς χρᾷ βοηθήμασιν;

ὁ δὲ καὶ πάνυ διετείνετο καὶ οὕτως ἔχειν ἔφασκε
354K καὶ ἀδύνατον ἄλλως. Ἆρ᾽ οὖν, ἔφην, | ἐθεάσω σύ ποτε
φλεγμαῖνον ἕλκος, ὄξει δριμυτάτῳ μετὰ Γλαυκίου
θεραπευθέν;

Οὐ μὲν οὖν, εἶπεν, ἀλλ᾽ εἰ τὸν Ἀνδρώνειόν τις

it cannot dry wounds completely, but it can "ripen" those that are inflamed, inasmuch as when he used a medication most inimical to the condition, after a day or two the amount of the ichors was immediately increased and the foul smell was intolerable.

The relatives did not agree to the Thessaleian touching the ear. Nevertheless he, through shamelessness coupled with obtuseness, thought it worthwhile not only to put in some of the "tetrapharmaceutical potency," but also to plaster the ear externally with a relaxing poultice. When the relatives drove him away, and fearing that the patient was at very great risk, I urged them to entrust his treatment to the Thessaleian for one more day. Obviously ichor was going to be much more on the next day and the smell was going to be very offensive. On that day, I attempted [to find out] if it were possible to persuade the Thessaleian ass, so that he might not distress all his patients, but might now be able to save some patients on some occasions, even if only a few, by abandoning the most amethodical sect. I therefore initiated the following discussion with him.

"Is it not" [I said,] "because the inflammation seems to you to be in the depths of the canal that you use the relaxing remedies?"

And he started to protest strongly and assert it was so and could not be otherwise. "Why then," I said, "did you ever see an inflamed wound treated with the very sharp vinegar with Glaucium?"[23]

"I did not," he replied, "but if someone were to use the

23 Despite the capital γ this is taken to be γλαύκιον, the substance also referred to at 954–55 and 1002 (see Galen XI.857, Dioscorides III.100 and vol. III, 954, note 10), rather than "after the manner of Glaucias"; see Boulogne (2009), pp. 307–8.

κυκλίσκον ὄξει δεύσας χρῶτο, τάχα δ' ἄν, ἔφη, καὶ σπασθείη.

Ἐὰν οὖν, εἶπον, ἄλλο τι φάρμακον οὐκ ὀλίγῳ τοῦ Ἀνδρωνείου σφοδρότερον ὄξει τις δριμυτάτῳ διεὶς χρήσηται καὶ ταῦτ' εἰς οὓς ἐγγὺς οὕτω μόριον ἐγκεφάλου καὶ μηνίγγων, ἆρ' οὐκ ὄντως σπασθήσεται κατὰ τὸν σὸν λόγον, εἴπερ γε φλεγμονή τις ὑπόκειται;

ταῦτ' ἐδόκει κἀκείνῳ καὶ τοῖς ἄλλοις τοῖς συμπαροῦσιν ἀληθῶς εἰρῆσθαι.

Ὅσον μὲν οὖν, ἔφην, ἐπὶ τῷ δεῖσθαι τὴν διάθεσιν τῶν μορίων ἐκτεθηλυσμένων ὑπὸ τῆς σῆς ἀγωγῆς ἐσχάτως ξηραινόντων φαρμάκων ἐχρησάμην ἂν ἤδη τοιούτῳ· νυνὶ δ' ἐπειδὴ κακὸν ἔθος εἴθισας τὰ μόρια πλείοσιν ἡμέραις, ἐπὶ τοὐναντίον αὐτὰ μετάγειν ἀθρόως οὐκ ἔτ' ἐγχωρεῖ. σοὶ μὲν γὰρ καὶ Θεσσαλῷ καταφρονεῖν ἔθους ἔξεστιν, ὥσπερ καὶ φύσεως μερῶν· ἡμῖν δ' οὐκ ἔξεστιν·

ἀλλὰ τῇ μὲν πρώτῃ τῶν ἡμερῶν ὄξει χρήσομαι μετὰ Γλαυκίου, τῇ δὲ δευτέρᾳ τῷ Ἀνδρωνείῳ, τῇ τρίτῃ
355K δ' ἐπί τι σφοδρότερον ἔτι καὶ αὐτοῦ τοῦ | Ἀνδρωνείου παραγενήσομαι φάρμακον, ᾧ χρησάμενος ἡμέραις τρισίν, ἢ καὶ τέτταρσιν, ἐάν μοι φαίνηται δεῖσθαι σφοδροτέρου φαρμάκου τὸ ἕλκος, οὐκ ὀκνήσω κἀκείνῳ χρήσεσθαι. Ἔξωθεν δ', ἔφην, ἐπιθήσω τῇ κεφαλῇ κατὰ τὸ τοῦ πεπονθότος ὠτὸς χωρίον οὐ μὰ Δί' ὡς σὺ κατάπλασμα χαλαστικόν, ἀλλά τι τῶν ξηραντικωτάτων φαρμάκων, ὁποῖόν ἐστι τὸ διὰ τῶν ἰτεῶν, ἢ καὶ

Andronian troche[24] moistened with vinegar, perhaps the patient would be caused to convulse."

"Therefore," I said, "if someone were to use some other medication a good deal stronger than the Andronian, moistened with very sharp vinegar, and introduced this into the ear, which is a part near to the brain and meninges, would this not be really convulsion-inducing, according to your argument, if in fact there is some underlying inflammation?"

This seemed to be a true statement, both to that man and to the others present.

"Therefore," I said, "on the basis that the condition of the parts which have been weakened by your treatment requires very drying medications, I would by now have already used such a medication. But now, because you have made the parts accustomed to a bad state over many days, it is no longer possible to change these to the opposite all of a sudden. For it is possible for you and for Thessaleians [generally] to think little of the state, just as you think little of the nature of the parts, but for me it is impossible."

On the first day, I shall use vinegar with Glaucium; on the second day, I shall use the troche of Andron while on the third day, I shall come to some medication stronger still than the troche of Andron itself, and having used that 355K for three or four days, if the wound seems to me to need a stronger medication, I shall not hesitate to use that. "However," I said, "I shall apply to the head externally, in the region of the affected ear, not, by Zeus, a relaxing poultice like you do, but one of the very drying medications, like the one that is derived from mushrooms, or I shall smear

24 See note 10 above.

GALEN

αὐτοῦ τοῦ Ἀνδρωνείου καταχρίσω μετ᾽ ὄξους, ἢ τῶν
τούτου τινὶ ξηραντικωτέρων. ἐπεὶ γὰρ ξηρὸν ἐσχάτως
ἐστὶ τὸ θεραπευόμενον μόριον, ἀναγκαῖον αὐτὸ ξηραίνειν ἐσχάτως· ἐνδείκνυται γὰρ ὥσπερ τὸ πάθος τοὐναντίον ἑαυτῷ πρὸς τὴν θεραπείαν, οὕτω τὸ μόριον ὅ
τί περ ἂν ὁμοιότατον ἑαυτῷ τυγχάνῃ. καὶ τοίνυν
ὥσπερ εἶπον, οὕτω καὶ ἔπραξα, καὶ ὁ ἄνθρωπος ὑγιάσθη, μὴ δεηθεὶς ἰσχυροτέρων ἄλλων φαρμάκων. ἐπ᾽
ἐνίων μέντοι δεηθέντων, οἷς ἐξ ἐνιαυτοῦ καὶ δυοῖν ἐτῶν
ἦν ἕλκη ἐν τῷ ὠτί, προσηνέγκαμεν φάρμακον ἁπάντων τῶν ῥηθέντων ἰσχυρότερον· ἔστι δὲ καὶ ἡ καλουμένη σκωρία τοῦ σιδήρου, κοπτομένη τε καὶ διακοσκι-
356K νομένη λεπτοτάτῳ κοσκίνῳ, κἄπειτα λειουμένη | μέχρι
τοῦ χνοώδης γενέσθαι· μετὰ δὲ ταῦτα σὺν ὄξει δριμυτάτῳ καθεψομένη μέχρι τοῦ σύστασιν σχεῖν μελιτώδη τε καὶ γλοιώδη. δῆλον δ᾽ ὅτι πολλαπλάσιον εἶναι
χρὴ τὸ ὄξος ἐν τῇ μίξει·

ἀλλ᾽ ὅπερ εἶπον ἤδη πολλάκις, ἡ μὲν εὐπορία τῆς
ὕλης τῶν φαρμάκων ἐξ ἑτέρων ὑπομνημάτων ἔσται
σοι· τὰ δὲ τῆς μεθόδου περαινέσθω. τὸ γάρ τοι τῆς
ἀκοῆς χωρίον, ἐπειδὴ ξηρότατόν ἐστι, διὰ τοῦτο τῶν
ἄκρως ξηραινόντων δεῖται φαρμάκων, ὧν οὐδὲν οὐδενὶ
τῶν ἄλλων μορίων ἁρμόττει προσφέρειν, ἀλλ᾽ ὀφθαλμῷ μὲν ἡλκωμένῳ τὸ διὰ τοῦ λιβανωτοῦ κολλύριον, εἰ
οὕτως ἔτυχε· μυκτῆρσι δὲ πολὺ μὲν ξηραντικώτερον ἢ
κατ᾽ ὀφθαλμούς, ἧττον δὲ ἢ κατ᾽ οὖς. ὥστε καὶ οἱ
κυκλίσκοι πάντες οἱ προειρημένοι χρηστοί, καὶ τὸ τοῦ

on the Andronian medication itself along with vinegar, or one of the medications more drying than this. Since the part being treated is exceedingly dry, it is necessary to dry it to the greatest degree because it indicates that, just as the affection may meet with what is opposite to itself in relation to treatment, so too the part should meet with what is most like itself." Accordingly, I did just as I said, and the man was restored to health and did not need other, stronger medications. Nevertheless, for some in whom there were ulcers in the ears for one or two years, I applied a medication stronger than all those mentioned. This is the so-called "dross of iron"[25] which has been pounded and sifted thoroughly with a very fine sieve and then ground down until it becomes a fine powder. After this, it is brayed with the strongest vinegar until it has the consistency of honey and is glutinous. It is clear that the vinegar must be in much greater proportion in the mixture.

356K

But what I have often said already is that an abundance of the material of the medications will be available to you from other treatises, so let the matter pertaining to the method be brought to a conclusion. Certainly, the region of the ear, since it is very dry, will require because of this the most extreme drying medications, which are unsuitable for application to any of the other parts, although for the ulcerated eye the collyrium of frankincense is suitable, if this happens to be the problem. Medications for the nostrils [should be] far more drying than those for the eyes, but less drying than those for the ears. As a result, all the pastilles previously mentioned are useful, as is the medica-

[25] See Galen, *De simplicium medicamentorum temperamentis et facultatibus*, XII.235–36K, and Dioscorides, V.94.

Μουσᾶ φάρμακον, ὅσα τ᾽ ἄλλα τοιαῦτα. τῶν δ᾽ ἐν τοῖς στόμασιν ἑλκῶν ὅσα μὲν ἱκανῶς πλαδαρὰ τῶν ξηραινόντων ἰσχυρῶς δεῖται φαρμάκων, οἷον τοῦ διφρυγοῦς αὐτοῦ τε καθ᾽ αὑτὸ καὶ μετὰ μέλιτος, οἴνου τε καὶ οἰνομέλιτος· ὡσαύτως δὲ καὶ τῆς ἴρεως καλουμένης· ἔτι τε τῆς ἀνθηρᾶς, ἤτοι ξηρῶν, ἢ μετὰ μέλιτος, ἢ 357K οἰνομέλιτος, | ἢ οἴνου. ἀγαθὸν δ᾽ εἰς τὰ τοιαῦτα καὶ τὸ τοῦ Μουσᾶ φάρμακον, ὅ τε τοῦ ῥοῦ χυλὸς ὀμφάκιόν τε καὶ ὅσα πέρ ἐστιν ἄλλα γενναίως ξηραίνοντα.

τί γὰρ δεῖ τὰς ὕλας ἐπέρχεσθαι; τὰ δ᾽ ἁπλούστερα τῶν ἐν τοῖς στόμασιν ἑλκῶν ἱκανὰ θεραπεύειν ἐστὶ καὶ τὰ μετρίως ξηραίνοντα φάρμακα· καθάπερ ταυτὶ τὰ διὰ τὸ συνεχὲς τῆς χρείας ὀνομασθέντα στοματικά· τό τε διὰ μύρων καὶ βάθου καρποῦ καὶ καρύων χλωρῶν λέμματος χυλοῦ, καὶ τούτων ἔτι μᾶλλον τὸ διὰ γλεύκους τε καὶ τῶν τῆς κυπαρίττου σφαιρίων. ὅσα δ᾽ ἱκανῶς ὑγρὰ τῶν ἐν τοῖς στόμασιν ἑλκῶν πλησίον ὀστῶν, ἔστι δὲ καὶ κίνδυνος διὰ τοῦτο καὶ αὐτὸ τὸ ὀστοῦν σφακελίσαι, σφοδροτάτων ταῦτα δεῖται φαρμάκων διὰ τὴν τῶν ὀστῶν οὐσίαν ξηρὰν οὖσαν. ὥστ᾽ ἐγὼ λειαίνων ἀεὶ τοὺς προειρημένους κυκλίσκους ἐπιτίθημι τὸ φάρμακον ξηρόν.

11. Εἴρηται δ᾽ ὀλίγον ἔμπροσθεν καὶ τῶν κατὰ κύστιν καὶ μήτραν ἔντερά τε καὶ πνεύμονα τὰ γένη τῶν φαρμάκων, οἷς ἐφ᾽ ἑκάστου χρηστέον ἐστί, ἀπὸ μὲν τῆς οὐσίας τῶν θεραπευσομένων μορίων τοῦ φαρ-
358K μάκου λαμβανομένου | μετὰ τοῦ προεπεσκέφθαι τὴν διάθεσιν, τοῦ τρόπου δὲ τῆς χρήσεως αὐτῶν ἀπό τε

tion of [Antonius] Musa,[26] and other such things. Those ulcers in the mouth that are excessively moist need medications that are strongly drying such as diphryx (baked clay), either by itself or with honey, wine or muscadell (a honey-wine blend). In like manner too, they need the so-called iris, and furthermore, its flowers, either dry or with honey, or melikraton, or wine. For such wounds, the medication   357K
of Musa is also good, and the juice of sumac, omphakion and other such things that are strongly drying.

But why is there need to go over the materials. The moderately drying medications are sufficient to treat the more straightforward wounds and ulcers in the mouth; i.e. medications such as those that are continually in use and are called "stomatics." These include the medications made from mulberry and blackberry fruit and the juice of the green skin of walnuts, and still more than these, the medication made from sweet, new wine and from the globular catkins of the cypress. With those wounds and ulcers of the mouth that are excessively moist and near bones, there is also a danger, because of this, of the bone itself necrosing, and these need very strong medications because the substance of the bones is dry, so that having ground up the previously mentioned troches, I always apply a medication that is dry.

11. The classes of medications for wounds in the bladder, uterus, intestines and lung, which we should use in each case, were spoken of a little earlier. The [choice of] medication is taken from the substance of the parts being treated along with the previously evaluated condition,   358K
while the manner of their use is taken from the conforma-

26 Antonius Musa was an Asclepiadian doctor in the last century BC; see *EANS*, p. 101.

τῆς διαπλάσεως ἅμα καὶ τῆς θέσεως. ἐντεῦθεν γὰρ
ἐπενοήθησαν ὠτεγχύται τε καὶ μητρεγχύται καὶ καθ-
ετῆρες καὶ κλυστῆρες. ἐντεῦθεν δὲ καὶ ὅτι τὰ μὲν κατὰ
γαστέρα καὶ θώρακα καὶ πνεύμονα συνιστάμενα τῶν
ἑλκῶν διὰ τῶν ἐσθιομένων καὶ πινομένων θεραπεύεσθαι χρή, τὰ δὲ κατ᾽ ἔντερα ὅσα μὲν ἐγγὺς τῇ
γαστρί, διὰ τῶν ἐσθιομένων καὶ πινομένων, ὅσα δ᾽
ἤδη κατωτέρω, διὰ τῶν ἐνιεμένων· οὔτε γὰρ τοῖς πλησίον τῆς γαστρὸς ἐπαναβῆναι δύναται τὸ διὰ τῆς
ἕδρας ἐνιέμενον οὔτ᾽ εἰς τὰ κάτω τοῦ στόματος λη-
φθέντος ἡ δύναμις ἀκραιφνὴς ἐξικνεῖσθαι. διὰ τοῦτο
καὶ ὅσα κατὰ θώρακα καὶ πνεύμονα συνίσταται τῶν
ἑλκῶν ἅμα τε δυσιατότερα τῶν κατὰ τὴν γαστέρα,
πόρρω γὰρ αὐτῶν ἡ θέσις, ὡς ἐκλύεσθαι τῶν φαρμάκων τὴν δύναμιν, ἅμα τε δι᾽ αὐτὸ τοῦτο πολὺ
σφοδροτέρων χρῄζει τῶν καταπινομένων φαρμάκων ἢ
εἰ αὐτοῖς τοῖς ἕλκεσιν ἄντικρυς προσεφέρετο. καὶ διὰ
τοῦτο ἐξεύρηται τοῖς ἰατροῖς, ἡνίκα ἐκκαθαίρειν δέον-
359K ται πῦον ἐκ θώρακος ἢ πνεύμονος, ἰσχυρότατά | τε καὶ
τμητικώτατα φάρμακα καὶ τοιαῦτα τὴν δύναμιν ὡς
παροξῦναι τὸ ἕλκος, εἴπερ ἦν ἐν τῇ γαστρί. καὶ μέν γε
τὸ διὰ τῶν βηχῶν ἐκκαθαίρειν αὐτὰ ἐκ τῆς διαπλάσεως ἐλήφθη τῶν μορίων, οὐδεμίαν ἐχόντων ἐκροὴν
τοιαύτην οἵαν μήτρα καὶ κύστις καὶ ἕδρα καὶ ὦτα καὶ
μυκτῆρες καὶ στόμα. διὰ τοῦτο καὶ ἡ γαστὴρ κατ᾽
ἄμφω τὰ μέρη καθαίρεσθαι δύναται· ἄνω μὲν διὰ τῶν
ἐμέτων, κάτω δ᾽ ὡς ἡ φύσις. ἀπὸ γάρ τοι τῶν μορίων
ὡς ὀργανικῶν αἱ τοιαῦται τῶν ἐνδείξεων, ὥσπερ αἱ

tion and position [of the parts being treated]. Hence, I have in mind ear syringes, uterine syringes, catheters and clysters. Hence also, it is necessary that those wounds and ulcers existing in the stomach, chest and lung are treated with things that are eaten or drunk, whereas those in the intestines that are near the stomach are treated by things that are eaten or drunk and those that are already below are treated by things inserted *per rectum*. For neither is what is inserted *per rectum* able to reach up to the intestines near the stomach nor is the potency of what is taken *per os* able to reach the intestines below unharmed. Because of this too, those wounds and ulcers that exist in the chest [wall] and lung are at once more difficult to cure than those in the stomach because their position is further onward, so as to dissipate the potency of the medications, and they require, for this very reason, much stronger medications to be drunk than those that are applied directly onto the wounds or ulcers themselves. And because of this, it was discovered by doctors that when they need to clear away pus from the chest [wall] or lung, the strongest and 359K most cutting medications [are required]—those that in respect to potency are such as would irritate the wound or ulcer if it were in the stomach. Indeed, the cleansing of these wounds through coughing is taken from the conformation of the parts, since they have no outlet of the kind [present in] the uterus, bladder, anus, ears, nostrils and mouth. Also because of this, the stomach can be purged in both directions; above by vomiting and below as is natural. For certainly there are such indications from the parts as

κατὰ τὸ ξηραίνειν ἐνδείξεις ἀπὸ τῆς οὐσίας αὐτῶν ὡς ὁμοιομερῶν.

ἀπὸ δέ γε τῶν παθῶν αἱ τοιαῦται πάλιν οἷον ἀπὸ τῶν ἑλκῶν, ἐπειδὴ περὶ τούτων ὁ λόγος ἦν, ὅτι τε ξηραντέον, ὡς ἔμπροσθεν ἐδείκνυτο, καὶ ὅτι τῶν κατὰ τὴν γαστέρα τὸ πῦον ἀποπλῦναι βουλομένοις οὐκ ἀκίνδυνον ἐμεῖν· ὥσπερ εἰ καὶ ἄλλως ἀπορρίπτοις ἐμπεπλασμένον αὐτῇ φλέγμα δι᾿ ὀξυμέλιτός τε καὶ ῥαφανίδων, ἀλλὰ βέλτιον ὑπάγειν κάτω· κίνδυνος γὰρ ἐμοῦντι καὶ σπαράξαι τὸ ἡλκωμένον καί τινα χυμὸν οὐ χρηστὸν ἐκ τῶν πλησίον ἐπισπᾶσθαι. διὰ τοῦτο δέ, 360K ὡς ἔφην, καὶ τὰ κατὰ πνεύμονα τῶν | ἑλκῶν δυσιατότατα· χωρὶς μὲν γὰρ τοῦ βήττειν οὐκ ἂν ἐκκαθαρθείη, βηττόντων δ᾿ ἐπιρρήγνυται. δι᾿ ἀλλήλων οὖν αὐτοῖς κυκλεῖται τὸ κακόν· ἐπιφλεγμαίνοντα γὰρ αὖθις τὰ σπαραχθέντα δεύτερον αὖθις δεῖται πεφθῆναί τε τὴν φλεγμονὴν ἀποκαθαρθῆναί τε τὸ πῦον, ὥστ᾿ ἐκ πάντων αὐτοῖς παρεσκευάσθαι τὸ δυσίατον, οὔτε τῶν οἰκείων ἕλκεσι φαρμάκων ψαῦσαι δυναμένων τοῦ ἕλκους, ὡς ἐν γαστρί, καὶ φθανόντων ἐκλύεσθαι μεταξύ, καὶ ὅτι κινεῖται κατὰ τὰς ἀναπνοὰς καὶ σπαράττεται κατὰ τὰς βήχας. ὥσθ᾿ ὅταν ἀγγεῖον ἐν πνεύμονι ῥαγῇ, γιγνώσκειν χρὴ σαφῶς εἰ μὴ παραχρῆμα πρὶν φλεγμῆναι κολληθείη, μετὰ ταῦτά γε ἀνίατον ἐσόμενον.

---

27 For the composition of oxymel (basically a honey-vinegar

organs, just as the indications that relate to drying are from the substance of these as *homoiomeres*.

Again, there are such [indications] from the affections, as for example from wounds and ulcers since the discussion is about these, that you must dry them, as was shown before, and that for those who wish to wash away the pus of the wounds and ulcers in the stomach, vomiting is not without danger, just as it is, differently, for those who wish to expel the phlegm adhering to it by oxymel and radishes.[27] But it is better to lead it downward, for there is a danger for the person vomiting of also tearing what is ulcerated, and for some humor which is not beneficial being drawn from the adjacent parts. Because of this, the wounds and ulcers in the lung are difficult to heal, as I said, in that they may not be purified other than by coughing, and yet when there is coughing, there is laceration.This sets up a vicious circle because those things that are lacerated, in turn become inflamed, which secondarily requires that the inflammation be "ripened" (brought to a head) and the pus cleared away, so that from all these things a difficulty of curing is produced in them since the medications that are proper for wounds cannot reach the wound as they can in the stomach, being dissipated beforehand in the process, and because they are changed by respiration and broken up by coughing. Consequently, whenever a vessel in the lung ruptures, it is necessary to know clearly whether or not it can be immediately conglutinated before it becomes inflamed [in that] after this, it will be incurable.

360K

mixture), see Dioscorides, V.22. On the use of radish (*Raphanus sativus*), see Dioscorides, II.137.

12. Ὅσα μέντοι τῶν ἑλκῶν ἐν ταῖς τραχείαις ἀρτηρίαις γίγνεται κατὰ τὸν ἔνδον αὐτῶν χιτῶνα, καὶ μάλισθ᾽ ὅσα τοῦ λάρυγγος πλησίον, ἢ καὶ κατ᾽ αὐτόν ἐστι, ταῦτα θεραπεύεται· καὶ ἡμεῖς οὐκ ὀλίγους ἰασάμεθα τῶν οὕτω καμνόντων. εὕρομεν δὲ μάλιστα τὴν θεραπείαν αὐτῶν ἐνθένδε κατὰ τὸν μέγαν τοῦτον λοι-

361K μόν, ὃν εἴη ποτὲ παύσεσθαι, πρῶτον | εἰσβάλλοντα. τότε νεανίσκος τις ἐνναταῖος ἐξήνθησεν ἕλκεσιν ὅλον τὸ σῶμα, καθάπερ καὶ οἱ ἄλλοι σχεδὸν ἅπαντες οἱ σωθέντες. ἐν τούτῳ δὲ καὶ ὑπέβηττε βραχέα. τῇ δ᾽ ὑστεραίᾳ λουσάμενος αὐτίκα μὲν ἔβηξε σφοδρότερον, ἀνηνέχθη δ᾽ αὐτῷ μετὰ τῆς βηχὸς ἣν ὀνομάζουσιν ἐφελκίδα. καὶ ἡ αἴσθησις ἦν τἀνθρώπῳ σαφὴς κατὰ τὴν τραχεῖαν ἀρτηρίαν τὴν ἐν τῷ τραχήλῳ πλησίον τῆς σφαγῆς ἡλκωμένου τοῦ μέρους. καὶ μέντοι καὶ διανοίξαντες αὐτοῦ τὸ στόμα κατεσκεψάμεθα τὴν φάρυγγα, μή που κατ᾽ αὐτὴν εἴη τὸ ἕλκος. οὔτ᾽ οὖν οὕτως ἐπισκοπουμένοις ἐφαίνετο πεπονθέναι, καὶ πάντως ἂν ἐδόκει τῇ διόδῳ τῶν ἐσθιομένων τε καὶ πινομένων αἴσθησις ἔσεσθαι τῷ κάμνοντι σαφής, εἴπερ ἕλκος ἦν αὐτόθι.

καὶ μέντοι καὶ δι᾽ ὄξους καί τινα διὰ νάπυος ἐδώκαμεν αὐτῷ προσενέγκασθαι βεβαιοτέρας ἕνεκα διαγνώσεως. οὔτ᾽ οὖν τούτων ἔδακνεν αὐτόν τι καὶ ἡ αἴσθησις ἦν ἐν τῷ τραχήλῳ σαφής· ἠρεθίζε τότε κατ᾽ ἐκεῖνο τὸ χωρίον ὡς ἐξορμᾶν εἰς βῆχας· συνεβουλεύομεν οὖν αὐτῷ ἀντέχειν καθ᾽ ὅσον οἷό[9] τ᾽ ἐστὶ καὶ

362K μὴ βήττειν. ἔπραττε δὴ τοῦτο. βραχύ τε γὰρ | ἦν τὸ ἐρεθίζον, ἡμεῖς τε τρόπῳ παντὶ συνεπράττομεν εἰς

12. Nevertheless, those ulcers that occur in the rough arteries (trachea and bronchial tree) in relation to their inner wall, and especially those near the larynx, or even in it, are treatable, and I have cured quite a few patients so affected. In particular, I discovered the cure of these ulcers here at the time of the great plague (would that it will at some point cease), which first came upon us. At that time, a 361K young man broke out in sores[28] over his whole body on the ninth day, just as did almost all the others who were saved. On that day there was also a slight cough. On the following day, immediately after he washed himself, he coughed more violently and brought up with the cough what they call a scab. And to the person, the sensation was clearly related to the rough artery in the neck near the throat (the trachea), this being the ulcerated part. However, I also opened his mouth and examined the pharynx in case the sore might be somewhere in this. But to those who examined him in this way he did not appear to have been affected, and anyway, to the patient the sensation would clearly seem to be in the passage of what is eaten and drunk, if the sore was in that spot.

Nevertheless, I gave him something made from vinegar and from mustard to take for the purpose of a more certain diagnosis. None of these things stung him and yet the sensation was clearly in the neck. At the time, he felt an irritation in that region to the point of exciting coughing. I then advised him to persevere as far as possible and not to cough, so he did this. And because the irritation was 362K slight, I helped in every way so the sore was brought to

28 *Helkos* is also used here.

---

9 B; καθόσον οἷός K

οὐλὴν ἀχθῆναι τὸ ἕλκος, ἔξωθεν μὲν ἐπιτιθέντες τι
τῶν ξηραινόντων φαρμάκων, ὕπτιον δὲ κατακλίναντες·
εἶτα διδόντες ὑγρὸν φάρμακον τῶν πρὸς ἕλκη τοιαῦτα
ποιούντων· καὶ τοῦτ᾽ ἐν τῷ στόματι κατέχειν ἀξιοῦντες, ἐπιτρέποντα βραχύ τι παραρρεῖν εἰς τὴν τρα-
χεῖαν ἀρτηρίαν. καὶ τοίνυν οὕτω πραττόντων αἰσθάνεσθαι σαφῶς ἔφασκε τῆς ἀπὸ τοῦ φαρμάκου στύψεως
περὶ τὸ ἕλκος, εἴτε κατὰ διάδοσιν γιγνομένης, εἴτε
καὶ αὐτοῦ τοῦ φαρμάκου περὶ τὸ ἕλκος δροσοειδῶς
παραρρέοντος εἰς τὴν ἀρτηρίαν καὶ παρηθουμένου. ἦν
δὲ οὐδ᾽ αὐτὸς ὁ κάμνων ἄπειρος τῆς ἰατρικῆς, ἀλλά τις
τῶν ἐκ τριβῆς τε καὶ γυμνασίας ἐμπειρικῶς ἰατρευόν-
των. αἰσθάνεσθαί τε οὖν ἔλεγε παραρρέοντος εἰς τὴν
ἀρτηρίαν τοῦ φαρμάκου καί ποτε καὶ βῆχα κινοῦντος,
ἀντεῖχε μέντοι πολλὰ μὴ βήττων. καὶ τοίνυν αὐτὸς
προθυμηθεὶς ἐν Ῥώμῃ μέν, ἔνθα περ ἐλοίμωξεν, ἄλλας τρεῖς ἡμέρας ἐπέμεινε μετὰ τὴν ἐννάτην· μετὰ
ταῦτα δ᾽ ἐνθεὶς ἑαυτὸν πλοίῳ κατέπλευσε μὲν πρῶτον
363K ἐπὶ τὴν θάλατταν διὰ τοῦ ποταμοῦ, τετάρτῃ | δ᾽
ὕστερον ἡμέρᾳ πλέων ἐν ταῖς Ταβίαις γίγνεται, καὶ
κέχρηται τῷ γάλακτι θαυμαστήν τινα δύναμιν ὄντως
ἔχοντι καὶ οὐ μάτην ἐπηγνημένῳ.

καί μοι δοκεῖ καιρὸς ἥκειν εἰπεῖν τι περὶ γάλακτος
χρήσεως οὐ τοῦ κατὰ τὰς Ταβίας μόνον, ἀλλὰ καὶ τοῦ
ἄλλου παντός. οὐδὲ γὰρ τοὺς ἐν Ἰταλίᾳ μόνῃ χρὴ
θεραπεύειν, ἀλλ᾽ ὅσον οἷόν τε τοὺς πανταχόθι. τῷ μὲν
οὖν ἐν ταῖς Ταβίαις γάλακτι πολλὰ συνετέλεσεν εἰς
ἀρετήν· αὐτό τε τὸ χωρίον ὑψηλὸν ὑπάρχον αὐτάρκως,

cicatrization, applying some of the drying medications externally while lying the patient on his back. Then, having given him a moist medication from among those made for such sores, and having required him to retain this in his mouth, I relied on it flowing into the rough artery (trachea) gradually. Further, while I was doing this, he said he clearly sensed the astringency from the medication around the sore, this occurring either by virtue of distribution or by the medication itself flowing in dewlike fashion around the sore into the [rough] artery and filtering through. Nor was the patient himself without experience of the medical art but was someone who practiced healing empirically by massage and exercise. He said he sensed the medication flowing into the [rough] artery and was sometimes also stirred to coughing, and yet he struggled hard not to cough. Therefore, although he was eager to be in Rome, where in fact plague was raging, he remained three more days after the ninth. After this, however, he put himself on board a ship and sailed first down the river to the sea. After 363K the fourth day sailing, he found himself among the Tabians and used milk, which has a truly wondrous potency and is not recommended without good reason.

It seems to me that the time has come to say something about the use of milk, not only among the Tabians, but also everywhere else. For it is not people in Italy alone we must treat, but as far as possible people everywhere. Many things contribute to the excellence in the milk among the Tabians. The region itself is sufficiently elevated, the ambi-

ὅ τε πέριξ ἀὴρ ξηρός, ἥ τε νομὴ τοῖς ζῴοις χρηστή.
ταύτην μέν γε ἀλλαχόθι τεχνήσεσθαι δυνατὸν ἐν
λόφῳ μετρίως ὑψηλῷ, φυτεύσαντας καὶ βοτάνας καὶ
θάμνους, ὁπόσοι γάλα χρηστόν τε ἅμα καὶ στῦφον
ἐργάσονται· λεχθήσεται δ᾽ αὐτῶν ὀλίγον ὕστερον
παραδείγματα. τὸν μέντοι πέριξ ἀέρα κατασκευάσαι
μὲν ὅμοιον ἀδύνατον, ἐκλέξασθαι δὲ τῶν ὄντων τὸν
ὁμοιότατον οὐκ ἀδύνατον. ὁμοιότατος δ᾽ ἂν εἴη ὁ ταῦτ᾽
ἔχων, ἅπερ ἐκείνῳ πάρεστιν· ὕψος μὲν τοῦ λόφου
μέτριον, ὁδὸς δ᾽ ἐπ᾽ αὐτὸν ἀπὸ τῆς θαλάττης εἰς
τριάκοντα στάδια καί τι πλέον οὐ πολλῷ. τὸ δὲ
364K χωρίον αὐτὸ τὸ ἐπὶ τῇ θαλάττῃ | αἱ Ταβίαι κατὰ τὸν
πυθμένα τοῦ κόλπου μάλιστά ἐστι τοῦ μεταξὺ Σουρ-
ρέντου τε καὶ Νεαπόλεως, ἐν τῇ πλευρᾷ μᾶλλον τῇ
κατὰ Σούρρεντον.

αὕτη δ᾽ ἡ πλευρὰ πᾶσα λόφος ἐστὶν εὐμεγέθης,
μακρός, εἰς τὸ Τυρρηνὸν ἐξήκων πέλαγος. ἐγκέκλιται
δ᾽ ἠρέμα πρὸς τὴν δύσιν ὁ λόφος οὗτος, οὐκ ἀκριβῶς
δ᾽ ἐπὶ τὴν μεσημβρίαν ἐκτέταται. οὗτος μὲν ὁ λόφος
ἄκλυστον[10] τοῖς ἀνατολικοῖς ἀνέμοις φυλάττει τὸν
κόλπον, Εὔρῳ καὶ ἀπηλιώτῃ καὶ βορρᾷ· συνάπτει δ᾽
αὐτῷ κατὰ τὸν μυχὸν τοῦ κόλπου λόφος ἕτερος οὐ
μικρός, ὃν ἔν τε τοῖς συγγράμμασιν οἱ παλαιοὶ Ῥω-
μαῖοι καὶ τῶν νῦν οἱ ἀκριβέστεροι Βεσούβιον ὀνομάζουσι. τὸ δ᾽ ἔνδοξόν τε καὶ νέον ὄνομα τοῦ λόφου
Βέσβιον[11] ἅπασιν ἀνθρώποις γνώριμον διὰ τὸ κάτωθεν ἀναφερόμενον ἐκ τῆς γῆς ἐν αὐτῷ πῦρ· ὅ μοι δοκεῖ
καὶ αὐτὸ μεγάλα συντελεῖν εἰς ξηρότητα τῷ πέριξ

88

ent air is dry, and the pasture for animals is good. In fact, it is possible to contrive the same elsewhere on moderately high ridges by planting plants and bushes of whatever sort will make for good and, at the same time, astringent milk. I shall mention examples of these things a little later. It is, however, impossible to make the ambient air similar, although it is not impossible to choose the most similar from those that do exist. These are the aspects that would be most similar to those present in that place: a moderate height of the ridge, a way to the ridge from the sea up to 30 stadia, or a little more (but not much more). The place itself (i.e. Tabiae) is toward the sea and is situated right at the head of the bay that lies between Sorrento and Naples, although more on the side toward Sorrento.

364K

This whole side is a ridge of good size, long and reaching out into the Tyrrhenian Sea. The ridge itself is inclined slightly toward the west and does not extend precisely toward the south. This ridge protects the bay against easterly winds so that is not flooded; namely the Euros, the east wind and the north wind. Joined to this, in the innermost part of the bay, is another ridge, by no means small, which the ancient Romans in their writings and those of the present time who are more accurate call Vesuvius. Now the famous and recent name of the ridge is Vesuvium and it is known to all men because of the fire inside it carried up from the earth below. And this seems to me to contribute greatly to the dryness of the ambient air. Apart from the

---

10 B (q.v. Cobet, *Mnem.* 12, 1884, 445); ἄκλειστον K
11 K; Λέσβον B

GALEN

ἀέρι· καὶ χωρὶς δὲ τοῦ πυρὸς οὔτε λίμνη τις ἐγγὺς οὔθ᾽
ἕλκος οὔτε ποταμὸς ἀξιόλογος οὐδαμόθι τοῦ κόλπου.
τῶν δ᾽ ἀρκτικῶν πνευμάτων ἁπάντων ἄχρι δύσεως
θερινῆς ὁ Βεσουβηνὸς λόφος πρόβλημ᾽ ἐστί, καὶ

365K πολλὴ τέφρα μέχρι τῆς θαλάσσης ἀπ᾽ | αὐτοῦ καθ-
ήκει, λείψανόν τι ἐν αὐτῷ κεκαυμένης τε καὶ νῦν ἔτι
καιομένης ὕλης. ταῦτα πάντα ξηρὸν ἐργάζεται τὸν
ἀέρα.

δύναιτ᾽ ἂν οὖν τις ἑτέρωθι τῆς οἰκουμένης ἐκλέξασθαι λόφον οὕτω ξηρόν, οὐ πόρρω θαλάττης,
οὔτε μέγαν ὡς ἐγκεῖσθαι ταῖς τῶν ἀνέμων εἰσβολαῖς,
οὔτε πάνυ ταπεινὸν ὡς τὰς ἐκ τῶν ὑποκειμένων πεδίων
ἀναθυμιάσεις ἑτοίμως δέχεσθαι. πεφυλάχθω δ᾽ αὐτὸν
ἐστράφθαι πρὸς ἄρκτον· οὕτω γὰρ ἂν ἀπεστραμμένος
εἴη τὸν ἥλιον. εἰ δὲ κἂν τοῖς εὐκράτοις τῆς ὅλης
οἰκουμένης ὁ λόφος ὑπάρχοι, καθάπερ ὁ κατὰ τὰς
Ταβίας ἐστίν, ἄμεινον μακρῷ. ἐν τοιούτῳ λόφῳ πόαι
μὲν ἔστωσαν ἄγρωστις καὶ λωτὸς καὶ πολύγονον καὶ
μελισσόφυλλον, θάμνοι δὲ σχῖνος καὶ κόμαρος καὶ
βάτος καὶ κισσὸς καὶ κύτισος, ὅσοι τ᾽ ἄλλοι τούτοις
ἐοίκασιν. οὕτω μέν σοι τὰ τοῦ λόφου παρεσκευάσθω.
τὰ δὲ ζῷα βόες μέν εἰσιν ἐν τῷ κατὰ Ταβίας, καὶ ἔστι
τούτου τοῦ ζῴου παχὺ τὸ γάλα, καθάπερ τὸ τῶν ὄνων
λεπτόν. ἐγὼ δ᾽ ἂν καὶ βοῦς καὶ ὄνους καὶ αἶγας ἀφείην
ἐπὶ τὰς νομάς, ὥστ᾽ ἔχειν χρῆσθαι γάλακτι παντί,
παχεῖ μὲν ἐκ τῶν βοῶν, λεπτῷ δ᾽ ἐκ τῶν ὄνων, μέσῳ δ᾽ |

366K ἀμφοῖν ἐκ τῶν αἰγῶν.

οἱ παλαιοὶ δὲ καὶ γυναῖκα θηλάζουσαν ἐφίστων
τοῖς τῇ φθόῃ κάμνουσι· κἀγὼ δὲ ἀποδέχομαι τὴν

90

fire, there is neither marshy lake, nor bog, nor a sizable river anywhere in the bay. However, the Vesuvian ridge is an obstacle to all the northern winds as far as the setting of the summer sun; much ash comes down from this even as far as the sea, and has in it some remnant of material that 365K has burned and is now still burning. All these things make the air dry.

It should, then, be possible for someone to pick out elsewhere in the Roman world a ridge that is dry like this, not far from the sea, nor so large as to be in the path of the onrushing winds, nor so low as to easily receive the vapors from the low-lying plains. Avoid it being turned toward the north for, in this way, it would have its back to the sun. And even if the ridge is in the temperate parts of the whole Roman world, as that at Tabiae is, it is better by far. Let the plants on such a ridge be dog's tooth grass, clover, knotgrass, mastich, arbutus, bramble, ivy, tree medick[29] and others that are similar to these. Let the things on the ridge be prepared for you in this way. The animals on the ridge at Tabiae are cows, and the milk of this animal is thick, just as that of asses is thin. I would let loose cows, asses and goats in the pastures so as to be able to use every kind of milk— thick from the cows, thin from the asses, and that midway between the two from the goats. 366K

The ancients also brought in a lactating woman for those suffering from consumption (*phthoe*), and I accept

[29] *Medicago arborea*; see, for example, Theophrastus, *De causis plantarum*, V.15.

91

γνώμην αὐτῶν, ὅτι τε τὸ οἰκεῖον ᾑροῦντο καὶ ὅτι πρὶν ψυγῆναι τῷ πέριξ ἀέρι. καί σοι τοῦτ᾿ ἔστω μέγιστον παράγγελμα γάλακτος χρήσεως ἐπὶ πάντων οἷς γάλακτος χρεία, αὐτίκα πίνειν ἀμελχθέν, τῷ ζῴῳ παρεστῶτα· προσεπεμβάλλειν δὲ καὶ μέλιτος, ὅτῳ τυροῦσθαι πέφυκεν ἐν τῇ γαστρί· εἰ δ᾿ ὑπελθεῖν αὐτό ποτε θᾶττον βουληθείης, καὶ ἄλων. ἐκεῖνος μέν γε οὖν ὁ νεανίας ἐκ τῆς λοιμώδους νόσου κατὰ τὴν ἀρτηρίαν ἕλκος ἔχων ὑγιὴς ἐγένετο καὶ ἄλλοι μετ᾿ αὐτὸν ὁμοίως. ἑτέρῳ δὲ μειρακίῳ περὶ ἔτος ὀκτωκαιδέκατον ἐκ κατάρρου πλείοσιν ἡμέραις γενομένου τὰ μὲν πρῶτα μετὰ βηχὸς αἷμα θερμὸν εὐανθὲς οὐ πολύ, μετὰ δὲ ταῦτα ἀνεπτύσθη τι καὶ τοῦ χιτῶνος αὐτοῦ μέρος, ὃς ὑπαλείφων ἔνδοθεν τὴν ἀρτηρίαν εἰς τὴν φάρυγγά τε καὶ τὸ στόμα διὰ τοῦ λάρυγγος ἀνεφέρετο. ἐδόκει δέ μοι τῷ τε πάχει τεκμαιρομένῳ καὶ τῇ τοῦ κάμνοντος αἰσθήσει τοῦ λάρυγγος ὑπάρχειν τὸ ἔνδοθεν σῶμα· 367K καὶ μέντοι καὶ ἐβλάβη τοὐντεῦθεν | εἰς τὴν φωνὴν ὁ ἄνθρωπος· ἀλλὰ καὶ οὕτως ἐν χρόνῳ μὲν πλείονι διεσώθη καὶ αὐτός.

οἱ δ᾿ ἐκ τοῦ λοιμοῦ ῥᾳδίως ὑγιάζεσθαί μοι δοκοῦσι τῷ προεξηράνθαι τε καὶ προκεκαθάρθαι σύμπαν τὸ σῶμα· καὶ γὰρ ἔμετός τισιν αὐτῶν ἐγένετο καὶ ἡ γαστὴρ ἅπασιν ἐταράχθη. καὶ οὕτως ἤδη κεκενωμένοις τοῖς σῴζεσθαι μέλλουσιν ἐξανθήματα μέλανα διὰ παντὸς τοῦ σώματος ἀθρόως ἐπεφαίνετο· τοῖς πλείστοις μὲν ἑλκώδη, πᾶσι δὲ ξηρά. καὶ ἦν εὔδηλον ἰδόντι τοῦ σεσηπότος ἐν τοῖς πυρετοῖς αἵματος εἶναι

their idea because they chose milk that was suitable and because it was prior to cooling by the ambient air. Make your most important precept regarding the use of milk be, in the case of all those for whom there is need of milk, to drink it immediately it is milked, placing [the patient] beside the animal, and adding some honey also, by which it is naturally curdled in the stomach. If you wish it to pass down quicker on occasion, add salt. In fact, that young man who had a sore in the trachea from the plague became healthy, and others after him likewise. Another young lad, around eighteen years of age, having suffered from catarrh for many days, expectorated for the first time blood that was hot and not very fresh with coughing, and after this, a part of the actual wall which was lining the trachea internally was expectorated and passed into the pharynx and mouth via the larynx. It seemed to me from the observed thickness and by the perception of the patient, that it was a body from within the larynx. And indeed, the man henceforth suffered damage to his voice, but he too was saved   367K over a longer time in this way.

Those who are easily restored to health from the plague seem to me to be dried and purged beforehand in respect of the whole body, for vomiting occurred in some of them and the stomach was disturbed in all of them. And in the same way, in those already purged who are going to be saved, exanthemata appear close together over the whole body, in the majority there are sores, while in all there is dryness. And it was quite clear to an observer that, when

τοῦτο λείψανον, οἷον τέφραν τινὰ τῆς φύσεως ὠθούσης ἐπὶ τὸ δέρμα, καθάπερ ἄλλα πολλὰ τῶν περιττῶν. οὐ μὴν ἐδέησέ γε πρὸς τὰ τοιαῦτα τῶν ἐξανθημάτων φαρμάκου· καθίστατο γὰρ αὐτόματα τρόπῳ τῷδε· τινῶν μέν, οἷς γε καὶ ἡλκώθη, τὸ ἐπιπολῆς ἀπέπιπτεν, ὅπερ ὀνομάζουσιν ἐφελκίδα· κἀντεῦθεν ἤδη τὸ λοιπὸν ἐγγὺς ἦν ὑγείας· καὶ μετὰ μίαν ἢ δύο ἡμέρας ἐπουλοῦτο. τινῶν δέ, οἷς οὐχ ἡλκώθη, τὸ μὲν ἐξάνθημα τραχύ τε καὶ ψωρῶδες ἦν, ἀνέπιπτε καὶ οἷόν τι λέμμα, 368K κἀκ τούτου πάντες | ὑγιεῖς ἐγίγνοντο. θαυμαστὸν οὖν οὐδὲν εἰ καὶ κατὰ τὸν πνεύμονα τοιούτων ἐξανθημάτων γεγενημένων ἐσώζοντο διὰ τὴν ξηρότητα τῶν ἑλκῶν. ὃν γὰρ ἂν ἐπὶ τῶν ἄλλων ἑλκῶν ἁπάντων ὁ πρόσθεν λόγος ἐδείκνυε σκοπὸν εἶναι τῆς θεραπείας, τοῦθ' ὑπῆρχεν ἤδη τοῖς ἐκ τοῦ λοιμοῦ γεγενημένοις. ἅπαντα γὰρ ἦν ξηρὰ καὶ τραχέα, τὰ μὲν πλεῖστα ψώρᾳ, τινὰ δ' αὐτῶν καὶ λέπρᾳ παραπλήσια. μαρτυρούσης οὖν τῷ λόγῳ τῆς πείρας καὶ τοῖς ἕλκεσιν ἕνα τοῦτον ἔχουσι σκοπὸν τῆς ἰάσεως τὸ ξηρανθῆναι, δύναιτ' ἄν τις σῴζειν παμπόλλους τῶν αἷμα πτυσάντων ἐκ πνεύμονος, ὥσπερ καὶ ἡμεῖς ἐσώσαμεν.

13. Ὅπως μὲν οὖν χρὴ θεραπεύειν οἷς ἀγγεῖον ἀξιόλογον ἐρράγη κατενεχθεῖσιν, ἢ μέγα βοήσασιν, ἢ βάρος ἀραμένοις ὑπὲρ τὴν δύναμιν, ἤ τινος ἐμπεσόντος τῷ θώρακι σκληροῦ καὶ μεγάλου ἔξωθεν σώματος, εἴρηται πρόσθεν. ὡς δ' ἄν τις κάλλιστα καὶ τοὺς ἐπὶ κατάρρου πτύσαντας αἷμα μεταχειρίζοιτο, 369K νῦν εἰρήσεται, σαφηνείας ἕνεκα | προχειρισαμένων

the blood had been putrefied in those with fever, there was this remnant like ash which nature forced out of the skin, just as with many other superfluities. In fact, there was no need of drying medications for such exanthemata for they spontaneously existed in the following manner: in some, in whom there was also ulceration, the surface fell off, which they call scabs, and henceforth what remained was already close to health, and after one or two days scarred over. In others, in whom there was not ulceration, the exanthem was rough and itchy, and fell off like something scaly, and from this all patients became healthy. It is no wonder, then, 368K if also when such exanthemata have occurred in the lung, [the patients] were saved due to the dryness of the sores. For what the previous discussion showed to be the objective of treatment in the case of all other ulcers and sores has already occurred in those with the plague because all the sores were dry and rough, the majority with scabs, and some of them also with what is like *lepra*. Therefore, since experience provides evidence for the theory, and [doctors] have this one objective of cure—to be dried—for ulcers and sores, it is possible for someone to save many of those spitting up blood from the lung, just as I too saved them.

13. I spoke earlier of how you should treat the rupture of a major vessel in those who are knocked down, or shout loudly, or lift a heavy weight beyond their capacity, or have some hard and large body fall on their chest externally. I shall speak now of how you should best manage those who are coughing up blood in catarrh. For the sake of clarity, I have chosen the treatment I carried out for a woman 369K

95

ἡμῶν ἦν ἐποιησάμεθα θεραπείαν ἐν Ῥώμῃ γυναικὸς
τῶν ἐν τέλει. αὐτὴ γὰρ ἀκηκουῖα λόγων τοιούτων
οἵους νῦν διεληλύθαμεν ὑπὲρ τῶν ἐκ πνεύμονος αἷμα
πτυσάντων, εἴτ' ἐκ κατάρρου τινὸς ἢ βηχὸς ἰσχυρᾶς
ὀλίγον τι πτύσασα διὰ νυκτὸς ἔπεμψεν αὐτίκα πρός
με παρέχειν ἑαυτὴν φάσκουσα πᾶν ὅ τι ἂν ἐθέλοιμι
πράττειν. ἠκηκόει δὲ κατὰ τοὺς ἔμπροσθεν χρόνους
ὡς εἰ μή τις αὐτίκα βοηθήσειε δραστικῶς πρὶν ἢ
φλεγμῆναι τὸ ἕλκος, οὐδὲν ἀνύσει, καὶ ὡς τοῦτ' ἂν εἴη
μάλιστα τὸ αἴτιον τοῦ διαφθείρεσθαι τοὺς ἀναβήξαν-
τας τὸ αἷμα. φλέβα μὲν οὖν τεμεῖν αὐτῆς οὐκ ἠξίωσα,
τεττάρων ἡμερῶν ἤδη διὰ τὸν κατάρρουν ὀλίγου δεῖν
ἀσίτου γεγενημένης. κλύσματι δὲ δριμεῖ κελεύσας
χρήσασθαι καὶ τρίψασθαί τε καὶ σκέλη καὶ τὰς χεῖρας ἐπὶ πλεῖστον ἅμα φαρμάκῳ θερμαίνοντι καὶ διαδήσας, εἶτα ξυρήσας τὴν κεφαλὴν ἐπέθηκα τὸ διὰ τῆς
κόπρου τῶν ἀγρίων περιστερῶν φάρμακον.

ὡρῶν δὲ μεταξὺ τριῶν γενομένων ἐπὶ τὸ βαλανεῖον
ἤγαγον καὶ λούσας ἄνευ τοῦ ψαῦσαι λίπους τῆς κεφα-
370K λῆς, εἶτα σκεπάσας αὐτὴν συμμέτρῳ πίλῳ | πρὸς τὴν
ἐνεστηκυῖαν ὥραν ἔθρεψα ῥοφήματι μόνῳ τῶν αὐστηρῶν ὀπωρῶν ἐπιδιδούς. εἶθ' ὕπνουν μελλούσῃ τὸ διὰ
τῶν ἐχιδνῶν ἔδωκα φάρμακον ὡς πρὸ τεττάρων μηνῶν
ἐσκευασμένον· ἔτι γὰρ ἔχει τὸν τοῦ μήκωνος ὀπὸν
ἰσχυρὸν τὸ τοιοῦτον, ἐν δὲ τοῖς κεχρονικόσιν ἐξίτηλος

---

[30] Although Galen devotes a whole section of *De simplicium
medicamentorum temperamentis et facultatibus* to the medicinal

of high status in Rome. Since she had heard those arguments I have now gone through in detail about those spitting up blood from the lung, when she expectorated a little [blood] during the night, either from catarrh or a strong cough, she immediately sent for me, saying she would submit herself to anything whatever I might wish to do. She had heard on previous occasions that, if someone does not provide an effective remedy immediately, before the wound becomes inflamed, he will accomplish nothing, and that this, particularly, would be the cause of death in those who coughed up blood. Therefore, I did not think it worthwhile to open one of her veins, since she had already gone four days almost without food due to the catarrh. However, I ordered the use of a sharp clyster and rubbing, and binding around of the legs and arms as much as possible, along with a heating medication. Then, having shaved her head, I applied the medication made from the excrement of wild
. pigeons.[30]

After an interval of three hours, I led her to the bath and washed her without touching her head with any oil. Then I covered her head with a well-fitting felt cloth and, according to the prevailing season, I nourished her with thick gruel alone, afterward giving her bitter fruits. Then, when she was about to go to sleep, I gave her the medication made from vipers[31] that had been prepared four months before, for such a medication still has the juice of the poppy in strong degree whereas, in medications that have been aged, the strength becomes less. Because of

370K

properties of the excrement of various creatures—including crocodiles, no less (XII.290–309K)—there is no mention of pigeons.

[31] See Dioscorides, II.18.

97

γίνεται, διὰ τοῦτο οὖν ὑπνῶδές τέ ἐστι καὶ ξηραίνει τὰ
ῥεύματα σὺν τῷ καὶ παχύνειν ἀτρέμα. παυσαμένου δὲ
τελέως τοῦ κατάρρου δῆλος μὲν ὁ πνεύμων ἦν ἔκ τε
τοῦ τῆς ἀναπνοῆς εἴδους καὶ τοῦ ψόφου τῆς γενομένης
ἅπαξ που βηχὸς ἐκκαθάρσεως δεόμενος. οὐ μὴν συν-
έπραξά γε εἰς τοῦτο κατὰ τὴν δευτέραν ἡμέραν, ἀλλ'
ἐφ' ἡσυχίας τε καὶ σιγῆς ἁπάσης τὴν γυναῖκα φυ-
λάξας καὶ τρίψας τὰ κῶλα καὶ διαδήσας ἐκέλευσα καὶ
τἆλλα πάντα μέρη τρίψασθαι πλὴν τῆς κεφαλῆς, ἔτι
γὰρ ἦν ἀπὸ τοῦ φαρμάκου θερμή. αὖθις δὲ εἰς τὴν
ἑσπέραν ἔδωκα τοῦ διὰ τῶν ἐχιδνῶν φαρμάκου μέγεθος, ὅσον γε ὁ παρ' ἡμῖν κύαμος. ἦν δὲ πολὺ τούτου
μεῖζον ὃ κατὰ τὴν προτέραν εἰλήφει. ὡς δὲ καὶ ταύτῃ
τῇ νυκτὶ καλῶς ἐκοιμήθη, τῇ τρίτῃ τῶν ἡμερῶν ἕωθεν
371K δοὺς μέλιτος ἑψημένου συχνόν, ἐφ' ἡσυχίας | ἐφύλαξα.
προηκούσης δὲ τῆς ἡμέρας ἅπαν ὁμοίως τρίψας τὸ
σῶμα τροφὴν ἐκέλευσα λαβεῖν πτισάνης χυλὸν σὺν
ἄρτῳ βραχεῖ. κἄπειτα τῇ τετάρτῃ τῶν ἡμερῶν ἕωθεν
μὲν τοῦ διὰ τῶν ἐχιδνῶν φαρμάκου ἀκμαίου κατὰ τὴν
ἡλικίαν ἅμα σὺν χρηστῷ μέλιτι προσέδωκα. τῇ κεφαλῇ δὲ τοῦ αὐτοῦ φαρμάκου πάλιν ἐπιθείς, ὃ θερμαίνει
τε καὶ ξηραίνει σφοδρῶς, εἶτα λούσας καὶ θρέψας
μετρίως, ἐκκαθαίρειν ἤδη σφοδρότερον ἐπεχείρησα
τὸν πνεύμονα τῇ πέμπτῃ τῶν ἡμερῶν. εἶτ' αὖθις καὶ
αὖθις ἐχρησάμην ἐκ διαλειμμάτων κατὰ μὲν τῆς κεφαλῆς τῇ συνήθει κηρωτῇ, τῇ διὰ τῆς θαψίας. τὴν δ'
ὅλην τοῦ σώματος ἐπιμέλειαν ἀναθρεπτικὴν ἐποιησάμην ἐν αἰωρήσεσι καὶ τρίψεσι καὶ περιπάτοις καὶ

this, then, it is sleep-inducing and dries the fluxes also with a gentle thickening. When the catarrh had completely stopped, it was clear from both the form and sound of the breathing that was occurring, that her lung was in need of clearing once and for all by coughing. I did not, in fact, accede to doing this on the second day, but directed the woman to keep herself quiet and be totally silent, and ordered rubbing and binding of the limbs, and rubbing of all other parts except her head, for this was still hot from the medication. Again, toward evening, I gave her an amount of the medication made from vipers, equal in size to a bean. The amount which she had taken on the previous day was far more than this. As she also slept well that night, early on the morning of the third day I gave her a large amount of boiled honey and kept her quiet. As the day pro-   371K
gressed, I ordered rubbing of the whole body in a similar way, and [for her] to take as nourishment the juice of ptisan with a little bread. And then, on the fourth day, early in the morning, I gave her the medication made from vipers as well, at its full strength, and because of her age, along with beneficial honey, I again applied to her head the same medication which heats and dries strongly, and then, after bathing and feeding her moderately, on the fifth day I attempted to clear out the lungs more vigorously. Then, at intervals, I repeatedly used on her head the customary salve from thapsia.[32] I provided total care and nourishment for the body with passive exercises, rubbings, perambulations, abstinence from baths and a moderate diet,

[32] See Dioscorides' detailed account of thapsia as a medication; IV.157.

GALEN

ἀλουσίαις καὶ διαίτῃ μετρίᾳ τε ἅμα καὶ εὐχύμῳ. αὕτη
καλῶς ἔσχεν ἡ γυνὴ μηδὲ δεηθεῖσα τοῦ γάλακτος.

ἄλλον δέ τινα νεανίσκον οὐκ ἐκ κατάρρου τῆς
βηχὸς ὁρμηθείσης, ἀλλ᾽ ἐπὶ ψύξει τῶν ἀναπνευστικῶν, αἷμα καὶ αὐτὸν πτύσαντα πλῆθος ὅσον ἥμισυ
κοτύλης, παραχρῆμα μὲν ἐφλεβοτόμησα καὶ δὶς τῆς
372K αὐτῆς | ἡμέρας ἐπαφεῖλον· εἶτα δὶς κατὰ τὴν ὑστε-
ραίαν· τρίψει δὲ τῶν κώλων καὶ δεσμῷ ἐν τῇ πρώτῃ
χρησάμενος, εἰς ἑσπέραν ἔδωκα τὸ διὰ τῶν σπερμάτων ἡμέτερον φάρμακον. ἐν δὲ τῇ δευτέρᾳ μετὰ τὴν
ἐπαφαίρεσιν ἐπέθηκα παντὶ τῷ θώρακι τὴν διὰ τῆς
θαψίας κηρωτήν, εἶτ᾽ ἄρας αὐτὴν εἰς ἑσπέραν, ὅπως
μὴ θερμήνειε περαιτέρω τοῦ δέοντος, ἐν τῇ τρίτῃ
πάλιν ἐπιθεὶς ὥραις που τρισὶν ἔλουσα τὸν ἄνθρωπον,
ἔθρεψά τε ταῖς τρισὶν ἡμέραις παραπλησίως, ῥοφή-
ματα μὲν ἐν ταῖς πρώταις δύο, τῇ τρίτῃ δὲ χυλὸν μὲν
πτισάνης προδούς, ἰχθὺν δέ τινα τῶν εὐπέπτων δοὺς
ἁπλῶς ἠρτυμένον. ἔδωκα δὲ καὶ τὸ διὰ τῶν σπερμάτων φάρμακον, ἐν δὲ τῇ δευτέρᾳ καὶ τῇ τρίτῃ
τῶν ἡμερῶν ὁμοίως εἰς ἑσπέραν· ὑπνοποιόν τε γάρ
ἐστι καὶ ἀνώδυνον καὶ ξηραντικόν. ἤδη δὲ καὶ τῶν
ἀναπνευστικῶν ἐν εὐκρασίᾳ τῇ κατὰ φύσιν ὄντων καὶ
τοῦ παντὸς σώματος κεκενωμένου καὶ μηδὲ μιᾶς ὑπο-
ψίας φλεγμονῆς ὑπολειπομένης κατὰ τὸ ἐρρωγὸς ἀγγεῖον, ἀνακαθαίρειν ἠρξάμην. εἶτα πίνοντα τὸ διὰ τῶν
ἐχιδνῶν φάρμακον ἀκμαῖον κατὰ τὴν ἡλικίαν ἐπὶ τὰς
373K Ταβίας ἐξέπεμψα. | πάντας οὕτως ἐγὼ τοὺς κατὰ τὴν
πρώτην ἡμέραν ἑαυτοὺς ἡμῖν ἐγχειρίσαντας ἰασάμην.

100

along with a healthy state of the humors (*euchymia*). This woman became well without the need for milk.

There was another young man who, when coughing was provoked not from catarrh but due to cooling of the respiratory organs, spat up blood, as much as half a *kotyle*.[33] I carried out phlebotomy immediately and withdrew blood 372K twice on the same day, and then twice on the following day. Then, having used rubbing and binding of the limbs on the first day, toward evening I gave my own medication made from seeds (*diaspermaton*). On the second day, after drawing off [blood], I applied a salve made from thapsia to the whole chest. Then, on the same evening, I removed it so he did not become heated more than was necessary. On the third day, after I had applied it over three hours, I bathed the man. On the first two days, I gave him beforehand thick porridge, while on the third I gave him juice of ptisan and one of the fish that are easily digested, prepared simply. And I gave the medication made from seeds (*diaspermaton*) similarly on the second and third days toward evening, for it is sleep-producing, pain-relieving and drying. When the respiratory organs were already in a *eukratic* state in accord with nature and the whole body had been evacuated, and no suspicion of inflammation remained in relation to the ruptured vessel, I began to purge thoroughly. Then, when he had drunk the medication made from vipers at full strength according to the age, I sent him off to Tabiae. I cured all those who entrusted 373K themselves to me on the first day in the same way.

[33] Approximately one quarter of a pint.

101

14. Οὐ μὴν τούς γε μετὰ δύο ἢ τρεῖς ἅπαντας, ἀλλ᾽ ἔνιοί τινες αὐτῶν ἀνίατον ἔσχον τὸ ἕλκος. ὅσοι δ᾽ ἔφθασαν οὕτω φλεγμήναντες ὡς πυρέξαι, τούτων οὐδεὶς ἐθεραπεύθη τελέως. ἀλλ᾽ ὅσοι μετὰ ταῦτα διὰ πάντων ὀρθῶς τῶν βοηθημάτων διεξῆλθον, ὡς ξηρανθῆναι τὸ ἕλκος, ἐκέρδησαν οὗτοι τοσοῦτον ὡς μήτ᾽ ἐπινέμεσθαι μήτε μεῖζον γίγνεσθαι, ξηρανθὲν δὲ καὶ τυλωθὲν ἐπιτρέψαι χρόνῳ πλέονι ζῆσαι τὸν ἄνθρωπον. μόνοι δὲ ἀνιάτως διακεῖσθαί μοι δοκοῦσι τῶν ἕλκος ἐν πνεύμονι ἐχόντων οἱ διὰ κακοχυμίαν τινὰ ἀναβρωθέντες, ὧν ἔνιοι καὶ τοῦ σιάλου φασὶν ὥσπερ ἅλμης αἰσθάνεσθαι. χρόνου μὲν γὰρ οἶμαι πάντως δεῖν μακροῦ πρὸς τὴν τῆς κακοχυμίας ἐπανόρθωσιν. ἀνάγκη δ᾽ ἐν τῷ χρόνῳ δυοῖν θάτερον, ἢ ξηραινόμενον τὸ ἕλκος τυλωθῆναι καὶ διὰ τοῦτο ἀνίατον γενέσθαι παντάπασιν, ἢ μὴ ξηραινόμενον αὐτό τε πρῶτον ἀποδειχθῆναι σηπεδονῶδες ἐπινέμεσθαί τε τὰ πέριξ καὶ
374K οὕτως ἐν χρόνῳ διασῆψαι τὸν πνεύμονα. | πολλοὶ δὲ τῶν ἐσχηκότων τοιαύτην κακοχυμίαν ἤδη βήττοντες ἐξ αὐτῆς, οὐ μέντοι γ᾽ ἐπτυκότες οὐδέπω τὸ αἷμα, τῆς παρ᾽ ἡμῶν τυχόντες προνοίας εἰς τέλος ὑγιάσθησαν.

χρὴ δὲ ἐν ἀρχῇ μὲν οὐδενὸς οὕτως ὡς τοῦ μήτε βήττειν αὐτοὺς φροντίσαι μήτ᾽ ἐκ τῆς κεφαλῆς καταρρεῖν τι εἰς τὸν πνεύμονα. καὶ γίνεται τοῦτο τρισὶ βοηθήμασι, καθάρσει μὲν πρώτῳ, δευτέρῳ δὲ τῷ διὰ τῶν σπερμάτων φαρμάκῳ, τρίτῳ δὲ τῇ προνοίᾳ τῆς κεφαλῆς. δεῖ δὲ τὸ μὲν καθαῖρον εἶναι μικτὸν ἐκ διαφερουσῶν δυνάμεων, οἷά πέρ ἐστι τὰ ἡμέτερα

14. Not all of these [patients] are in fact [cured] after two or three days; some of them have an ulcer that is incurable. None of those who have inflammation beforehand such that they are febrile is cured completely. But those who, after this, have gone through all the remedies properly so that the ulcer is dried are benefited to the extent that it neither spreads nor becomes larger, and being dried and made hard allows the person to live for a longer time. The only patients among those who have an ulcer in the lung who seem to me to be in an incurable state are those who have suffered erosion due to some *kakochymia*, some of whom say they perceive their saliva like brine. There is no doubt, I believe, that it requires a long time for the *kakochymia* to be corrected. During this time one of two things is inevitable: either the ulcer, having been dried, is made hard and, because of this, becomes completely incurable, or if it is not dried, it is first shown to be putrefying and then it spreads all around, and over time causes the lung to putrefy in the same way. However, many of those who had such a *kakochymia* and were already coughing from this, but who had not yet expectorated blood were, with some forethought on my part, completely restored to health.

374K

In the beginning, you should consider nothing as more important than that they neither cough nor have anything flow down from the head to the lung. And this occurs with three remedies: first by purging, second by the medication made from seeds (*diaspermaton*), and third by giving careful thought to the head. The purgative needs to be mixed from things having different potencies, like, for example,

103

καταπότια δι' ἀλόης καὶ σκαμμωνίας καὶ κολοκυν-
θίδος καὶ ἀγαρικοῦ καὶ βδελλίου καὶ κόμμεως Ἀρα-
βικοῦ συγκείμενα, πρὸς τὸ πλέονας ἰδέας ἐκκαθαίρειν
τῶν περιττωμάτων. ἱκανὸν δὲ ὠφελῆσαι καὶ τὸ χωρὶς
κόμμεως· ὕστερον δ' ἄν, εἰ δεήσει, καὶ τοῖς τὰ μέλανα
καθαίρουσι χρήσαιο· τὴν δὲ τῆς κεφαλῆς πρόνοιαν
ἐπιτρέψαι τῇ διὰ θαψίας κηρωτῇ. ταῦτα μὲν ἐν ἀρχῇ
ποιητέον ἐστίν, ἐφεξῆς δὲ ἀνατρέφειν χρηστῶς εὐχύμοις ἐδέσμασι καὶ τρίψεσι καὶ περιπάτοις καὶ λουτροῖς. εἰρήσεται δὲ ἡ περὶ τὰ τοιαῦτα μέθοδος ἐν τῷ
375K προσήκοντι | χωρίῳ τῆς συγγραφῆς, ὁποῖοι οὗτοι
μάλιστα δέονται τοῦ γάλακτος, ἀμελήσαντες δὲ πάντων ἀνιατότατοι γίνονται· περὶ δὲ φλεβοτομίας αὐτῶν
ὧδε χρὴ γινώσκειν. ὅσοι μὲν ἂν δόξωσιν ὀλίγον ἔχειν
αἷμα, καθ' ὃν εἴρηκα τρόπον εἰς εὐχυμίαν τινὰ προσ-
αγαγὼν αὐτούς, φλεβοτομῆσαί τε δυνήσῃ καὶ πάλιν
ἀναθρέψαι, καὶ πάλιν καθᾶραι καὶ αὖθις ἀναθρέψαι,
καὶ αὖθις, εἰ δεήσειε, φλεβοτομῆσαι, καὶ μάλισθ'
ὅσοις ἐστὶν οἷον ἰλύς τις μοχθηρὰ καὶ παχεῖα τὸ
σύμπαν αἷμα· τοὺς δ' ἰσχυρούς τε καὶ πολυαίμους
εὐθὺς ἐξ ἀρχῆς φλεβοτομεῖν.

15. Οὐδὲν ὧν εἶπον οὔτε νῦν οὔτε πρόσθεν ὧν ἂν ἢ
αὐτὸς ἐξεῦρον, ἢ κατὰ τὴν Ἱπποκράτους ὁδὸν ἐχρησάμην, ἀβασάνιστόν ἐστιν οὐδ' ἄκριτον, ἀλλ' ὑπὸ τῆς
πείρας ἅπαντα κέκριται, τὸν μὲν τῆς ἀποτυχίας κίνδυνον ἡμῶν ὑπομεινάντων, τὸν καρπὸν δὲ τῆς χρήσεως ἐξόντων ἁπάντων οἷς ἂν μέλῃ τῶν ἔργων τῆς
τέχνης. ἄλλοι μὲν γάρ εἰσιν οἱ ἀληθεῖς, ἄλλοι δὲ οἱ

our own little pills compounded from aloes, scammony, colocynth, agaric, bdellium and gum arabic, made up to purge the many kinds of superfluities. What lacks the gum arabic is also sufficient to be of benefit. Later, perhaps, you might if necessary also use those things that purge the black [bile], and give consideration to the head through a salve made from thapsia. These are the things you must do at the beginning. The next step is to nurture the patient properly with *euchymous* foods, massage, walking and bathing. I shall speak of the method relating to such things in the appropriate part of the book—that is, of what sort 375K they are who particularly require milk, and those who, being neglectful of all things, become incurable. Here it is necessary to know about phlebotomy for these patients. Those who would seem to have little blood when you bring them to a *euchymia* in the way I have described, you will be able to phlebotomize and again feed up, and again purge, and again feed up, and again, if necessary, phlebotomize. This applies particularly to those whose blood in its entirety is of bad quality and thick like slime. On the other hand, phlebotomize those who are strong and full of blood right from the start.

15. None of the things that I said, either now or previously, which I myself discovered, nor any of those I used according to the Hippocratic way, is untried or untested. They have all been tested by experience in which I have borne the risk of failure, so that all those who intend to practice the tasks of our craft will enjoy the benefit of their use. For some arguments are true while some are sophisti-

τῶν σοφιστῶν λόγοι. καὶ τί δεῖ λέγειν περὶ τῶν
376K σοφιστῶν, ὅπου καὶ τῶν ἀξιολόγων ἰατρῶν | ἔνιοι ὅλα
γράψαντες βιβλία περὶ αἵματος ἀναγωγῆς; ἄλλα μέν
τινα πολλὰ καὶ μικρὰ καλῶς διεξῆλθον, οὐδενὸς δὲ
τῶν μεγίστων ἐμνημόνευσαν, οὐκ ἐννοοῦντες ὡς μὲν
διὰ τὰ πολλὰ καὶ σμικρὰ τὰ καλῶς παρ' αὐτῶν παρη-
νημένα μακρὸς ὁ θάνατος αὐτοῖς τοῖς κάμνουσι γενή-
σοιτο, διὰ δὲ τρία ταῦτα σῴζονται πάντες οἱ σωθῆναι
δυνάμενοι, διὰ φλεβοτομίαν καὶ κάθαρσιν καὶ τὰ
ῥωννύντα τὴν κεφαλὴν φάρμακα. εἰρήσεται δὲ τελεώτερον ἐν τοῖς ἑξῆς περὶ φλεβοτομίας, ὅταν ὑπὲρ
τῆς τῶν κακοχύμων θεραπείας διερχώμεθα. μὴ τοίνυν
ὥσπερ ἔνιοι τῶν ἰατρῶν ἀπὸ τῶν μικρῶν ἄρχεσθαι
βοηθημάτων, μηδ' ὅπερ ἐκεῖνοι λέγουσιν, ἐννοεῖν ὡς
ἀποπειρατέον ἐστὶ τούτων πρότερον, εἶτ', εἴπερ ἀνύει
μηδέν, ἐπιτίθεσθαι τοῖς μείζοσιν. ἐπὶ μὲν γὰρ τῶν
ἀκινδύνων παθῶν ἀληθὴς ἡ τοιαύτη δόξα, τεθνηξο-
μένου δὲ πάντως τοῦ κάμνοντος, ἐὰν εἰς ἀρχὴν κατα-
στῇ φθόης, ἀλογώτατόν ἐστιν ἀπὸ τῶν μικροτέρων
ἄρχεσθαι.

τά τε γὰρ ἄλλα πάντα καλῶς ὑφ' Ἱπποκράτους
377K εἴρηται καὶ σὺν αὐτοῖς ὁ ἀφορισμὸς ὅδε· | εἰς δὲ τὰ
ἔσχατα νοσήματα αἱ ἔσχαται θεραπεῖαι εἰς ἀκριβίην
κράτισται. τί ποτ' οὖν Ἐρασιστράτῳ δόξαν ἀργός τε
καὶ ῥάθυμος ἐν ἀρχῇ τῶν τοιούτων παθῶν γενόμενος,
ἐπιμελὴς αὖθίς ἐστιν, ὅτε οὐδὲν ὄφελος; φλεβοτομίᾳ
γὰρ οὐδ' ὅλως ἐπ' οὐδενὸς τοῦ αἵματος ἀναγωγῇ
χρῆται, λογισμῷ φαυλοτάτῳ πιστεύσας· ἐκάθηρέ τε

cal. But why must we talk about those who are sophistical when there are also some notable doctors who have writ- 376K ten whole books about the bringing up of blood?[34] But although they cover many of the other, minor matters well, they fail to mention any of the major ones, not realizing that, because of the many minor things which they rightly recommend, the major thing—death—will befall the actual patients. All those who can be saved are saved through these three things: phlebotomy, purging and medications that strengthen the head. A more comprehensive statement will be made about phlebotomy in what follows when I go over the treatment of the *kakochymias*. Therefore, do not, as some doctors do, begin with the small remedies, nor, as those doctors say, think that you must make a trial of these first, and then, if they accomplish nothing, try those that are major. For although in the case of nondangerous affections such a notion is true, when the patient is altogether at the point of dying, if in the first place he is suffering from consumption (*phthoe*), it is absolutely irrational to begin with the smaller remedies.

And as everything else was well stated by Hippocrates, so with these matters the *Aphorisms* has the following: "for extreme diseases extreme treatments are the strong- 377K est to the highest degree."[35] Why ever, then, being of the same opinion as Erasistratus, would a doctor be inactive and indifferent at the start of such affections and be attentive again when there is no benefit? For he did not use phlebotomy at all in the case of anyone bringing up blood, relying on this completely foolish reasoning. Nor did he

[34] Presumably a reference to Erasistratus' *Expectoration of Blood*. [35] See Hippocrates, *Aphorisms* I.7.

οὐδένα, καθάπερ οὐδὲ τὴν κεφαλὴν ἐξήρανεν. ἅπερ εἴ
τις ἐξέλοι τῶν κινδυνευόντων ἁλῶναι φθόῃ, κἂν τἆλλα
πάντα διέλθοι καλῶς, οὐδέν, οἶμαι, πλέον ἀνύσει.
φλεβοτομίαν γὰρ παραιτεῖται, δεσμοῖς μόνοις χρώμενος ἐπὶ τῶν κώλων ὑπὲρ ἀντισπάσεως ἵν᾽, ὡς αὐτός
φησι, εἰς τοὺς τῆς φλεγμονῆς καιροὺς ἔχωμεν αὐταρ-
κες αἷμα καὶ μὴ διὰ τὴν ἔνδειαν αὐτοῦ τρέφειν ἀναγ-
καζώμεθα τὸν ἄρρωστον. ἀλλ᾽ ὦ γενναιότατε, φαίη τις
ἄν, οἶμαι, πρὸς αὐτόν, Ἐὰν φθάσῃ φλεγμήνας πνεύμων ἐπ᾽ ἀγγείου ῥήξει, μηκέτι ἔλπιζε τοῦτον ἰαθή-
σεσθαι δι᾽ οὓς ὀλίγον ἔμπροσθεν εἶπον λογισμούς·
ὥστ᾽ οὐδὲν ἔτι σου χρῄζει, προδοθεὶς ἐν ἀρχῇ. παραπλήσιον γάρ τι ποιεῖς κυβερνήτῃ δι᾽ ἀμέλειαν περι-
τρέψαντι τὸ πλοῖον, εἶτ᾽ ἐγχειρίζοντι σανίδας τῶν
378K πλωτήρων | τινὶ καὶ συμβουλεύοντι διὰ ταύτης πορί-
ζεσθαι τὴν σωτηρίαν.

ἀλλ᾽ ἴσως Ἐρασίστρατος ἡγεῖτο τῶν ἐξ ἀνάγκης
ἑπομένων τραύμασιν εἶναι τὴν φλεγμονήν, ἀγνοεῖ δ᾽,
εἴπερ οὕτω γιγνώσκει, μεγάλα· πάρεστι γοῦν θεάσασθαι τῷ βουλομένῳ μυρίους τῶν ὁσημέραι μονομα-
χούντων ἄνευ φλεγμονῆς τραύματα μέγιστα κολλη-
θέντας· ὥστε τῇ δευτέρᾳ ἢ τῇ τετάρτῃ τῶν ἡμερῶν ἐν
ἀσφαλείᾳ πάσῃ καθεστηκέναι. παμπόλλους δὲ καὶ
ἡμεῖς ἐθεραπεύσαμεν ἀγγεῖον ἐν πνεύμονι ῥήξαντας
ἐκ καταπτώσεως ἢ ἐκ πληγῆς ἢ βοῆς, πρὶν φλεγμῆναι
τὸ ῥαγέν. εἰ δὲ καὶ τοῦτόν φησι φλεγμῆναι τὸν πνεύμονα, τοῖς ἑαυτοῦ μάχεται δόγμασιν, ἅμα μὲν φλεγμαίνειν λέγων κύριον σπλάγχνον ἐγγυτάτω κείμενον

108

purge, just as he did not dry the head. If someone were to remove the dangers of succumbing to consumption (*phthoe*), even if all the other things were to come out well, nothing, I think, would accomplish more [than phlebotomy]. But he rejects phlebotomy, using bindings of the limbs only for the purpose of revulsion so that, as he himself says, we may have sufficient blood for the times of inflammation and not be compelled, due to the lack of this, to nourish the sick person. But, my most noble fellow, someone might, I think, say to you: "If, because of rupture of the vessel, the lungs are already inflamed, do not hold out hope any longer that this man will be cured because of those reasons I stated a little earlier. So he no longer needs you, having been lost from the start. What you are doing is similar to a ship's captain who, when his vessel capsizes due to negligence, puts his trust in one of the navigable planks and recommends providing safety through this."   378K

But perhaps Erasistratus thought inflammation was one of those things that necessarily follow wounds. If, however, he did think like this, he was very ignorant indeed. In fact, it is within the power of anyone who wishes to see in a very large number of those who engage in single combat every day (i.e. gladiators), major wounds conglutinated without inflammation so that, by the second or fourth day, they are completely safe. I have also treated very many [patients] who had ruptured a vessel in the lung from a fall, blow, or shout, before what was ruptured became inflamed. If, however, he also says this lung is inflamed, he is disputing his own opinions, since he is saying, at one and the same time, that an important organ lying very near the

καρδίας, ἅμα δ' ἀπύρεκτον φαίνεσθαι τὸν ἄνθρωπον,
ἔτι τε λυθείσης τῆς φλεγμονῆς, μηδὲν ἀναπτυσθῆναι.
ἀντακουσάτω τοιγαροῦν παρ' ἡμῶν ὡς οὔτε κολληθῆναι τὸ ἕλκος οἷόν τ' ἐστίν, ἐὰν ὁ πνεύμων φλεγμαίνῃ·
καὶ πάντως πυρέξει φλεγμήναντος· ἀναπτύσει θ' ἑξῆς
σὺν βηχὶ πυώδη πτύσματα, λυομένης τῆς φλεγμονῆς.

379K ἐὰν τοίνυν μήτε πυρέξῃ μήτε βήξῃ μήτε ἀναπτύσῃ | τι
τῶν ἐφ' ἕλκει καὶ φλεγμονῇ ἰχώρων, τίνα νοῦν ἔχει
φλεγμῆναι φάναι τούτου τὸν πνεύμονα; τοῦτό τε οὖν
τὸ βοήθημα μέγιστον ὑπάρχων οὐκ ὀρθῶς παρ' αὐτοῦ
κατέγνωσται, τούτῳ τ' ἐφεξῆς ἡ κάθαρσις ἀμνημόνευ-
τος ἔαται, καὶ φάρμακον ἀγωνιστικὸν οὐδὲν εἴρηται
τῷ ἀνδρὶ κατ' οὐδὲν γένος τῶν προειρημένων. ἀλλ' ἐάν
τε κεφαλὴ πέμπῃ τὸ ῥεῦμα, φυλαχθήσεται πέμπουσα·
ἐάν τε διὰ δυσκρασίαν αὐτῶν τῶν ἀναπνευστικῶν
γένηται ἡ βήξ, καὶ αὐτὴ φυλαχθήσεται.

παραπλήσιος γοῦν ἐστιν ὁ Ἐρασίστρατος ἀγαλ-
ματοποιῷ, τὰ μὲν ἄλλα πάντα κοσμήσαντι, τυφλὸν δ'
ἐργασαμένῳ τὸ ἄγαλμα. τί γὰρ ὄφελος τοῦ λοιποῦ
κάλλους, ὀφθαλμῶν μὴ παρόντων; εἶτα[12] τῶν τηλικού-
των ἀνδρῶν μέγιστα σφαλλομένων ὁ θαυμασιώτατος
Θεσσαλός, οὐδ' ἐννοήσας τὴν τέχνην, ἀξιώσει μεθ-
οδικὸς ὀνομάζεσθαι· καὶ νῦν ὁρῶμεν σχεδὸν ἅπαντας
τοὺς ἀπ' αὐτοῦ φλεβοτομοῦντας ἄλλους τε πολλοὺς
τῶν καμνόντων, οἷς οὐχ ὅπως ὠφέλιμον, ἀλλὰ καὶ
βλαβερὸν ἐχρῆν ὑπειλῆφθαι τὸ βοήθημα, φυλατ-
τόντων γε αὐτῶν τὰς οἰκείας ὑποθέσεις· οὐχ ἥκιστα δὲ

380K καὶ τοὺς αἷμα πτύσαντας, εἴτε | σὺν ἐμέτοις εἴτε σὺν

110

heart is inflamed and that the person appears to be free of
fever, and further, that when the inflammation is resolved,
nothing is expectorated. For that very reason, let him lis-
ten to me—it is impossible for the wound to be conglutinated if the lung is inflamed. Assuredly, when the lung is
inflamed, the patient will be febrile, and then, while the inflammation is resolving, he will expectorate purulent spu-
tum with coughing. Therefore, if the person is not febrile,
does not cough, and does not expectorate any of the ichors   379K
associated with a wound that is inflamed, how does he
make sense of the claim that this person's lung is inflamed?
And so this, which is the best remedy, is wrongly condemned by him, and next, purging is allowed to pass un-
mentioned, and no medication, in any of the previously
mentioned classes, is said to be "fit for the contest." But if
the head sends the flux, it will be preserved by sending it,
and if a cough occurs due to a *dyskrasia* of the respiratory
organs themselves, this too will be maintained.

In fact, Erasistratus is like a sculptor who, having put
everything else in order, makes the statue eyeless. What is
the good of the rest being beautiful if eyes are not present?
Next, when so many men are very greatly in error, the most
egregious is Thessalus who, although ignorant of the art,
thinks himself worthy to be called methodical. And now
we see almost all those who follow him carrying out phlebotomy on many other patients in whom it is not only not
beneficial to take up the remedy but harmful, which they
do, in fact, in order to stand by their own theories. Not
least [do they do this] also in those spitting up blood, either
with vomiting or coughing, whenever they are strong in   380K

___

12 B K; postea si diis placet KLat

111

βηξίν, ὅταν ὦσιν ἰσχυροὶ τὴν δύναμιν. εἶτα πῶς ἀλλήλοις ὁμολογεῖ ταῦτα, τὸ φλεβοτομεῖν τινα τῶν αἷμα πτυσάντων, αὐτοὺς ἐν τοῖς ἰδίοις συγγράμμασι γράφοντας τοῖς στεγνοῖς πάθεσιν ἐπιτήδειον εἶναι τὸ βοήθημα; μὴ τοίνυν ἔτι μηδὲ Μεθοδικοὺς ἑαυτοὺς ὀνομαζέτωσαν, ἀλλ᾽ Ἐμπειρικούς, οἳ παρέντες τὸν λόγον, ὃν ἐνόμιζον ὑπάρχειν ὑγιῆ, τῇ πείρᾳ χρῶνται πρὸς τὴν τῶν βοηθημάτων εὕρεσιν. ἆρ᾽ οὖν ἐν τούτοις μὲν ἐναργῶς ἐξελέγχονται μηδὲν μήτε μεθόδῳ μήθ᾽ ὅλως λόγῳ τινὶ πράττοντες, ἐν οἷς δὲ τὰ μέρη φασὶν ὑπάρχειν ἄχρηστα πρὸς τὴν τῆς θεραπείας εὕρεσιν, οὐ πολὺ μᾶλλον; καὶ μὴν εἴ τις ἀναμνησθείη τῶν εἰρημένων ἡμῖν ἐπὶ ὤτων καὶ μυκτήρων καὶ ὀφθαλμῶν καὶ στόματος καὶ θώρακος καὶ πνεύμονος, ἔτι τε μήτρας καὶ κύστεως καὶ τῶν κατὰ τὴν γαστέρα, τοῦ παντὸς ἁμαρτάνοντες φανοῦνται.

τοιοῦτος γοῦν καὶ ὁ τῷ Μακεδονικῷ φαρμάκῳ χρώμενος ἐπ᾽ αἰδοίου φλεγμαίνοντος, ἅμα τῷ χαλαστικῷ καταπλάσματι, τῷ συνήθει τούτῳ τῷ δι᾽ ἄρτου καὶ ὑδρελαίου σκευαζομένῳ. καί τις ἄλλος αὐτῷ παρα-
381K πλήσιος ἐφ᾽ | ἕδρας ἡλκωμένης τοῖς αὐτοῖς χρώμενος· ἀλλ᾽ ἐν τοῖς περὶ τῶν φλεγμονῶν λόγοις ὑπὲρ τῶν τοιούτων ἐροῦμεν. ἕλκη δὲ χωρὶς φλεγμονῆς ἐν αἰδοίῳ καὶ ἕδρᾳ καταπλάσματος μὲν οὐδενὸς δεῖται, φαρμάκου δ᾽ ἐπουλοῦντος, οὐ μὰ Δία τοιούτου τὴν φύσιν οἷον εἰς οὐλὴν ἄγει τὰ κατὰ τὰς σάρκας ἕλκη, ἀλλ᾽ εἰς τοσοῦτον ξηροτέρου τὴν δύναμιν εἰς ὅσον ἐστὶ καὶ τὰ μόρια ξηρότερα σαρκός. καὶ τό γε θαυμασιώτερον,

terms of capacity. How, then, will these things agree with others when they write, in their own particular treatises, that phlebotomy is a suitable remedy for the constricting affections? Therefore, let them no longer call themselves Methodics but rather Empirics, since they bypass the theory they know to be sound and use experience for the discovery of remedies. Are they not, then, in these matters, much more clearly refuted when acting neither on the basis of method nor, to speak generally, any theory, they say, in these, the parts are useless for the discovery of treatment? And if someone were to call to mind what I said in the case of the ears, nostrils, eyes, mouth, chest and lung, and besides these, the uterus, bladder and those things pertaining to the stomach, he will see they are manifestly mistaken in every respect.

In fact, such a person also uses the Macedonian medication[36] in the case of inflammation of the penis together with a relaxing poultice prepared in the customary way with bread and water mixed with oil. And another, similar to him, uses these same things in the case of ulceration in the anal region. But I shall speak about such matters in the discussions about inflammation. Wounds and ulcers in the penis and anal region don't require a poultice but a cicatrizing medication, although not, by Zeus, one of such a nature that it leads wounds and ulcers of the flesh to a scar, but one which, in terms of capacity, is drier to the same degree as the parts are drier than flesh. What is more to be

381K

36 There is no other reference to the Macedonian medication in the Kühn index.

αὐτῶν τῶν ἐν αἰδοίῳ συνισταμένων ἑλκῶν ἐπὶ μᾶλλον
δεῖται ξηραίνεσθαι τά τε τοῦ καυλοῦ σύμπαντος ὅσα
τε ἐκτὸς αὐτοῦ κατὰ τὸ πέρας ἐστίν, ὃ προσαγο-
ρεύουσι βάλανον. ἧττον δὲ τούτων χρῄζει ξηραινόντων φαρμάκων ὅσα τῆς πόσθης ἐστὶν ἕλκη, κἀκ
τούτων ἔθ' ἧττον ὅσα κατὰ τοῦ λοιποῦ δέρματος ὃ
περὶ σύμπαν ἐστὶ τὸ αἰδοῖον.

  ὑγρὸν οὖν ποθ' ἕλκος ἐπὶ τῆς βαλάνου τῶν ἀμεθόδων ἰατρῶν τις, τούτων δὴ τῶν Θεσσαλείων, μὴ
δυνάμενος ἰάσασθαι τούτοις τοῖς ἐπουλωτικοῖς ὀνομαζομένοις φαρμάκοις, εἰς συμβουλὴν ἡμᾶς παρεκάλε-
σεν. ἀκούσας οὖν ὅτι πολὺ ξηροτέρου δεῖται φαρμάκου τὸ μόριον τοῦτο, παρ' ὅσον ἐστὶ καὶ φύσει
382K ξηρότερον, | ἠπίστει μὲν τὸ πρῶτον· ὡς δ' ὑπ' ἀνάγκης
εἰς τὸ χρήσασθαί τινι τῶν ἡμετέρων ἀφίκετο, τρισὶ
μὲν ἡμέραις ὑγιάσθη τὸ ἕλκος· εὔδηλος δὲ ἦν ὁ ἰατρὸς
οὐ τοσοῦτον χαίρων ἐπὶ τῷ θεραπεῦσαι τὸν ἄνθρωπον
ὅσον ἀνιώμενος ἐπὶ τῷ πονηρᾷ συντετράφθαι Δογμάτων αἱρέσει. τὸ γάρ τοι διὰ χάρτου κεκαυμένου, τοῦτο
δὴ τὸ σύνηθες ἡμῖν, ἰᾶται τὰ τοιαῦτα τῶν ἑλκῶν,
ὥσπερ γε καὶ ἄνηθον κεκαυμένον ὁμοίως ἐπιπαττό-
μενον, καὶ κολοκύνθη δὲ ξηρὰ κεκαυμένη κατὰ τὸν
αὐτὸν τρόπον· ἄλλα τε πολλὰ τῶν ὄντως ἰσχυρῶς
ξηραινόντων φαρμάκων. ὅσα δ' ἄνικμα τῶν τοιούτων
ἑλκῶν ἐστι καὶ πρόσφατα, τούτοις καὶ ἡ ἀλόη μόνη
φάρμακον ἀγαθόν· ἐπιπάττεται δὲ χνοώδης ξηρά. αὕτη δὲ καὶ τὰ κατὰ τὴν ἕδραν ἕλκη τὰ ξηρὰ θεραπεύει
καλῶς. ὁμοιοτάτην δ' αὐτῇ δύναμιν ἔχει καὶ καδμεία

114

wondered at is that the actual wounds that exist in the penis need to be dried still more, both those that are in the whole shaft and those that are beyond the end of it in what they call the glans. Wounds and ulcers that are in the foreskin have less need of drying medications than these, and the need is even less in the remaining skin which surrounds the whole penis.

On one occasion one of the amethodical doctors, in fact one of the Thessaleians, called me to consult on a moist ulcer of the glans when he could not effect a cure with these so-called cicatrizing medications. Therefore, when he heard that this part needed a much more drying medication because it was, by nature, more dry, he was at first incredulous, yet when, of necessity, he came to use one of our [medications], the ulcer was made healthy in three days. Clearly the doctor was not so much delighted with the treatment of the man as he was distressed by being aligned with the terrible sect of Dogmatics. For the medication made from burned papyrus, which is what I customarily use, certainly cures such ulcers, just as burned dill sprinkled on also does, and burned dry colocynth in the same way, and many other of the really strongly drying medications. For some ulcers of this sort, which are without moisture and are recent, aloes alone is a good medication. Sprinkle it on as a fine powder. This also treats dry ulcers of the anal region well. Cadmia washed with wine

382K

δι᾽ οἴνου πεπλυμένη ξηρά. καὶ ταύτης ἐγγύς ἐστιν ἡ
καλουμένη λιθάργυρος. εἶθ᾽ ἑξῆς ἡ μολύβδαινα. πάντων δ᾽ αὐτῶν ἀνωδυνώτατόν τε καὶ οὐδενὸς ἧττον
δραστήριον ὁ πομφόλυξ ἐστίν. εἰ δ᾽ ὑγρότερα τύχοι
πίτυός τε φλοιὸς αὐτὸς καθ᾽ ἑαυτόν, ὅ τε λίθος ὁ αἱμα-
383K τίτης ὀνομαζόμενος. | εἰ δὲ καὶ βάθος αὐτοῖς συνείη
τι, μετὰ τὸ ξηρᾶναι τῶν εἰρημένων τινί, μάννης αὐτοῖς
τοσοῦτον μικτέον ὅσον αὔταρκες εἰς γένεσιν σαρκός.

τούτων τε οὖν οὐδὲν οὐδὲ ὄναρ ἴσασιν οἱ ἀπὸ τῆς
ἀμεθοδωτάτης αἱρέσεως ἅπαν ἕλκος ἡγούμενοι τῆς
αὐτῆς δεῖσθαι θεραπείας, ἐν ᾧπερ ἂν ᾖ τοῦ ζῴου
μορίῳ· καὶ πρὸς τούτοις ἔτι, μηδ᾽ ὅπως χρὴ ῥάπτειν
αὐτῶν ἔνια, καθάπερ εἰ τύχοι τὰ κατ᾽ ἐπιγάστριον·
ὑπὲρ ὧν ἐν τῷ μετὰ ταῦτα λόγῳ λεχθήσεται σὺν τοῖς
ὑπολοίποις ἅπασι.

and dried has a very similar capacity to this. And the so-called litharge is close to this; then, next in order, there is molybdena. Of all these, pompholyx is the least painful and no less efficacious. If the wound or ulcer happens to be more moist, the bark of the pine tree by itself and the stone which is called hematite [are suitable].[37] If some depth is 383K present in these, after drying with one of the previously mentioned [medications], you must mix manna with them to a degree that is sufficient for the generation of flesh.

Those of the most amethodical sect, who think that every wound or ulcer requires the same treatment regardless of what part of the organism it might be in, do not know any of these things even in a dream, and as well as this, they don't know how it is necessary to suture some wounds should they occur in the epigastrium. I shall say more about these wounds and ulcers, along with all those remaining, in the book following this one.[38]

[37] The Latin terms have been used for these five medications. The references to Dioscorides are as follows: cadmia, V.84; litharge (*Argenti spuma*), V.102; molybdena, V.100; pompholyx, V.85; and hematite, V.144 and 145. For the first four Peter English has, respectively, Oare of Brasse, Litharge, Oare of Lead, and Nil.

[38] The closure of abdominal wounds is dealt with in detail in Book 6, chapter 5.

ΒΙΒΛΙΟΝ Ζ

384K 1. Οἶδ᾽ ὅτι μηκύνειν δόξω τισὶν ἓν γένος ἔτι νοσήματος ἐξηγούμενος, ὅπως χρὴ θεραπεύεσθαι μεθόδῳ. προσήκει δ᾽ αὐτοὺς οὐκ ἐμοὶ τοῦ μήκους ἐγκαλεῖν, ἀλλὰ τοῖς ἃ μηδ᾽ ὅλως ἔγνωσαν ὑφ᾽ Ἱπποκράτους ὀρθῶς εἰρημένα διαβάλλειν ἐπιχειρήσασιν, οὓς οὐδ᾽ ἕλκος ἰᾶσθαι καλῶς ἐπισταμένους ἔδειξα, μή τοί γε δὴ τῶν ἄλλων τι τῶν μειζόνων. αὐτὸ μὲν οὖν τὸ γένος τοῦ πάθους οὗ τῶν εἰδῶν ἕν ἐστι τὸ ἕλκος, εἴτε συν-
385K εχείας λύσιν εἴθ᾽ ἑνώσεως ὀνομάζοι τις, οὐ | διοίσει. δέδεικται δὲ ὅτι καθ᾽ ἕκαστον μέλος τοῦ σώματος ἡ μέθοδος τῆς θεραπείας αὐτοῦ τὰς μὲν ἐξ αὐτοῦ τοῦ πάθους ἐνδείξεις φυλάττει κοινάς, ἐπικτᾶται δὲ ἀπὸ τοῦ μέρους ἄλλοτε ἄλλας. ἐν μὲν γὰρ τοῖς σαρκώδεσι μορίοις ἐπιγιγνόμενον ὀνομάζεται μὲν ἕλκος, ἔχει δὲ τὸν μὲν κοινὸν σκοπὸν ἁπάντων τῶν παρὰ φύσιν ἀναίρεσιν ἑαυτοῦ, τὸν δ᾽ ὡς διαθέσεως, τὸν διὰ τῶν ἐναντίων, τὸν δὲ ὡς διαιρέσεως ἕνωσιν· ἡ γάρ τοι τοῦ νοσήματος τοῦδε γένεσις ἐν τῇ διαιρέσει τῆς ἑνώσεώς ἐστι. καὶ διὰ τοῦτο κατὰ μὲν τὸ ὀστοῦν ὀνομάζεται κάταγμα, κατὰ δὲ τὰς νευρώδεις ἶνας σπάσμα, κατὰ

118

BOOK VI

1. I know that I shall seem to some to be dragging out un-   384K
duly my account of how we must treat by method one class
of diseases. It is not, however, appropriate for them to accuse me of prolixity. Nevertheless, these men, whom I
showed did not know how to cure a wound or ulcer properly, let alone anything else more major, attempt to take issue with me about those things which were correctly stated
by Hippocrates but which they don't understand at all.
Someone may call the actual class of affection, of which
the wound or ulcer is one of the kinds, either dissolution of
continuity or of union—it makes no difference which. It   385K
has been demonstrated that in each part of the body the
method of treatment of this [affection] preserves the indications common to the affection itself, and gains in addition different indications at different times from the part.
For when it occurs in the fleshy parts it is called a wound
(ulcer, sore) and has, as its general objective, removal of all
things contrary to its own nature; as a condition, by means
of opposites, and as a division, by means of union, for certainly the genesis of this disease lies in the division of
union. For this reason it is called a fracture in bone, a rup-

GALEN

δὲ τοὺς μῦς ἕλκος τε καὶ ῥῆγμα· προείρηται δὲ αὐτῶν
ἡ διαφορά. νεύρῳ μέντοι καὶ ἀρτηρίᾳ καὶ φλεβὶ ταὐτὸν τοῦτο τὸ πάθος ἐγγινόμενον οὐδὲν ἴδιον ὄνομα
κέκτηται· συγχρῆται δ᾽ ἐνίοτε μὲν τῇ τοῦ ἕλκους, ἔστι
δ᾽ ὅτε τῇ τοῦ τραύματος ἢ ῥήξεως προσηγορίᾳ.

ἀλλ᾽ ὅτι βραχύ τι χρὴ φροντίζειν ὀνομάτων
ὅτῳ περ σπουδὴ τῶν πραγμάτων αὐτῶν ἐξευρίσκειν
τὴν ἐπιστήμην, εἴρηταί μοι πολλάκις. ἐπὶ ταύτην
386K οὖν σπεύδωμεν | ἀμελήσαντες τῶν ὀνομάτων· οὐδὲ
γὰρ ἐκ τῆς τούτων ἀκριβοῦς θέσεως, ἀλλ᾽ ἐκ τῶν
προσηκόντων ἰαμάτων οἱ κάμνοντες ὀνίνανται. πάλιν
οὖν ἀναμνηστέον ὡς ὁ τολμηρότατος Θεσσαλὸς
οὐδεμίαν εἰπὼν μέθοδον ἑλκῶν ἰάσεως οἴεται πάσας
εἰρηκέναι. τὸ γάρ, εἰ οὕτως ἔτυχε, τὸ μὲν κοῖλον ἕλκος
δεῖσθαι σαρκώσεως, τὸ δ᾽ ἁπλοῦν κολλήσεως, οὐδὲ
τοὺς ἰδιώτας λανθάνει, πῶς δ᾽ ἄν τις εὕροι μεθόδῳ
φάρμακα δι᾽ ὧν ἤτοι σαρκωθήσεται τὸ κοῖλον, ἢ
κολληθήσεται τὸ ἁπλοῦν ἕλκος, οὐκέτι οὐδεὶς ἰδιώτης
ἐπίσταται· καὶ τοῦτό ἐστιν ὃ πρόκειται σκοπεῖσθαι
τοῖς ἰατροῖς· κἂν τούτῳ βελτίων ἐστὶν ἕτερος ἑτέρου.
καὶ γὰρ εὑρήσει φάρμακα καὶ τοῖς εὑρημένοις ὀρθῶς
χρήσεται, καθάπερ ἐδείκνυμεν ἐν τοῖς ἔμπροσθεν, ὁ
γεγυμνασμένος ἐν τῇ θεραπευτικῇ μεθόδῳ.

2. Ἵνα γὰρ ἤδη τινὸς ἐχώμεθα τῶν ἀκολούθων τοῖς
ἔμπροσθεν, ὑποκείσθω τις ἥκων πρὸς ἡμᾶς νενυγμέ-

1 The term *neuron* and its cognates cover a number of distinct
structures—nerves, sinews, tendons, even blood vessels. On this,

120

ture in nerve fibers,[1] and a wound or tear in muscles. The differentiation of these was spoken of previously. However, when this same affection has occurred in a nerve, artery or vein, it has not acquired a specific name. Sometimes the term *helkos* (wound, ulcer) is used, sometimes *trauma* (wound, trauma), and sometimes *rhexis* (rupture).

But what I have often said is that we must give little thought to names so that we may hasten to discover the knowledge of the matters themselves. Let us press on toward this, paying scant attention to names because patients 386K are benefited not by the precise application of names, but by the appropriate remedies. Once again, we must recall the most overweening Thessalus who, although he states no method of cure for wounds and ulcers, thinks he has mentioned them all. Thus, the wound or ulcer that is hollow, should this occur, has need of enfleshing; the simple wound of conglutination. Nor is this opaque to laymen, although how someone might discover by method medications through which either the hollow wound or ulcer will be filled with flesh or the simple wound will be conglutinated, no layman as yet knows. This is what is being put forward for doctors to consider. Even in this, one [doctor] is better than another, for if he is practiced in my method of treatment, he will discover medications, and he will use the medications discovered correctly, as I showed before.

2. So that we may now follow up one of the consequences of those things previously considered, let us assume that someone comes to us having been pierced by a

see section 6 on terminology in the Introduction and particularly Galen, *De ossibus ad tirones*, II.719K. In what shortly follows, Galen is clearly referring to nerves.

GALEN

νος αὐτὸ μόνον τὸ δέρμα βελόνη· τοῦτον τὸν ἄνθρωπον, εἰ μὲν εὐελκὴς εἴη, κἂν χωρὶς φαρμάκου, γυμνὸν
387K   ἔχοντα τὸ | μέλος, ἐπὶ τὰς συνήθεις ἀπολύσῃς πρά-
ξεις, οὐδὲν πείσεται φαῦλον· εἰ δὲ δυσελκὴς εἴη, πρῶτα μὲν ὀδυνήσεται, μετὰ ταῦτα δὲ ἤδη καὶ σφύξει καὶ
φλεγμανεῖ τὸ μέρος. ὁ μὲν οὖν Ἐμπειρικὸς ἐξ ἀνακρί-
σεως εἴσεται τὴν φύσιν τοῦ ἀνθρώπου· ἡμεῖς δὲ κἀκ
ταύτης μὲν ἅπασί γε χρώμενοι τοῖς ἐκ τῆς πείρας
εὑρισκομένοις, οὐδὲν δὲ ἧττον καὶ ἐξ ὧν ἔχομεν γνωρισμάτων εὐχύμου καὶ κακοχύμου φύσεως, εὐαισθή-
του τε καὶ δυσαισθήτου, πληθωρικῆς τε καὶ συμμέτρως ἐχούσης χυμῶν, εὖ εἰδότες ὡς ὁ μὲν πληθωρικός,
ἢ κακόχυμος, ἢ εὐαίσθητος, ἤ τινα τούτων, ἢ πάντα
ἔχων, φλεγμαίνει· ὅτῳ δ' ὑπάρχει τἀναντία, δεινὸν
οὐδὲν πείσεται. καὶ ταῦτα προγνόντες, οὐ μὰ Δία τὸ
κολλητικὸν ἐπιθήσομεν φάρμακον, ὁποῖα τὰ πλεῖστα
τῶν καλουμένων ἐναίμων ἐστίν, ἃ προσφάτοις ἐπιτί-
θενται τραύμασιν, ἀλλὰ μαλακόν τε καὶ παρηγορικὸν
καὶ ἀνώδυνον. ἔνθα μὲν γὰρ ἡ διαίρεσις ᾖ μεγάλη,
σπουδὴν χρὴ ποιεῖσθαι ξηραντικωτέροις φαρμάκοις
εἰς σύμφυσίν τε καὶ ἕνωσιν ἄγειν τὰ χείλη τοῦ ἕλ-
κους· |
388K   ἔνθα δ' ἐκ βελόνης ἢ γραφείου[1] διαίρεσις, ἑνὸς
μόνου χρὴ φροντίζειν τοῦ μὴ φλεγμῆναι. κἂν τῷδε
δῆλον ὡς ὁ γεγραμμένος ὑπὸ Θεσσαλοῦ σκοπὸς τῶν
ἐναίμων ἑλκῶν οὐδὲν ἡμᾶς οὐδέπω διδάσκει πλέον οὗ
καὶ τοῖς ἰδιώταις μέτεστιν. οὐ γὰρ ὃ χρὴ ποιεῖσθαι
γιγνώσκειν μέγα, φύσει γε ὑπάρχον ἅπασιν ἀνθρώ-

122

needle in the skin itself alone. This person, should he be someone who heals well, will suffer only slightly, even without medication, if having the limb uncovered, you re- 387K lease him to his normal activities. If, however, he is some-one who does not heal well, he will first suffer pain, and af-ter this the part will presently throb and become inflamed. The Empiric will know by examination the nature of the man. But we, from this [same examination], by using all those things discovered by experience, but no less also from the signs we have of a good- or bad-humored na-ture, and of a normal or disordered sensation, and of an abundance or balance of humors, know well that someone who is *plethoric*, or *kakochymous*, or of normal sensation, or is some or all of these things, develops inflammation, whereas someone who is the opposite will suffer nothing dire. When we know these things in advance we shall apply something mild, soothing and pain-relieving, and not, by Zeus, a conglutinating medication, like the majority of the so-called hemostatics which are applied to fresh wounds. When the division is great, we must act quickly, using more drying medications to bring the margins of the wound to coaptation and union.

Where the division is [made] by a needle or stylus,[1] it is 388K necessary to consider one thing alone—that it does not be-come inflamed. Even in this case it is clear that the indica-tor prescribed by Thessalus for bloody wounds teaches us nothing more than what is available even to laymen. It is not knowing what we should do that is important (for in fact that comes naturally to all men), but knowing by what

---

[1] B; γραφίου K

GALEN

ποις, δι' ὧν δ' ἄν τις αὐτὸ ποιήσειεν ἐπίστασθαι τῶν
τεχνικῶν ἐστι. καὶ γὰρ εἰ ναῦν τις μέλλει πήξεσθαι
καλῶς, ἴσμεν δήπου καὶ ἡμεῖς ἅπαντες οὐκ ὄντες
ναυπηγοί, ποῦ μὲν χρὴ τάξαι τὰ πηδάλια, ποῦ δὲ τὴν
πρύμναν καὶ τὴν πρῷραν, ἕκαστόν τε τῶν ἄλλων· ἀλλ'
οὐδὲν ἡμῖν πλέον πρόσεστιν ἀγνοοῦσι πῶς αὐτὰ
δημιουργήσομεν. οὕτω δὲ καὶ οἰκίαν κατασκευαζό-
μενός τις οὐκ ἀγνοεῖ δήπουθεν οὔθ' ὡς πρῶτα χρὴ
θέσθαι τὰ θεμέλια τῶν τοίχων, οὔθ' ὡς ἐπ' αὐτοῖς καὶ
ἀκλινεῖς ἐγεῖραι τοὺς τοίχους, οὔθ' ὡς κατὰ τοῦτον
πῆξασθαι τὴν ὀροφήν, οὔθ' ὡς θύρας τε καὶ θυρίδας,
ἕκαστόν τε τῶν ἄλλων μερῶν τῆς οἰκίας ἐν οἰκείῳ
389K τάξαι χωρίῳ· ἀλλ' οὐδὲν | τοῦτο πλέον οὐδὲν εἰς οἰκίας
κατασκευήν, ἄχρι περ ἂν ἀμαθὴς ὑπάρχων οἰκοδο-
μικῆς, ἀγνοεῖ δημιουργεῖν αὐτά.

μόνῳ τοίνυν τῶν ἁπάντων ἀνθρώπων σοφωτάτῳ
ἀρκεῖ Θεσσαλῷ πρὸς ἐπιστήμην τεχνικὴν ὃ χρὴ ποιῆσαι γινώσκειν. ἀλλ' ἡμεῖς καὶ διὰ τῶν ἔμπροσθεν
ἐδείξαμεν, ὡς ἀρχὴ μέν τίς ἐστι τοῦτο τῶν κατὰ τὰς
τέχνας πράξεων, οὔπω μὴν ἴδιον οὐδὲν αὐτῶν μόριον,
ἀλλ' ἔτι κοινὸν ἰδιώταις ἅπασιν· αἱ γὰρ ἐνδείξεις αἱ
πρῶται κατὰ πᾶσαν τέχνην φύσει πᾶσιν ἀνθρώποις
ὑπάρχουσιν. ὥστε εἴπερ εἰσὶν ἱκαναὶ τεχνίτας αὐταὶ
ποιεῖν, οὐδὲν κωλύει καὶ ναυπηγεῖν ἡμᾶς καὶ τεκταίνε-
σθαι καὶ δύνασθαι πάντας ὑποδήματά τε κατασκευάζειν καὶ ἱμάτια καὶ οἰκίας ἀρχιτεκτονεῖν τε καὶ κιθα-
ρίζειν καὶ ῥητορεύειν. ἀλλ' οὐχ οὕτως ἔχει τὸ ἀληθές·
οὐδ' ὁ γινώσκων ὅτι τῷ τρωθέντι μορίῳ τὴν κατὰ

124

means someone might do this, which is to know the technicalities. For if we intend to build a ship properly, although we are not shipwrights we obviously all know where we should place the rudder, where the poop and the prow, and each of the other things. But nothing more is added by us, who do not know how we will fashion these things. In this way too, someone building a house is not, of course, unaware that it is first necessary to lay the foundation stones of the walls, or that it is necessary to erect the walls on these foundations without inclination, or that it is necessary to fix the roof in relation to these, and to put doors and windows in place, and to arrange each of the other parts of the house in their proper position. But he contributes 389K nothing more to the construction of a house who, whilever he remains unschooled in house building, doesn't know how to fashion them.

Therefore, only for Thessalus, that wisest of all men, is it enough to know what he should do in the matter of technical knowledge. However, I have also shown by what has gone before that what constitutes a principle of the practices in the crafts is not yet a specific part of them, but is still common [knowledge] to all laymen, for the indications that are primary in every craft come naturally to all men. So if these are sufficient to make [people] craftsmen, nothing prevents us from all being shipwrights or joiners, or all being able to make sandals and clothes, or build houses, or play the cithara, or practice oratory. But really it is not like this. A doctor is not someone who knows that he must bring about a natural union in the wounded part, but one

125

φύσιν ἔνωσιν ἐκποριστέον ἰατρός ἐστιν, ἀλλ' ὁ δι' ὧν
ἐκπορισθήσεται· καίτοί γε οὐδὲ τοῦτό γ' αὐτὸ μόνον
ἱκανόν, ἐὰν ἀγνοῇ τις ὅπως χρηστέον αὐτοῖς ἐστιν,
ἀλλ' ὁ τὴν ὁδὸν ἅπασαν ἐπιστάμενος τῆς θεραπείας,
390K ἄχρι τοῦ τυχεῖν | τοῦ σκοποῦ, μόνος οὗτός ἐστιν ὁ
γινώσκων ἰᾶσθαι.

συμβαίνει τοιγαροῦν γε τοῖς ἀμεθόδοις Θεσσα-
λείοις ἐξ ὧν μείζω τολμῶσιν ἢ δύνανται μηδὲ τῶν
δυνατῶν ἐφικνεῖσθαι. γραφείῳ γοῦν τινος ἔναγχος εἰς
τὴν χεῖρα πληγέντος, ὡς διαιρεθῆναι μὲν ὅλον τὸ
δέρμα, νυχθῆναι δέ τι τῶν ὑποκειμένων αὐτῷ νεύρων,
ἐν ἀρχῇ μὲν ἐπέθηκεν ἔμπλαστόν τι φάρμακον ὁ
σοφώτατος Θεσσάλειος, ᾧ πολλάκις ἐπὶ μεγίστων
τραυμάτων εὐδοκίμει χρώμενος. ᾤετο γὰρ, οἶμαι, τῆς
αὐτῆς δεῖσθαι θεραπείας ἅπαν τραῦμα. φλεγμονῆς δὲ
ἐπιγενομένης, ἐπὶ τὸ δι' ἀλεύρου πυρίνου κατάπλασμα
μεταβάς, ἐν τούτῳ τε σήπων τὸν ἄνθρωπον, ἀπέκτει-
νεν ἐντὸς τῆς ἑβδόμης ἡμέρας. οὐδ' ἀριθμήσασθαι
δυνατὸν ὅσοι σπασθέντες ἀπέθανον, εἰς τὰς παιωνεί-
ας χεῖρας ἐμπεσόντες αὐτῶν· ἵνα δηλαδὴ σῴζηται τὸ
Θεσσάλειον θέσπισμα, πᾶν ἕλκος ἔναιμον ὁμοίως
θεραπευθῆναι, μηδὲν τῶν πεπονθότων μορίων συνεν-
δεικνυμένων. ἀλλ' οὐχ ἥ γε ὄντως μέθοδος εἰς τοσοῦτον ἀμέθοδός ἐστιν εἰς ὅσον ἡ Θεσσάλειος· εὑρίσκειν δὲ δύναται καὶ νῦν ἔτι μετὰ τοσούτους τε καὶ
391K τηλικούτους | ἰατροὺς οὐ μόνον φάρμακα βελτίω τῶν
ἔμπροσθεν, ἀλλὰ καὶ τὸν σύμπαντα τῆς θεραπείας
τρόπον. οὐδεὶς γοῦν χρώμενος εὐθὺς ἐξ ἀρχῆς οἷς

who knows by what means this will be brought about. And yet this is not enough by itself if a person does not know how he should use these means. Rather, it is the one who knows the whole course of treatment up to the point of attaining the objective who alone knows how to cure.   390K

For that very reason then, what happens to the amethodical Thessaleians is that they dare to take on more than they are able [to do], and yet they are not even able to bring to fruition those things they can [do]. At any rate, when someone was recently wounded in the hand by a stylus such that the whole skin was divided, and one of the nerves underlying it was pierced, the most sapient Thessaleian initially applied a certain emplastic medication which he often used successfully in very severe wounds, for he believed, I think, that every wound required the same treatment. When inflammation supervened, he changed to the poultice made from barley meal, whereupon the man became septic and [as a result] he killed the patient within seven days. It is impossible to calculate how many people have died from convulsions when they have fallen into the "healing" hands[2] of the Thessaleians. In order that the Thessaleian oracular saying is clearly preserved [let me state it thus]: "every wound that is filled with blood is treated in the same way," since no contributing indication is taken from the affected parts. But no method is so truly amethodical as the Thessaleian method. Even now, after so many and such notable doctors, it is still possible to discover not only medications that are better than those previously [available], but also a whole manner of treatment. At all events, nobody who   391K

[2] Literally, "Paeonian hands." Paeon alludes to Apollo, the divine founder of medicine and father of Asclepius.

127

ἡμεῖς εὕρομεν φαρμάκοις ἐπὶ νεύρων τρώσεσιν ἐσπά-
σθη. κατενόησα γὰρ ὡς, ἐπειδὰν νυγῇ νεῦρον, ἀναγκαῖον αὐτῷ διὰ περιττὸν τῆς αἰσθήσεως ὀδυνᾶσθαί τε
μειζόνως ἢ τἆλλα καὶ φλεγμαίνειν ἐξ ἀνάγκης, εἰ μή
τις ἐξεύροι ἄκος τῶν ὀδυνῶν, ἐπίσχοι τε τὴν γένεσιν
τῆς φλεγμονῆς.

εὔλογον οὖν ἐφαίνετό μοι τὴν μὲν τοῦ δέρματος
τρῶσιν ἀκόλλητον φυλάττειν, ὅπως ἐκρέοιεν δι᾽ αὐτῆς
οἱ ἐκ τῆς τοῦ νεύρου τρώσεως ἰχῶρες, ἀπέριττον δ᾽
ἐργάζεσθαι τὸ σύμπαν σῶμα καὶ τοῦ μηδεμίαν ὀδύνην ἐπιγίνεσθαι τῷ τετρωμένῳ μορίῳ, μεγάλην ποιεῖσθαι φροντίδα. καὶ τοίνυν ἐξεῦρον οὐκ ὀλίγα φάρμακα κατὰ τοῦ τραύματος ἐπιτιθέμενα πρὸς ἄμφω
θαυμαστῶς παρεσκευασμένα, τό τε τὴν ὀδύνην ἰᾶσθαι
καὶ τὴν ἐκροὴν ἀναστομοῦν τοῖς ἰχῶρσιν. ἀσφαλέστερον δ᾽ ἂν εἴη καὶ προσανατέμνειν τὸ δέρμα, τὸ δ᾽
ὅλον σῶμα διὰ φλεβοτομίας ἐκκενοῦν, ἐρρωμένης τῆς
392K δυνάμεως· | εἰ δ᾽ εἴη κακόχυμον, ἐκαθαίρειν αὐτίκα.
τὸ δὲ τὰς ἄλλας φλεγμονὰς ἱκανῶς παραμυθούμενον
ὕδωρ θερμὸν πολεμιώτατον ἡγεῖσθαι νεύρου τρώσει,
γεγενημένης γε τῆς οὐσίας αὐτοῦ κατὰ ψύξιν τε καὶ
πῆξιν ἐξ ὑγροτέρας ὕλης· ἅπασαι γὰρ αἱ τοιαῦται
συστάσεις ὑπὸ τῶν θερμαινόντων τε καὶ ὑγραινόντων
ἅμα διαλύονταί τε καὶ σήπονται. τοῦ μὲν οὖν ὕδατος
τοῦ θερμοῦ παντάπασιν ἀπέσχον τοὺς οὕτω τρωθέν-
τας· ἐλαίῳ δὲ καταιονᾶν ἄμεινον εἶναι κρίνας θερμῷ·
διότι καὶ τοῦτο ἀποδέδεικταί μοι, ψυχρὸν μὲν προσ-
φερόμενον ἐμπλάττειν, θερμὸν δὲ διαφορεῖν.

128

used the medications I discovered on wounds of the nerves right from the start suffered convulsions. For I observed that, whenever a nerve is pierced, because of the excess of sensation in it, it is inevitable that pain is felt more than in other parts, and it inevitably becomes inflamed, unless someone finds a cure for the pain and prevents the generation of inflammation.

Therefore, it seemed reasonable to me to keep a wound of the skin unconglutinated so that the ichors from the nerve wound might flow out through the wound itself, but also to render the whole body free of superfluity, and to make absolutely sure that nothing painful supervenes in the wounded part. Therefore, I discovered quite a few medications which, when applied to the wound, surprisingly provided both; that is, they cured the pain and created an opening for the outflow of ichors. However, it may be safer to incise the skin as well and to clean out the whole body by phlebotomy, if the capacity is strong. If, on the other hand, the body is *kakochymous*, purge immediately. Although warm water may be sufficient to assuage other inflammations, it is considered most inimical for a wound of a nerve whose substance has been generated through the cooling and congealing of a more watery material. All such structures are, at one and the same time, dissolved and putrefied by things that are heating and moistening. Therefore, I keep away from warm water altogether in those who are wounded in this way, judging fomentation with warm oil to be better because this too is something I have shown. If it is applied cold it causes blockage whereas, if the oil is applied warm, it causes dispersion.

392K

129

ἔφυγον μὲν οὖν καὶ τούτου τό τε ὠμοτριβὲς ὀνομαζόμενον καὶ ὅλως τὸ στῦφον, εἱλόμην δὲ τὸ λεπτο-
μερέστατον, οἷόν πέρ ἐστι τὸ Σαβῖνον· ἄμεινον δὲ εἴη,
εἰ καὶ δυοῖν, ἢ τριῶν ἐτῶν εἴη· διαφορητικώτερον γὰρ
τοῦτο τοῦ προσφάτου· τὸ δ' ἔτι παλαιότερον, ἀεὶ μὲν
καὶ μᾶλλον ἑαυτοῦ γίνεται διαφορητικώτερον, ἀνώδυνον δὲ ἧττον. ἐν δὲ τοῖς φαρμάκοις ἐθέμην σκοπόν,
ὧν ἡ δύναμις λεπτομερής τέ ἐστι καὶ θερμὴ συμ-
393K μέτρως καὶ ξηραίνειν ἀλύπως ἱκανή· | μόνη γὰρ αὕτη
δύναιτ' ἂν ἐκ βάθους ἕλκειν ἰχῶρας ἄνευ τοῦ συντεί-
νειν καὶ δάκνειν τὸ μόριον. ἐχρησάμην δὲ πρῶτον
μὲν τῇ τερμινθίνῃ ῥητίνῃ, καθ' ἑαυτήν τε καὶ βραχὺ
προσμίξας εὐφορβίου, καθ' ἑαυτὴν μὲν ἐπί τε παίδων
καὶ γυναικῶν καὶ ὅλως τῶν ἁπαλοσάρκων, σὺν εὐφορβίῳ δὲ ἐπὶ τῶν σκληροσάρκων. οὕτω καὶ προπόλει
καθ' ἑαυτήν τε καὶ σὺν εὐφορβίῳ μαλάττων αὐτήν·
εἴπερ δ' εἴη σκληροτέρα, σύν τινι τῶν λεπτομερῶν
ἐλαίων. καὶ σαγαπηνῷ δὲ ἐχρησάμην κατὰ τὰ σκληρὰ
σώματα μετ' ἐλαίου τε καὶ τῇ τερεβινθίνῃ μιγνύς· καὶ
ὀποπάνακί τε κατὰ τὸν αὐτὸν τρόπον, ὡς καὶ ὁ Κυρη-
ναῖος ὀπὸς ὀνήσειν, εἴ τις δι' αὐτοῦ σκευάσειε φάρμακον ἐπίπαστον ἐμπλαστόν, ὁποῖον ἡμεῖς τὸ δι' εὐφορβίου συνεθήκαμεν, ἀλλ' οὐδέπω τοῦθ' ἡμῖν ἐκρίθη τῇ
πείρᾳ, καθάπερ τὰ ἄλλα σύμπαντα. χρὴ γὰρ ἡγεῖσθαι μὲν τὴν ἐκ τῆς ἀληθινῆς μεθόδου τῶν ἰασομένων

130

Of oils, then, I also avoided that which is called "omotribes"[3] and what is altogether astringent, although I did choose the most fine-particled oil like the Sabine oil; and this is better if it is two to three years old because this is more dispersing than it is when new. If it is even older, it is invariably more dispersing but less anodyne. I established as an objective in medications that their potency be fine-particled and moderately heating, and particularly that they dry sufficiently without pain, because it is this potency alone that can draw ichors from the depths without contracting or stinging the part. I first used the resin of turpentine, both by itself and mixed with a little euphorbium: by itself, in children and women and, in general, those with soft flesh but [mixed] with euphorbium in those with hard flesh. I also used propolis in the same way, either by itself or after softening it with euphorbium, or if it is too hard, with one of the fine-particled oils. However, I also used sagapine in hard bodies, both with oil, and also mixed with turpentine. [I would also use] opopanax[4] in the same way, as the Cyrenian sap would be beneficial, if someone were to prepare from this a medication to be sprinkled or daubed on, of the sort I compounded from euphorbium. But I have not yet evaluated this through experience, as I have all the others, because it is necessary to carry out the discovery of things that will be curative as a result of true

393K

[3] Under Elaion Omotribes (Oil of Unripe Olives), Dioscorides (I.29) has, "Which is also called Omphacinum is best for the use in health. Of this that is reckoned the best which is new not biting, of a sweet smell" (Goodyer's translation, p. 25).

[4] The gum of *Oponopax hispidus*, having the common name Hercules' woundwort, has a wide range of uses listed by Dioscorides (III.55).

εὕρεσιν, ἐπιμαρτυρεῖν δ' αὐτῇ καὶ τὴν πεῖραν εἰς
ἀκριβεστέραν πίστιν.

ἀπὸ γοῦν τῆς τοιαύτης μεθόδου καὶ τὸ θεῖον τὸ
394K ἄπυρον, ὅταν γε² | μὴ ᾖ³ λιθῶδες ἀλλ' ἱκανῶς ὑπάρχῃ
λεπτομερές, ἠλπίσαμεν ὀνήσειν τοὺς τὰ νεῦρα τρωθέντας εἰς τοσοῦτον ἐλαίῳ λεπτομερεῖ μιγνύντες, ὡς
ἐργάσεσθαι γλοιῶδες τὸ μικτὸν ἐξ ἀμφοῖν. ἐπὶ δὲ τῶν
ἰσχυροτέρων τε καὶ σκληροτέρων σωμάτων, εἰ καὶ
μελιτῶδες ἐργάσαιο τῇ συστάσει, καὶ οὕτως ὀνήσεις,
ἐκρίθη γὰρ καὶ τοῦτο τῇ πείρᾳ. καὶ τὴν πεπλυμένην δὲ
τίτανον ἐπενοήσαμεν ὁμοίως μιγνύντες ἐλαίῳ προσφέρειν αὐτοῖς· καὶ μᾶλλόν γε ὀνίνησιν, ἐπειδὰν θαλάττῃ πλυθῇ· καλλίστη δὲ ἡ πλύσις ὥρᾳ θέρους ἐν
τοῖς ὑπὸ κύνα καύμασιν· εἰ δὲ καὶ δὶς αὐτὴν ἢ τρὶς
ἐκπλύναις, ἄμεινον ἐργάσῃ τὸ φάρμακον· ὅπως δὲ
χρὴ σκευάζειν τὰ τοιαῦτα διὰ τῶν περὶ τῆς φαρμάκων
συνθέσεως ὑπομνημάτων⁴ εἴρηται. νυνὶ δὲ ἀρκέσει μοι
λεχθῆναι τοσοῦτον, ὡς καὶ ταῦτα καὶ τἆλλα πλείω
φάρμακα πρὸς νεύρων τρώσεις ἐξεῦρον, οὔτε τῶν
διδασκάλων τινὰ θεασάμενος χρώμενον οὔτε ἀνα-
γνούς που γεγραμμένα κατά τι θεραπευτικὸν σύγγραμμα τῶν πρεσβυτέρων, ἢ τούτων δή τι τῶν ἐπιγε-
γραμμένων αὐτοῖς δυνάμεων ἐξ ἰδίας σοφίας, ἀλλ' ἐξ
395K αὐτῆς τῶν πραγμάτων | τῆς⁵ φύσεως ἐνδεικτικῶς ὁρμη-
θείς, ὅπερ ἐστὶν ἴδιον ἰατροῦ μεθόδῳ χρωμένου.

² K; γε om. B   ³ B; ᾖ om. K   ⁴ διὰ τῶν περὶ τῆς
φαρμάκων συνθέσεως ὑπομνημάτων nos; διὰ τῆς τῶν περὶ
φαρμάκων συνθέσεως ὑπομνημάτων B, K   ⁵ K; τῆς om. B

method, appealing to this, and to experience, for a more secure belief.

Anyway, on the basis of such a method, I also hoped that brimstone which is unburned, at least whenever it is 394K not stony but is sufficiently fine-particled, would help those with wounds of the nerves when I mixed it to such a degree with fine-particled oil as to make the mixture [prepared] from both viscous. In the case of bodies that are stronger and harder, if you also make it like honey in consistency, you will help in this way. This too was proven by experience. I also thought to apply washed chalk to them, having mixed it with oil in the same way; this is, in fact, particularly helpful when it is washed with seawater, the washing being best in the burning heat of the dog days of summer. Also, you will make the medication better if you wash it two or three times. How you should prepare such things is stated by way of my treatises on the composition of medications.[5] For the moment, however, I have said enough about such a medication, as I have discovered both these and other medications, more in number, for wounds of nerves, which I neither saw any of my teachers use, nor read about anywhere in a manual of treatment written by my predecessors, nor indeed in any of those things written by them about potencies derived from specific knowledge. Rather, I have taken my starting point probatively from the very nature of the matters, which is characteristic of the 395K doctor who uses method.

---

[5] De compositione medicamentorum secundum locos (XII.378–1007K and XIII.1–361K) and De compositione medicamentorum per genera (XIII.362–1058K).

Θεσσαλὸς δὲ ὁ θαυμασιώτατος οὐδὲν φάρμακον
εὑρὼν ἐγνῶσθαί φησι τὴν ὕλην αὐτῶν ἐκ πολλοῦ. καὶ
μὴν οὐκ ἔγνωσταί τι παμπόλλων φαρμάκων ὧν ἡμεῖς
εὑρήκαμεν, οὐ μόνον Θεσσαλῷ καὶ τοῖς πρὸ αὐτοῦ
πᾶσιν ἰατροῖς, ἀλλ᾽ οὐδὲ τοῖς μετ᾽ αὐτὸν ἄχρι δεῦρο.
ὑπαχθεὶς γοῦν ἐγὼ πρός τινά ποτε τῶν σηπομένων
ὑπὸ Θεσσαλείων ἰατρῶν ἀμεθόδων καὶ θεασάμενος
αὐτὸν μέλλοντα καταπλάττεσθαι τῷ δι᾽ ἀλεύρου πυρίνου καταπλάσματι, μηδὲν ἐν τῷ παραχρῆμα φάρμακον ἔχων, ᾔτησα κονίαν στακτήν, ἑωρακὼς ἐκ γειτό-
νων τοῦ κάμνοντος πηλοποιοῦ, ἑψήσας δὲ δι᾽ αὐτῆς
ἄλευρον κρίθινον, οὐχ ὡς ἐκεῖνοι δι᾽ ὑδρελαίου τὸ
πύρινον ἑψοῦσιν, ἐπέθηκα. καὶ αὖθις ὀρόβινον ἄλευρον ὁμοίως ἑψήσας, ἐπὶ φλεγμαίνοντός τε καὶ σηπο-
μένου τοῦ νεύρου, διὰ τὰς καλὰς αὐτῶν θεραπείας
ἐπιθείς, ἔπαυσα ταύτης τῆς σηπεδόνος τὸν ἄνθρωπον.

ἀλλὰ περὶ μὲν τῶν φλεγμαινόντων καὶ σηπομένων
396K   ἕτερος μὲν | λόγος ἐκδέχεται μακρὸς ἐν ἰδίῳ καιρῷ τῆς
πραγματείας. τῆς δὲ τοῦ νεύρου τρώσεως ἀρκεῖ τὰ
εἰρημένα φάρμακα. καὶ πάμπολλοί γε τῶν εὑρημένων
ἡμῖν ἑνὶ χρῶνται, μάλιστα τῷ δι᾽ εὐφορβίου καὶ
κηροῦ καὶ φρυκτῆς ῥητίνης ἐμπλαστῷ φαρμάκῳ· εἶτα
οὐκ οἶδ᾽ ὅπως ἔνιοι μὲν ἔμιξαν αὐτῷ Σινωπίδος, ἔνιοι
δ᾽ ὤχρας, ὑπὲρ τοῦ χρωσθὲν ἐκ τίνων σύγκειται λαν-

---

6 The red earth of Sinope, a town on the Black Sea littoral of
Paphlagonia, is described by Dioscorides (V.111) in part as follows: "It is gathered in Cappodocia in certain dens, but it is puri-

The most remarkable Thessalus, although he discovered no medication, said he had known their material for a long time. And yet, some of the very many medications which I have discovered were not known, and this applies not only to Thessalus and all the doctors before him, but also to those who came after him up to the present time. At all events, on one occasion, when I was sent for to attend someone suffering putrefaction due to the amethodical Thessaleian doctors, and saw him about to have a poultice made from barley meal applied, because I had no medication myself on the spot, I asked for myrrh powder, as I had seen a potter among the patient's neighbors. I boiled up the barley gruel with the myrrh powder, not as those men do who boil wheat in water mixed with oil, and I applied [it]. And again, in a similar fashion, I boiled up meal from bitter vetch for the inflammation and putrefaction of the nerve, and when I applied these, due to their therapeutic virtues, I put an end to the man's putrefaction.

But on the matters of inflammation and putrefaction another lengthy discussion is taken up at an appropriate 396K time in the treatise. When there is a wound of a nerve, the aforementioned medications suffice. Very many [doctors], in fact, use one of those discovered by me, particularly the emplastic medication made from euphorbium, beeswax and roasted pine resin. Some (and I do not know how) mixed Sinopian earth[6] with it, others ocher, compounding it from such things for the purpose of concealing those

fied and carried to ye city Sinope and sold whence also it had the sirname. It hath a drying emplasticall faculty, for which cause also it is mixed with vulnerarie plasters" (Goodyer, p. 638). See also Galen, *De compositione medicamentorum per genera*, XIII.785K.

θάνειν, ἵνα μὴ δοκῇ εἶναι τὸ ἐμὸν φθονοῦντες· οὐ μὴν
ὑπ' ἐμοῦ γε ἐξ ἀρχῆς οὕτως συνετέθη. τινὲς δ' ἂν ἴσως
αὐτῷ μίξειαν ἁλὸς ἄνθος, ἤ τι τοιοῦτον ἕτερον ὃ καὶ
τὴν χροιὰν ὑπαλλάξει καὶ τὴν δύναμιν οὐ βλάψει·
δύναιτο δ' ἄν τις ὑπ' ἀγνοίας μῖξαί τι καὶ τοιοῦτον,
ὅπερ βλάψει τὴν δύναμιν. ἀλλ' ἡμεῖς γε συνεθήκαμεν
αὐτὸ διὰ κηροῦ καὶ ῥητίνης τερμινθίνης καὶ πίττης
καὶ εὐφορβίου, τοῦ μὲν κηροῦ βάλλοντες μέρος ἕν,
τῆς τερμινθίνης δὲ καὶ τῆς πίττης ἑκατέρας ἥμισυ·
ὥστε τὸ ἐξ ἀμφοῖν ἴσον εἶναι τῷ κηρῷ. δύναιτο δ' ἂν
ποτε καὶ πλέον γίγνεσθαι τὸ ἐξ ἀμφοῖν τοῦ κηροῦ·
δύναιτο δ' ἂν καὶ τὸ ἕτερον αὐτῶν μόνον μίγνυσθαι τῷ
κηρῷ.

397K   καὶ | μέντοι καὶ μὴ παρούσης τερμινθίνης ἥ τ' ἐκ
τῶν κεραμίων ὑγρὰ καὶ ἡ φρυκτὴ καλῶς ἂν μιχθεῖεν·
οὐδὲν δὲ ἧττον αὐτῶν καὶ ἐλατίνη μιγνύοιτ' ἄν. ἡ δὲ
στροβιλίνη μόνον τῶν σκληρῶν σωμάτων ἐστὶ φάρ-
μακον· ἐπὶ δὲ τῶν μαλακωτέρων οὐ χρὴ μιγνύειν. εἰ
μὲν οὖν ὑγρὰ ῥητίνη μιγνύοιτο, κόψας καὶ διασείσας
τὸ εὐφόρβιον ἀναμίγνυε τοῖς ἄλλοις τακεῖσιν· ἔστω
δὲ ὁ σταθμὸς αὐτοῦ τὸ δωδέκατον μέρος τοῦ κηροῦ καί
ποτε καὶ πλέον, εἰ βούλοιο ποιεῖν ἰσχυρότερον· εἰ δὲ
ξηρά, καθάπερ ἡ φρυκτὴ δεήσεται βραχέος ἐλαίου τὸ
εὐφόρβιον. ὥστ' ἔγωγε τηνικαῦτα λεῖον αὐτὸ σὺν
ἐλαίῳ καὶ γλοιῶδες ἐργαζόμενος ἐψυγμένοις τοῖς ἄλλοις μετὰ τὸ τακῆναι ξύσας μιγνύω, πολλάκις δὲ καὶ
ὕδατος ἔμιξα σκευαζομένῳ τῷ φαρμάκῳ τοσοῦτον
ὅσον ἐν αὐτῷ τῷ τήκεσθαι τὰ μιγνύμενα δαπανηθῆναί

components used in its synthesis, acting out of jealousy so that it did not seem to be my medication; nor was it compounded like this by me from the beginning. Some may, perhaps, mix flower of salt or some other such thing with it, which will both change the color and not harm the potency. Someone also may be able, through ignorance, to mix the sort of thing that will harm the potency. But I compounded it with wax, turpentine resin, pitch and euphorbium, putting in one part of wax and half each of turpentine resin and pitch, so that the amount of both [combined] is equal to that of the wax. Sometimes, also, it may be possible for the amount of both to be greater than the wax. It might also be possible for one of these alone to be mixed with the wax.

However, if turpentine is not available, you can mix potters' liquid or roast pine satisfactorily; no less, you could also mix cankerwort. But the pine cone is a medication for hard bodies only. In the case of those that are softer, you should not mix this. Therefore, if you do mix moist pine resin, when you have pounded and vigorously shaken the euphorbium, mix it with those other dissolved [materials]. Let the weight of this be a twelfth part of wax, and sometimes even more if you wish to make it stronger. If it is dry, like the resin, the euphorbium will need a little oil. As a result, under these circumstances, I myself, after I have made it soft and glutinous with oil, mix it with the other cooled components, grinding it after they have been dissolved. Frequently, I also mixed water with the prepared medication to the extent that those things mixed were "consumed" by the dissolving itself and were able to be-

397K

τε δύναται καὶ ἀφανισθῆναι τελέως. ἀλλὰ τὰ μὲν
τοιαῦτα σύμπαντα τῆς περὶ φαρμάκων συνθέσεως[6]
πραγματείας ἐστὶν οἰκειότερα. νυνὶ δ᾽ ἀρκεῖ, καθάπερ
ἔμπροσθεν ἐποιήσαμεν, ὁποῖον εἶναι χρὴ τὸ γένος

398K τῶν φαρμάκων δηλώσαντας, ὀλίγα προστιθέναι | τῶν
κατὰ μέρος ἕνεκα παραδείγματος. ἄφθονον γὰρ ἁπάντων τὴν ὕλην ἐν τῇ περὶ τῶν ἁπλῶν φαρμάκων ἔχεις
πραγματείᾳ· σκευάζειν δ᾽ ὡς ἐν τοῖς[7] περὶ συνθέσεως
αὐτῶν[8] εἴρηται. νῦν οὖν ἀρκεῖ τοῦτο εἰπόντι μόνον ἐπ᾽
ἄλλα μεταβαίνειν.

ἡ τοῦ τετρωμένου νεύρου θεραπεία φαρμάκων δεῖται, θερμασίαν μὲν ἐγειρόντων χλιαράν, ξηραινόντων
δ᾽ ἱκανῶς, καὶ τῇ φύσει δὲ τῆς οὐσίας ἑλκτικῶν τε ἅμα
καὶ λεπτομερῶν. οὐ μόνον δὲ εἰς εὕρεσιν φαρμάκων τε
καὶ διαιτημάτων ἡ θεραπευτικὴ μέθοδος ἐπιτήδειός
ἐστιν, ἀλλὰ καὶ εἰς τὴν τῶν εὑρημένων χρῆσιν. ὥσπερ
γὰρ ἐν τοῖς ἔμπροσθεν ἐδείκνυμεν οὐδὲ τῷ τυχόντι
φαρμάκῳ τῶν πρὸς ἕλκη καλῶς δύνασθαι χρήσασθαι
τὸν ἄνευ μεθόδου μεταχειριζόμενον αὐτά, τὸν αὐτὸν
τρόπον ἔνεστι καὶ νῦν ἐπιδεικνύναι. γνωρίζειν γὰρ εἰς
ὅσον ἤ τί γε ἐξήρανεν, ἢ ἐθέρμηνεν, ἢ παρηγόρησεν,
ἢ ἠρέθισε τὴν διάθεσιν ἡ τοῦ φαρμάκου προσφορά,
μόνος ὁ κατὰ μέθοδον ἰατρεύων ἱκανός ἐστιν. οὗτος δ᾽
αὐτὸς οἶδε μόνος ἤ τί γ᾽ ἐπιτεῖναι τὴν ἐξ ἀρχῆς
χρῆσιν, ἢ ἐκλῦσαι.

τὸ γοῦν δι᾽ εὐφορβίου φάρμακον ἐπιθείς τις παρα-

---

[6] B; συνθέσεως om. K   [7] B; τῇ K recte fort.
[8] B; αὐτῶν om. K

138

come completely invisible. But all such things belong more appropriately to the treatise on compound medications. It is enough for the moment to put forward a few of the medications individually by way of example, just as I did before, after I have made clear what kind the class of medications should be, for you have material in abundance on all these in the treatise on simple medications; the preparation is as described in the treatises on the compounding of these.[7] Stating this now is enough to allow me to move on to other things. 398K

The treatment of the wounded nerve requires medications which excite a lukewarm heat, which dry sufficiently, which have a drawing effect by virtue of the nature of their substance, and at the same time are also fine-particled. My method of treatment is not only of value for the discovery of medications and regimens, but also for the use of those that are already discovered. For just as I showed in what has gone before, so is it possible now also to demonstrate in the same way that for someone who practices these things without method it is impossible to make proper use of some chanced-upon medication for wounds. Only a person who practices medicine according to method is capable of knowing to what degree and why the application of the medication either dried, heated, soothed or irritated the condition. This very person is the only one who knows whether to extend the use from the beginning, or to end it.

At all events, someone, after he applied the medication

[7] For details of Galen's work on simple medications and the two on compound medications, see Book 3, note 14.

399K χρῆμα νευροτρώτῳ, | πολλάκις δὲ ἄρα πρόσθεν εὐδο-
κιμοῦντος αὐτοῦ πεπείρατο, τριταῖον ἐπεδείκνυέ μοι τὸ
τετρωμένον μόριον, ὀδυνώμενόν τε ἅμα καὶ φλεγμαῖνον, ἀπορῶν ὅπως οὐδὲν ἤνυσε τὸ φάρμακον. ἠρόμην
οὖν τὸν τετρωμένον εἰ θερμασίας τινὸς οἷον ἐξ ἡλίου
πραέος ᾔσθετο, τῇ πρώτῃ τῶν ἡμερῶν ἐπιβληθέντος
αὐτοῦ· ὁ δὲ ἀπεκρίνατο μηδενὸς ᾐσθῆσθαι τοιούτου.
πάλιν οὖν ἠρόμην τὸν ἰατρὸν ἐκ πόσου τε χρόνου τὸ
φάρμακον ἐσκευασμένον ἔχει καὶ τίνας αὐτῷ θεραπεύσειεν. ὁ δὲ τὸν μὲν χρόνον ἐνιαυτοῦ πλέονα, τοὺς
δὲ θεραπευθέντας ἔλεγε δύο μὲν παῖδας, ἐν δὲ μει-
ράκιον εἶναι. πυθομένου δέ μου καὶ περὶ τῆς τοῦ
μειρακίου σχέσεως τῆς κατὰ τὸ σῶμα, λευκὸν ἔφρα-
σεν αὐτὸ καὶ μαλακὸν εἶναι. τούτων ἀκούσας ἐγὼ καὶ
συνεὶς ἐλλείπειν ὡς πρὸς τὴν τοῦ παρόντος νεανίσκου
κρᾶσιν τὸ εὐφόρβιον, αἰτήσας τε τὸ ἐμπλαστὸν φάρμακον καὶ τὸ εὐφόρβιον ἀνελόμενος, ὅσον ἤλπιζον
αὐτάρκως αὐτῶν ἀλλήλοις μιχθήσεσθαι, μαλάττειν
μὲν ταῖν χεροῖν ἐκέλευσα τὸ φάρμακον, ἀκριβῶς δὲ
λειῶσαι τὸ εὐφόρβιον. ὡς δὲ ταῦτα ἐγένετο, μιγνύειν

400K αὐτὰ πρὸς ἄλληλα | κελεύσας, εἶτ᾽ αὐτὸς ἔλαιον Σαβῖ-
νον ἠρέμα παλαιὸν θερμήνας συμμέτρως, κατήντλη-
σα τὸ τετρωμένον μόριον, ἀνακείρας τε βραχὺ με-
μυκυῖαν αὐτοῦ τὴν ὀπὴν τῆς νύξεως, ἐπέθηκα τὸ
φάρμακον, ἀσιτῆσαι προστάξας τὸν ἄνθρωπον· ἐς
ἑσπέραν τε πάλιν ἐκέλευσα τὸν ἰατρὸν ἐπιλῦσαί τε καὶ
χρήσασθαι κατὰ τὸν αὐτὸν τρόπον ἐλαίῳ καθ᾽ ὃν ἐμὲ
χρώμενον ἐθεάσατο. τούτων οὖν γενομένων ἀνώδυνόν

made from euphorbium immediately to a wounded nerve
(this was a man who had tried this often before, since it had   399K
a good reputation) on the third day, when he showed me
the wounded part, which was painful as well as inflamed,
was puzzled as to how the medication had accomplished
nothing. So I asked the person who had suffered the
wound whether, on the first day, when it was applied, he
felt a certain warmth as if from mild sunshine. He replied
that he had felt nothing of the sort. So, in turn, I asked the
doctor how long it had been since the medication had been
prepared and whom he had treated with it. He said the
time was more than a year and those treated were two children and one adolescent. When I also inquired about the
state of the body of the adolescent, he replied that it was
white and soft. After hearing of these people and realizing
that euphorbium was left out on the basis of the *krasis* of
the youth in question, I asked for both the emplastic medication and the euphorbium, and taking up as much of
these as I anticipated would be mixed sufficiently with
each other, I gave orders to soften the medication manu-
ally and to triturate the euphorbium thoroughly. After
these things occurred, I gave orders to mix them with each
other. Then I myself gently warmed aged Sabine oil to a   400K
moderate degree and poured it over the wounded part,
and having slightly cut open his puncture wound that had
closed up, I applied the medication and enjoined the man
to fast. I directed the doctor to open up [the wound] again
toward evening and to use oil in the same way he saw me
using it. After these things happened, the affected part ap-

GALEN

τε ἅμα καὶ ἀφλέγμαντον ἐφάνη κατὰ τὴν ὑστεραίαν
τὸ πεπονθὸς μόριον, ἔγνωσάν τε πάντες οἱ παραγενόμενοι τῷ ἔργῳ τῷδε τὸ πολλάκις ἡμῖν λεγόμενον
ἀληθέστατον ὑπάρχειν, ὡς οὐδέν τι μέγα δύναιτο τὰ
φάρμακα χωρὶς τῶν χρωμένων ἐπιτηδείως αὐτοῖς.

τοῦτο δ' αὐτὸ τὸ χρῆσθαι δεξιῶς, εὔδηλον δήπου
μεθόδῳ γιγνόμενον, οἵαν ἡμεῖς ἤδη πολλάκις ἐν τοῖς
πρὸ τοῦδε διήλθομεν ὑπομνήμασιν, ἐπὶ τὸ θερμὸν καὶ
ψυχρὸν καὶ ξηρὸν καὶ ὑγρὸν ἁπάσης χρήσεως φαρμάκων ὡς ἐπὶ κανόνας ἀναφερομένης. ἣν οὐ μόνον οἱ
ἀμέθοδοι Θεσσάλειοι μεταχειρίζεσθαι καλῶς ἀδυνα-
τοῦσιν, ἀλλὰ καὶ οἱ περὶ τὸν Ἐρασίστρατόν τε καὶ
401Κ τοὺς ἄλλους ἰατροὺς ὅσοι τὰ | στοιχεῖα τοῦ σώματος
ἢ οὐδ' ὅλως ἐζήτησαν, ἢ ἄλλ' ἄττα τῶν εἰρημένων
ἔθεντο. περὶ μὲν δὴ τούτων ἅλις.

3. Εἰ δὲ μὴ ὑγιείη τὸ νεῦρον, ἀλλὰ τρωθείη σαφῶς
τῇ τομῇ, σκέπτεσθαι τὴν τρῶσιν ὁποία τις ἐγένετο·
πότερον ἐγκαρσία τις, ἢ κατὰ τὸ τοῦ νεύρου μῆκος·
ὁπόσον τέ τι τοῦ προκειμένου διῄρηται δέρματος.
ὑποκείσθω δὴ πρότερον ἀνεπτύχθαι πολὺ τοῦ δέρματος, ὡς γυμνὸν φαίνεσθαι τὸ νεῦρον, ὄρθιον οὐκ ἐγ-
κάρσιον διῃρημένον. οὐ χρὴ τούτῳ τῷ νεύρῳ τῶν
εἰρημένων φαρμάκων οὐδὲν προσφέρειν, ὅσα δι' εὐφορβίου καὶ τῶν οὕτω δριμέων ἔμπλαστα γίνεται·
γεγυμνωμένον γὰρ οὐκ οἴσει τὴν δύναμιν αὐτῶν,
ὥσπερ ὅτε διὰ μέσου τοῦ δέρματος ἔφερεν. ἄριστον
οὖν τηνικαῦτα τὴν πεπλυμένην τίτανον ἐλαίῳ πλείονι
δεύσαντα χρῆσθαι· ἀγαθὸν δὲ καὶ τὸ διὰ τοῦ πομ-

142

peared free of pain and without inflammation on the next day, and all those who were present knew by this action that what I often said was absolutely true—medications are able to achieve nothing of significance without those who use them properly.

Of course, this skillful use quite clearly occurs by means of method of the kind I already went over thoroughly and often in the books prior to this one, i.e. a method directed at the hot, cold, dry and moist of every use of medications with a view to the formulation of rules. It is not only the amethodical Thessaleians who are unable to practice this properly, but also the followers of Erasistratus and those other doctors who neither attempted to seek the elements 401K of the body at all nor applied certain other of the things spoken of. But enough, surely, on these matters.

3. If the nerve is not pierced but is clearly wounded by the cut, consider what sort of wound it is; whether it is oblique or along the length of the nerve, and how much of the overlying skin is divided. Let us assume, first, that the skin is cut open to a marked degree so that the nerve appears exposed, and that the division is straight and not oblique. You should not apply any of the previously mentioned medications to this nerve, such as those emplastics made from euphorbium or from things that are similarly bitter because, when the nerve has been exposed, it will not bear the potency of these substances in the way this is borne when the skin is intact. Best, then, under these circumstances, that you moisten washed chalk using much oil. Also good is the medication made from pompholyx[8]

8 Pompholyx is listed as zinc oxide in LSJ. Dioscorides (V.85) has a detailed entry on its preparation; see Goodyer, pp. 624–26. For oil of roses (rhodinon, rosaceum), see Dioscorides, I.53.

φόλυγος φάρμακον ἐν πλέονι ῥοδίνῳ τακέν· ἄμεινον
δὲ εἰ καὶ τὸ ῥόδινον εἴη καὶ τοὔλαιον ἄναλον· σκοπὸς
γάρ σοι γιγνέσθω τῆς θεραπείας ἁπάσης ἐπὶ νεύρου
γεγυμνωμένου τὸ ξηραίνειν ἀδηκτότατα· πάνυ δὲ |
402K   ὀλίγιστα φάρμακα τὸ τοιοῦτο πέφυκε δρᾶν.

ἔστω τοίνυν ἡ τίτανος τηνικαῦτα πολλάκις ὕδατι
χρηστῷ πεπλυμένη ὥρᾳ θέρους· πεπλύσθω δέ, ὡς εἴρηται, καὶ ἡ πομφόλυξ, ὥσπερ οὖν καὶ πέπλυται κατὰ
τὸ σύνθετον φάρμακον ᾧ συνήθως χρώμεθα, πολλὰ
καὶ ἄλλα πεπλυμένα δεχόμενον. ὅσα γὰρ ἐκ μετάλλων
ἐστί, πεπλύσθαι χρὴ πάντα μέλλοντά γε ξηραίνειν
ἀδήκτως. ἀγαθὸν δὲ καὶ τὸ διὰ μέλιτος ἐμπλαστὸν
φάρμακον, ὅταν ἐκ καλλίστου μέλιτος ᾖ. λύειν δὲ καὶ
τοῦτο χρὴ ῥοδίνῳ τά τε ἄλλα καλλίστῳ καὶ τῶν ἁλῶν
ἥκιστα μετέχοντι· καὶ τὸν κηρὸν δὲ χρὴ πεπλύσθαι
τὸν εἰς τὰ τοιαῦτα φάρμακα βαλλόμενον· εἰ δὲ καὶ
τερμινθίνης τι ῥητίνης μιγνύοιτο, πεπλύσθαι χρὴ καὶ
ταύτην· ἔτι δὲ μᾶλλον, εἰ ἄλλης ἡστινοσοῦν. ἐξ ἁπάντων γὰρ ὧν ἂν πλύνῃς φαρμάκων οἱ δριμεῖς καὶ
δακνώδεις ἰχῶρες ἐκκλύζονταί τε καὶ ἀπορρύπτονται.

εἰ δ᾽ ἰσχυρὸς ὁ τετρωμένος εἴη καὶ σὺν τούτῳ καὶ
ἀπέριττος τὸ σῶμα, δυνατὸν ἐπ᾽ αὐτοῦ χρῆσθαι καὶ
τῶν ἰσχυροτέρων φαρμάκων ἐνίοις· ὥσπερ ἐγώ ποτε
403K   τοιαύτης | τρώσεως γενομένης ἐν καρπῷ ἐπὶ νεανίσκου
φιλοσοφοῦντος, εὐέκτου τε τἆλλα καὶ κατωπτηκότος
ἐν ἡλίῳ θερινῷ τὸ σῶμα τὸν Πολυείδου τροχίσκον
ἀνεὶς σιραίῳ, κἄπειτα χλιάνας ἐφ᾽ ὕδατος θερμοῦ,
βάψας ἐν αὐτῷ μοτοὺς ἐπέθηκα. καὶ γὰρ καὶ τούτου

144

dissolved in a rather large volume of rosaceum. And it is better if the rosaceum and the oil are free of salt. In every treatment of an exposed nerve let your aim be to dry in the least stinging way, although this is something that, overall, very few medications are able to do by nature. 402K

Therefore, under these circumstances, let the chalk be frequently washed with wholesome water in the summertime. Let the pompholyx also be washed, as I said, just as it has also been washed in the compound medication which I customarily use, and which contains numerous other washed [components]. Those medications made from metals should all be washed, if the intention is to dry without stinging. Also good is the emplastic medication made from honey, whenever the latter is of the best quality. It is also necessary to dissolve this in oil of roses which is the best in other respects and contains the least salt. However, it is necessary to wash the beeswax that is put into such medications and, if some turpentine is mixed in, it is necessary to wash this too, and still more so if you [put in] anything else at all, because the sharp and stinging ichors are washed away and cleansed thoroughly by all the medications which you wash.

If the wounded person is strong and is also without superfluity in the body, it is possible to use some of the stronger medications on him, as I once did when such a 403K wound occurred on the wrist of a young philosophy student who was otherwise in a good bodily state and was sunburned by the summer sun. I dissolved a troche of Polyeidos in new wine, then warmed this with hot water and, having dipped pledgets in it, applied it. For we must

145

χρὴ μάλιστα φροντίζειν ἀεί, τοῦ μηδὲν τῶν ψαυόντων
τῆς τρώσεως ψυχρὸν εἶναι, ἐπειδὴ τὸ πεπονθὸς μόριον
αἰσθητικώτατόν ἐστι καὶ τῇ κυριωτάτῃ τῶν ἀρχῶν
συνεχές, ἔτι τε τῇ κράσει ψυχρότερον· ἐξ ὧν ἁπάντων
ἑτοίμως μὲν ὑπὸ τοῦ ψυχροῦ βλάπτεται, διαδίδωσι δὲ
ἐπὶ τὸν ἐγκέφαλον τὸ πάθος. εἰ δὲ καὶ τῶν εἰς μῦς
καθηκόντων εἴη, καὶ σπασμοὺς ἐπικαλεῖται ῥᾳδίως·
ἐδείχθησαν γὰρ οἱ μύες ὄργανα τῆς κατὰ προαίρεσιν
κινήσεως. οὕτω δὲ καὶ ἐπὶ τῶν τενόντων προσδόκα
γενήσεσθαι διὰ τὰς αὐτὰς αἰτίας. ἀλλὰ τό γε προειρημένον φάρμακον ἐπιθεὶς τῷ νεανίσκῳ κατά τε τοῦ
τραύματος αὐτοῦ καὶ μέντοι καὶ τῶν ὑπερκειμένων
μερῶν, οὐκ ὀλίγον ἐπιλαμβάνων αὐτῷ τὰ κατὰ
μασχάλας καὶ τράχηλον καὶ κεφαλήν, ἐλαίῳ θερμῷ
διέβρεχον ἅπαντα συνεχῶς. ἀφεῖλον δὲ καὶ τοῦ αἵμα-
404K τος αὐτοῦ σχάσας τὴν | φλέβα κατὰ τὴν πρώτην ἡμέ-
ραν εὐθέως. ἐν οὖν τῇ τετάρτῃ τῶν ἡμερῶν οὗτος ὁ
νεανίσκος ἔσχε καλῶς, ὥστε ῥυσσὸν καὶ βραχὺ καὶ
προσεσταλμένον φαίνεσθαι τὸ ἕλκος. οὐ μὴν ἀλλὰ
καὶ προμηθέστερον ἐδόκει μηδὲν νεωτεροποιεῖν ἄχρι
τῆς ἑβδόμης, μεθ᾽ ἣν τελέως ἦν ὑγιής.

ἐλαίῳ δ᾽ αἰονᾶν οὐ χρὴ τὸ τοιοῦτον ἕλκος, καὶ
μάλισθ᾽ ὅταν ὡς νῦν εἴρηται θεραπεύηται· τῇ γὰρ τοῦ
τροχίσκου δυνάμει τὸ ἔλαιον ἐναντίον ἐστὶ καὶ ῥυπαί-
νει τὸ ἕλκος. οὐ γὰρ δήπου ταὐτόν ἐστιν ἢ γυμνῷ τῷ
νεύρῳ προσφέρειν τοὔλαιον, ἢ διὰ προβεβλημένου
τοῦ δέρματος. ἀποπλύνειν οὖν αὐτοῦ τοὺς ἰχῶρας,
ἔριον μαλακὸν ὑπαλείπτρῳ περιελίττοντας· ἀποβρέ-

146

always give particular thought to this—that none of those things that touch the wound are cold, since the affected part is highly sensitive and in continuity with the most important of the principal parts, which is even colder in *krasis*. From all these factors, the nerve is readily injured by cold, and transmits the affection to the brain. If also [the nerve] is one of those that pass down to muscles, it readily evokes spasms because the muscles were shown to be the organs of voluntary motion. Expect the same thing to occur in the case of the tendons, and for the same reasons. But after I applied the previously mentioned medication in the young man, both to the wound itself and, of course, to the overlying parts, including the axilla, neck and head to no small extent, I kept moistening them all continuously with warm oil. I also withdrew blood, having immediately opened a vein on the first day. On the fourth day, this young 404K man was doing well, so that the wound appeared wrinkled, small and contracted. Nevertheless, it seemed more prudent not to attempt anything new until the seventh day, after which he was completely restored to health.

You should not moisten such a wound with oil, especially when it is treated as I just now described, for the oil is inimical to the potency of the troche and contaminates the wound. To apply the oil to the exposed nerve is not, of course, the same as applying it through the intervening skin. Therefore, wash away its ichors, wrapping soft wool around a spatula. If you wish, soak the wool so that you do

χειν δὲ εἰ βούλοιο τὸ ἔριον, ὅπως μὴ ξηρῷ ψαύσῃς τὸ
ἕλκος, ἀρκέσει τὸ σίραιον, ὅπερ καὶ ἕψημα καλοῦσι
παρ᾽ ἡμῖν ἐπὶ τῆς Ἀσίας συνήθως. ἐν τούτῳ βάψας τὸ
ἔριον, εἶτ᾽ ἀποθλίψας, οὕτως ἀπομάττειν τὸ ἕλκος.
ἔστω δὲ καὶ αὐτὸ τοῦτο χλιαρὸν ἐν ταῖς πρώταις
μάλιστα ἡμέραις. εἰ δὲ πάντα κατὰ γνώμην περαίνοιτο, καὶ οἴνῳ γλυκεῖ βρέχειν ἀκίνδυνον, ἄδηκτος δὲ
405K   ὁ γλυκὺς οἶνος ἔστω παντάπασιν, οἷος ὁ Θηραῖός | τε
καὶ Σκυβελίτης ἐστὶ καὶ μετ᾽ αὐτοὺς ὁ καρύϊνος ὀνο-
μαζόμενος· ὅσοι δὲ γλυκεῖς τε ἅμα καὶ κιρροὶ τῶν
οἴνων εἰσίν, ὥσπερ ὁ Φαλερῖνος, ἀνεπιτήδειοι, δριμεῖς
γὰρ ἅπαντες οἱ τοιοῦτοι καὶ πέρα τοῦ μετρίου θερμοί·
ἤδη δὲ εἰς οὐλὴν ἰόντων τῶν ἑλκῶν, ὅσοι λευκοὶ τῶν
οἴνων εἰσὶν ἀκριβῶς καὶ λεπτοὶ καὶ ὀλιγοφόροι καὶ
ἄνοσμοι, βελτίους τῶν γλυκέων εἰσίν. ὕδατος δὲ χρῆσιν ἀεὶ φεῦγε τρωθέντος νεύρου, καθάπερ γε καὶ
καταπλάσματος χαλαστικοῦ. τῇ δὲ εἰρημένῃ τοῦ
τροχίσκου χρήσει πλησίον ἥκει τὸ διὰ τῆς χαλκίτεος
φάρμακον, ᾧ συνήθως χρώμεθα· τετῆχθαι δὲ καὶ
τοῦτο χρὴ θέρους μὲν ῥοδίνῳ, χειμῶνος δὲ Σαβίνῳ·
γέγραπται δ᾽ ἐν τῷ πρώτῳ περὶ φαρμάκων συνθέσεως.
ὁ δὲ Πολυείδου τροχίσκος, ἢ κυκλίσκος, ἢ ὅπως ἄν τις

---

[9] This refers to the Roman province of Asia in western Asia
Minor.   [10] This is defined in LSJ as "must boiled down to
one third part"; see Hippocrates, *Regimen*, 2.52.

[11] An oil made from walnuts; see *De simplicium medicamentorum temperamentis et facultatibus*, XI.871K, and Dioscorides,
I.41.

not touch the wound with anything dry; newly boiled-down wine which those among us in Asia[9] are accustomed to call "hepsema,"[10] will suffice. Dip the wool in this, then squeeze it out, and in this way wipe the wound clean. Let this newly decocted wine also be lukewarm, particularly on the first days. If all these things happen according to plan, there is no danger in irrigating it with sweet wine as well, but this must be altogether nonpungent like the Theran and the Scybellite are, and after these, that which is called 405K "caryinus."[11] Wines that are both sweet and tawny orange like the Falernian are unsuitable because they are all pungent and warm beyond the average.[12] When wounds are already forming a scar, wines that are genuinely white and thin, cannot bear much water (i.e. are weak), and are odorless are better than those that are sweet. When a nerve is wounded, always avoid the use of water, just as you would in fact also avoid the use of a relaxing poultice. In the previously mentioned use of the troche, the medication made from copper comes close to that which I customarily use. In summer, this should be dissolved in oil of roses, and in winter, in Sabine oil—I have written about this in the first book on the composition of medications.[13] The troche of Polyeides or lozenge,[14] or whatever one might wish to call

[12] On the Theran and Scybellite wines, see A. Dalby, *Empire of Pleasures* (London, 2000), pp. 137–38, and on the Falernian, pp. 142–44.

[13] The only reference in the Kühn index to this medication for nerve wounds is to this passage.

[14] For the composition of these troches, see the list of medications in the Introduction (section 10).

ὀνομάζειν ἐθέλοι, γιγνώσκεται σχεδὸν ἅπασι· καὶ εἰ
μὴ παρείη, τὸν Ἄνδρωνος, ἢ τὸν Πασίωνος ἀντ᾽ αὐτοῦ
περιλαμβάνειν, ἢ τὸν ἡμέτερον ἰσχυρότερον ἁπάντων
αὐτῶν ὑπάρχοντα. προείρηται δὲ ὅτι τοῖς μὲν ἰσχυροῖς σώμασι τὰ ἰσχυρὰ φάρμακα, τοῖς δὲ ἀσθενέσι
τὰ μαλακὰ προσφέρειν χρή· ταῦθ᾽ εὑρίσκει μὲν ἡ
406K ἀληθὴς | μέθοδος, ἐπισφραγίζεται δὲ ἡ πεῖρα.

Θεσσαλὸς δὲ ἅμα τοῖς ἑαυτοῦ σοφισταῖς ἐφ᾽ ὑψηλοῦ θρόνου καθήμενος ἐν κριομύξοις⁹ ἀνδράσιν, ὡς ὁ
Κερκίδας φησίν, εὐδοκιμήσει, κατασκευάζων τῷ λόγῳ
παντὸς ἕλκους προσφάτου τὴν αὐτὴν εἶναι θεραπείαν,
οὐδεμίαν ἔνδειξιν ἐκ τῆς τοῦ μορίου φύσεως λαμ-
βάνουσαν. εἷς δέ τις τῶν ὑπὸ τῆς σοφίας αὐτοῦ
κεκομισμένων θαυμαστὴν ἐξεύρισκε θεραπείαν τῶν
νευροτρώτων· αὐτίκα γὰρ ὅλα διέκοπτεν ἐξαίφνης αὐτά, μηδὲ προειπών τι τῷ τρωθέντι· καίτοι κἀνταῦθα
προεδίδου τὴν αἵρεσιν. ἐχρῆν γὰρ ἢ καὶ τοὺς μύας
τρωθέντας καὶ τὰς ἀρτηρίας καὶ τὰς φλέβας καὶ πᾶν
ὁτιοῦν ἄλλο διακόπτειν ὅλον, ἢ μηδὲ τὸ νεῦρον. ἢ
οὕτως ἂν ὃ φεύγουσι πράττοντες ἁλίσκοιντο,¹⁰ διάφορον ἐκ τῶν διαφερόντων μορίων ἔνδειξιν θεραπείας
λαμβάνοντες. τούτων μὲν οὖν ἀπαλλαγῶμεν ἤδη, τῆς
δ᾽ ἐγκαρσίας τρώσεως τῶν νεύρων μνησθῶμεν, ἐφ᾽ οἷς
καὶ ὁ κίνδυνος τοῦ σπασθῆναι μείζων, τῆς μὲν φλεγμονῆς ἐκ τῶν τετρωμένων ἰνῶν εἰς τὰς ἀτμήτους
διαδιδομένης, τοῦ σπασμοῦ δὲ διὰ τὰς ἀτμήτους

it, is known to almost everyone. If this is not available, get hold of that of Andron or Pasion instead of it, or my own, which is stronger than all of these. I said before that it is necessary to apply strong medications to strong bodies and soft medications to weak bodies. These are things the true 406K method discovers and experience confirms.

Thessalus, together with his own sophists, seated on a lofty seat among men who are drooling sheep, as Cercidas says,[15] will be held in high esteem as he fabricates his argument that the treatment of every recent wound is the same, taking no indication from the nature of the part [involved]. One particular individual among those carried away by his wisdom discovered an amazing treatment for wounded nerves. He would immediately and suddenly cut through them all without saying anything beforehand to the person who was wounded, although here also he betrayed his sect. He should have either completely cut through wounded muscles, arteries, veins and everything else whatsoever, or not cut through the nerve. Otherwise, such people would be caught in the act of doing what they eschew—that is, taking a different indication of treatment from the different parts. Therefore, let me now set these matters aside and make mention of the oblique wounds of nerves in which the danger of spasm is greater, since inflammation is being passed from the wounded fibers to those that are undivided, and spasm occurs due to these

[15] Cercidas (ca. 290–220 BC) of Megalopolis was a Cynic philosopher and poet; see *OCD*, p. 312.

---

[9] B, K; κριομύξαις conj. Cobet, *Mnem.* 12, 1884, 445, *recte fort.*   [10] B; ἁλίσκονται K

407K γιγνομένου. τὰ μὲν δὴ τῆς θεραπείας τοῦ | ἕλκους καὶ
τούτοις τὰ αὐτά· κενοῦν δ' αἵματος ἀφειδέστερον αὐτῶν καὶ λεπτότερον ἢ κατ' ἐκείνους διαιτᾶν, ἐν ἡσυχίᾳ
τε καὶ στρωμνῇ μαλακῇ συνέχειν τὸν ἄνθρωπον ἐλαίῳ
θερμῷ δαψιλῶς χρώμενον ἐπὶ μασχαλῶν καὶ τραχή-
λου καὶ τενόντων καὶ συνδέσμων καὶ κεφαλῆς. εἰ δὲ
τῶν ἐν σκέλει νεύρων εἴη τὸ τμηθέν, ὥσπερ ἐπὶ χειρὸς
μασχάλας, οὕτως ἐνθάδε βουβῶνας ἐλαίῳ πολλῷ τέγ-
γειν, ἐπαναβαίνειν δὲ δι' ὅλης τῆς ῥάχεως ἐπὶ τράχη-
λόν τε καὶ κεφαλήν. ἡ δὲ θλάσις ἡ τῶν νεύρων, ὅταν
μὲν ἅμα τῷ δέρματι θλασθέντι τε καὶ ἑλκωθέντι γίγνηται, φαρμάκων δεῖται τῶν μὲν τοῦ ξηραίνειν σκοπὸν ἐχόντων κοινόν, συνάγειν δέ πως καὶ σφίγγειν
δυναμένων τὰ διὰ τὴν θλάσιν ἀλλήλων ἀφεστῶτα
μόρια. τὰ δὲ χωρὶς τοῦ θλασθῆναι τὸ δέρμα συν-
εχέστατα κατανλεῖν ἐλαίῳ θερμῷ διαφορητικῷ, τὴν
δ' ὅλην τοῦ σώματος ἐπιμέλειαν ὁμοίαν ποιεῖσθαι.

ἅπαξ οὖν εἶδον τοῦτο γιγνόμενον, ἰασάμην τε
ταχέως αὐτὸ διὰ τῆς καταντλήσεως. ἅμα μέντοι τῷ
δέρματι πάνυ πολλάκις ἐθεασάμην νεῦρα θλασθέντα,
καὶ διὰ τὸ συνεχὲς τοῦ συμπτώματος οἱ ἀθληταὶ τῇ
408K πείρᾳ | διδαχθέντες ἔχουσι κατάπλασμα τὸ δι' ὀξυμέ-
λιτος καὶ τοῦ τῶν κυάμων ἀλεύρου· καὶ ἔστιν ὄντως
ἀγαθὸν φάρμακον. εἰ δ' ὀδύνη τις συνείη τῇ θλάσει,
πίττης ὑγρᾶς μιγνύειν ἕψοντάς τε καλῶς ἐπιτιθέναι
θερμόν· ὅταν δὲ ξηραντικώτερον ἐθέλῃς γίγνεσθαι,
μιγνύειν ὀρόβων ἀλεύρου. καὶ εἰ μᾶλλον βούλοιο
ξηραίνειν, ἴρεως τῆς Ἰλλυρίδος ἢ μυρίκης. ἡ δὲ τοῦ

undivided fibers. Surely, those things that pertain to the treatment of wounds [in general] are also the same for these: to drain the blood of these patients quite liberally, to establish a thinner diet than for other [patients], to restrain the person in quietude and on a soft bed, and to use warm oil copiously to the axillae, neck, tendons, ligaments and head. If the cut involves the nerves in the leg, just as with the axillae in the case of the arm, moisten the groins in the same way with copious oil, extending up along the whole spine to the neck and head. Bruising of the nerves, when it occurs in conjunction with skin that is bruised and wounded, requires medications that have the common objective of drying and being able somehow to coapt and compress the parts separated from each other by the bruising. As for nerves [that are affected] without bruising of the skin, bathe them very frequently with warm oil that is dispersing, and make your whole care of the body similar.

On one occasion, when I saw this happening, I cured it quickly by fomentation. However, very frequently I saw nerves bruised along with the skin and, because of the common occurrence of the symptom, athletes have learned from experience to keep a poultice made from oxymel and bean meal [handy], and this is a really good medication. If any pain is present with the bruising, mix liquid pitch, and when you have boiled it well, apply it hot. When you want the poultice to be more drying, mix meal of bitter vetch, and if you wish to dry more, Illyrian iris or tamarisk.[16] The benefit for the whole body with these

407K

408K

16 For these two medications, see Dioscorides, I.1 and I.116 respectively.

παντὸς σώματος ἐπιμέλεια καὶ τούτοις κοινή. δια-
κοπέντος δὲ ὅλου τοῦ νεύρου κίνδυνος μὲν οὐκέτι
οὐδείς, ἀνάπηρον δ' ἔσται τὸ μόριον. ἡ δὲ ἴασις κοινὴ
τοῖς ἄλλοις ἕλκεσιν, ἣν μόνην ἴσασιν οἱ Θεσσάλειοι.
περὶ μὲν δὴ τῶν νεύρων ἱκανὰ καὶ ταῦτα. τὰς δὲ
ἐπιγινομένας αὐτοῖς φλεγμονὰς ὅπως χρὴ θεραπεύειν
ἐν τῷ τῶν φλεγμονῶν λόγῳ διαιρησόμεθα.

4. Ὁμοιοτάτην δὲ τοῖς τένουσιν ἔχοντες ἰδέαν οἱ
σύνδεσμοι θεραπείας ἰσχυροτάτης ἀνέχονται, διά τε
τὸ μὴ περαίνειν ἐπὶ τὸν ἐγκέφαλον ἀναίσθητοί τε
εἶναι· τὰ μὲν γὰρ νεῦρα σύμπαντα τὰ μέν, ἄντικρυς ἐξ
409K αὐτοῦ πέφυκε | τοῦ ἐγκεφάλου· τὰ δὲ διὰ τοῦ νωτιαίου.
καὶ μέν γε καὶ οἱ τένοντες, ἐπειδὴ σύνθετον αὐτῶν
ἐπεδείκνυμεν εἶναι τὴν οὐσίαν ἐκ νεύρου καὶ συν-
δέσμου, καθ' ὅσον μὲν νεύρου μετέχουσι, κατὰ τοσοῦτον ἐξ ἐγκεφάλου πεφύκασιν, οὐ μὴν ἧττόν γε τῶν
νεύρων αὐτῶν ἐπιφέρουσι τοὺς σπασμούς. οἱ σύνδεσμοι δὲ ἐξ ὀστοῦ πεφυκότες, ὅσοι μὲν στρογγύλοι,
νεύροις ἐοίκασι, πολὺ δὴ τῇ σκληρότητι διαφέροντες·
ἀλλὰ τῷ λευκοί τε εἶναι καὶ ἄναιμοι καὶ ἀκοίλιοι,
διαλύεσθαί τε εἰς ἶνας, ἡ ὁμοιότης αὐτῶν ἐστι πρός τε
τὰ νεῦρα καὶ τοὺς τένοντας. ἔνθα μὲν οὖν στρογγύλοι
σύνδεσμοι καὶ τένοντές εἰσι, τοῖς ἀπείροις ἀνατομῆς
ὡς νεῦρα φαντάζονται, καὶ μάλισθ' ὅταν ἀγνοῶσιν ὅτι
καὶ σκληρότεροι πολὺ τῶν νεύρων εἰσίν· ἔνθα δὲ
πλατεῖς, ἐνταῦθ' ὅτι μὲν νεύρων διαφέρουσιν ἐπίστανται, καίτοι γε οὐδὲ τοῦτο πάντες· οὐ μὴν γιγνώ-
σκουσί γε διακρίνειν αὐτοὺς ἀπ' ἀλλήλων. ἀλλὰ σύ γε

agents is generalized. If the whole nerve is cut through, there is no longer any danger, although the part will be paralyzed. The cure is common to other wounds and is the only one the Thessaleians know. These considerations are sufficient regarding nerves. I shall distinguish how you should treat inflammation supervening in them in the discussion on inflammations.[17]

4. Since ligaments are very like tendons in form, they tolerate the strongest treatment because they do not come to an end in the brain and are without sensation. For all nerves are by nature direct from the brain itself or from the   409K spinal cord. Furthermore, tendons, since I showed their substance to be compounded from nerve and ligament, arise from the brain to the degree that they partake of nerve, and bring about spasms no less than the nerves themselves. The ligaments, however, arise from bone and are round like nerves, although they certainly differ greatly in hardness. But in their being white, bloodless, solid, and resolvable into fibers, they have a similarity to both nerves and tendons. In some places, then, ligaments and tendons are round and are imagined to be nerves by those ignorant of anatomy, especially when they do not know that they are much harder than nerves. In some other places, however, they are flat, so that here these people know that they differ from nerves. And yet not everyone knows this. What they do not know is how to distinguish them from one another. But you, since you know the

[17] There is no extant work on inflammation per se. The treatment of inflammation is dealt with in Book 13 of the present work, which of course postdates the present book by a considerable time.

τὴν φύσιν ἑκάστου τῶν τριῶν μορίων ἐπιστάμενος, ἔτι τε διάπλασιν καὶ θέσιν ἣν ἔχουσιν ἐν ἅπαντι τῷ σώματι καθ᾿ ἕκαστον μέλος, ἐπειδάν ποτε τύχῃ κατ᾿ ἐκεῖνο τραῦμα γενόμενον, ἑτοίμως | γνωριεῖς εἴτε νεῦρόν ἐστι τὸ τετρωμένον, εἴτε σύνδεσμος, εἴτε τένων.

410K

περὶ μὲν δὴ τῆς τῶν νεύρων καὶ τενόντων θεραπείας προείρηται. σύνδεσμος δὲ τρωθεὶς ὁ μὲν ἐξ ὀστοῦ διήκων εἰς ὀστοῦν ἀκινδυνότατός ἐστι, καὶ πάντῃ ξηραίνων αὐτὸν ὁποίοις βούλοιο φαρμάκοις οὐδὲν βλάψεις τὸν ἄνθρωπον. ὁ δ᾿ εἰς μῦν ἐμφυόμενος ὅσον ἀκινδυνότερός ἐστι τένοντος καὶ νεύρου, τοσοῦτον τῶν ἄλλων συνδέσμων σφαλερώτερος, ἢν μὴ χρηστῶς θεραπεύηται. τούτων οὐδὲν οἷοί τέ εἰσι διαπράξασθαι μεθόδῳ τῶν ἰατρῶν ὅσοι μήτε τὴν ἀπὸ τῶν μορίων ἔνδειξιν εἰς τὰς τῶν ἑλκῶν ἰάσεις χρήσιμον εἶναι συγχωροῦσιν, οὔθ᾿ ὅσοι τοῦτο μὲν ὁμολογοῦσιν, ἀγνοοῦσι δὲ τὴν ἑκάστου φύσιν, ἥτις, ὡς ἐδείκνυμεν, ἐκ τῆς τῶν στοιχείων γίνεται κράσεως. ἀλλ᾿ οὗτοι μὲν εἰ καὶ μηδὲν ἄλλο, τά γ᾿ ἐκ τῆς ὀργανικῆς κατασκευῆς τῶν μορίων ἐνδεικτικῶς λαμβανόμενα γιγνώσκουσι·

οἱ δ᾿ ἀπὸ τοῦ Θεσσαλοῦ καὶ ταῦτ᾿ ἀγνοοῦσιν, οἷον αὐτίκα διαιρεθέντος ἐπιγαστρίου μέχρι τοσούτου βάθους ὡς προπεσεῖν ἔντερον, ὅπως τε χρὴ καταστεῖλαι τοῦτο· καὶ ἢν | ἐπίπλους προπέσῃ, πότερον ἀποκόπτειν ἢ οὐκ ἀποκόπτειν αὐτὸν χρή· καὶ πότερον βρόχῳ διαλαμβάνειν ἢ μή· καὶ εἰ ῥάπτειν ἢ μὴ ῥάπτειν τὸ τραῦμα· καὶ ῥάπτοντας ὅτῳ χρὴ τρόπῳ ῥάπτειν. οὐδὲν τούτων ἐπίστανται, οὐδὲ γὰρ ἡμεῖς ἂν

411K

nature of each of the three structures, and in addition, the conformation and position which they have in every body in relation to each limb, will readily know when at some time a wound might happen to that limb, whether it is a   410K
nerve that has been wounded, or a ligament, or a tendon.

I have previously spoken about the treatment of nerves and tendons. When a ligament which extends from bone to bone is wounded, there is absolutely no danger, and when you dry it completely with whatever medications you might wish, you will not harm the person. A ligament which is inserted into a muscle is less dangerous [when wounded] than a tendon or nerve to the degree that it is more dangerous than other ligaments, if it is not treated properly. Those doctors who do not accept that the indication from the parts is useful for the cure of wounds are not able to accomplish any of these things by method. Nor are those who do agree on this point but do not know the nature of each [structure], a nature which, as I showed, arises from the *krasis* of the elements. But the latter, even if they know nothing else, do at least know those things taken indicatively from the organic construction of the parts.

The followers of Thessalus, however, don't even know the following: when, for example, the epigastrium is divided to such a depth that the intestine prolapses, how they should restore it; and if the omentum should pro-   411K
lapse, whether they should cut it off or not; and whether they should tie it off with a ligature or not; and whether to suture the wound or not; and if they are suturing, how they should do this. They know none of these things, nor would

ἔγνωμεν, εἰ μὴ δι᾽ ἀνατομῆς ἐμάθομεν ἁπάντων τῶν τῇδε μορίων τὴν φύσιν, ἣν καὶ διελθεῖν ἀναγκαῖόν ἐστιν οὐ σαφηνείας μόνης ἕνεκα τῶν λεχθησομένων, ἀλλὰ καὶ πίστεως. δέρμα μὲν ἔξωθεν ἁπάντων προβέβληται τελευτῶν εἰς ὑμένα. μεθ᾽ ὃ κατὰ μὲν τὴν μέσην χώραν ἀπονευρώσεις μυῶν τέτανται διτταὶ δίκην ὑμένων· ἃς οὐδ᾽ ὅτι δύο εἰσὶν οἱ πολλοὶ τῶν ἀνατομικῶν ἐπίστανται. καὶ γὰρ καὶ συμπεφύκασιν ἀλλήλαις, ὡς ἔργον εἶναι χωρίσαι, καὶ λεπτότητος εἰς ἄκρον ἥκουσιν. ἐφεξῆς δὲ ταῖσδε δύο μύες ὄρθιοι σαρκώδεις ἀπὸ τοῦ στήθους ἐπὶ τὰ καλούμενα τῆς ἥβης ὀστᾶ καθήκουσι.

ταῦτα μὲν δὴ σύμπαντα τὰ εἰρημένα συμφυῆ τ᾽ ἐστὶ καὶ οἱ τὰς καλουμένας γαστρορραφίας ὅπως χρὴ ποιεῖσθαι γράψαντες, ἐπιγάστριον ὀνομάζουσι τὸ συγκείμενον | ἐξ αὐτῶν. ὅσον δ᾽ ἐφεξῆς τῷδε, καλεῖται μὲν ὑπ᾽ αὐτῶν περιτόναιον, οἰομένων ἓν ἁπλοῦν ἀσύνθετον εἶναι σῶμα· τὸ δὲ οὐχ οὕτως ἔχει, σύγκειται γὰρ ἐκ δυοῖν σωμάτων, ἀναίμων μὲν ἀμφοῖν καὶ νευρωδῶν, ἀλλὰ τὸ μὲν ἕτερον αὐτῶν ἀπονεύρωσίς ἐστι μυῶν ἐγκαρσίων, τὸ δ᾽ ἕτερον ὑμὴν ἀκριβῶς λεπτός, οἷόν περ τὸ ἀράχνιον, ὅπερ δὴ τὸ περιτόναιον ὄντως ἐστί. τοιοῦτον μὲν ἐν τοῖς μέσοις ἑαυτοῦ τὸ ἐπιγάστριον· ὅσον δὲ ἀποκεχώρηκεν ἑκατέρωσε πρὸς τὸ πλάγιον ὡς ἐπὶ τέσσαρας δακτύλους, τοῦτο ἐφεξῆς τῷ δέρματι τοὺς λοξοὺς ἔχει μῦς· προτέρους μὲν τοὺς ἀπὸ τοῦ θώρακος καταφερομένους, δευτέρους δὲ τοὺς ἀπὸ τῶν λαγόνων ἀναφερομένους· εἶτα ἐπὶ τούτοις τὸν ἐγκάρ-

412K

I have known had I not learned through anatomy the nature of all the parts here, which it is also essential to go over, not only for the sake of clarity in regard to what will be said, but also for the sake of proof. The skin covers everything externally and ends in a membrane. After this (i.e. the skin), two tendinous muscles (aponeuroses) extend like membranes,[18] which the majority of anatomists do not even know are two. Moreover, they are united with each other so it is a difficult task to separate them, and they come to be extremely thin. After these [aponeuroses], there are two muscles, straight and fleshy, extending from the breast to the so-called "bones of youth" (i.e. the pubic bones).

Now all these things spoken of are united and those who have described how one should carry out the so-called "gastrorraphies"[19] term what is composed of them the epigastrium. What is next in turn after this, they call the peritoneum, thinking it to be one simple, uncompounded body. But this is not the case, for it is compounded from two bodies, both of which are bloodless and sinewy. However, one of them is an aponeurosis of obliquely running muscles while the other is strictly a thin membrane, like a spider's web, which is in fact the true peritoneum. Such is the epigastrium in its central parts, whereas what extends away from this to either side toward the flanks to a breadth of approximately four fingers has beneath the skin the [two] oblique muscles; the first extends down from the chest, and the second extends up from the flanks. Then,

412K

---

[18] Linacre (p. 304) adds in parentheses, "which the Greeks call aponeuroses."   [19] This term is still in use, albeit with a more restricted meaning—i.e., suture of a perforation of the stomach.

σιον, ἐφ᾽ ὧν τὸ περιτόναιον. ἀκινδυνότερόν τε οὖν ἐστι
τοῦτο τὸ χωρίον τοῦ μέσου, μηδεμίαν ἔχον ἀπονεύρω-
σιν· αἵ τε γαστρορραφίαι κατὰ τὸ μέσον εἰσὶ δυσ-
μεταχείριστοι· καὶ γὰρ καὶ προπίπτει τὰ ἔντερα ταύτῃ
μάλιστα καὶ δυσκάθεκτά ἐστιν ἐν τούτῳ τῷ τόπῳ· τὸ
γὰρ σφίγγον αὐτὰ καὶ προστέλλον οἱ ὄρθιοι μύες
413K   ἦσαν οἱ σαρκώδεις, οὓς ἐκ | τοῦ θώρακος ἐπὶ τὰ τῆς
ἥβης ὀστᾶ καθήκειν ἔφαμεν. ὅταν οὖν τις τούτων
τρωθῇ διὰ διττὴν αἰτίαν, ἀναγκαῖόν ἐστι προπίπτειν
ἔντερον, ἐκ μὲν τῶν πλαγίων μερῶν ὑπὸ τῶν ταύτῃ
μυῶν σφιγγόμενον, ἐκ δὲ τῶν μέσων οὔτε τὸν μῦν
ἐρρωμένον ἔχον, ἐπιτήδειόν τε χώραν εἰς πρόπτωσιν.
εἰ δὲ καὶ μεῖζον εἴη τὸ τραῦμα, πλείω τε προπίπτειν
ἀναγκαῖόν ἐστιν ἔντερα καὶ χαλεπώτερον καταστέλ-
λεσθαι. καθ᾽ ἕτερον δ᾽ αὖ τρόπον αἱ βραχεῖαι τρώσεις
δυσμεταχείριστοι. ἢν γὰρ μὴ παραχρῆμά τις εἰς τὴν
ἑαυτοῦ χώραν ἐμβάλλῃ τὸ προπεπτωκός, εἰς ὄγκον
αἴρεται πνευματούμενον· ὥστ᾽ οὐκ ἔτι οἷόν τε δι᾽ ὀπῆς
στενῆς ἐμβάλλειν αὐτό. βέλτιον οὖν ἐν ταῖς τοιαύταις
τρώσεσι τὸ σύμμετρον τραῦμα. ταυτὶ μὲν δὴ προεπί-
στασθαι χρή· ὡς δ᾽ ἄν τις ἄριστα μεταχειρίζοιτο τὰς
τοιαύτας τρώσεις, ἐφεξῆς σκεπτέον.

ὅτι μὲν γὰρ οὐκ ἀρκεῖ τὸ Θεσσάλειον παράγγελμα
τὸ κολλᾶν τοῖς ἐναίμοις φαρμάκοις αὐτάς, οὕτως
ἡγοῦμαι πρόδηλον ὑπάρχειν ὡς οὐδένα λαθεῖν τῶν
414K   ἐχόντων νοῦν. ἐπεὶ τοίνυν προηγεῖσθαι | μέν χρεών
ἐστιν εἰς[11] τὴν οἰκείαν χώραν ἀποτίθεσθαι τὰ προ-
πεπτωκότα ἔντερα, δεύτερον δὲ ἐπὶ τόδε ῥάψαι τὸ

beneath these, is the transversus muscle, and deep to these the peritoneum. Therefore, this part is less dangerous than the middle since it has no aponeurosis. Gastrorraphies in the middle are hard to manage. Certainly, the intestines prolapse here especially and are hard to hold in at this place, for what encloses and covers them are the rectus muscles which are fleshy and which I said pass down from 413K the chest to the pubic bones. Therefore, whenever one of these is wounded, the intestines inevitably prolapse for two reasons: because they are compressed from the sides toward the center by the muscles, and because the muscle in the middle is not strong and it is a place that predisposes to prolapse. If the wound is even larger, the intestines inevitably prolapse more and are more difficult to reduce. In another way again, small wounds are hard to manage. This is because, unless someone immediately replaces the intestines that have prolapsed, they become inflated and expand into a swelling, so that it is no longer possible to replace them through a narrow opening. Therefore, in such wounds, one of moderate size is better. Certainly, this is something you must know beforehand. What you must consider next is how best to manage such wounds.

It is obvious to anyone with any intelligence that the Thessaleian precept of conglutinating these wounds with blood-stanching (hemostatic) medications is inadequate. Therefore, it is first necessary to give thought to restoring 414K the prolapsed intestines to their proper place, second to

---

11 B; μὲν χρὴ ἐς K

ἕλκος, εἶθ᾽ ἑξῆς τρίτον ἐπιθεῖναι τὸ φάρμακον, εἶτ᾽ ἐπ᾽
αὐτῷ[12] τέταρτον ὅπως μὴ συμπάθῃ τι τῶν κυριωτέρων
προνοεῖσθαι,[13] περὶ τοῦ πρώτου ῥηθέντος ἤδη σκοπῶμεν.[14] οὐσῶν δὲ, ὡς εἴρηται, τριῶν ἐν τοῖς τραύμασι
κατὰ μέγεθος διαφορῶν, ἀφ᾽ ἑκάστης αὐτῶν πειρᾶσθαι χρὴ λαβεῖν οἰκείαν ἔνδειξιν. ἔστω δὴ πρότερον
μικρὸν οὕτως ὡς τὸ προπεσὸν ἔντερον ἐμφυσηθὲν
μηκέτι οἷόν τε εἶναι καταστέλλειν, ἆρ᾽ οὐκ ἀναγκαῖον
ἐνταῦθα δυοῖν θάτερον, ἤτοι τὴν φῦσαν ἐκκενοῦν ἢ τὸ
τραῦμα μεῖζον ἐργάζεσθαι; βέλτιον οὖν οἶμαι τὸ πρότερον, ἐάν περ οἷόν τε ᾖ τυχεῖν αὐτοῦ. πῶς δ᾽ ἄν τις
τύχοι μᾶλλον; εἰ τὴν αἰτίαν ὑφ᾽ ἧς ἐκφυσᾶται τὸ
ἔντερον ἐκποδὼν ποιησαίμεθα. τίς οὖν ἐστιν αὕτη; ἡ
ἐκ τοῦ περιέχοντος ἀέρος ψύξις· ὥστε καὶ ἡ ἴασις ἐν
τῷ θερμῆναι. σπόγγον οὖν χρὴ μαλακὸν ὕδατι θερμῷ
βρέξαντας, εἶτ᾽ ἐκπιέσαντας ἐκθερμῆναι τούτῳ τὸ ἔντερον. εὐτρεπιζέσθω δὲ ἐν τῷ τέως οἶνος αὐστηρὸς
415K θερμός, | καὶ γὰρ θερμαίνει μᾶλλον ὕδατος καὶ ῥώμην
ἐντίθησι τῷ ἐντέρῳ. εἰ δὲ καὶ τούτῳ χρησαμένων ἔτι
διαμένει τὸ ἔντερον ἐμπεφυσημένον, ἐπιτέμνειν τοῦ
περιτοναίου τοσοῦτον ὅσον δεῖται τὸ προπεπτωκός.
ἐπιτήδεια δ᾽ ἐστὶν εἰς τὴν τοιαύτην τομὴν τὰ καλούμενα συριγγοτόμα. τὰ δ᾽ ἀμφήκη τῶν μαχαιρίων, ἢ
κατὰ τὸ πέρας ὀξέα παντὶ τρόπῳ φευκτέα.

σχῆμα δ᾽ ἐπιτήδειον τῷ κάμνοντι πρὸς μὲν τοῖς

[12] B; αὐτό K   [13] K; προνοήσασθαι B, recte fort.
[14] B (cf. tentemus KLat); σκοποῦμεν K

suture the wound, next and third to apply the medication, and fourth after this, to consider how to avoid one of the more important [parts] becoming sympathetically affected.[20] Let us now examine the first point mentioned. Because, as I said, there are three different types of wound in terms of size, you should attempt to take the proper indication from each of them. First, suppose the wound is small so that it is no longer possible to reduce the prolapsed intestines once they have become inflated. Is it not critical here to do one of two things—either empty the flatus or enlarge the wound? The former is better, I think, if it can be done. But how can someone achieve this other than by removing the cause due to which the intestines are inflating? What, then, is this cause? It is the coldness of the surrounding air. Consequently the cure lies in heating. Therefore, it is necessary to wet a soft sponge with warm water, then having squeezed it out, to warm the intestines thoroughly with it. In the meantime prepare warm, astringent wine because this heats more than water and puts strength into the intestines. If, when you have used this, the intestines still remain inflated, cut as much of the peritoneum as the prolapsed intestine requires. The so-called syringotomes are suitable for such a cut. Knives that are double-edged or sharp at the point must be avoided at all costs.

415K

A suitable position for the patient is to be tilted upward

[20] For a consideration of Galen's thinking on "sympathetic affections," see R. E. Siegel, *Galen on Sense Perception* (Basel: S. Karger, 1970), pp. 187–89.

GALEN

κάτω μέρεσι τῆς τρώσεως γεγενημένης τὸ ἀνάρροπον,
πρὸς δὲ τοῖς ἄνω τὸ κατάρροπον. εἷς δ' ἐπ' ἀμφοῖν
σκοπός, ὡς μὴ βαρύνοιτό τι πρὸς τῶν ἄλλων ἐντέρων,
τὸ προπεπτωκός. ὥστε καὶ τοῦ σκοποῦ τοῦδε κατὰ μὲν
τὰ δεξιὰ μέρη τοῦ τραύματος γεγονότος, ἐπὶ θάτερον
ῥέπειν· εἰ δὲ ἐκ τῶν ἀριστερῶν ᾖ, πρὸς τὴν δεξιὰν
ἐπικλίνεσθαι πλευρὰν ὑψηλότερον ἀεὶ ποιοῦντας τὸ
τετρωμένον μόριον. τοῦτο μέντοι καὶ τοῖς μεγάλοις
καὶ τοῖς μέσοις ἕλκεσι συνοίσει· κοινὸς γὰρ ἁπάντων
ὁ σκοπός. αἱ δ' ἀποθέσεις τῶν ἐντέρων εἰς τὴν οἰκείαν
χώραν, ὅταν ἐπὶ τοῖς μεγάλοις γίγνωνται τραύμασιν, |
416K ὑπηρέτου δέονται δεξιοῦ. χρὴ γὰρ αὐτὸν ὅλον ἔξωθεν
καταλαβόντα τὸ τραῦμα ταῖς ἑαυτοῦ χερσὶν εἴσω
προστέλλειν τε καὶ σφίγγειν, ὀλίγον ἑκάστοτε τῷ
ῥάπτοντι προγυμνοῦντα· καὶ μέντοι καὶ τὸ ῥαφὲν αὐτὸ
μετρίως προστέλλειν, ἄχρι περ ἂν ὅλον ἀκριβῶς
ῥαφῇ.

τίς δ' ἂν εἴη τρόπος ἐπιτήδειος εἰς τὰ τοιαῦτα τῆς
καλουμένης γαστρορραφίας ἐφεξῆς λέγωμεν. ἐπειδὴ
συμφῦσαι χρὴ τῷ περιτοναίῳ τὸ ἐπιγάστριον, ἀρ-
κτέον μὲν ἀπὸ τοῦ δέρματος ἔξωθεν εἴσω διαπείροντα
τὴν βελόνην. ἐπειδὰν δὲ τὸ δέρμα καὶ τὸν μῦν τὸν
ὄρθιον ὅλον διεξέλθοι,[15] τὸ παρακείμενον ὑπερβαί-
νοντα περιτόναιον, ὠθεῖν αὐτὴν ἔσωθεν ἔξω διὰ τοῦ
λοιποῦ ἀντικειμένου περιτοναίου, κἄπειτ' ἐντεῦθεν
ἔσωθεν ἔξω διαπείρειν τὸ ἕτερον ἐπιγάστριον. διεξελ-
θούσης δὲ τελέως αὐτῆς, ἔξωθεν εἴσω τὸ ἐπιγάστριον
τοῦτο διαιροῦντας, εἶτα τὸ παρακείμενον αὐτῷ περι-

164

when the wound has occurred in the lower parts, and to be tilted downward when it has occurred in the upper parts. The objective in both cases is a single one—to insure that what has prolapsed is not weighed down by the other intestines. So in accordance with this aim, when the wound has occurred on the right side, tilt the body to the left whereas, if it has occurred on the left side, tilt the body to the right, always making the wounded part higher. This will certainly be beneficial for the large and medium-sized wounds because the objective is common to all these. Restoration of the intestines to their proper place, whenever prolapses occur in large wounds, requires a dexterous assistant. He 416K should take hold of the whole wound externally with his hands inside protecting and compressing, leaving a small part exposed on each side for the one who is suturing, and further, protecting to a moderate degree what is sutured until the whole wound is completely sutured.

Let me next state what might be a suitable technique of so-called gastrorraphy for such wounds. Since it is necessary to unite the epigastrium with the peritoneum, we must begin from the skin, driving the needle through from without inward. Once the needle has passed right through the skin and the whole rectus muscle, bypass the associated peritoneum and drive it through the contralateral peritoneum, and then, in turn, through the contralateral epigastrium from within outward. When the needle has passed through this completely, drive it through this [same contralateral] epigastrium from without inward, again bypassing the associated peritoneum, and coming to the

---

15 K; διεξέλθη B, *recte fort*.

GALEN

τόναιον ὑπερβαίνοντας, ἐπί τε τὸ ἀντικείμενον ἐλ-
θόντας ἔσωθεν ἔξω τοῦτο διακεντεῖν, ἅμα δ' αὐτῷ καὶ
τὸ πλησίον ἐπιγάστριον ἅπαν· εἶτ' αὖθις ἀπὸ τούτου
πάλιν ἀρξαμένους συρράπτειν αὐτὸ τῷ ἀντικειμένῳ |
417K περιτοναίῳ· κἄπειτα διεκβάλλειν διὰ τοῦ πλησίον
δέρματος, ἐκεῖθεν δ' αὖ πάλιν εἴσω διείρειν, συρρά-
πτοντας αὐτὸ τῷ ἀντικειμένῳ περιτοναίῳ, διεκβάλ-
λοντάς τε διὰ τοῦ πλησίον δέρματος. εἶτ' αὖθις καὶ
αὖθις ἐργάζεσθαι ταῦτα, μέχρι περ ἂν ὅλον ὁμοίως
ῥάψωμεν τὸ τραῦμα.

διάστημα δὲ τῶν ῥαφῶν ὅσον μὲν ἐπὶ τῷ σφίγγεσθαι τὰ ὑποκείμενα βραχύτατον εἶναι χρεών· ὅσον δ'
ἐπὶ τῷ τὸ μεταξὺ τῶν ῥαφῶν δέρμα διαμένειν ἀσύρ-
ρηκτον οὐ χρηστὸν τὸ βραχύ. φεύγων οὖν ἑκατέρου
τὴν ὑπερβολὴν ἀμφοῖν αἱρεῖσθαι τὸ μέτριον. ἤδη δὲ
καὶ τοῦτο κοινόν πως ἁπάντων ἑλκῶν, ὥσπέρ γε καὶ
αὐτοῦ τοῦ ῥάμματος ἡ σύστασις. τὸ μὲν γὰρ σκληρό-
τερον χρὴ ῥήσσειν τὸ δέρμα, τὸ δὲ μαλακώτερον αὐτὸ
φθάνει ῥηγνύμενον. οὕτω δὲ καὶ τὸ μὲν ἐγγυτάτω τῶν
ἄκρων χειλῶν διαπείρειν τὴν βελόνην, τὸ λοιπὸν τοῦ
δέρματος ὀλίγιστον ὂν, ἀναγκάζεται καὶ βιάζεται

[21] This is the first of three techniques described by Galen for
closing a full-thickness abdominal wound. Two points of terminology are important. First, the term "gastrorrhaphy" has changed in
meaning and is now applied specifically to repair of the stomach, a
usage especially associated with the nineteenth-century German
surgeon Theodor Billroth. It is notable, however, that the 1933
OED still lists only the ancient meaning, which is repair of the an-

opposite (i.e. original ipsilateral) peritoneum, piercing this from within outward and along with it, the adjacent epigastrium in its entirety. Then, beginning again from this once more, suture it with the opposite (i.e. original contralateral) peritoneum and then pass the needle through the adjacent skin. From there, once again draw it through inwardly, suturing it together with the oppositely lying peritoneum, and pass [the needle] inward suturing this [epigastrium] with the opposite (i.e. original ipsilateral) peritoneum, passing the needle outward through the adjacent skin. Do this repeatedly until you suture the whole wound in similar fashion [Figures 2a–f].[21]

417K

For the purpose of restraining the underlying structures, what is required is that the interval between the sutures is kept very short. However, for the purpose of maintaining the skin between the sutures unbroken, a short interval is not good. Therefore, avoid the extreme in each case and choose the mean. And this is, in a way, a general rule for all wounds, just as the consistency of the suture material itself is also, because that which is too hard must break the skin, whereas that which is too soft is prone to breaking itself. Similarly, if you drive the needle through very near the skin edges, because the skin that is left is very

terior abdominal wall, as in Galen, citing two eighteenth-century references. Second, "epigastrium" here means the abdominal wall as a whole. We are very grateful to Mathias Witt for his suggestions on these matters and for kindly allowing us to reproduce the included figures. Interested readers are referred to his 2009 work, *Weichteil- und Viszeralchirurgie bei Hippokrates* (Berlin: Walter de Gruyter). This section of Galen is more or less exactly reproduced in Paul of Aegina's Book VI, section 52.

ῥήγνυσθαι, τὸ δὲ ἐπὶ πλεῖστον ἀποχωρεῖν τοῦδε πολὺ
τοῦ δέρματος ἀκόλλητον ἀπολείπει.

418K   ταυτὶ μὲν οὖν εἰ καὶ πάντων | ἑλκῶν ἐστι κοινά,
μάλιστα αὐτὰ φυλακτέον ἐν ταῖς γαστρορραφίαις.
αὐτὰς δὲ τὰς γαστρορραφίας ἤτοι γε ὡς προείρηται
ποιητέον, ἐστοχασμένου τοῦ συμφῦσαι τῷ περιτοναίῳ
τὸ ἐπιγάστριον, ἐπειδὴ μόγις αὐτῷ συμφύσεται, νευρῶδες ὑπάρχον, ἢ ὡς ἔνιοι, συνάγοντας ἀλλήλοις τὰ
κατὰ φύσιν οἰκεῖα, περιτοναίῳ μὲν περιτόναιον, ἐπιγάστριον δ' ἐπιγαστρίῳ. ἔσται δὲ τοῦτο κατὰ τόνδε
τὸν τρόπον. ἀπὸ τοῦ πλησίον ἡμῶν ἐπιγαστρίου διεκβάλλειν χρὴ τὴν βελόνην ἔξωθεν εἴσω δι' αὐτοῦ
μόνου· κἄπειθ' ὑπερβάντας ἄμφω τὰ χείλη τοῦ περιτοναίου, πάλιν ἀντεπιστρέψαι τὴν βελόνην ἔξωθεν
εἴσω, δι' ἀμφοτέρων τῶν χειλῶν τοῦ περιτοναίου·
κἄπειτ' αὖθις ἀντεπιστρέφοντας ἔσωθεν ἔξω διεκβάλλειν κατὰ τὸ ἀντικείμενον ἐπιγάστριον. οὗτος ὁ
τρόπος τοῦ κοινοῦ καὶ προχείρου, καθ' ὃν διὰ τῶν τεττάρων χειλῶν ἐπιβολῇ μιᾷ διεκβάλλουσι τὴν βελόνην, διαφέρει τῷ κατακρύπτειν ὅλον ἀκριβῶς ἔνδον
τοῦ ἐπιγαστρίου τὸ περιτόναιον.

    ἑξῆς δὲ περὶ τῶν φαρμάκων εἴπωμεν. εἴη δ' ἂν
δήπου καὶ ταῦτα τῆς αὐτῆς ὕλης τοῖς ἐναίμοις ὀνο-
419K   μαζομένοις, | ἃ κἂν τοῖς ἄλλοις μέρεσι συμφύειν τὰ

---

[22] The third method is the continuous through-and-through
suture still used today.

little, it is forced and compelled to break whereas, if you go away from the edges to a marked extent, much of the skin is left ununited.

Therefore, even if these same factors are also common to all wounds, we must guard against them in gastrorraphies in particular. We must carry out gastrorraphies themselves either in the manner previously described, having as our aim the uniting of the epigastrium with the peritoneum, since the latter, being sinewy, will only unite with itself with difficulty. [An alternative method] which some use is to bring those things of the same kind in terms of nature together with one another—that is, peritoneum to peritoneum and epigastrium to epigastrium. This will be done in the following way. We must drive the needle from without inward through the epigastrium nearest to us, and through this alone. Then, having bypassed both the margins of the peritoneum (i.e. ipsi- and contralateral), we must turn the needle around 180 degrees and pass it back again from without inward through both the margins of the peritoneum. Then turning the needle around again through 180 degrees, we pull it from within outward through the oppositely lying epigastrium. This method differs from the one that is common and easy in which [doctors] drive the needle through the four margins with a single stroke by putting away the whole peritoneum completely and precisely within the epigastrium [Figures 3a–c and 4].[22]

Let me speak next about the medications. Presumably these would also be of the same material as those that are called "hemostatics" which I showed in the earlier sec-

418K

419K

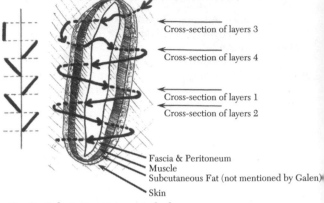

Cross-section of layers 3

Cross-section of layers 4

Cross-section of layers 1

Cross-section of layers 2

Fascia & Peritoneum
Muscle
Subcutaneous Fat (not mentioned by Galen)
Skin

Fig. 2a. Galen's First Suture Method

In the first method of abdominal wall suturing the initial sutures are
mattress stitches. What follow are tailor's stitches, alternating overhand on
either side. There is a peculiarity in the initial stitch in that the outgoing
stitch passes through all layers of the abdominal wall, while the fascia
together with the peritoneum is not picked up with the needle by the
ingoing stitch. In this way there is an artificial discarding of tissues because
the layers of the abdominal wall penetrated from the peritoneal side are
pulled ventralward in relation to the opposite side of the wound, so the
wound edge of the fascia will be opposed to the muscle tissue of the
opposite side (see Figs. 2e & 2f). Similar displacements occur with the
penetrating tailor's stitch (see Figs. 2c & 2d). To the left of Fig. 2a is a
diagram in which the completed wound suture is represented by solid lines
indicating the stitches visible on the skin and interrupted lines those below
the level of the skin.

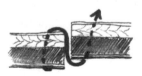

Fig. 2c. Cross-section of layers 1

Fig. 2d. Cross-section of layers 2

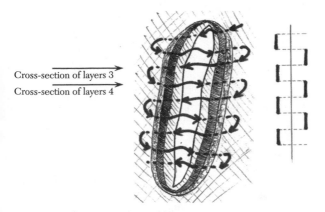

Cross-section of layers 3

Cross-section of layers 4

Fig. 2b. A Variant of Galen's First Suture Method

With this, not only the first two stitches but also the following ones are placed with a mattress technique, creating a corresponding continuous mattress suture. The suture technique follows in an analogous way to that commented on for Fig. 2a. That is to say, with this method the fascia is also bypassed with the ingoing stitch as a result of which the aforementioned displacement again occurs. The needle tip is accordingly always turned away from the bowel so the danger of bowel perforation is slight. To the right of the diagram is a schematic view of the completed suture.

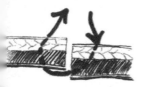

g. 2e. Cross-section of layers 3

Fig. 2f. Cross-section of layers 4

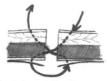

Fig. 3b, Galen's "Flaschenzugnaht"/
"block and tackle" suture (sagittal section)

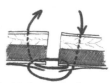

Fig. 3c, Modern "Flaschenzugnaht"/
"block and tackle" suture (sagittal section)

Fig. 3a. Galen's Second Suture Method

With this suture method, the fascia is approximated separately from the rest of the abdominal wall. We are dealing here with a two-layered suture. This type of continuous stitch is a modified "block and tackle" stitch as is still used by modern surgeons (Galen doesn't mention whether the continuation is by an overhand or mattress suture—accordingly the representation of the completed stitch is given in Figs. 3a and 3b while in 3a continuation as a mattress suture is illustrated). Galen achieves the separate approximation of the fascia by taking the needle out above the fascia and turning it through 180 degrees so the fascia is therefore approximated separately in the opposite stitch direction. In this way the point of the needle only penetrates a thin layer (fascia plus peritoneum) while pointed towards the bowel so the risk of bowel perforation would be less than with method 4. After the approximation of the fascia, the needle is turned through 180 degrees to penetrate the opposite side above the fascia in the same way as was done with the ingoing stitch of the abdominal wall. Fig. 3c shows a modern "block and tackle" suture which differs from Galen's technique only in the direction of the needle (i.e., without turning around).

Fig. 4. Galen's Third Suture Method

In this we are dealing with a widely used method. Both in- and out-going stitches penetrate all the layers of the abdominal wall. Galen doesn't actually comment on the continuation of the suture, but in theory it can be applied as a furrier's stitch (as shown, a representation of the completed suture, is next to it), or as a continuous mattress suture (the representation of the completed suture is given on the right of Fig. 4). Because the needle is turned towards the bowel during the 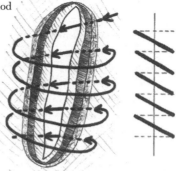 ingoing stitch through all the layers, the danger of bowel perforation is higher with the ingoing stitch compared to other methods (taking into account the fact that the patient is not relaxed). If you take the furrier's stitch as a starting point, this technique would relate most closely to the way a modern surgeon closes the fascial layer. Nowadays, the skin would be closed separately on the grounds of prevention of infection.

Figures 2–4. The methods of surgical closure of the anterior abdominal wall (gastrorrhaphy) as described by Galen in Book 6. *From* Mathias Witt, *Weichteil- und Viszeralchirurgie bei Hippocrates* (Berlin: Walter de Gruyter, 2009). His captions have been translated from the German and abridged.

GALEN

τραύματα διὰ τῶν ἔμπροσθεν ὑπομνημάτων ἀπεδεί-
ξαμεν. ἡ δ' ἔξωθεν ἐπίδεσις ἔτι δὴ καὶ μᾶλλον ἐπὶ
τούτων ἀναγκαία. τὸ δὲ δὴ τέταρτον ἔτι τῆς θεραπείας
μέρος οὐ σμικρῷ τινι τῶν ἄλλων ἀποκεχώρηκεν·
ἐλαίῳ γὰρ χρὴ θερμῷ συμμέτρως ἔριον ἁπαλὸν δεύ-
οντας ὅλον ἐν κύκλῳ περιλαμβάνειν τὸ μεταξὺ βουβώνων τε καὶ μασχαλῶν. ἄμεινον δὲ καὶ διὰ κλυστῆρος ἐνιέναι τοῖς ἐντέροις ἕτερον τοιοῦτον. εἰ δέ τι καὶ
αὐτῶν τῶν ἐντέρων τρωθείη, τὰ μὲν ἔξωθεν, ὡς εἴρηται, πάντα πράττειν ὡσαύτως, τὸ δὲ ἐνιέμενον οἶνος
αὐστηρὸς μέλας χλιαρὸς ἔστω καὶ μᾶλλον εἰ διατρω-
θείη σύμπαν εἰς τὸν εἴσω πόρον. εὐίατα μὲν οὖν τὰ
παχέα τῶν ἐντέρων, δυσιατότερα δὲ τὰ λεπτά. παν-
τάπασι δὲ ἀνίατος ἡ νῆστις διά τε τὸ πλῆθος καὶ τὸ
μέγεθος τῶν ἀγγείων καὶ τὸ λεπτὸν καὶ νευρῶδες τοῦ
χιτῶνος· ἀλλὰ καὶ τὴν χολὴν ἀκραιφνῆ πᾶσαν ἐκδέχεται τὸ ἔντερον τοῦτο, καὶ πάντων ἐστὶν ἐγγυτάτω
τοῦ ἥπατος. γαστρὸς δὲ τὰ μὲν κάτω τὰ σαρκώδη |
420K θεραπεύειν τολμᾶν· ἐγχωρεῖ γὰρ καὶ τυχεῖν, οὐ μόνον
ὅτι παχύτερα ταῦτ' ἐστίν, ἀλλὰ καὶ τοῖς ἰωμένοις
φαρμάκοις εὐπετὴς ἡ ἕδρα κατὰ τήνδε τὴν χώραν
αὐτῆς· τὰ δ' ἐν τῷ στόματι καὶ τῷ στομάχῳ τῇ
παρόδῳ μόνη ψαύει τῶν πεπονθότων· τοῖς δὲ ἐν τῷ
στόματι καὶ τὸ περιττὸν τῆς αἰσθήσεως ἐναντιοῦται
πρὸς τὰς ἰάσεις.

ὅπως δ' ἐπιχειρεῖ γαστέρα τετρωμένην ὁ Ἱπποκράτης ἰᾶσθαι, σὺν καὶ τοῖς ἄλλοις ὀλεθρίοις τρώμασι,
παρ' ἐκείνου μανθάνειν ἄμεινον. ἐγὼ γὰρ οὐχ ὑπὲρ τοῦ

174

tion unite wounds, even in other parts.[23] External bandaging is, however, even more necessary in these [abdominal wounds]. [This], the fourth part of the treatment, is to no small extent different from the others, for it is necessary to wind soft wool, moistened with moderately warm oil, in a circle around the whole body between the groins and the armpits. It is better to also insert another such thing into the intestines through a clyster. And if some of the intestines themselves are also wounded, in those that are external, do everything in the way I described, but let what is inserted be wine that is astringent, dark and warm, particularly if the whole intestine has been pierced through to the internal channel (lumen). The thick intestines are easy to cure whereas the thin intestines are more difficult. The jejunum is altogether incurable due to the number and size of its vessels and the thinness and sinewy nature of its wall. But also [this part of] the intestines receives all the bile unmixed and it is nearest of all [the parts of the intestines] to the liver. Be confident in your treatment of the lower parts of the stomach which are fleshy. It is possible to   420K
achieve this because these are not only thicker, but also because its position in this region is favorable to the curative medications. The medications in the mouth and esophagus make contact with the affected parts by their passage alone, while for those in the opening (cardiac orifice) of the stomach, the excess of sensation acts in opposition to the cures.

How Hippocrates attempts to cure a stomach that has been wounded, along with the other deadly wounds, is better learned from the man himself. It is not so that no

[23] See chapter 2 of the present book (387K).

μηδένα τοῖς Ἱπποκράτους συγγράμμασιν ὁμιλεῖν ἐπὶ
τήνδε τὴν πραγματείαν ἧκον, ἀλλ᾽ ὅτι μοι δοκεῖ πρῶτος μὲν ἐκεῖνος ὁδῷ χρήσασθαι προσηκούσῃ, μὴ
μέντοι γε ἅπαντα συμπληρῶσαι· καὶ γὰρ ἀδιόριστά
τινα τῶν ὑπ᾽ αὐτοῦ γεγραμμένων ἐστὶν εὑρεῖν καὶ
ἐλλιπῆ καὶ ἀσαφῆ· διὰ τοῦτο γοῦν ἐγὼ προὐθυμήθην
ἅπαντά τε σαφῶς διελθεῖν, ἐξεργάσεσθαί τε τὰ χωρὶς
διορισμῶν εἰρημένα καὶ προσθεῖναι τὰ λείποντα. προ-
γυμνασάμενος οὖν τις ἐν τοῖς ἡμετέροις ἐπὶ τὴν
τῶν Ἱπποκράτους συγγραμμάτων ἀνάγνωσιν ἴτω· καὶ
τότε τὸ Περὶ τῶν ἑλκῶν βιβλίον ἀναγνώτω τοῦ ἀνδρὸς |
421K   τό τε Περὶ τῶν ὀλεθρίων τρωμάτων. ἕξει δὲ εἰς αὐτὰ
μεγίστας ἀφορμὰς ἐκ τῶνδε τῶν ὑπομνημάτων· ἔτι τε
γνώσεται βεβαίως ὅτι μήτε τις τῶν ἀπὸ τῆς ὀνόματι
μὲν σεμνῷ κεκοσμηκυίας ἑαυτὴν αἱρέσεως τῆς Μεθο-
δικῆς, ἔργῳ δὲ ἀμεθοδωτάτης ὀρθῶς ἕλκος ἰᾶσθαι
δυνατός ἐστι, μήτε τῶν ἄλλων Λογικῶν, ὅσοι χωρὶς
τοῦ γνῶναι τὰ τῶν ὁμοιομερῶν στοιχεῖα μεταχειρίζεσθαι τὴν τέχνην δικαιοῦσιν. ἀπορήσουσι γὰρ καὶ
οὗτοι λογικῆς θεραπείας ἐπὶ τῶν ἁπλῶν τοῦ ζῴου
μορίων, ἀπὸ τῶν ὀργανικῶν μόνων ἐνδείξεις λαμβάνοντες. ὀλίγον οὖν ἔτι πρὸς τοὺς Μεθοδικοὺς εἰπόντες
ὑπὲρ τῶν κατὰ τὴν γαστέρα τραυμάτων ἐφ᾽ ἕτερόν τι
μεταβησόμεθα.

---

[24] The first is *Peri helkon*, generally rendered "Ulcers" in English (for a translation, see Potter, 1995, LCL, *Hippocrates*, vol.

one familiarizes himself with the Hippocratic writings that I have come to this present treatise, but because that man seems to me to have been the first to use an appropriate path. Nevertheless, he does not seem to me to have brought everything to completion, for among his writings it is possible to find things that are undefined, deficient or obscure. Anyway, it was because of this that I was keen to go over everything in detail clearly, both to finish off those things stated without definitions and to supply those things that are missing. Then, once someone has become practiced initially in my works, let him proceed to the reading of the Hippocratic treatises, and then let him read that man's book *On Wounds(Ulcers)* and the book *On Fatal Injuries*.[24] He will have the best possible starting point for those books from these treatises [of mine]; and further, he will know with certainty that none of those from the Methodic sect, which has honored itself with an imposing name but in practice is completely amethodical, is able to cure a wound properly. Nor are any of the other Rationalists who, without knowing the elements of the *homoiomeres*, think themselves fit to practice the art. These men too will be at a loss for a rational treatment in the case of the simple parts of the organism, since they take indications from the organic parts alone. When I have said a little more with regard to the Methodics on wounds in the abdomen, I shall pass on to something else.

421K

VII); it is about rather more than ulcers in the presently understood sense; see ἕλκος in the section on diseases in the Introduction (section 9). There does not seem to be an extant work corresponding to the second treatise mentioned.

τρωθέντος τοῦ περιτοναίου προπίπτει πολλάκις
ἐπίπλους, οὔτ᾽ εἰ κύριον ὑπάρχει μόριον, οὔτ᾽ εἰ μὴ
κύριον, οὔτ᾽ ἐξ ὧντινων σύγκειται γινωσκόμενον αὐτοῖς, οὔθ᾽ ἥντινα ἔχων ἐνέργειαν ἢ χρείαν. ἐὰν οὖν
πελιδνὸν καὶ μέλαν γένηται τὸ προπεσὸν αὐτοῦ μέρος,
ὅ τι ποτὲ πράξουσιν ἐπ᾽ αὐτοῦ καλὸν ἀκοῦσαι· πότερον ἀποτεμοῦσιν, ἢ καταθήσουσιν ἔσω τοῦ περιτο-
ναίου; πάντως μὲν γὰρ ἤτοι γ᾽ ἐκ πείρας ἡ γνῶσις ἢ |
422K ἐξ αὐτῆς τοῦ μορίου τῆς φύσεως ἡ ἔνδειξις αὐτοῖς
ἔσται. καίτοι φεύγουσί γ᾽ ἑκατέρας· τὴν μὲν ἐκ τῆς
πείρας γνῶσιν ἀνατεινόμενοι τὸ σεμνὸν δὴ τοῦτο τῆς
αἱρέσεως αὐτῶν ὄνομα, τὴν Μέθοδον· τὴν δ᾽ ἐκ τῆς
φύσεως τοῦ μορίου, διότι μήτε τὴν οὐσίαν αὐτοῦ
γινώσκουσι, μήτε τὴν ἐνέργειαν ἢ τὴν χρείαν, ἀποστάντες ὡς ἀχρήστου τῆς ἀνατομῆς. ὥστε οὐκ ἴσασιν
εἴτε τῶν ἀναγκαίων ἐστὶν εἰς τὸ ζῆν εἴτε τῶν οὐκ
ἀναγκαίων μέν, οὐ μὴν ἀκύρων γε παντάπασιν· ἀλλ᾽
οὐδ᾽ εἰ συμπάσχειν αὐτῷ τι[16] μέλλει μέρος[17] κύριον, ἢ
μὴ συμπάσχειν· οὐδ᾽ εἰ τῶν ἀγγείων κατ᾽ αὐτὸ δύνα-
ταί τι δι᾽ αἱμορραγίας ἀποκτεῖναι τὸν ἄνθρωπον· οὐδ᾽
εἰ μετὰ τὴν ἀποτομὴν τοῦ μελανθέντος, εἰ βρόχῳ
διαληφθείη τὸ ὑγιές, ὑπὲρ τοῦ μηδεμίαν αἱμορραγίαν
γενέσθαι, τοῦτ᾽ αὐτῷ κίνδυνον οἴσει τινά· καὶ γάρ τοι
νευρώδης φαίνεται κατά γε τὴν πρόχειρον φαντασίαν.
ὥστε εἰ μή τις ἀκριβῶς εἰδείη τὴν φύσιν αὐτοῦ, μή
ποτ᾽ ἂν θαρρήσῃ βρόχῳ χρήσασθαι, φόβῳ σπασμοῦ.
τούτων οὐδὲν οἱ θαυμασιώτατοι γιγνώσκοντες Μεθο-

178

When the peritoneum has been wounded, the omentum often prolapses. However, they (i.e. the Thessaleians) do not know whether it is an important part or not, nor from what things it is composed, nor what function or use it has. Therefore, if the prolapsed part of it becomes livid or black, it would be good to hear at some time what they will do with it. Will they excise it or will they replace it within the peritoneum? At all events, for these matters knowledge will either be from experience or the indication   422K
will be from the actual nature of the part. And yet they shrink from both. In respect of the knowledge derived from experience, they actually exalt this as the imposing name of their sect—that is, the Methodic. In respect to the knowledge gained from the nature of the part, because they know neither its substance nor function nor use, they reject anatomy as useless. So they don't know which things are necessary for life and which are not. Nor, indeed, do they know which things are altogether unimportant. But nor [do they know] if some important part will suffer a sympathetic affection with it (i.e. the omentum) or not, nor if any of the vessels in it can, through hemorrhage, kill the person. Nor do they know whether cutting off what is blackened and preserving the healthy part by a ligature so that no hemorrhage occurs will carry some danger with it. Certainly the omentum has a "neural" appearance, at least to ordinary perception. So unless someone knows its nature precisely, he would never have the courage to use a ligature for fear of spasm (convulsion). Since the most

16 K; τί B
17 B (cf. pars KLat); μέλος K

423K δικοί, τί ποτε | πράξονται¹⁸ ἐπίπλου μελανθέντος οὐκ
ἔχουσι φάναι.

ἀλλ᾽ οὐχ ἡμεῖς γε παραπλησίως ἐκείνοις ἄπρακτοι
καθεδούμεθα· γινώσκοντες δὲ τὴν μὲν χρείαν αὐτοῦ
μικρόν τι τῷ ζώῳ συντελοῦσαν, τὴν δὲ οὐσίαν ἐξ
ὑμένος τε λεπτοῦ καὶ ἀρτηριῶν καὶ φλεβῶν συγκει-
μένην, ἰδόντες δὲ καὶ τούτων τὰς ἀρχὰς ἀπὸ μεγίστων
οὔσας ἀρτηριῶν καὶ φλεβῶν, αἱμορραγίαν μὲν εὐλα-
βησόμεθα, συμπάθειαν δὲ νεύρων οὐ φοβηθησόμεθα·
καὶ διὰ τοῦτο βρόχῳ τε διαληψόμεθα τὸ πρὸ τοῦ
μελανθέντος ἀποτεμοῦμέν τε τὸ μετὰ τὸν βρόχον ἐν
τῷ κάτω πέρατι τῆς προειρημένης γαστρορραφίας,
ἐκκρεμεῖς τοῦ βρόχου τὰς ἀρχὰς ποιησάμενοι, πρὸς
τὸ κομίσασθαι ῥαδίως αὐτάς, ὅταν ἀποπτυσθῶσιν
ἐκπυήσαντος τοῦ τραύματος.

5. Περὶ μὲν οὖν τῶν ἄλλων μορίων τοῦ σώματος
αὐτάρκως εἴρηται, περὶ δὲ τῶν ὀστῶν ὑπόλοιπον ἂν
εἴη λέγειν. ἐγγίνεται γὰρ δὴ καὶ τούτοις τὸ προκείμενον ἐν τῷ λόγῳ νόσημα, τὸ καλούμενον ὑφ᾽ ἡμῶν
ἑνώσεώς τε καὶ συνεχείας λύσις. ὄνομα δὲ ἴδιον αὐτῷ
κεῖται κατὰ ταῦτα τὰ μόρια κάταγμα, σχεδὸν πᾶσιν
424K ἀνθρώποις ὅσοι γε τὴν | Ἑλλάδα γλῶσσαν ἐπίσταν-
ται σύνηθες· ἄπαγμα δὲ τῶν ἰατρῶν ἴδιον ὄνομά ἐστι
τοῖς πολλοῖς ἀνθρώποις ἀηθές· εἰώθασι δὲ οὕτω προσ-
αγορεύειν, ὅταν ἀποκαλισθῇ τι πέρας ὀστοῦ καθ᾽ ὃ
διαρθροῦται μάλιστα. καὶ μὲν δὴ καὶ τῶν καταγμάτων
ὅσα τελέως διέστησε τὰ μέρη τοῦ κατεαγότος ὀστοῦ,
καυληδὸν γεγονέναι φασίν. εὔδηλον δὲ ὡς ἐγκάρσιος

amazing Methodics know none of these things, they are 423K
not able to say what they will ever do if the omentum becomes black.

But we, at least, shall not uselessly take our seats (as
teachers) like those men, since we know its (i.e. the omentum's) use contributes little to the organism, and its substance is composed of a thin membrane, arteries and veins.
And since we have seen also that the origins of these are
from the major arteries and veins, we will show due cau-
tion about hemorrhage, whereas we will not be afraid of a
sympathetic affection of nerves. Because of this, we will
take up with a ligature that which lies proximal to what is
blackened, and we will excise that which is distal to the ligature in the lower margin of the previously mentioned
gastrorraphy after we have made the ends of the ligature
suspended for the purpose of removing them easily whenever they are cast out once the wound suppurates.

5. I have said enough about the other parts of the
body. What remains is to speak about the bones. For the
disease previously put forward in the discussion, which I
have called dissolution of union or continuity, also befalls
them. The specific name applied to this in relation to these
parts is fracture (*katagma*) which is customary to almost all
men, or at least to those who know the Greek language. 424K
However, *apagma*, a name restricted to doctors, is unfa-
miliar to most men. They (the doctors) are accustomed
to apply the term whenever some margin of the bone is
fractured across, particularly when the fracture involves
a joint. Furthermore, those fractures where the parts of
the fractured bone are completely separated, they say are

---

18 K; πράξουσιν B

ἡ τοιαύτη διαίρεσις, ὥσπερ ἑτέρα κατὰ τὸ μῆκος
μᾶλλόν ἐστιν, οὐ διακόπτουσα παντάπασιν ἀπ' ἀλλήλων τὰ μόρια τῶν οὕτω παθόντων, ἀλλ' οἷον σχίζουσα
κατ' εὐθυωρίαν, ἣν ὀνομάζειν αὐτοῖς ἔθος ἐστὶ σχι-
δακηδόν· ἔνιοι δὲ τῶν νεωτέρων ἰατρῶν ἄχρι τοσούτου
φιλοτιμοῦνται πάσας τῶν καταγμάτων τὰς διαφορὰς
ἰδίοις ὀνόμασιν ἑρμηνεύειν, ὥστε καὶ ῥαφανηδόν τι
καὶ ἀλφιτηδὸν γίνεσθαί φασιν, οὐκ ἀρκούμενοι τῷ
λόγῳ δηλῶσαι τὸ πολυειδῶς συντετριμμένον ὀστοῦν.
οὐ μὴν Ἱπποκράτης γε τοιοῦτος, ἀλλ' ὡς ἔνι μάλιστα
τοῖς συνηθεστάτοις ὀνόμασι χρώμενος, ἑρμηνεύειν
οὐκ ὀκνεῖ λόγῳ καὶ ταύτας τῶν καταγμάτων τὰς
425K διαφορὰς καὶ τούτων οὐδὲν | ἧττον ὅσα κατὰ τὰ τῆς
κεφαλῆς ὀστᾶ γίγνονται. καὶ εἴπερ οὕτως αὐτῷ περὶ
πάντων τῶν παθῶν ὁ λόγος ἐξείργαστο, σύντομος ἂν
ἡμῖν ἡ προκειμένη πραγματεία ἐγεγένητο.

νυνὶ δ' ἐπειδὴ τῶν πλείστων ἐπιδείξας τὴν ὁδὸν
οἵαν δεῖ[19] ποιεῖσθαι, τὰ κατὰ μέρος ἀνεξέργαστα
παρέλιπεν, ἀναγκαῖον ἡμῖν ἐν ἐκείνοις χρονίζειν· οὐ
μὴν οὐδὲ ταῦθ' ὑπερβῆναι δίκαιον, ἀλλ' ὅσα μὲν
εἴρηται τελέως Ἱπποκράτει διὰ κεφαλαίων ὑπομνῆ-
σαι, προσθεῖναι δ' ἐνίοις ἀπόδειξιν ἄνευ πίστεως ὑπ'
αὐτοῦ ῥηθεῖσι· καί τι καὶ διορίσασθαι τῶν ἀδιορί-
στων, καὶ τάξαι τῶν ἀτάκτων, καὶ σαφέστερον ἑρμη-
νεῦσαί τι τῶν ἀσαφεστέρων ὑπ' ἐκείνου γραφέντων.
ἀλλ' εἰ μὲν τὰς ῥήσεις αὐτοῦ ἐφ' ἑκάστου παρα-

19 K; χρή B

"stalklike." It is quite clear that such a division is transverse, just as in another case, when the fracture is more longitudinal, the parts of the affected [bones] are not broken apart from each other altogether but are, as it were, split longitudinally; their custom is to call this "splinterlike." Some of the younger doctors are ambitious to the extent that they give specific names to all the differentiae of fractures, so that they also say there are "radishlike" and "comminuted" [fractures], not being satisfied with the term to signify the bone that is shattered into pieces of many kinds. At least Hippocrates was not like this in that, although as far as possible he used the most customary names, he did not hesitate to explain by discussion these differentiae of fractures, and no less than these, those fractures that occur in the bones of the head. If the discussion about all the affections had been carried out by him in the same way, the matter I just now set forth would have been brief.

425K

For the present, since he set out the kind of path it is necessary to take for most things but left the individual matters unresolved, it is essential for me to take some time over those. It is certainly not right to pass them over. But I shall make mention of them under the chief points of those matters stated completely by Hippocrates. However, for some matters which were stated by him without proof, I shall add a demonstration. Further, I shall define those matters stated without definition, bring order to those not ordered, and explain more clearly those things that were written not so clearly by him. But if I were to place my own

183

γράφοιμι, μῆκος ὑπομνημάτων ἐξηγητικῶν ὁ λόγος
ἕξει· καί τις ἴσως ἡμῖν ἐγκαλέσει μακρολογίαν εὐλόγως τῶν οὐκ εὐλόγως ἤδη μεμψαμένων ἐπὶ τῷ τρίτῳ
καὶ τετάρτῳ γράμματι· κατ᾽ ἐκεῖνα μὲν γὰρ ἀναγκαῖον
ἐγένετό μοι πολλὰς ἐκ τοῦ Περὶ τῶν ἑλκῶν βιβλίου
ῥήσεις Ἱπποκράτους παραθεμένῳ δεικνύναι τοῖς ἄλ-
λοις ἰατροῖς | ἅπασιν ὁποῖόν τι τὸ κατὰ μέθοδον
ἑλκῶν ἐστιν ἰάσεις γράφειν. ἐν δὲ τῷ Περὶ τῶν
καταγμάτων τίς οὕτως ἔμπληκτος ὃς οὐκ ἐπαινεῖ τὴν
διδασκαλίαν ὡς σαφῆ τε ἅμα καὶ τελεώτατα πᾶσαν
ἐξειργασμένην; εἰ δὲ καὶ φαίη τις εἰς ἢ δύο μὴ
θαυμάζειν τὸ γράμμα, πρὸς τοὺς τοιούτους μάλιστα
μὲν ἐν καιρῷ τὸ τοῦ ποιητοῦ λεχθείη, Τούσδε δ᾽ ἔα
φθινύθειν ἕνα καὶ δύο.

τίς οὖν ἐστιν ἡ τῆς καταγμάτων ἰάσεως ἀληθής τε
καὶ ὄντως μέθοδος, ἐξ αὐτῶν τῶν πραγμάτων τῆς
φύσεως ἐνδεικτικῶς λαμβανομένη, λέγειν ἂν εἴη καιρὸς ἀρχὴν τῶν λόγων τήνδε θεμένους. ἐπειδὴ λέλυται
τῆς συνεχείας τὰ τοῦ κατεαγότος ὀστοῦ μόρια, σκοπὸς μὲν τῆς θεραπείας αὐτοῖς, ὁ γοῦν πρῶτος, ἡ
ἕνωσις· εἰ δ᾽ οὗτος ἀδύνατος φαίνοιτο, διὰ τὴν τῶν
πεπονθότων μορίων ξηρότητα, δεύτερος ἄλλος ὁ τῆς
δι᾽ ἑτέρου κολλήσεως ἀπόκειται σκοπός· ὃς εἰ μηδ᾽
αὐτὸς εὑρίσκοιτο δυνατός, ἀνίατον ἐροῦμεν εἶναι τὸ
πάθος. ὅτι μὲν οὖν ἀδύνατόν ἐστι συμφῦναι τὸ οὕτως
σκληρὸν ὀστοῦν, οἷον ἐπὶ νεανίσκων τ᾽ ἐστὶ καὶ μει-
ρακίων καὶ ἀνδρῶν καὶ πολὺ δὴ μᾶλλον ἐπὶ γερόντων,
εὔδηλον | δήπου παντί. μόνον δ᾽ ἐγχωρεῖ μαλακὸν

426K

427K

writings alongside his statements in each case, the discussion would have the length of my exegetical treatises[25] and someone will, perhaps, accuse me of prolixity, and reasonably so—one of those who has already reproached me unreasonably for this in the third and fourth books. For in those it was necessary for me to provide many statements from the Hippocratic work *On Wounds (Ulcers)* to show all other doctors what sort of thing it is to write about cures of 426K wounds according to method. In the work *On Fractures*, who is so capricious that he does not approve the teaching as lucid and, at the same time, also describing everything very comprehensively? If, however, one or two were to say that the work is not remarkable, for such people particularly what was said by the poet is apposite—"permit those one or two to perish."[26]

Therefore, this may be an appropriate time to speak of what the true method for the cure of fractures really is when taken indicatively from the nature of the actual matters, beginning the discussions as follows. When the parts of the fractured bone have suffered a loss of continuity, the object of treatment for them, or at least the primary one, is union. If, however, this seems impossible due to the dryness of the affected parts, the other, secondary objective of conglutination by some other means is held in reserve. If this objective is itself not found to be possible, we shall say the affection is incurable. That it is impossible for hard bone to unite in this way, as for example in young men, youths and men, and especially the elderly is, I presume, 427K quite clear to everyone. It is only possible for bone that is

[25] *In Hippocratis librum De fracturis commentarii III*, XVIIIB.318–628K.   [26] *Iliad*, 2.346.

ἱκανῶς ὀστοῦν, οἷον ἐπὶ τῶν παιδίων ἐστί, δέξασθαι
σύμφυσιν. ὅτι δὲ καὶ δι᾽ ἑτέρας οὐσίας οἷον διὰ
κόλλης ἐγχωρεῖ κολληθῆναί τε καὶ δεθῆναι πρὸς ἄλληλα τὰ διεστῶτα μόρια τοῦ κατεαγότος, ἐνθένδε
μάλιστα ἄν τις ἐλπίσειεν. ἕκαστον τῶν τοῦ ζῴου
μορίων τὴν ἰδίαν τροφὴν ὁμοίαν ἔχειν ἑαυτῷ δέδεικται. εἴπερ οὖν ἀληθὲς τοῦτο, καὶ ἡ τῶν ὀστῶν
οἰκεία τροφὴ παχυτάτη καὶ γεωδεστάτη τῶν ἄλλων
ἁπασῶν ἔσται τῶν καθ᾽ ὅλον τὸ ζῷον. οὔκουν ἀδύνατον ἐξ αὐτῆς ταύτης τῆς οἰκείας τροφῆς περιττόν τι
τοῖς τοῦ κατάγματος ἐπιπηγνύμενον χείλεσι, δι᾽ ἑαυτοῦ μέσου κἀκεῖνα κολλῆσαι. καὶ δὴ καὶ φαίνεται
γιγνόμενον οὕτως· καὶ μαρτυρεῖ τῇ λογικῇ τοῦ πράγματος ἐλπίδι καὶ ἡ πεῖρα.

σκεπτέον οὖν ἐφεξῆς, ὅτῳ μάλιστ᾽ ἄν τις τρόπῳ
τουτὶ τὸ σῶμα τὸ τοῖς κατάγμασιν ἐπιτρεφόμενον,
ὅσον τε καὶ οἷον χρὴ γεννήσειεν. ὅτι μὲν γὰρ οὔθ᾽
ὁπόσον ἔτυχεν οὔθ᾽ ὁποῖον, ἀλλὰ συμμετρίας τινὸς εἰς
ἄμφω δεόμενον, ἄντικρυς δῆλον. ἥτις δὲ ἡ συμμετρία,
κατά τε τὸ ποιὸν αὐτὸ καὶ κατὰ τὸ ποσὸν ἔτυχεν |
428K εὑρόντας, ἐφεξῆς χρὴ ζητῆσαι τὸν τρόπον ᾧ μάλιστ᾽
ἄν τις ἑκατέρου τυγχάνῃ. ζητῆσαι δὲ οὐδὲν ἧττον
ἀναγκαῖόν ἐστι καὶ τὸν καιρὸν ἐν ᾧ ταῦτ᾽ ἐργάζεσθαι
χρή· πότερον εὐθέως ἅμα τῷ γενέσθαι τὸ κάταγμα,
καθάπερ ἐπὶ τῶν τραυμάτων αὐτίκα τὴν σύμφυσιν
ἐποιοῦμεν, ἢ μοχθηρὸς μὲν ὁ τοιοῦτος καιρός, ἕτερον
δέ τινα βελτίω χρὴ ζητεῖν. εὕροιμεν δ᾽ ἂν καὶ αὐτὸ
τοῦτο πρὸς τῆς τοῦ πράγματος φύσεως, ὥσπερ καὶ
τἄλλα πάντα, διδασκόμενοι.

sufficiently soft, as it is for example in the case of children, to accept union. Here, particularly, someone might expect that due to another substance, for example, due to glue, it is possible for the separated parts of the fracture to be conglutinated and bound to each other. It has been shown that each of the parts of the organism has the specific nutriment similar to itself. Therefore, if this is true, the specific nutriment of the bones will be very thick and earthlike compared to that for all other parts of the whole organism. It is not, therefore, impossible for some excess from this same specific nutriment to have congealed within the margins of the fracture and, through its own agency, to conglutinate them. Furthermore, it clearly does occur in this way, and experience also bears witness to the rational expectation of the matter.

Next, we must consider in what way especially someone may produce this particular body which provides nutriment for the fractures—that is, how much and what kind he should produce. For it is patently obvious that it is not as much or whatever kind you might like, but a certain moderation in both that is required. And what this moderation actually is, we should discover in terms of quality and quantity, and next in order we should seek the way in which, particularly, someone might hit upon each of these. No less is it essential to also seek the proper time at which these things should be effected—whether it is straightaway, at the very time the fracture occurs, just as in the case of wounds we immediately effected the union, or whether, such a time being difficult, we should seek another, better time. We would discover this particular thing, just as we would all others, when we have learned about the nature of the matter. 428K

GALEN

τίς οὖν ἡ τοῦ πράγματος φύσις; ὀστοῦν κατεαγὸς
ἅμα τινὶ τῶν εἰρημένων ὀλίγον ἔμπροσθεν διαφορῶν.
ἴδωμεν οὖν εἴ τι παρ' ἑκάστης τῶν διαφορῶν εἰς τὴν
θεραπείαν αὐτοῦ λαβεῖν ἐστιν, ἀπὸ τῆς καυληδὸν
ὀνομαζομένης ἀρξάμενοι. παραλλάττει δ' ἐπὶ ταύτης
ἀλλήλων τὰ μέρη τοῦ κατεαγότος ὡς μὴ κεῖσθαι κατ'
εὐθύ. δῆλον οὖν ὅτι κατ' εὐθὺ χρὴ πρότερον αὐτὰ
ποιήσαντα τῶν ἐφεξῆς τι πράττειν. ἕξει δὲ δήπου τὴν
τοιαύτην θέσιν, ἐὰν ἐπὶ τἀναντία παράγηται τὰ ἐξ-
εστῶτα παραδείγματι χρωμένοις τῷ ὑγιεῖ· παρ' ἐκείνου γὰρ ἀκριβὴς ἡ ἔνδειξίς ἐστι τῆς μεταθέσεως
429K   αὐτῶν. ἐγχωρεῖ γέ τοι καὶ πρόσω τοῦ | κώλου καὶ
ὀπίσω καὶ τῇδε καὶ τῇδε γενέσθαι τὴν μετάστασιν.
ὅσα μὲν οὖν ὀπίσω μᾶλλον μετέστη, ταῦτ' ὠθεῖν χρὴ
πρόσω μετὰ τοῦ τὸ ἕτερον μέρος τοῦ κατεαγότος
ἀντωθεῖν μετρίως εἰς τοὐναντίον· ὅσα δὲ πρόσω, ταῦτ'
εἰς τοὐπίσω μὲν αὐτὰ τὸ δὲ ὑπόλοιπον μέρος ἄγειν
ἠρέμα πρόσω. κατὰ ταὐτὰ δὲ καὶ τῶν εἰς τὸ δεξιὸν
μέρος ἐξεστώτων ἡ ἴασις εἰς τὸ ἀριστερὸν ἀγομένων·
ὥσπερ καὶ τῶν ἐπὶ τὸ ἀριστερὸν εἰς τὸ δεξιόν, ἀεὶ τοῦ
ἑτέρου μέρους ἀντωθουμένου μετρίως. ἀλλ' ἐν τῷ τὴν
ἐναντίαν ὁδὸν ἄγειν ἑκάτερον οὐ σμικρὸς κίνδυνός
ἐστι θραυσθῆναί τινας ἐξοχὰς αὐτῶν. οὐ γὰρ δὴ λεῖόν
γε τὸ πέρας ἑκατέρου τοῦ μέρους ἐστίν, ὡς ἐπὶ τῶν
ἀποπριομένων γίνεται.

What, then, is the nature of the matter? [It is] a fractured bone together with one of the differentiae spoken of a little earlier. Let us see, therefore, if it is possible to take something toward its treatment from each of the differentiae, making a start from [the type of fracture] that has been called the "stalklike" fracture.[27] In this, the parts of the fracture are not in alignment with each other so they do not lie straight. It is clear that the first thing we must do is to make them straight, and then do whatever else is next. It (i.e. the fracture) will, of course, have such a position if the displaced fragments are brought around (i.e. reduced) to the opposite position by us, using [what obtains in] health as the model, for from that there is the precise indication of their change of position. It is possible, in fact, for 429K the change of position of the limb to be either forward or backward, here or there. Therefore, those that are displaced more posteriorly, we should push anteriorly and, along with this, push the other part of the fracture moderately in the opposite direction. Those that are displaced anteriorly, we must push posteriorly and draw the other part gently forward. In the same way, the cure of those fragments displaced to the right is to draw them more to the left, just as with those fragments displaced to the left, it is to draw them to the right, always applying a moderate counterthrust to the other part. But in drawing each [fragment] to the opposite path, there is no small danger of what is prominent in each of the parts being fragmented. Certainly the end of each part is not smooth, as occurs in the case of things that are sawn off.

[27] We have taken this to be a transverse fracture of the shaft of a long bone.

καὶ μὴν εἴπερ αὗται θραυσθεῖεν, οὐκ ἂν ἀκριβῶς
ἔτι συναρμοσθεῖεν πρὸς ἄλληλα τοῦ κατεαγότος
ὀστοῦ τὰ πέρατα, διὰ διττὴν αἰτίαν. αὐτά τε γάρ,
οἶμαι, τὰ θραύσματα μεταξὺ τῶν συναγομένων μερῶν
ἱστάμενα διακωλύσει ψαύειν ἀλλήλων τὰ συναρμοτ-
τόμενα, καὶ εἴπερ ἄρα δυνηθείη τι πρὸς τοὔκτὸς τῶν
430K   ὀστῶν ἐκπεσεῖν, οὐδ' οὕτως ἀκριβὴς | ἡ ἁρμονία
γενήσεται τῶν διαπλαττομένων, ὡς τὴν ἀρχαίαν ἔνωσιν μιμήσεσθαι· μόνως γὰρ ἂν ἐκείνη γένοιτο τῶν
ἐξοχῶν ἁρμοσθεισῶν ἐν ταῖς κοιλότησιν. εἰ δ' ἅπαξ
ἀπόλοιντο θραυσθεῖσαι, κενὴν χώραν ἀναγκαῖον ἀπο-
λειφθῆναι μεταξὺ τῶν ἁρμοζομένων ἀλλήλοις ὀστῶν,
εἰς ἣν ἰχῶρας ἀθροιζομένους, εἶτ' ἐν τῷ χρόνῳ σηπο-
μένους ὅλον ἅμ' ἑαυτοῖς τὸ κῶλον ἀνάγκη διαφθεῖραι.
διὰ ταῦτα μὲν οὖν χρὴ τὴν παραγωγὴν τῶν ὀστῶν
ποιεῖσθαι διεστώτων. αὐτὸ δὲ δὴ τοῦτο πάλιν οὐχ οἷόν
τε καλῶς ἐργάσασθαι χωρὶς ἀντιτάσεως. χρὴ τοίνυν
ἤτοι διὰ τῶν χειρῶν, εἰ μικρὸν εἴη τὸ κῶλον, ἢ διὰ
βρόχων περιβαλλομένων, ἢ καὶ σὺν αὐτοῖς ὀργάνων
οἵων Ἱπποκράτης ἡμᾶς ἐδίδαξε τὴν ἀντίτασιν ποιεῖσθαι τῶν ὀστῶν· ἐπειδὰν δὲ ἱκανῶς διαταθῇ καὶ
μηκέτι κίνδυνος ἀλλήλοις ἐνερείδειν τῷ παράγεσθαι,
κατ' εὐθὺ θέμενον ἀνεῖναι τοὺς βρόχους, ἐπιτρέψαντα
τοῖς μυσὶ συνάγειν εἰς ταὐτὸ τὰ διεστῶτα· συνεφάπτε-
σθαι δὲ ἐν τούτῳ ταῖς χερσὶ καὶ αὐτὸν εἴ τί που
παραλλάττοι[20] σμικρόν,[21] ἐπανορθούμενόν τε καὶ δια-
πλάττοντα.

[20] B; παραλλάττει K   [21] B; μικρόν K

Furthermore, if they are fragmented, the ends of the fractured bone may no longer be precisely fitted together with each other for two reasons: either because there are broken fragments themselves, I think, lying between the parts being brought together, which will prevent the parts being coapted from touching each other, or if some part of the bones is able to fall out externally, so the means of joining of the bones being reduced will not occur precisely in such a way as to reproduce the original union. That would only occur when the protrusions are fitted together in the concavities. If the fragments are destroyed once and for all, an empty space is necessarily left between the bones being coapted with each other; ichors collect in this and when over time there is putrefaction, it is inevitable that the whole limb breaks down along with them. Because of these factors it is necessary to bring about coaptation of the bones that are separated. It is impossible to do this properly without contrary distraction (antistasis). Accordingly, we ought to carry out distraction of the bones, either with our hands if the limb is small, or with ligatures surrounding [the limb], or with the very instruments of the sort Hippocrates taught us about.[28] When the bones are sufficiently distracted and there is no longer any danger of them overriding by being brought alongside one another, and you have placed them in alignment, release the ligatures and allow the separated bones to come together to the same place, relying on the agency of the muscles. In this, lay hold with your hands and, if something small in some way slips aside, correct and mold it.

430K

28 See Hippocrates, *On Joints*, 386, and *Mochlicus*.

μετὰ δὲ ταῦτα δεύτερος ἂν εἴη σκοπὸς ἀτρεμεῖν |
431K ἀκριβῶς τὸ κῶλον, ὡς μηδὲν κινοῖτο τῶν διαπλα-
σθέντων· ἀναγκαῖον γὰρ ἐν τῷδε παραλλάττειν αὖθις.
εἰ μὲν οὖν ἐπιτρέψαις αὐτῷ τῷ κάμνοντι προνοεῖσθαι
τῆς ἡσυχίας αὐτοῦ, περὶ μὲν τὸν τῆς ἐγρηγόρσεως
χρόνον ἴσως ἂν τοῦτο πράξειεν, ὑπνώττων δ' ἐξ ἀνάγκης κινήσει τὸ κῶλον. ὅπως οὖν ἀεὶ φυλάττοιτο τῶν
διαπλασθέντων ἡ θέσις, οὐ μόνον ὑπνώττοντος, ἀλλὰ
καὶ πρὸς ἄφοδον ἀνισταμένου τἀνθρώπου, κἂν ταῖς
μεθυποστρώσεσιν ἀσφαλεῖ δεσμῷ περιλαβεῖν χρὴ τὸ
κάταγμα, σφίγγειν ἀκριβῶς δυναμένῳ πρὸς ἄλληλα
τὰ μέρη τοῦ κατεαγότος ὀστοῦ. ἀλλ' ἐπεὶ τῶν δεσμῶν
ὁ μὲν χαλαρὸς ἐπιτρέπει κινεῖσθαι τοῖς ὀστοῖς, ὁ δ'
ἰσχυρὸς σφιγγόμενος ὀδύνην ἀπεργάζεται, πειρατέον
ἀπολαύοντα τῆς ἀφ' ἑκατέρου χρείας, φυλάττεσθαι
τὴν βλάβην. γένοιτο δ' ἂν τοῦτο φευγόντων τὰς ἀμετρίας.

μήτε οὖν εἰς τοσοῦτον σφίγγειν ὡς ἤδη καὶ θλίβειν, μήθ' οὕτως ἐκλύειν τὴν σφίγξιν ὡς χαλαρὸν
ἐργάζεσθαι. εἰ μὲν οὖν ἰσοπαχὲς εἴη ἕκαστον τῶν
μορίων, ὁ πλατύτατος ἂν ἐπίδεσμος ἄριστος ὑπῆρχεν,
ὅλον[22] ὁμαλῶς τε καὶ συνεχῶς ἐκ παντὸς μέρους
432K περιλαμβάνων | τὸ κατεαγός. ἐπεὶ δ' οὐχ οὕτως ἔχει,
τῷ μὲν θώρακι περιβάλλειν ἐγχωρεῖ τὸν πλατύτατον,
οὔτε δ' ἐπὶ τῶν κώλων οὔτ' ἐπὶ κλειδός, ἀλλ' ἐπὶ τῶν
τοιούτων ὁ στενὸς ἀμείνων, ὡς ἂν μήτε ῥυτιδούμενος
ἅπαντί τε μορίῳ ψαύων τοῦ δέρματος ᾧ περιελήλικται·
ἀλλ' οὗτός γε οὐκ ἀσφαλής, ὀλίγαις λαβαῖς συνέχων
τὸ κατεαγός. ὅσον οὖν εἰς ἀσφάλειαν αὐτοῦ διὰ τὴν

After this, the second objective is to keep the limb com-   431K
pletely still so that none of the reduced parts moves, because if this does happen, there is inevitably overlapping
again. If you allow the patient himself to make provision
for his rest, although during the time of wakefulness he
may perhaps do this, when he is sleeping he will unavoid-
ably move the limb. Therefore, in order that the position of
the reduced fragments is constantly maintained, not only
when the patient is sleeping, but also when he is standing
up to urinate, and during changes of bedding, it is necessary to surround the fracture with a secure bandage which
can compress the parts of the fractured bone against one
another accurately. But when slackening of the bandages
allows the bones to move, or the force of the compressing [bandage] produces pain, you must attempt to guard
against the harm while gaining the benefit from the use of
each. This would occur should we avoid the extremes.

Therefore, neither bind to such a degree as to compress, nor release the binding in such a way as to make it
loose. If each of the parts is of equal thickness, the widest
binding is best, surrounding the fracture in its entirety
evenly and continuously on every side. But this is not [al-   432K
ways] the case: in the chest it is possible to bind around a
very wide bandage, whereas it is not possible in the limbs
or the collarbone. In the case of such parts, a narrow binding is better as it does not create wrinkles at every part
where it touches the skin around which it is bound. But
this is not in fact secure, if it encompasses the fracture with
[only a] few turns of the bandage. What it lacks in terms of
the security of it due to its narrowness, is made up for by

___

22 B; ὅλων K

στενότητα λείπει, τοῦτ' ἐκ πλήθους τῶν περιβολῶν καὶ τῆς ἐπὶ τὸ ὑγιὲς νομῆς προστιθέναι.

ἀλλ' ἐπεὶ τῶν ἐπιδέσμων, ὅσοι πιλοῦσί τε καὶ σφίγγουσιν ἀλύπως τὴν σάρκα, φύσιν ἔχουσιν ἐκθλίβειν μὲν ἐξ ἐκείνων τῶν μορίων οἷς πρώτοις ἐπιβάλλονται τοὺς χυμούς, ἐναποτίθεσθαι δὲ καὶ στηρίζειν εἰς ἅπερ ἐτελεύτησαν, εὔλογον οἶμαι βάλλεσθαι μὲν τὴν ἀρχὴν τῶν ἐπιδέσμων ἐπ' αὐτὸ τὸ κάταγμα· νέμεσθαι δὲ ἐντεῦθεν ἐπὶ τὸ λοιπὸν κῶλον. ὁ γὰρ ἔμπαλιν ἐπιδῶν ἀπὸ τῶν ὑγιεινῶν ἐπὶ τὸ πεπονθὸς ἐκθλίβει τὸ αἷμα. καὶ μὴν εἴπερ ἀπὸ μὲν τοῦ πεπονθότος ἄρχεσθαι προσήκει, τελευτᾶν δὲ ἐπὶ τὸ ὑγιές, οὐ μόνον ἀβλαβὴς ἡ ἐπίδεσις εἰς ἅπερ εἴπομεν ἡ τοιαύτη 433K γένοιτ'[23] ἄν, ἀλλὰ καὶ | χρηστόν τι προσεκπορίζουσα· φλεγμονὴν γὰρ οὐδεμίαν ἐάσει συστῆναι περὶ τὸ κάταγμα· χρὴ δέ, οἶμαι, καὶ τούτου φροντίζειν ἐν τοῖς μάλιστα. κίνδυνος γὰρ ἐπί τε ταῖς ἀντιτάσεσιν, ἃς διαπλάττοντες τὸ κῶλον ἐποιούμεθα, καὶ πρὸ τούτων, ὅτι τὰ πλεῖστα τῶν ἐργαζομένων αἰτίων τὸ κάταγμα τὴν περιβεβλημένην τοῖς ὀστοῖς σάρκα πρότερον ἀδικεῖ πιλοῦντα καὶ θλῶντα, ὡς μεγάλας ἀκολουθῆσαι φλεγμονάς. οὐ θαυμάσαιμι δ' ἂν οὐδ' εἰ καὶ τοῖς ὀστοῖς αὐτοῖς τοῖς κατεαγόσιν ἀνάλογόν τι φλεγμονῇ συμπίπτει. καὶ γὰρ καὶ φαίνεται τά γε μὴ καλῶς ἰαθέντα σαφῶς ὑγρότερα τῶν κατὰ φύσιν, ὅταν ἅμα τραύματι τὸ κάταγμα γενόμενον ὑπὸ τὴν τῆς ὄψεως ἥκῃ διάγνωσιν. οὐ μὴν οὐδὲ ὁ σφάκελος ἐξ ἄλλης προφάσεως ὁρμᾶται, φθορὰ καὶ αὐτὸς ὑπάρχων ὅλης

the number of turns of the bandage and by the distribution extending toward the healthy part.

But since those bandages that compress and bind the flesh painlessly are of a nature to squeeze out the humors from those parts to which they are first applied, and to store them up and establish them in those parts where they end up, it is, I think, reasonable to place the start of the bandages on the fracture itself, and from here to distribute them to the rest of the limb. For someone who bandages in the opposite way, from the healthy parts to what is affected, squeezes out the blood. Furthermore, since it is appropriate to begin from what has been affected and end at what is healthy, not only would such binding be noninjurious in those ways I spoke of, but it would also supply some- 433K thing useful besides, for it will allow no inflammation to exist around the fracture, which is, I think, also something we should consider, especially in these [fractures]. There is a danger when we realign the limb, due to the distractions which we carry out and prior to these, because the majority of the causes which bring about the fracture first harm the flesh that surrounds the bones, causing compression and bruising such that major inflammation follows. Nor would I be surprised even if something analogous to inflammation were to befall the fractured bones themselves. It is also obvious that those things that are not cured properly are more moist than is natural whenever the fracture occurring along with a wound comes under the diagnosis of what is visible. Indeed, gangrene surely arises from no other cause, since this too is a destruction of the whole sub-

---

23 K; γίγνοιτ' B

τῆς οὐσίας αὐτῶν. οὔκουν ἀμελεῖν χρὴ τοῦ τὸ περιττὸν ὑγρὸν ἐκθλίβεσθαι πάντων τῶν περὶ τὸ κάταγμα μορίων. ἄρχεσθαί τε οὖν ἀπὸ τοῦ πεπονθότος καὶ δὶς ἢ τρὶς ἐν κύκλῳ περιβαλόντα νέμεσθαι τοὐντεῦθεν ἐπὶ τὸ ὑγιές. ὁ γὰρ οὕτως ἐπιδῶν ἀπείργει τε τὴν ἐκ τῶν

434K ὑγιῶν εἰς τὸ πεπονθὸς ἐπιρροὴν τοῦ | αἵματος, ἐκθλίβει τε τὸ φθάσαν ἠθροῖσθαι κατ᾽ αὐτό.

διττῶν δὲ ὄντων χωρίων τῶν δυναμένων ἐνδέξασθαί τι παρὰ τῶν πεπονθότων καὶ πέμψαι τὰ μὲν ὑπερκείμενα διά τε τὸ πλῆθος καὶ τὸ μέγεθος ἱκανώτερα πρὸς ἄμφω, τὰ δὲ ἄκρα διὰ τὰ ἐναντία, βραχὺ μέν τι καὶ δέξασθαι καὶ πέμψαι δύναται, πολὺ δὲ οὔτε χορηγῆσαί ποτε τοῖς πεπονθόσιν οὔτε παρ᾽ ἐκείνων λαβεῖν. ὥστε διὰ ταῦτα δύο ἐπιδεύματα πρῶτα ποιησάμενος ὁ Ἱπποκράτης, τῷ μὲν ἑτέρῳ προτέρῳ τό τε περιεχόμενον ἐν τοῖς πεπονθόσιν ἐκθλίβει πρὸς τὰ ὑπερκείμενα, τό τε ἐξ ἐκείνων ἐπιρρέον ἀνείργει.²⁴ τῷ δὲ ὑπολοίπῳ κατὰ μὲν τὰς πρώτας ἐπιβολάς, ἃς κατ᾽ αὐτοῦ τε τοῦ κατάγματος, ἔτι τε τῆς ἐπὶ τὰ κάτω νομῆς τῶν ὀθονίων ποιεῖται, τὸ μὲν ἐκθλίβει πρὸς αὐτά, τὸ δὲ ἀνείργει, κατὰ δὲ τὰς ἑξῆς ἁπάσας, ἡνίκα ἐκ τῶν κατωτάτω μερῶν ἄνω παλινδρομεῖν ἀξιοῖ τὸν ἐπίδεσμον, ὡς εἰς ταὐτὸ τῷ προτέρῳ τελευτῆσαι τοῖς ὑπερκειμένοις ἐκθλίβων, τὰς ἐξ αὐτῶν ἐπιρροὰς κωλύει. φρουρεῖ τ᾽ οὖν ἅμα καὶ στηρίζει τὸ κατεαγός, ἀφλέγμαντόν τε φυλάττει τὰ πρῶτα δύο τῶν ὀθονίων·

²⁴ K; ἀπείργει B, recte fort.

stance of the bones. We must not, therefore, neglect to squeeze out the excess moisture from all the parts around the fracture. Start from the affected part and, when you have encircled this two or three times, extend [the bandage] from here to what is healthy. Bandaging in this way restricts the influx of blood from the healthy parts to the affected part, and squeezes out what has already gathered in the latter.

434K

However, since there are two places that are able to receive something from the affected parts and transmit it to the overlying parts due to their great number and size, they are more adequate in both respects, whereas for the opposite reasons the farthest parts are able to receive and transmit little, being able neither to provide much at any time to the affected parts, nor to receive from them. Consequently, for this reason Hippocrates made two primary bindings: with the first one he squeezed out what was contained in the affected parts toward the overlying parts and restricted the inward flow from those. With the remaining binding, in the first applications, which he made on the fracture itself, he then continued the distribution of the linen bandages downward, on the one hand squeezing out with these and on the other restricting [inflow]. However, in all those [bandages] that follow, since he thinks it a good idea for the bandage to run back upward again from those parts that are lowest so as to end in the same place as the first bandage, he squeezes out from the overlying parts and prevents the inflows from these. The first two of the linen bandages therefore protect and support the fracture and keep it free of inflammation, but they are not in fact suf-

197

435K οὐ | μὴν ἱκανά γε μόνα τὰ εἰρημένα πρὸς ἀμφοτέρας τὰς χρείας.

ὅθεν εἰς μὲν τὴν φρουρὰν ἐπικουρίαν αὐτοῖς ὁ Ἱπποκράτης ἐξεῦρε τὴν τῶν σπληνῶν ἐπιβολήν, ἅμα καὶ τοῖς ἔξωθεν ὀθονίοις, ἃ τούτους αὐτοὺς στηρίζει. πρὸς δὲ τῷ μὴ φλεγμῆναι φαρμάκῳ τινὶ κελεύει χρῆσθαι τῶν ἀφλεγμάντων, οἷά πέρ ἐστιν ἡ ὑγρὰ κηρωτή. ταῦτά τε οὖν ἅπαντα τοῖς εἰρημένοις λογισμοῖς εὑρέθη καὶ πρὸς τούτοις ἔτι τὸ τῆς ἀποθέσεως σχῆμα, διττῷ καὶ τοῦτ᾽ ἐνδείξεως ὑποπῖπτον τρόπῳ· προτέρῳ μὲν ἐκ τῶν κοινῶν ἐννοιῶν ἠρτημένῳ ψιλῷ, ἑτέρῳ δὲ ἐκ τῆς τῶν θεραπευομένων ὀργάνων κατασκευῆς. ὁ μὲν οὖν πρότερος ἀνωδυνώτατον αἱρεῖσθαι συμβουλεύει σχῆμα, πρός τε τῷ μὴ φλεγμαίνειν τὰ μόρια καὶ διαμένειν ἀτρέμα ἐπ᾽ αὐτοῦ μόνου δύνασθαι τὸν κάμνοντα χρόνῳ παμπόλλῳ. ὁ δὲ δεύτερος τρόπος ὁ ἐκ τῆς τῶν θεραπευομένων φύσεως ὁρμώμενος ἀρτηρίας καὶ φλέβας καὶ νεῦρα καὶ μῦς ὡς εὐθύτατα κεῖσθαι κελεύει. καὶ δὴ καὶ ὁμολογοῦσιν ἀλλήλοις οἱ τρόποι· τό τε γὰρ 436K εὐθύτατον ἑκάστου μορίου σχῆμα μάλιστά | ἐστιν ἀνώδυνον, ὅ τι δ᾽ ἂν ἑτέρου μᾶλλον ἀνώδυνον ᾖ, κατὰ φύσιν μάλιστά ἐστι τῷ κώλῳ· χειρὶ μὲν τὸ καλούμενον ἐγγώνιον, σκέλει δὲ τὸ μικροῦ δεῖν ἐκτεταμένον.

εἰς ἀνωδυνίαν δὲ οὐ μόνον τὸ κατὰ φύσιν ἑκάστου σχῆμα τῶν ὀργάνων, ἀλλά τι καὶ τὸ ἔθος ἔοικε συντελεῖν. αὕτη μὲν οὖν σοι καὶ ἡ τοῦ σχήματος εὕρεσις, ἐν ᾧ χρὴ διαφυλάττειν τὸ κῶλον· ἡ δ᾽ αὐτὴ κατὰ τὴν ἀντίτασίν τε καὶ διάπλασιν. ἄμεινον γὰρ ἐν εὐθυτάτῳ

ficient alone for what has been mentioned in respect to   435K
both uses.

From this, Hippocrates discovered as a protective aid
for fractures the application of linen compresses along
with the linen bandages externally which support these.[29]
So that there was no inflammation, he ordered the use of
one of the anti-inflammatory medications like, for example, the moist salve. All these things were discovered on
the basis of the previously mentioned considerations, as
was the form of the setting in addition to these. This [form]
also falls under a twofold manner of indication: in the first
by a mere dependence on the common concepts, and in
the second, by the constitution of the organs being treated.
The first suggests the choice of the most pain-free position,
as a result of which the parts do not become inflamed, and
the patient remains without movement on his own for a
very long time. The second way, which takes its origin from
the nature of those things being treated, demands that
arteries, veins, nerves and muscles lie in the straightest
possible position. Moreover, the [two] ways are in accord
with each other because the best form of each part is the   436K
most pain-free, and what is more pain-free than anything
else is most natural for the limb. For the arm, it is the so-
called "position of angulation" and for the leg, it is almost
stretched out.

However, for purposes of being pain-free, not only does
the natural form of each of the organs (limbs), but also custom seem to contribute something. The actual discovery of
the position in which you must maintain the limb is for you
[to make]. The same applies to distraction and conforma-

[29] See Hippocrates, *Fractures*, 4–5.

καὶ ἀνωδυνωτάτῳ διατείνειν καὶ διαπλάττειν τὸ κῶλον·
πολὺ δὴ μᾶλλον ἐν τῷ αὐτῷ σχήματι τὴν ἐπίδεσίν τε
καὶ τὴν ἡσυχίαν τοῦ μορίου ποιεῖσθαι προσήκει. τὸ
γὰρ ἐν ἄλλῳ μὲν ἐπιδεῖν, ὑπαλλάττειν δὲ εἰς ἕτερον,
οὐκ ὀδύνην μόνον, ἀλλὰ καὶ διαστροφὴν ἐργάζεται
τοῖς ὀστοῖς. εἰ γάρ τι μεμνήμεθα τῶν ἐν τοῖς Περὶ
μυῶν κινήσεως ὑπομνήμασιν εἰρημένων, ἀναγκαῖόν
ἐστιν ὑπαλλαττόντων τὸ σχῆμα τινὰς μὲν τῶν μυῶν
ἐντείνεσθαί τε καὶ οἷον σφαιροῦσθαι συναγομένους,
ἐνίους δ᾽ ἐκλύεσθαί τε καὶ χαλᾶσθαι. ἔνθα μὲν οὖν
ἐντείνονται, θλίβεσθαί τε πρὸς τῶν ἐπιδέσμων αὐτοὺς
437K ἀναγκαῖόν ἐστιν ὀδυνᾶσθαί τε διὰ τὴν | θλίψιν· ἔνθα
δὲ ἡ τάσις ἐκλύεται, χαλαρὰν μὲν ἐν ἐκείνῳ τῷ μέρει
τὴν ἐπίδεσιν, ἀστήρικτον δὲ γίνεσθαι τὸ κάταγμα. διὰ
ταῦτ᾽ οὖν ἅπαντα διατείνειν τε καὶ διαπλάττειν ἐπιδεῖν
τε καὶ ἀποτίθεσθαι καθ᾽ ἓν σχῆμα τὸ ἀνωδυνώτατον,
ἕλοιτ᾽ ἄν τις. εἰς μὲν δὴ τὴν πρώτην ἐνέργειαν ἣν
ἐνεργοῦμεν ἐν τοῖς κατεαγόσιν, οὐδὲν ἔτι λείπει παράγγελμα.

λύειν δὲ κελεύει διὰ τρίτης ὁ Ἱπποκράτης, ὅπως
μήτ᾽ ἄση τις γίγνοιτο, μήτε κνῆσις ἀήθως σκεπα-
σθέντι τῷ μορίῳ, μήτ᾽ ἐπὶ πλέον αἱ διαπνοαὶ κωλύ-
οιντο τοῦ φθάσαντος ἐστηρίχθαι κατὰ τὸ κάταγμα· δι᾽
ἃς οὐ μόνον ἀσωδῶς κνήσεσθαι συμβαίνει τισίν,
ἀλλὰ καὶ διαβρωθέντος ὑπὸ τῆς τῶν ἰχώρων δριμύτη-
τος ἐνίοτε τοῦ δέρματος ἕλκωσιν γενέσθαι. κατ-
αντλεῖν οὖν ὕδατος εὐκράτου τοσοῦτον, ὅσον ἱκανόν
ἐστι διαφορῆσαι τοὺς τοιούτους ἰχῶρας. ἢν δὲ δὴ καὶ

tion. It is better to extend and realign the limb in the straightest and most pain-free way. It is much more appropriate to fashion the binding and place the limb at rest in the same position. To bind in some other way, or to change to another [position] not only causes pain but also distortion of the bones. For if we recall any of those things said in the treatise *On the Movement of Muscles*,[30] it is essential, when we are changing the position, to stretch out some of the muscles, and as it were, to bring them to contraction, but to release and relax others. Then, when they are stretched out, it is essential to compress them with bandages and for there to be pain due to the pressure. When, however, the tension is released and the bandage is loose in that part, the fracture becomes unstable. Therefore, because of all these factors, you should choose to extend, realign, bandage and set [the limb] in the one position that is most pain-free. Certainly, as regards the primary function, which we carry out in fractures, no further example is lacking. 437K

Hippocrates directs [us] to release [the bandage] every two days so that neither any distress occurs, nor unwonted itching in the covered part, nor more importantly that transpirations interfere with the previously firm fixation in the fracture. Due to these transpirations, not only does it happen to some patients that they become nauseatingly itchy but also, when there is erosion due to the sharpness of the ichors, ulceration of the skin sometimes occurs. Therefore, pour on *eukratic* water in an amount sufficient to carry away such ichors. If you also do the same thing

[30] *De motu musculorum*, IV.367–464K.

αὖθις ὁμοίως πράξῃς, ἑβδόμη μὲν ἂν εἴη μετὰ τὴν πρώτην ἀρχὴν ἡμέρα· φαίνοιτο δ' ἂν ἤδη μηδενός γε ἐμποδὼν γενομένου πάντ' ἀφλέγμαντα, καὶ αὐτοῦ γε τοῦ κατὰ φύσιν ἐνίοτε μᾶλλον ἰσχνά. τότε οὖν ἐγχωρεῖ νάρθηκάς τε περιτιθέναι καὶ λύειν διὰ πλέονος· |

438K ἔμπροσθεν μὲν γὰρ ὅθ' ὁ τῆς φλεγμονῆς ἐπεκράτει σκοπός, οὐκ ἦν ἀσφαλὲς θλίβειν τοῖς νάρθηξι· νυνὶ δὲ ἐπειδὴ πέπαυται μὲν αὕτη, στηρίζεσθαι δὲ χρὴ τὸ κάταγμα, καλῶς ἄν τις αὐτοῖς χρῷτο. καὶ μὲν δὴ καὶ διὰ πλέονος ἐγχωρεῖ λύειν, ὡς ἂν μηκέτι χρῃζόντων τῶν μερῶν ἀποκρίνειν ἰχῶρας. ἀλλὰ καὶ ἡ πώρωσις ἄμεινον ἂν οὕτως γίγνοιτο· χρὴ γάρ, ὡς ἔμπροσθεν εἴρηται, παγῆναί τι τῆς οἰκείας τροφῆς τῶν ὀστῶν, ἵνα γένηται πῶρος. οὔκουν χρὴ τοῦτ' ἀποκλύζειν τῶν χειλῶν τοῦ κατάγματος, ἢ διαφορεῖν ἐκτός, ὅπερ αἱ συνεχεῖς λύσεις ἐργάζονται. οὐ μὴν οὐδ' οὕτω πολὺ διαλείπειν, ὡς μηδὲ γνῶναι πῶς προσχωρεῖ τὸ κάταγμα.

πολλάκις γοῦν ἐθεασάμην ἀκριβῶς ξηρανθέντων τῶν ὀστῶν δυσχερῶς γιγνομένην τὴν πώρωσιν. ἐπαντλεῖν οὖν χρὴ τοῖσδε μέτριον τῷ πλήθει, διὰ τρίτης ἢ τετάρτης ἡμέρας ὕδωρ θερμόν, ὅρον ἔχοντα τοῦ παύσασθαι τῆς καταιονήσεως τὴν εἰς ὄγκον ἐρυθρὸν ἔπαρσιν τῶν σαρκῶν. ἐπὶ μὲν δὴ τούτων πρὶν ἄρχεσθαι καθίστασθαι παύσασθαι χρή, καθάπερ ἐφ' ὧν διαφορῆσαί τι βουλόμεθα, μὴ παύεσθαι πρὸ τοῦ

439K συμπεσεῖν | τὸν ἐκ τῆς αἰονήσεως ὄγκον. ἐφ' ὧν δ' ὑγρότης ἐστὶ πλείων καὶ διὰ ταύτην οὐ πωροῦται τὰ κατεαγότα, ξηραίνειν αὐτήν, ὡς ἔμπροσθεν εἴρηται,

again, it should be on the seventh day from the outset. If there seems to be no contraindication, everything appearing now to be free of inflammation, and [the limb] itself sometimes seems rather thinner than normal, it is possible at that time to put splints around it and to release them after a longer interval. Previously, when the objective was to   438K
prevail over the inflammation, it was not safe to compress [the limb] with the splints. Now, when inflammation has ceased and it is necessary for the fracture to be set, one may use these successfully. Furthermore, it is permissible to release [the bandages] for a longer interval as the parts no longer need to get rid of ichors. As well, callus formation would occur better in this situation. It is necessary, as I said before, for some of the proper nutriment of the bones to be made solid for a callus to be generated. We should not, therefore, wash this away from the margins of the fracture or disperse externally what the repeated loosenings create. Certainly, not to leave a significant interval in this way is not to know how a fracture unites.

Often I observed that precisely when the bones were dried, callus formation occurred with difficulty. We should, therefore, pour warm water over these fractures in moderate amount during the third and fourth days, taking as a limit for stopping the fomentation, the raising up of the flesh to a red swelling. In the case of these [swellings], it is necessary to stop before they begin to become established, just as with those things we wish to disperse it is necessary not to stop before the swelling from the fo-   439K
mentation occurs. In these cases, however, the moisture is greater, and because of this moisture, the fractured parts do not form callus. Attempt to dry this, as I said before,

πειρᾶσθαι δι᾽ ἐπιδέσεως προσηκούσης καὶ καταντλήσεως, ἤτοι γ᾽ ἐλαχίστης παντάπασιν ἢ πλείστης. ἡ
μὲν γὰρ ἐλαχίστη, πρὶν ἐπιρρυῆναί τι παυομένη, τά τε
πρόχειρα τῶν ὑγρῶν διαφορεῖ, τά τε ἐν τῷ βάθει
διαχεῖ μετρίως· συμφέρει δὲ οὕτω κεχύσθαι τοῖς ὑπὸ
τῆς ἐπιδέσεως ἐκθλίβεσθαι μέλλουσιν. ἡ δὲ πλείστη
διαφορεῖ πλέον ἢ ἕλκει. εὔδηλον δὲ δήπου καὶ ὡς ἐπὶ
μὲν τῆς ἐκθλιβούσης ἐπιδέσεως ἧττον χρὴ πιέζειν τὰ
τελευταῖα τῶν ὀθονίων, ἐπὶ δὲ τῆς ἀνατρεφούσης,
οὔθ᾽²⁵ ἧττον ταῦτα, καὶ συμπάσας τὰς περιβολὰς
χαλαρωτέρας ποιεῖσθαι.

ἐπεὶ δὲ οὐ μόνον ἐν ᾧ χρὴ καιρῷ συμπράττειν ταῖς
πωρώσεσιν, ἀλλὰ καὶ ὅπως ἐξεύρηται, λείποιτ᾽ ἂν ἔτι
περὶ τῆς συμπάσης διαίτης εἰπεῖν. ὅτι μὲν οὖν ἐν
ἀρχῇ λεπτοτάτως χρὴ διαιτᾶν ἐν τῇ τῶν φλεγμονῶν
οἰκείᾳ θεραπείᾳ λεχθήσεται· λεχθήσεται δὲ καὶ ὡς
φλέβα χρὴ τέμνειν, ἐνίοτε δὲ καὶ διὰ γαστρὸς ἐκ-
440K κενοῦν | τὰ περιττά. κατὰ δὲ τὸν τῆς πωρώσεως καιρὸν
ἀνατρέφειν χρὴ τὸ σῶμα σιτίοις εὐχύμοις καὶ τροφί-
μοις, ἐξ ὧν εἴωθε γεννᾶσθαι χυμὸς οὐ μόνον χρηστός,
ἀλλὰ καὶ γλίσχρος, ἐξ οἵου χρὴ μάλιστα γίγνεσθαι
τὸν πῶρον· ἐκ μὲν γὰρ τῆς ὀρρώδους καὶ λεπτῆς
ὑγρότητος οὐδ᾽ ἂν γεννηθείη τὴν ἀρχήν, ἐκ δὲ τῆς
παχείας μέν, ἀλλὰ ψαθυρᾶς καὶ ἀλιποῦς, γεννηθείη
μὲν ἂν οὐ βραδέως, ἀλλ᾽ ἐν τῷ χρόνῳ ξηραινόμενος
κραυρότερος καὶ διὰ τοῦτο εὔθραυστος γίνεται. μεγέ-

²⁵ Β; οὐχ Κ

through an appropriate binding and thorough fomentation, either very little altogether, or a very great amount. Before ceasing to flow in, very little disperses those fluids that are present and dissolves to a moderate extent those in the depths. It is expedient for those things that are going to be squeezed out by the bandage to be dissolved in this way. A very great amount disperses more than it draws. It is also quite clear, of course, that in the case of the compressing bandage, we must have less pressure in the ends of the linen bandages, whereas in the case of that which is nourishing, not less, and we must make all these coverings looser.

Since it is not only necessary to help in callus formation at the appropriate time, but also to discover how [to do this], it still remains [for me] to speak about regimen as a whole. I shall speak about the fact that it is necessary to put the patient on a very light diet at the beginning in the specific treatment of the inflammations. I shall say too that it is necessary to open a vein and sometimes to evacuate the 440K superfluities through the stomach as well. At the time of callus formation, it is necessary to nourish the body with *euchymous* and nutritious foods from which a humor is customarily generated which is not only useful but also viscous—the kind from which, particularly, the callus should arise. For the callus would not even be generated initially from a serous and thin fluid in the way it would be from a fluid that is thick, but it would be noncohesive and nonfatty, and it would not be generated slowly, but in time it would be dried out and become more friable and, because of this, easily broken. In magnitude, the callus must be of

θει δὲ ἔστω τηλικοῦτος, ὡς ἀσφαλῆ τε ἅμα δεσμὸν εἶναι τοῖς ὀστοῖς καὶ μὴ θλίβειν τοὺς μῦς· ὁ μὲν γὰρ μικρότερος ἢ χρὴ σφαλερὸς τοῖς ὀστοῖς, ὁ δὲ μείζων τοῦ προσήκοντος ὀδυνηρὸς τοῖς μυσί.

προσέχειν οὖν ἀκριβῶς αὐτῷ τὸν νοῦν, ἤτοι προτρέπειν ἐνδεῶς γιγνόμενον, ἢ κωλύειν ἀμέτρως αὐξανόμενον. ἐξ ὧν δ' ἐργάσῃ ταῦτα, τό τε ποσὸν τῆς αἰονήσεώς ἐστι καὶ τὸ ποιὸν καὶ τὸ ποσὸν τῶν ἐδεσμάτων, αἵ τε ἔξωθεν ἐπιτιθέμεναι κατ' αὐτοῦ δυνάμεις φαρμάκων. περὶ μὲν οὖν αἰονήσεως καὶ διαίτης προείρηται. τῶν δὲ φαρμάκων ὅσα μὲν ἐμπλαστικὰ ταῖς
441K οὐσίαις ἐστὶ μετρίως θερμαίνοντα, | προτρέπει τε καὶ συναύξει τοὺς πώρους· ὅσα δὲ διαφορητικά, καὶ τοὺς ὄντας ἤδη μεγάλους καθαίρει. μήτε δὲ ἀνατρέφειν ἐθέλων μήτε καθαίρειν τὸν πῶρον, ἀλλ' ὡς αὐτὸς ὡρμήθη δέχεσθαι, φαρμάκων κέχρητο τῶν ἐναίμων ὀνομαζομένων· ἃ δὴ μετρίως ξηραίνοντα καὶ τὴν πῆξιν ἐργάζεται μετρίαν τοῦ πώρου. περὶ μὲν δὴ τῶν ἐγκαρσίων καταγμάτων ἱκανὰ καὶ ταῦτα·

περὶ δὲ τῶν παραμήκων τὰ μὲν ἄλλα σύμπαντα τὰ αὐτά, πιέζειν δὲ χρὴ μᾶλλον ταῦτα κατ' αὐτὸ τὸ κάταγμα καὶ προστέλλειν εἴσω τὸ ἀφεστηκός. ὅσα δὲ ἐθραύσθη πολυειδῶς καὶ μάλιστα σὺν ἕλκεσιν, ὅπερ ὡς τὰ πολλὰ φιλεῖ γίγνεσθαι, σπλῆνα σκεπαρνηδὸν ὡς Ἱπποκράτης ἐκέλευσε περιβάλλειν, οἴνῳ τέγγοντας αὐστηρῷ τε καὶ μέλανι καὶ μάλιστ' ἐν θέρει.

such a size as to be secure and, at the same time, a bond for the bones without compressing the muscles, for what is smaller than it ought to be is insecure for the bones and what is bigger than is appropriate is painful for the muscles.

Therefore, direct your attention precisely to this; either to promote it if it is deficient or to inhibit it if it is increasing excessively. You will achieve these aims from the following: the amount of the fomentation, the quality and quantity of the foods, and the potencies of the medications applied externally to it. Fomentation and regimen were previously spoken of. Those medications that are emplastic in their substances are moderately heating; they promote and help 441K to increase calluses. Those medications that are dispersing also reduce those calluses that are already large. When you wish neither to build up nor reduce the callus but to accept it as it has arisen, make use of what are termed "hemostatic" medications which are moderately drying and bring about a solidification of the callus that is within the proper limits. This is enough about the transverse fractures.

Concerning the oblique fractures, all the same considerations apply. However, it is necessary to compress these more at the site of the fracture itself and to push what is displaced back in. Those which are broken into many pieces (i.e. comminuted), and particularly those with wounds (i.e. compound), as a large number are wont to be, surround with an oblique compressing bandage as Hippocrates directed,[31] moistening it with wine that is astringent and dark, especially in summer. Someone who uses

31 See Hippocrates, *On Fractures*, 24.

σήπεται γὰρ εἴτε ἐλαίῳ τις εἴτε κηρωτῇ χρῷτο, δεό-
μενά γε ξηραίνεσθαι μειζόνως ἢ τἆλλα, διότι καὶ
μεῖζον τὸ πάθημα. καὶ τοίνυν καὶ τὰ φάρμακα πάντα
τῶν ξηραινόντων ἔστω, καθάπερ ἐκεῖνος ἐκέλευσεν,
ἅμα καὶ τὸ μέτρον ἐπ᾽ αὐτοῖς ὁρίσας· ἐκ γὰρ τῶν
442K ἐναίμων ὀνομαζομένων | εἶναι κελεύει τὰ τοιαῦτα τῶν
φαρμάκων. εἰ δὲ καὶ κατ᾽ ἀρχάς τις εὐθέως αὐτοῖς
χρῷτο, τοιοῦτον εἶναι συνεβούλευε τὸ ἔναιμον, ὃ σύν-
τροφόν ἐστιν ἐπιτέγξει. τὰ μὲν οὖν ἄλλα σύμπαντα
ταῖς ἐκείνου ποιητέον ὑποθήκαις, οὐκ ἐν τοῖς εἰρημένοις μόνον, ἀλλὰ κἂν γυμνωθὲν τὸ ὀστοῦν ἀπο-
πρίσεως δέοιτο,[26] κἂν παρασχίδας τινὰς ἤτοι γ᾽
αὐτὸν ἀφαιρεῖν ἢ τῆς φύσεως ἐξωθούσης συμπράττειν προσήκει.

περὶ δὲ τῶν ὑποτιθεμένων σωλήνων τοῖς σκέλεσιν,
ὅσα μὲν Ἱπποκράτης ἠπόρησεν, εἴτε χρηστέον εἴτε
μὴ χρηστέον, ἅπαντές τε γινώσκουσιν, ἐπαινεῖν τε
χρὴ τοὺς λογισμούς. ὃ δὲ τοῖς νεωτέροις εὕρηται
γλωσσόκομον,[27] ᾧ χρώμεθα μάλιστα κατὰ τὸν τῆς
πωρώσεως καιρόν, ἐπαινῶ μὲν εἴπερ τι καὶ ἄλλο τῶν
τοιούτων μηχανημάτων, οὐ μὴν ἔοικέ γε Ἱπποκράτης
γινώσκειν αὐτὸ καίτοι γε οὐκ ἀμελὴς εἰς τὴν τῶν
χρησίμων ὀργάνων εὕρεσιν. ἀλλὰ τό γε τοῖς σκέλεσιν
ὑποτιθέμενον ὄργανον ὀρθῶς τοῖς μετ᾽ αὐτὸν ἐξεύρηται, δι᾽ ἑνὸς ἄξονος ἐπὶ τελευτῇ τοῦ μηχανήματος ἐν

[26] K; δέηται B
[27] K (cf. autem -ττ- ad 443.14, 16); γλωττοκόμιον B

either oil or a salve brings on putrefaction, since these fractures need to be dried more than others do because the affection is greater. Therefore, also let all the medications be of the drying kind, just as Hippocrates directed when, at the same time, he defined the amount for these. Moreover, 442K he directed that such medications be from those termed "hemostatics." Also, if someone uses them right from the outset, he recommended that it be a hemostatic of the kind which will moisten. You must, then, do all the other things according to his instructions, not only in those [fractures] mentioned, but also where the bone, having been exposed, needs to be sawn off; and where there are some fragments, either take them away or, if it is appropriate, [the doctor] himself should assist their natural extrusion.

Regarding the proposed cylindrical box splints for the legs, Hippocrates had some doubt as to whether we should use them or not. Everyone knows them, and we must commend his reasoning.[32] What was invented by more recent doctors is the *glottocomon* [Figure 5] which I use especially at the critical time of callus formation, and which I recommend, even if it is another one of such contrivances. Hippocrates in fact did not seem to know of this, and yet he was by no means negligent regarding the discovery of useful instruments. But anyway, the instrument discovered by his successors, when applied correctly under the legs, brings about a distraction in the distal parts through one

32 See Galen, *In Hippocratis librum De fracturis commentarii III*, XVIIIB.499K.

Figure 5. The *glottocomon* as described by Galen. The upper of the two smaller figures shows the leg bandages applied in numerical sequence to the tibial fracture to dispel ichors from the fracture site. The lower small figure shows the *solenic* splint with a hole for the heel.

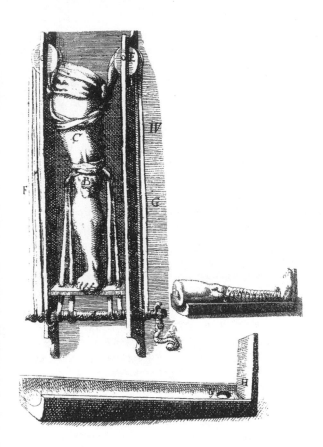

443K τοῖς κάτω μέρεσι τεταγμένου τὴν ἀντίτασιν | ἐργα-
ζόμενον ὅλῳ τῷ κώλῳ ἐντάσεσι δινταῖς. ὀνομάζεται δὲ
αὐτῶν ἡ μὲν κατ' εὐθὺ τείνουσα τὸ κῶλον εὐθύπορος, ἡ
δ' ἄνω μὲν πρότερον, εὐθὺς δὲ κάτω μεταληπτική·
γίνονται δὲ ἀμφότεραι διὰ περιθέσεως βρόχων. ἄριστος δ' εἰς τοῦτο βρόχος ὁ ἐκ δυοῖν διαρτῶν. οὗτος
οὖν ἐν μὲν τοῖς κάτω τοῦ κατάγματος μέρεσι τοῦ
κώλου περιτιθεὶς τὴν εὐθύπορον ἐργάζεται τάσιν, τῶν
σκελῶν αὐτοῦ περιτεθέντων τῷ ἄξονι· ἐν δὲ τοῖς ἄνω
μέρεσι τοῦ κατάγματος ὁ αὐτὸς οὗτος βρόχος εἰ
περιβληθείη, τὴν μεταληπτικὴν ἀποτελεῖ τάσιν, ἀπ-
αχθέντων αὐτοῦ τῶν σκελῶν, ἄνω μὲν πρότερον, εἶτ'
αὖθις κάτω· περιβληθῆναι γὰρ χρὴ καὶ ταῦτα τῷ
ἄξονι. τὴν περιαγωγὴν δὲ καὶ τὴν οἷον καμπὴν τῶν
σκελῶν τοῦ βρόχου, τὴν ἐκ τῶν ἄνω μερῶν εἰς τὸ
κάτω, διὰ τροχιλιῶν γίγνεσθαι χρὴ κατὰ τὰς πλευρὰς
τοῦ γλωττοκόμου τεταγμένων. ἔξεστι δὲ καὶ σωλῆνα
καλεῖν τὸ τοιοῦτον μηχάνημα καὶ μετὰ προσθήκης
σωλῆνα μηχανικὸν ἢ γλωττόκομον μηχανικόν. ἀλλὰ
περὶ μὲν ὀργάνων ἐπὶ πλέον ἐροῦμεν ἐν ταῖς τῶν
ἐξαρθρημάτων ἰάσεσιν· ὡσαύτως δὲ καὶ περὶ τῆς |
444K ἐπιδέσμων ποικιλίας ἐν ἐκείνοις λελέξεται.

νυνὶ δὲ ἐπειδὴ περὶ τοῦ μηχανικοῦ σωλῆνος ἐμνη-
μόνευσα, πολλὴν χρείαν παρεχομένου τῷ σκέλει καὶ
κατὰ τὴν ἄλλην μὲν ἀπόθεσιν, ἔτι δὲ μᾶλλον ἐπειδὰν
ἀλλάττῃ τὴν κοίτην ὁ κάμνων, ἢ ἀποτρίβηταί τι τῶν
κατὰ γαστέρα, καλῶς ἂν ἔχοι πρὸς τῷ μηδὲν ἔτι
ἐνδεῖν εἰς τὸν ὑπὲρ τούτων λόγον, ἐπαινέσαι πολὺ

212

axle placed at the end of the apparatus by way of a twofold   443K
tension for the whole limb. The terms for these [two com-
ponents] are the *euthuporic*, which stretches the limb out
straight, and the *metaleptic*, which first stretches it upward
and then immediately downward, both occurring by way of
the encirclement of ligatures. The best ligature for this
is that which is doubly suspended. When this is placed
around the parts of the limb below the fracture, it brings
about the straight (*euthuporic*) tension, having had its own
limbs placed around the axle. If this same ligature is placed
around the parts above the fracture, it effects the reverse
(*metaleptic*) traction when its own limbs have been led
away, first upward and then downward again, for it is necessary for these also to have been put around the axle. The
going around and the winding, as it were, of the limbs of
the ligature, which is from the parts above to those below,
must occur through a system of pulleys arranged on the
sides of the *glottocomon*. It is also possible to call such a
device a *solen* [either alone] or with the adjective, i.e.
solenic device or *glottocomic* device.[33] But I shall say more
about instruments in relation to the cures of dislocations.
Similarly, more will also be said about the variety of ban-   444K
dages in those considerations.

For the present, since I have made mention of the
solenic device, which provides great service to the leg in
relation to repositioning and, more particularly, whenever
the patient changes his bed or when he rejects some of the
contents of the stomach, it may be right that there is nothing still lacking in the discussion about these instruments

[33] See Hippocrates, *On Fractures*, 16.

μᾶλλον ἐκεῖνο τὸ γλωττόκομον, οὗ καὶ τὴν ἑτέραν τῶν
πλευρῶν καὶ τὸ σανίδιον ἐφ᾽ ὃ τὸν πόδα στηρίζουσι,
κινούμενα κατασκευάζουσιν, ὥστε ἁρμόζειν ἅπαντι
μεγέθει κώλου. περὶ μὲν δὴ τῶν ἄλλων καταγμάτων
ἀρκεῖ καὶ ταῦτα μετὰ τῶν ὑφ᾽ Ἱπποκράτους εἰρημένων
γιγνώσκεσθαι.

6. Περὶ δὲ τῶν ἐν τῇ κεφαλῇ γέγραπται μέν που καὶ
περὶ τούτων Ἱπποκράτει βιβλίον ὅλον, ἅπαντα δι-
δάσκοντι ὅσα χρὴ πράττειν ἐπ᾽ αὐτῶν· καὶ ἡμεῖς δὲ
ὅταν τήνδε τὴν πραγματείαν ἐπιτελέσωμεν, ἐπιθησό-
μεθα ταῖς ἐξηγήσεσι τῶν συγγραμμάτων αὐτοῦ. κατὰ
δὲ τὸ παρόν, ἐπειδὴ τά τε προσευρημένα τοῖς ὑπ᾽
ἐκείνου λεχθεῖσιν ἐν τοῖσδε τοῖς ὑπομνήμασι προσ-
445K   τίθεμεν, ὅσα τε ἀδιορίστως εἶπε διοριζόμεθα, | πρῶτον
μὲν ἀναγκαῖόν ἐστι μνημονεῦσαι τῶν κοίλων ἐκκο-
πέων, οὓς καὶ κυκλίσκους ὀνομάζουσιν· εἶθ᾽ ἑξῆς τῶν
φακωτῶν· καὶ μετ᾽ αὐτοὺς τῶν στενῶν ξυστήρων· εἶτα
περὶ τῆς τῶν φαρμάκων χρήσεως εἰπεῖν τι.

τῶν τοίνυν καταγμάτων τοῦ κρανίου τινὰ μὲν ἄχρι
τῆς διπλόης ἐξήκει, τινὰ δὲ ἄχρι τῆς ἔνδον ἐπιφανείας
τῶν ὀστῶν· καὶ τινὰ μὲν ἁπλαῖ ῥωγμαί, τινὰ δὲ θλά-
σεις, τινὰ δὲ ἕδραι τῶν πληξάντων εἰσίν. αἱ μὲν οὖν
ἁπλαῖ ῥωγμαὶ μέχρι τῆς διπλόης διασχοῦσαι τῶν νῦν
εἰρημένων ξυστήρων χρήζουσι τῶν στενῶν. εἶναι δ᾽
αὐτοὺς χρὴ πολλοὺς μὲν τὸ πλῆθος, ἀνίσους δὲ τὸ
μέγεθος, ὡς μηδέποτε ἀπορεῖν τοῦ χρησιμωτάτου τῷ
ἔργῳ· κἄπειτα γυμνωθέντος κατὰ τὰ εἰθισμένα τοῦ
πεπονθότος ὀστοῦ, χρῆσθαι πρώτῳ μὲν τῷ πλατυ-

other than to praise much more that *glottocomon* of which they make one of the sides and the base, on which they fix the feet, movable so that it adapts to every size of leg. Regarding the other fractures, it is enough to know these things along with those stated by Hippocrates.

6. Of course, a whole book has been written about fractures in the head by Hippocrates, who teaches us all those things we must do for them.[34] And whenever I finish this particular work, I shall add my own explanations of his writings. For the present, since I add in this treatise the things discovered other than those mentioned by that man, let me define those things he stated vaguely. First, it is necessary to call to mind the hollow knives which they also call *cyclisci*; next are those with a lentiform guard; and after these, the narrow raspatories [Figures 6 and 7]. Then it is necessary to say something about the use of medications.

445K

Thus, of the fractures of the cranium, there are those that extend to the diploe, those that extend to the internal surface of the bones, those that are simple fractures, those that are comminuted, and those that are the "seats" of the things that inflicted the blow (i.e. depressed fractures). The simple fractures divide as far as the diploe, and of the fractures listed, they need the narrow raspatories mentioned just now. It is necessary for there to be a considerable number of these raspatories of various sizes so there is never a lack of one that is especially useful in its action. When the bone is laid bare in the customary manner for affected bone, use first the broader raspatory, and second

[34] See Hippocrates, *On Wounds of the Head* (*Vulnera capitis*). An English translation by E. T. Withington may be found in LCL, *Hippocrates* vol. III, pp. 1–51.

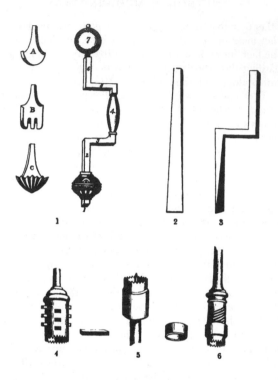

Figure 6. Instruments used in the treatment of skull fractures: a trephine (1), two raspatories (2, 3), and three modioli (4–6). From Francis Adams, trans., *The Genuine Works of Hippocrates* (New York: William Wood, 1886), vol. 1, plate 2.

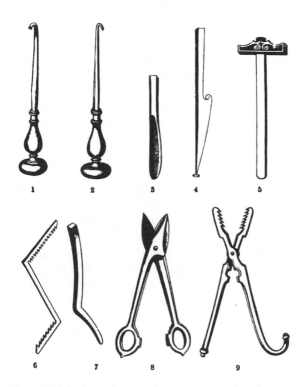

Figure 7. A further selection of instruments used in the treatment of skull fractures: in sequence, three raspatories, a lenticular knife, a mallet, a lever for raising depressed fragments, a meningophylax, and bone-holding forceps. From Francis Adams, trans., *The Genuine Works of Hippocrates* (New York: William Wood, 1886), vol. 1, plate 3.

τέρῳ, δευτέρῳ δὲ τῷ μετ' αὐτὸν στενωτέρῳ, κἄπειθ'
οὕτως ἑξῆς τοῖς ἄλλοις ἄχρι τοῦ στενωτάτου· τούτῳ δὲ
χρηστέον ἐπ' αὐτῆς τῆς διπλόης. εἶτ' ἰᾶσθαι χρὴ
ξηροῖς φαρμάκοις αὐτίκα τε καὶ μέχρι τέλους· ἃ δὴ
καὶ δι' αὐτὸ τοῦτο προσηγόρευται κεφαλικά. σύγκει-

446K   ται δὲ διά τε[28] τῆς | Ἰλλυρίδος ἴρεως καὶ τοῦ τῶν
ὀρόβων ἀλεύρου καὶ μάννης ἀριστολοχίας τε καὶ
πάνακος ῥίζης φλοιοῦ, καὶ ἁπλῶς εἰπεῖν ἁπάντων ὅσα
ῥύπτειν πέφυκεν ἄνευ τοῦ δάκνειν. αὕτη γὰρ ἡ γεννω-
μένη σάρξ, ἔργον τῆς φύσεώς ἐστιν· ὥστε οὐδὲν εἰς
τοῦτο τῆς ἰατρικῆς ὁ κάμνων δεῖται. τὸ δὲ τὴν γεννω-
μένην συμφύεσθαί τε καὶ περιφύεσθαι πᾶσι τοῖς τῶν
ὀστῶν μέρεσιν ἐκ τοῦ μήτε ῥύπον ἐπ' αὐτοῖς εἶναι μήτ'
ἐλαιώδη τινὰ λιπαρότητα γίγνοιτ' ἂν μάλιστα. τοῦτ'
οὖν ἐστιν ὃ παρὰ τῶν ἰατρῶν οἱ κάμνοντες ἴσχουσιν
εἰς τὴν σάρκωσιν, ὡς ξηρὸν μὲν ἅπαν εἶναι τὸ χωρίον,
ἕκαστον δὲ μέρος τῶν πεπονθότων ὀστῶν ἀκριβῶς
καθαρόν. ταῦτα μὲν οὖν κοινὰ πάντων τῶν καταγμάτων, ὅταν ἄρχηται σαρκοῦσθαι.

    τῶν δ' ἄχρι μήνιγγος διασχόντων, εἰ μὲν εἴη μόνη
ῥωγμή, τοῖς εἰρημένοις ξυστῆρσι χρηστέον· εἰ δὲ
μετὰ θλάσεώς τινος, ἐκκόπτειν χρὴ τὸ τεθλασμένον,
ἤτοι διὰ τρυπάνων ἐν κύκλῳ πρότερον κατατιτρῶντα,
κἄπειθ' οὕτω χρώμενον τοῖς ἐκκοπεῦσιν, ἢ διὰ τῶν
κυκλίσκων εὐθὺς ἐξ ἀρχῆς. ἡ μὲν οὖν διὰ τρυπάνων
ἐνέργεια σφαλερά, διὰ τὸ πολλάκις ἅψασθαί τινας, |

    [28] K; τε om. B, recte fort.

218

after this, one that is narrower, and then, in like manner, the others in order right up to the narrowest. You must use this for the diploe itself. Then it is necessary to cure with drying medications straightaway and right to the end; those which, for obvious reasons, they call "cephalics."[35] These are compounded from Illyrian iris, meal of bitter vetch, manna, aristolochia and the root and bark of opopanax—in short, from all those things which, by their nature, cleanse without stinging. The generated flesh itself is a work of Nature, so the patient has no need of the medical art for this. However, the uniting and adhering of the generated flesh to all the parts of the bones occurs particularly from there being neither contamination on them nor any oily fat. This, then, is what patients receive from their doctors as regards the growth of flesh—that the place is altogether dry and that each part of the affected bones is completely clean. These things are common to all fractures whenever the growth of flesh begins.

446K

Of those divisions that extend as far as the meninges, if it is a fracture alone, you must use the aforementioned raspatories. If it is combined with some crushing, it is necessary to excise what has been crushed, either perforating first in a circle with trephines and then, in like manner, using the knives, or perforating with the *cyclisci* immediately from the outset. The operation with trephines is dangerous because, when the trephine plunges in too suddenly, it

[35] This term was still being listed in the *OED* from 1933 as referring to remedies which cure or relieve disorders of the head.

447K ἀθροώτερον αὐτοῦ βαπτισθέντος, τῆς μήνιγγος τῆς
σκληρᾶς τῆς ὑποτεταμένης τοῖς ὀστοῖς. ἡ δὲ διὰ τῶν
κυκλίσκων οὐδ᾽ αὐτὴ παντάπασιν ἄμεμπτος, σείει
γὰρ ἐπὶ πλεῖστον τὴν κεφαλὴν ἡσυχάζειν δεομένην.
ἀρκεῖ δέ μοι κἀνταῦθα μεγάλων μὲν οὐσῶν τῶν
ῥωγμῶν καὶ τῶν ὀστῶν τῶν κατεαγότων ἰσχυρῶς
κεκινημένων τοῖς κυκλίσκοις χρῆσθαι· βραχείαις γὰρ
ἐπιβολαῖς χώραν παρέξεις τοῖς φακωτοῖς ἐκκοπεῦσιν·
ἰσχυρῶν δὲ κατὰ τὰ πλεῖστα τῶν ὀστῶν ὑπαρχόντων
κατατιτρῶναι τοῖς τρυπάνοις αὐτά. τινὲς δὲ ὑπὲρ τοῦ
μηδέποτε ἁμαρτεῖν ἀβάπτιστα τρύπανα κατεσκευ-
άσαντο. καλοῦσι δὲ οὕτως αὐτὰ διὰ τὸ μὴ βαπτίζε-
σθαι· περιθεῖ γὰρ ἐν κύκλω περιφερής τις ἴτυς, προὔ-
χουσα μικρὸν ὑπεράνω τοῦ κατὰ τὸ τρύπανον ὀξέος
πέρατος. εἶναι δὲ δήπου χρὴ πολλὰ καὶ ταῦτα πρὸς
ἅπαν πάχος κρανίου παρεσκευασμένα· τοῖς μὲν γὰρ
παχυτέροις τὸ μακρότερον ἁρμόσει τρύπανον· ὀνομάζω δὲ οὕτως ᾧ μεῖζόν ἐστι τὸ μεταξὺ τοῦ τ᾽ ἄκρου
πέρατος καὶ τῆς κυκλοτεροῦς ἴτυος· τοῖς δὲ λεπτο-
μερέσι τὸ μικρότερον· ἔστι δὲ δήπου καὶ τούτῳ τὸ
448K μεταξὺ τοῦ πέρατός τε | καὶ τῆς ἴτυος ἔλαττον.

ἔνιοι δὲ τούτων, εἴτε δειλοτέρους χρὴ λέγειν, εἴτ᾽
ἀσφαλεστέρους, ταῖς καλουμέναις χοινικίσιν ἐχρή-
σαντο. σὺ δὲ εἰ μήτε ἀφύλακτος εἶναι μέλλοις τῶν
ὄντων σφαλερῶν μήτε πέρα τοῦ προσήκοντος φοβε-
ρός, ἄριστόν σοι τοῖς κυκλίσκοις χρῆσθαι, κατὰ μὲν
τὴν πρώτην ἐπιβολὴν τοῖς πλατυτέροις, ἑξῆς δ᾽ αὐτῶν
τοῖς στενωτέροις, ἄχρι περ ἂν ἐπὶ τὴν παχεῖαν μήνιγ-

often contacts parts of the hard meninges underlying the 447K
bones. Nor is the operation with the *cyclisci* altogether
without hazard, for it shakes the head too much when it
needs to be still. However, when the fractures are large
and the bones which are fractured have been moved excessively, I find it suitable here to use the *cycliscus*, for you
will provide a place in the small sutures with the lenticular
knives. But when the bones are particularly strong, perfo-
rate them with the trephines. Some equip themselves with
nonplunging trephines so they never make a mistake. They
name them in this way because they don't plunge in, there
being an outer rim which runs around in a circle and is
raised up slightly above the sharp edge of the trephine. Of
course, there must also be many of these trephines prepared, to take into account every thickness of the skull. For
those skulls that are thicker, the larger trephine is adequate. I term them thus based on how much more the distance is between the sharp edge [of the trephine] and the
encircling [protective] rim. In those skulls that are thinner,
the smaller [trephine] suffices. There is, of course, less
space in these between the [sharp] edge and the [protec- 448K
tive] rim.

Some doctors, whom we should call either too timid or
too cautious, use the so-called "crown trephine."[36] But if
you are going to be neither careless about their being slip-
pery, nor fearful beyond due measure, it is best for you
to use the *cyclisci*; in the first application, those that are
broader, and next after these, those that are narrower, until
you come to the thick membrane (dura mater). However,

[36] Celsus provides a description of this instrument, which he
calls a *modiolus* (8.3.1).

γα κατέλθῃς. οὐ μὴν οὐδ' ἐν κύκλῳ πᾶν ὅσον ἐκκο-
πῆναι χρὴ τοῖς κυκλίσκοις γυμνωτέον, ἀλλὰ κατ'
ἐκεῖνο μάλιστα τὸ μέρος, ἔνθα τὸ κάταγμα βιαιότερον
αὐτοῦ γέγονε. πρὸς μὲν γὰρ αὖ τοῖς ἄλλοις καὶ ἡ
μῆνιγξ ἀφίσταται τάχιστα τῶν ἰσχυρῶς παθόντων
ὀστῶν, ὥστε οὐδεὶς ἔσται κίνδυνος ἅψασθαι τῆς ἀφ-
εστώσης ἤδη. ἢν δ' ἅπαξ ἔν τι γυμνώσῃς μέρος,
ὑποβαλὼν ἐκκοπέα, τὸ μὲν φακοειδὲς ἐπὶ τῷ πέρατι
προὔχον ἀμβλὺ καὶ λεῖον ἔχοντα, τὸ δὲ ὀξὺ κατὰ τὸ
μῆκος ὄρθιον· ὅταν στηρίξῃς κατὰ τῆς μήνιγγος τὸ
πλατὺ τοῦ φακοειδοῦς, ἐπικρούων τῇ μικρᾷ σφύρᾳ,
διαιρεῖν οὕτω τὸ κρανίον.

σύμβαίνει γὰρ ἐπὶ ταῖς τοιαύταις ἐνεργείαις πάντα
449K ὅσων χρήζομεν. ἡ μέν γε μῆνιγξ, | οὐδ' ἂν νυστάζων
τις ἐνεργῇ, τρωθῆναι δύναται, τῷ πλατεῖ μέρει μόνῳ
τοῦ φακοειδοῦς ὁμιλοῦσα· καὶ ἢν προσέχηται δέ που
τῷ κρανίῳ, καὶ ταύτης τὴν προσάρτησιν ἀλύπως
ἀποσπᾷ τὸ περιφερὲς πέρας τοῦ φακοειδοῦς· ἕπεται δὲ
ἐξόπισθεν αὐτῷ ποδηγοῦντι, διακόπτων τὸ κρανίον ὁ
ἐκκοπεὺς αὐτός. ὥστε οὔτε ἀκινδυνότερον οὔτε θᾶττον
ἐνεργοῦντα τρόπον ἕτερον ἀνατρήσεως εὑρεῖν ἐγχωρεῖ. μάλιστα δ' αὐτὸν ἐπαινέσεις ἐν τοῖς σφοδροτά-
τοις κατάγμασιν· ἃ δή τινες τῶν νεωτέρων ἰατρῶν
ἐγγεισώματά τε καὶ καμαρώσεις ὀνομάζουσι, τὰ μὲν
ἐγγεισώματα τῷ μέσῳ σφῶν αὐτῶν ἐρείδοντα κατὰ

---

37 Linacre simply transliterates the Greek terms, which LSJ
defines as follows: "fracture of the skull, such that one part slips in

what must be cut away in a circle is not everything that has been stripped bare by the *cycliscus*, but that part especially where the fracture has occurred more violently. Apart from other things, the thick membrane (dura mater) very quickly separates from the bones that are severely affected, so there will be no danger of touching the already separated membrane. Once you lay one part bare, insert beneath it the knife which has a lentiform guard that is blunt and smooth projecting at the margin, but sharp along its straight edge [Figure 7]. When you fix the flat surface of the lentiform guard in relation to the thick membrane (dura mater), tap it with a small hammer and, in this way, divide the cranium.

What happens in such operations is that we require instruments of every kind. The thick membrane (dura mater) cannot, in fact, be injured, even if the person operating is half asleep, because it only comes into contact with the flat part of the lentiform [guard]. And if somehow it stays close to the cranium, the rounded margin of the lentiform guard harmlessly strips away the attachment of this membrane. The knife itself, as it cuts through the cranium, follows behind this [guard] as it leads. As a consequence, it is impossible to discover any other method of trepanning that is less dangerous or quicker. You will especially commend it in the very severe fractures which some younger doctors call depressed and elevated.[37] Depressed fractures exert pressure downward on the thick membrane (dura mater) in their central part. In contrast, in the elevated fractures, because they have this same central part ele-

449K

under the bone like a cornice" and "arched fracture (opposite to depressed fracture)."

Figure 8. The procedure of trephination. A pair of bone-holding forceps is also shown. From Johannes Scultetus, *Armamentarium Chirugicum* (1656), and Vido Vidius, Chirurgia (1544).

τῆς μήνιγγος, τὰ δὲ καμαρώματα τουτὶ μὲν ὑψηλὸν
ἔχοντα, καθ᾽ ἃ δὲ πρῶτον ἀπὸ τῶν ὑγειῶν τὸ πεπονθὸς
ἤρξατο χωρίζεσθαι ταῖς τοῦ κατάγματος ῥωγμαῖς,
εἴσω μᾶλλον ἀπεληλυθότα καὶ κατὰ τῆς μήνιγγος
ἐστηριγμένα. τάχιστα γὰρ ἐπὶ τῶν τοιούτων ἐκκόπτε-
ται τὸ πεπονθὸς ἅπαν, ἑτοιμότερον ὑποδυομένου τοῦ
φακωτοῦ πέρατος ἐν τοῖς ἐπὶ πλεῖστον ἐξεστηκόσι τοῦ
κατὰ φύσιν. ἀλλὰ καὶ διὰ τῆς ὀστάγρας ἀνατείνοντες
450K ἢ ἀνακλάσαντες ἔνια | τῶν ἰσχυρῶς συντετρημένων
ὀστῶν, κατ᾽ ἐκεῖνο μάλιστα τὸ μέρος ὑποδῦναι τὸν
φακωτὸν παρασκευάσομεν. ἕπεται δὲ τούτου γενομένου πάντα ἑξῆς ὧν δεόμεθα σὺν ἀσφαλείᾳ τε ἅμα καὶ
τάχει· καὶ σχεδόν, ὡς ἄν τις εἴποι, κατὰ μὲν τἆλλα
σύμπαντα καλῶς εἴρηται τοῦτο δὴ τὸ πολυθρύλητον,
Ἀρχὴ δὲ τὸ ἥμισυ παντός· ἐνταυθοῖ δ᾽ οὐχ ἥμισυ τοῦ
παντός, ἀλλ᾽ ἤτοι πᾶν ἢ ὀλίγου δεῖν ἅπαν ἔχοις ἂν
ἐνθεὶς τὸν φακωτόν.

αὕτη μὲν οὖν ἀρίστη χειρουργία τῶν ἐν τῷ κρανίῳ
καταγμάτων· ὁπόσον δὲ ἐκκόπτειν χρὴ τοῦ πεπονθότος, ἐφεξῆς σοι δίειμι. τὸ μὲν ἰσχυρῶς συντριβὲν
ὅλον ἐξαίρειν· εἰ δ᾽ ἀπ᾽ αὐτοῦ τινες ἐπὶ πλέον ἐκτεί-
νοιντο ῥωγμαί, καθάπερ ἐνίοτε φαίνεται συμβαῖνον,
οὐ χρὴ ταύταις ἕπεσθαι μέχρι πέρατος, εὖ εἰδότας ὡς
οὐδὲν βλάβος ἀκολουθήσει διὰ τοῦτο, τῶν ἄλλων
ἁπάντων ὀρθῶς πραχθέντων. ἡμεῖς γοῦν οὐχ ἅπαξ
οὐδὲ δίς, ἀλλὰ πάνυ πολλάκις τοῦτο ποιήσαντες ἐτύ-
χομεν τοῦ τέλους. ἡ δ᾽ ἔνδειξις ἡμῖν κἀνταῦθα εἰκότως
τῶν ποιητέων ἐκ τῆς φύσεως τῶν πεπονθότων μορίων

vated, it is those parts of the fracture where what is affected first begins to be separated from the healthy bone that are displaced inwardly more and settle on the thick membrane (dura mater). In such fractures, very quickly cut away everything that is affected, since the edge of the knife is more readily slipped under in the bones that are more displaced from their normal position. But also, by lifting up and turning back some of the severely frag- 450K mented bones with the bone-holding forceps, in that part particularly we shall make preparation to slip the lentiform knife underneath. When this happens, all those things we need to do in succession follow with safety, and at the same time with swiftness. One might almost repeat that dictum which has been well said in regard to all other matters: "Beginning is half of the whole."[38] Here, however, you would not have half of the whole, but either the whole or only slightly less than the whole after you have inserted the lentiform knife.

This, then, is the best surgical treatment for fractures in the cranium. Next, I shall describe for you how much of what is affected you should cut away. Lift away everything that is severely shattered. If some fractures extend beyond this still further, as obviously happens sometimes, it is not necessary to follow these to the limit, since you know well that no harm will ensue because of this, provided everything else has been done properly. At any rate, when I have done this, not once only or twice but very many times, I have achieved the goal. Our indication here of the things to be done reasonably is satisfactorily furnished by the nature

[38] Hesiod, *Works and Days*, 40, and Plato, *Republic*, 5.466c.

εὐπορήθη· ἦν γὰρ ἐπὶ τῶν ἄλλων καταγμάτων ἐπίδε-
451K σιν | ὡς ἀφλεγμαντοτάτην ἐξεῦρεν ὁ λόγος, ἐπὶ κεφα-
λῆς οὐχ οἷόν τ᾽ ἐστὶ ποιήσασθαι. ὥστε οὔτε ἀπο-
στρέψαι τὸ ἐπιρρέον οὔτε τὸ περιεχόμενον ἐν τοῖς
πεπονθόσιν ἐκθλῖψαι δυνατόν· ὧν χωρὶς οὐδὲ τῶν
ἄλλων ὀστῶν οὐδὲν ὑγιὲς φυλαχθῆναι δύναται. ὑποκείσθω γοῦν βραχίων ἄχρι τοῦ μυελοῦ ῥωγμὴν ἐσχη-
κώς, εἶτα μηδεὶς αὐτὸν ἐπιδείτω νόμῳ καταγματικῷ,
πᾶσα δήπουθεν ἀνάγκη τοὺς ἰχῶρας οὐκ ἔξω μόνον
ὑπό τε τὸ δέρμα καὶ τοὺς μῦς ἀθροιζομένους, ἀλλὰ καὶ
τῷ μυελῷ περιεχομένους, ἐκεῖνόν τε πρῶτον καὶ μάλιστα σῆψαι καὶ σὺν αὐτῷ τὸ σύμπαν ὀστοῦν, ὅπου γε
καὶ νῦν ἁπάντων προσηκόντως γιγνομένων, ἔστιν ὅτε
τοιαῦτα συμπίπτει.

πῶς οὖν οὐχὶ καὶ μᾶλλον ἂν ἐπὶ κεφαλῆς ταῦτα
συμβαίνοι, μήτε καταγματικὴν ἐπίδεσιν ἐγχωρούσης
δέξασθαι, τῶν τε ἰχώρων ἐπὶ τὸ κάταντες φερομένων
ὡς ἀθροίζεσθαι πάντας ἐπὶ τῆς μήνιγγος; ἐπὶ μὲν οὖν
τῶν ἄλλων καταγμάτων ἡ ἐπίδεσις ὅταν ὀρθῶς γένηται, τοσούτου δεῖ περιττὴν ὑγρότητα συγχωρεῖν ὑπο-
τρέφεσθαι[29] κατὰ τὸ πεπονθὸς ὀστοῦν, ὥστε καὶ τοῦ
κατὰ φύσιν ἰσχνότερον ἀποφαίνειν τὸ χωρίον. ἐπὶ δὲ |
452K τῆς κεφαλῆς ὁ μὲν διὰ τῆς ἐπιδέσεως τρόπος οὐχ οἷός
τ᾽ ἐστὶ ξηραίνειν οὕτω τό τε κατεαγὸς ὀστοῦν αὐτὸ καὶ
τὰ πέριξ, ὡς μήτε φλεγμῆναι μήθ᾽ ὅλως ἐργάσασθαί
τινα ἰχῶρα· φάρμακόν τε οὐδὲν οὐδ᾽ ἐπὶ τῶν ἄλλων
μερῶν ἄνευ τῆς ἐπιδέσεως ἱκανὸν εἰς ὅσον εἴρηται
ξηρὸν καὶ ἀπέριττον ἐργάσασθαι τὸ κατεαγός. ἀναγ-

of the affected parts. Although, in the case of the other fractures, theory shows bandaging to be the least condu- 451K cive to inflammation, it cannot be done in the case of the head. As a result, it is neither possible to prevent influx nor to squeeze out what is contained in the affected parts; without this it is not possible for the health of the other bones to be preserved. Anyway, let us assume that there has been a fracture of the arm extending to the marrow and that nobody bandages it in the conventional way for a fracture. It is, of course, absolutely inevitable that the ichors are collected, not only externally under the skin and muscles, but are also contained in the marrow, and cause putrefaction in that first and foremost, and with it, in the whole bone. And even when everything happens as it should, such things sometimes occur.

How, then, would these things not occur even more in the case of the head, since it cannot receive the conven- tional bandaging for fractures, and when the ichors are carried downward so they all collect on the membrane (dura mater)? Therefore, in the case of other fractures, when the bandaging occurs properly, so far does it prevent excess moisture building up in the affected bone that the place also appears thinner than normal. In the case of the 452K head, however, the manner of the bandaging is such that it is not possible to dry the fractured bone itself or those things around it in this way, so that there is neither inflam- mation nor, in general, the creation of any ichor. Nor, in the case of the other parts, is there any medication which is sufficient to make the fracture dry and free of superfluity to the extent stated without bandaging. It is, then, neces-

---

29 B; ἐπιτρέφεσθαι K

καῖον οὖν ἡμῖν γίγνεται γυμνῶσαί τι τοῦ κατάγματος,
ἵν᾽ ἔχωμεν ἀπομάττειν καὶ ἀποπλύνειν ἀπὸ τῆς μήνιγγος τοὺς ἰχῶρας, ἐπειδὰν ὅ τε τῆς φλεγμονῆς παύση-
ται καιρός, ἀκριβῶς τε ᾖ ξηρὰ πάντα, σαρκῶσαι καὶ
συνουλῶσαι τὸ χωρίον. οὐκ ἔστιν ὁ λόγος οὗτος
ψιλός, οἷον οἱ μηδὲν ἐπ᾽ αὐτῶν τῶν ἔργων εἰδότες
σοφισταὶ ζητοῦσι, διὰ τί τὰ τῆς κεφαλῆς οὐ πωροῦται
κατάγματα· πωροῦται μὲν γὰρ ὦ βέλτιστοι· καὶ ὑμεῖς
οὕτως ἐστὲ ληρώδεις ὥστε τῶν οὐκ ὄντων ὡς ὄντων
λέγειν αἰτίας.

οἶδα γοῦν ποτε τὸ βρέγματος ὀστοῦν συντριβέν·
τὸ δ᾽ ἐφεξῆς αὐτῷ τὸ τοῦ κροτάφου καλούμενον, ἐν ᾧ
τὰς λεπιδοειδεῖς ἐπιβολὰς εἶναι συμβέβηκεν, ἄχρι
τοῦ[30] πλεῖστον μεγίστην σχὸν[31] ῥωγμὴν ἧς ἡμεῖς |
453K ὅλως μὴ προσαψάμενοι, μόνον δ᾽ ἐκκόψαντες τὸ τοῦ
βρέγματος ἰάσαμεν τὸν ἄνθρωπον, ὡς καὶ νῦν ἔτι ζῆν
ἐξ ἐτῶν πολλῶν. εἰ δὲ καὶ τὸ τοῦ βρέγματος εἰάσαμεν,
οὕτως ἐσάπη ἂν θᾶττον ἢ κατὰ τοῦθ᾽ ὑποκειμένη
μήνιγξ ἢ ἐπωρώθη τὸ κάταγμα. ὡς εἴ γε μηδεὶς ἐκ τῶν
πεπονθότων ἰχὼρ ἐντὸς ἔρρει, περιττὸν ἦν ἐκκόπτειν
ὀστοῦν. ἐκεῖνοι μὲν οὖν ὡς σύνηθες αὐτοῖς ἐστι φλυα-
ρείτωσαν, ἐγὼ δ᾽ ἐφ᾽ ἑτέρου κατάγματος ὁμοίως γεγονότος ἐνενόησα τὸ μὲν ὑψηλὸν ὀστοῦν ἐᾶσαι, τὸ δ᾽ ἐν
τοῖς πλαγίοις ἐκκόψαι πρὸς ὑπόρρυσιν τῶν ἰχώρων·
εἶτ᾽ ἐννοήσας τό τε πάχος αὐτοῦ καὶ τὴν σκληρότητα
βέλτιον ᾠήθην εἶναι τὸ μετέωρον ὀστοῦν ἐξελεῖν μᾶλλον ἢ τῆς ὑπορρύσεως φροντίζων σεῖσαι σφοδρῶς

[30] B; τοῦ om. K   [31] B; ἔχον K

230

sary for us to expose part of the fracture, so that we are able to wipe clean and wash off the ichors from the thick membrane (dura mater), and when the critical time of inflammation has passed and everything is completely dry, enflesh and cicatrize the place. This is not just an empty discussion, such as the sophistical [doctors], who know nothing of the matters themselves, pursue about why fractures of the head do not unite by callus formation for, my very good friends, they do unite by callus formation, and you are such fools to speak in this way of things that are not causes being causes.

Anyway, I know of one occasion when the frontal bone was shattered. The bone next to this is the one called the temporal in which the squamous suture happens to be. That bone, which had a very large fracture extending to the greatest extent, I did not touch at all, and cutting away the frontal bone only, I cured the man so that now after many years he is still alive. If, however, I had let the frontal bone remain, the thick membrane (dura mater) underlying it would have putrefied faster than the fracture would have been brought to callus formation. For if no ichor from the affected parts flows in, it is redundant to cut away the bone. Therefore, let those men talk nonsense, as is their wont. I, however, in another fracture that had occurred in a similar way, thought to leave the elevated bone and to cut away the bone at the sides to effect drainage of the ichors. Then, after estimating its thickness and hardness, I thought it better, when giving consideration to the drainage, to take away the elevated bone rather than to

453K

τὸν ἐγκέφαλον. ἐνενόησα δὲ καὶ ὡς εἰς τὰ πλάγια
γεννηθείσης ὀπῆς μεγάλης τοῖς ὀστοῖς ἐξίσχειν
ταύτῃ συμβήσεται τὸν ἐγκέφαλον. ἔστι δὲ δήπου πολ-
λαχόθι κατὰ τὰ πλάγια καὶ νεύρων τις ἔκφυσις ἐπι-
καίρων· ἐν δὲ τοῖς ὑψηλοῖς τῆς κεφαλῆς οὐδὲ σμικρό-
τατον οὐδαμόθεν νεῦρον ἐκφύεται. διὰ ταῦτα μὲν δὴ
ἀπέστην ἐκκόπτειν τὸ πλάγιον ὀστοῦν τῆς κεφαλῆς. |

454K ἐπωρώθη δὲ ἀεὶ θεραπευόντων ὡς χρή.

καὶ σχεδὸν ἔτι τοῦθ᾽ ἡμῖν ὑπολείπεται διασκέψα-
σθαι, τίς ἡ τῶν φαρμάκων τε καὶ ὅλης τῆς μετὰ τὴν
ἀνάτρησιν ἐπιμελείας ἀγωγὴ βελτίστη πασῶν ἐστιν·
ἆρά γε ἡ πραοτάτη καὶ παρηγορικωτάτη, καθάπερ
νῦν ὑπὸ τῶν πλείστων γίνεται, ἢ ἡ ταύτης ἐναντιω-
τάτη, διὰ τῶν ἰσχυρότατα ξηραινόντων φαρμάκων,
οἷον καὶ Μέγης ὁ Σιδώνιος ἐπαινεῖ. καί τις ἡμέτερος
πολίτης ἐχρῆτο διὰ παντός, ὡς καὶ τὴν Ἶσιν ἐπονο-
μαζομένην εὐθέως ἐπιθεῖναι γυμνωθείσῃ τῇ μήνιγγι
ἔμπλαστρον καὶ κατὰ ταύτης ἔξωθεν ὀξύμελι. πρεσβύτης δὲ ἦν οὗτος ἱκανῶς τρίβων τὰ τοιαῦτα τῆς
τέχνης· οὐ μὴν οὔτε ἄλλον τινὰ χρώμενον εἶδον, οὔτε
αὐτὸς ἐτόλμησα χρήσασθαι. τοσοῦτο μόνον ἔχω μαρτυρεῖν τῷ Εὐδήμῳ, τοῦτο γὰρ ὁ πρεσβύτης ἐκαλεῖτο,
ὡς ἐσῴζοντο μᾶλλον οἱ ὑπ᾽ ἐκείνου θεραπευόμενοι τῶν
παρηγορικῶς ἀγομένων. ἐπεχείρησα δ᾽ ἄν ποτε καὶ

---

39 Meges of Sidon (10 BC–30 AD) was a student of Themison.
He emigrated to Rome and gained fame as a surgeon, particularly
for bladder stone and fistulae; see *EANS*, p. 538.

shake the brain violently. I also thought that, if a large opening were created in the bones at the sides, what would happen would be that the brain would extrude through at this point. Of course, at many places in the sides there are also the origins of important nerves, whereas at no point in the upper parts of the head is there the origins of even the smallest nerve. For these reasons I refrained from cutting away the bone at the side of the head. Always, when the   454K treatment is as it should be, there is callus formation.

It just remains for me still to consider what the best form of medication is and the whole care after the trephining. That is, whether it is the most mild and paregoric, as now used by the majority, or the complete opposite of this, by means of those medications that are very strongly drying—medications of the sort Meges the Sidonian also praises.[39] And a fellow citizen of mine always used it, immediately applying what is also called "Isis"[40] as an emplastic to the exposed membrane (dura mater) and oxymel over this externally. This was a man old enough to be practiced in such aspects of the craft. I have not seen anyone else use this, nor have I myself been brave enough to use it. This much only am I able to testify regarding Eudemus,[41] for this was what the old man was called: that more of those people treated by him were saved than those who were treated with paregorics. I, too, would have endeavored to

[40] A compound medication mentioned several times in *De compositione medicamentorum per genera*; see, for example, XIII.736, 747, 774, and 794K.

[41] Probably Eudemus the Methodic (first century AD); see *EANS*, pp. 307–8.

αὐτὸς δι' ἐμαυτοῦ πειραθῆναι τῆς τοιαύτης ἀγωγῆς, εἰ
διὰ παντὸς ἐν Ἀσίᾳ κατέμεινα· διατρίψας δ' ἐν Ῥώμῃ
455K τὰ πλεῖστα τῷ τῆς πόλεως | ἔθει συνηκολούθησα,
παραχωρήσας τοῖς χειρουργοῖς καλουμένοις τὰ πλεῖστα τῶν τοιούτων ἔργων. τὴν μέντοι φύσιν τοῦ πράγματος ἐπισκοπούμενος ἐννοῶ τοιοῦτόν τινα διορι-
σμόν, ὑπὸ μακρᾶς πείρας ἡμετέρας μαρτυρούμενον· ὁ
καλούμενος ἀκουστικὸς πόρος, οὐ μόνον ἄχρι τῆς
σκληρᾶς ἐξικνεῖται μήνιγγος, ἀλλὰ καὶ αὐτοῦ τοῦ
νεύρου ψαύει τοῦ καθήκοντος ἐξ ἐγκεφάλου εἰς αὐτόν.
κείμενος δ' οὕτως ἐγγύς, ὅμως ἰσχυροτάτων ἀνέχεται
φαρμάκων, ὡς καὶ πρόσθεν εἴρηται. θαυμαστὸν οὖν
οὐδέν, εἰ καὶ μετὰ τὰς ἀνατρήσεις ἡ μῆνιγξ ἡ παχεῖα,
πρὶν ἀξιολόγως φλεγμῆναι, τοῖς ἰσχυροτάτοις χαίρει
φαρμάκοις, ὡς ἂν φύσει ξηρὰ τὴν οὐσίαν ὑπάρχουσα.

234

make trial of such a treatment for myself at some time if I had remained throughout in Asia. However, as I have spent most of my time in Rome, I have followed the cus- 455K tom of the city, yielding the majority of such activities to those who are called surgeons. Nevertheless, having given due consideration to the nature of the matter, I conceive of a certain such distinction, borne out by my long experience. The so-called porus acousticus not only extends up to the thick membrane (dura mater) but also contacts the nerve itself which passes down to it from the brain. Although it lies so close, it nonetheless tolerates the strongest medications, as I also said before. It is not, then, to be wondered at if the thick membrane (dura mater), after trepannations, and before it is significantly inflamed, welcomes the strongest medications, as it is, by nature, dry in substance.

456K 1. Τὴν θεραπευτικὴν μέθοδον, ὦ Εὐγενιανὲ φίλτατε,
πάλαι μὲν ὑπηρξάμην γράφειν Ἱέρωνι χαριζόμενος,
ἐπεὶ δὲ ἐξαίφνης ἐκεῖνος ἀποδημίαν μακρὰν ἀναγκα-
σθεὶς στείλασθαι, μετ' οὐ πολὺν χρόνον ἠγγέλθη
τεθνεώς, ἐγκατέλιπον κἀγὼ τὴν γραφήν. οἶσθα γὰρ
ὡς οὔτε ταύτην οὔτε ἄλλην τινὰ πραγματείαν ἔγραψα
τῆς παρὰ τοῖς πολλοῖς ἐφιέμενος δόξης, ἀλλ' ἤτοι
φίλοις χαριζόμενος ἢ γυμνάζων ἐμαυτόν, εἴς τε τὰ
παρόντα χρησιμώτατον γυμνάσιον εἴς τε τὸ τῆς λή-
θης γῆρας, ὡς ὁ Πλάτων φησίν, ὑπομνήματα θησαυ-
457K ρισόμενος.[1] ὁ γάρ τοι τῶν πολλῶν ἀνθρώπων ἔπαινος
εἰς μὲν χρείας τινὰς ἐπιτήδειον ὄργανον ἐνίοτε γίγνεται τοῖς ζῶσιν, ἀποθανόντας δὲ οὐδὲν ὀνίνησιν, ὥσπερ οὐδὲ τῶν ζώντων ἐνίους. ὅσοι γὰρ ἥσυχον εἵλοντο
βίον, ὠφελημένοι μὲν ἐκ τῆς φιλοσοφίας, αὐτάρκη δ'
ἔχοντες τὰ πρὸς τὴν τοῦ σώματος θεραπείαν, τούτοις
ἐμπόδιον οὐ σμικρόν ἐστιν ἡ παρὰ τοῖς πολλοῖς δόξα,
περαιτέρω τοῦ προσήκοντος ἀπάγουσα τῶν καλλί-
στων αὐτούς. ὥσπερ ἀμέλει καὶ ἡμᾶς οἶσθα πολλάκις
ἀνιωμένους ἐπὶ τοῖς ἐνοχλοῦσιν οὕτω συνεχῶς ἐνίοτε

---

[1] Κ; θησαυριζόμενος Β

236

BOOK VII

1. My dearest Eugenianus, a long time ago I began to write 456K
the *Method of Medicine* as a favor for Hiero. Then suddenly he was forced to spend a long period abroad. Soon
after, he was reported to have died, whereupon I abandoned the writing. For you know that I wrote neither this
nor any other treatise to advance my popular reputation. [I
write] either for the gratification of friends or so that I
might exercise myself as the most useful practice in the
present matter and as a laying by of notes against the for-
getfulness of old age (as Plato said).[1] Let me tell you, the 457K
praise of many is sometimes a beneficial instrument for
certain purposes to those who are living. However, for
those who are dead, it has no benefit at all, just as it does
not for some of those who are living. Those who choose a
quiet life, those who derive benefit from philosophy and
are self-sufficient when it comes to the care of the body,
find a reputation among the many to be no little hindrance,
drawing them further away from a concern with the things
that are best. Also, you are not unaware that I am often distressed by those things that are disturbing, and sometimes

[1] Plato, *Phaedrus* 276d3 (not quoted verbatim).

χρόνον ἐφεξῆς πολύν, ὡς μηδ᾽ ἅψασθαι δυνηθῆναι
βιβλίου. ἐγὼ δὲ οὐκ οἶδ᾽ ὅπως εὐθὺς ἐκ μειρακίου
θαυμαστῶς, ἢ ἐνθέως, ἢ μανικῶς, ἢ ὅπως ἄν τις
ὀνομάζειν ἐθέλῃ, κατεφρόνησα μὲν τῶν πολλῶν ἀνθρώπων δόξης, ἐπεθύμησα δὲ ἀληθείας καὶ ἐπιστή-
μης, οὐδὲν εἶναι νομίσας οὔτε κάλλιον ἀνθρώποις οὔτε
θειότερον κτῆμα. διὰ ταῦτ᾽ οὖν οὐδ᾽ ἐπέγραψά ποτε
τὸ ἐμὸν ὄνομα τῶν ὑπ᾽ ἐμοῦ γεγραμμένων βιβλίων
οὐδενί· παρεκάλουν δ᾽, ὡς οἶσθα, καὶ ὑμᾶς μήτ᾽ ἐπαινεῖν με παρὰ τοῖς ἀνθρώποις ἀμετρότερον, ὥσπερ
458K εἰώθατε, | μήτ᾽ ἐπιγράφειν τὰ συγγράμματα. κατὰ
ταῦτ᾽ οὖν ἅπαντα καὶ ἡ Θεραπευτικὴ μέθοδος ἐγκατ-
ελείφθη μοι, τὰ μὲν κεφάλαια τῶν εὑρημένων διὰ
βραχέων ὑπομνημάτων ἐμαυτῷ γράψαντός μου, διεξο-
δικὴν δ᾽ οὐδεμίαν ἔτι προσθέντος διδασκαλίαν. νυνὶ
δ᾽ ἐπειδὴ καὶ σὺ καὶ ἄλλοι πολλοὶ τῶν ἑταίρων,
ἅπερ ἐθεάσασθέ με πολλάκις ἐπὶ τῶν νοσούντων ἔργῳ διαπραττόμενον, ἐν ὑπομνήμασιν ἔχειν ἀξιοῦτε,
προσθήσω τὸ λεῖπον ἔτι τῇδε τῇ πραγματείᾳ.

2. Ἐν μὲν οὖν τῷ τρίτῳ καὶ τετάρτῳ καὶ πέμπτῳ
καὶ ἕκτῳ τῶνδε τῶν ὑπομνημάτων τὸ κοινὸν νόσημα
τῶν ὁμοιομερῶν τε καὶ ὀργανικῶν μορίων ὅπως χρὴ
θεραπεύειν ἐγεγράφειν· ἀρξάμενος μὲν οὖν ἀπ᾽ αὐτοῦ,
διότι τε σαφέστερόν ἐστι τῶν ἄλλων ἐναργῆ τε τὸν
ἔλεγχον ἔχει τῶν ἔμπροσθεν ἁπάντων σχεδὸν ἰατρῶν
ὅσοι μεθόδῳ μὲν ἐπαγγέλλονται μεταχειρίζεσθαι τὴν
τέχνην, ἀμέθοδοι δέ εἰσιν ἐν ταῖς θεραπείαις. πλὴν

continue to be so over a long period, so I have been unable to touch the book.[2] Remarkably, from my youth, and I do not know how—whether being inspired or crazy, or whatever you might wish to call it—I have despised the opinion of the majority and have set my heart on truth and knowledge, thinking no possession to be better or more divine for men. Because of this, I did not at any time attach my name to any of the books I wrote and I used to urge you also, as you know, neither to praise me too immoderately among men, as you are wont to do, nor to attach my name   458K
to the treatises. For all these reasons I abandoned the *Method of Medicine* after I had written, in brief notes for myself, the chief points of the things I had discovered without as yet putting forward any detailed teaching. Now, however, since you and many others among my pupils, who often observed me carrying out my work among the sick, think it worthwhile to have this [material] in a treatise, I shall add what remains to this particular work.

2. Thus, in the third, fourth, fifth and sixth [books] of this treatise, I had written how it is necessary to treat the disease common to the *homoiomeres* and the organic parts (i.e. dissolution of continuity). I began with this because it is more obviously apparent than the other diseases, and provides a clear refutation of almost all those previous doctors who profess to practice the craft by method and yet are without method in their treatments. Apart from

2 Vivian Nutton (*per litt.*) has pointed out that this remark about distress is somewhat at odds with Galen's claim of imperturbability in the face of discomfiting events as expressed in such works as *De indolentia*, *De moribus*, and *De propriorum animi cuiuslibet affecturum dignotione et curatione*.

γὰρ Ἱπποκράτους τοῦ πάντων ἡμῖν τῶν καλῶν παρα-
σχόντος τὰ σπέρματα, τῶν ἄλλων οὐδεὶς οὐδ᾽ ἐπεχεί-
459K ρησε τὸ γένος τοῦτο τοῦ νοσήματος | ἰάσασθαι μεθ-
όδῳ. δεῖται δὲ καὶ τὰ Ἱπποκράτους αὐτοῦ γεωργῶν
ἀγαθῶν, οἳ σπεροῦσί τε αὐτὰ καὶ αὐξήσουσι καὶ
τελειώσουσι προσηκόντως. καὶ ὅτι πρὸ ἡμῶν οὐδεὶς
τοῦτ᾽ ἔπραξεν, ἀλλ᾽ οἱ πλεῖστοί γε καὶ προσδιέφθει-
ραν αὐτοῦ τὰ σπέρματα, νομίζω σαφῶς ἐπιδεδεῖχθαι
τοῖς προσεσχηκόσι τὸν νοῦν. τοῦτο μὲν οὖν τὸ γένος
τοῦ νοσήματος, εἴτε συνεχείας λύσιν, εἴθ᾽ ἑνώσεως,
εἴθ᾽ ὁπωσοῦν ἄλλως ὀνομάζειν τις ἐθέλοι, καλεῖν ἐπ-
ετρέψαμεν ὡς ἑκάστῳ φίλον. οὐ γὰρ ὑπὲρ ὀνομάτων
σπουδάζομεν οὔτε κατὰ ταύτην τὴν πραγματείαν οὔτε
καθ᾽ ἑτέραν τινὰ τῶν ἰατρικῶν, ἀλλ᾽ ὅπως ἂν μάλιστα
τοῦ τέλους τῆς τέχνης τυγχάνωμεν.

ἐπὶ δὲ τὸ πρῶτον ἁπάντων ἐπανερχώμεθα νῦν τῷ
λόγῳ νόσημα, τὸ τοῖς ὁμοιομερέσιν ἐγγινόμενον, ὧν
πρῶτόν εἰσιν αἱ κατὰ τὸ ζῷον ἐνέργειαι. δέδεικται γὰρ
ἐν ἄλλαις πραγματείαις ὅπως ἕκαστον τῶν ὀργανικῶν
μορίων ἐνεργεῖν τι λέγομεν, οἷον ὀφθαλμὸν ὁρᾶν, ἢ
βαδίζειν σκέλος. εἶναι γὰρ οὔτε ὅλου τοῦ σκέλους,
ἀλλὰ τοῦ μυώδους ἐν αὐτῷ γένους τὸ κῦρος τῆς
ἐνεργείας· οὔτε[2] ὀφθαλμοῦ τὸ[3] βλέπειν, ἀλλὰ τοῦ
460K κρυσταλλοειδοῦς· | ἐπί τε τῶν ἄλλων ἁπάντων ὀρ-
γάνων ἀνάλογον. ἐδείχθη δὲ ἐν τῷ δευτέρῳ τῆσδε τῆς
πραγματείας καὶ ὡς ἀναγκαιότατόν ἐστι τὸ γένος τοῦ

---

[2] K; οὔτ᾽ B   [3] B; τι K

240

Hippocrates, who supplied for us the seeds of all that is good, none of the others attempted to cure this class of disease by method. However, the [seeds] of Hippocrates himself need good husbandmen to sow them and increase them, and bring them to fruition in the proper way. Also, I think I have clearly shown to those who are paying attention, that nobody before me did this. Rather, the majority even destroyed his seeds. This class of disease, then, I leave to each person to call as he pleases, whether it be dissolution of continuity, or of union, or whatever else he might wish to name it. For I am not concerned about names, either in this treatise, or in any other of my medical treatises. I am concerned particularly about how best we might attain the objective of our craft.

    Let me now return to the disease which, in the discussion, is the primary disease that befalls the *homoiomeres*, the functions of which are primary in the organism. For I have shown in other treatises how we say each one of the organic parts functions—for example, how the eye sees and the leg walks. The principal component of function does not belong to the whole leg but to the muscle class in it, nor is seeing a function of the eye but of the crystalline lens,[3] and analogously in the case of all the other organs. It was also shown in the second [book] of this treatise that what is most essential is the class of this disease, and that

459K

460K

[3] The anatomical terms pertaining to the eye are somewhat confusing, particularly when we attempt to relate them to current ideas of the anatomy of the eye. Galen's concept of the structure of the eye is set out in detail in *De usu partium*; see M. T. May (1968), vol. 2, pp. 463–503 and frontispiece.

νοσήματος τούτου, καὶ ὡς οὐδεὶς αὐτὸ λογικῶς ἰάσαιτο τῶν ἀγνοησάντων τὰ πρῶτα στοιχεῖα, καὶ ὡς οὐδὲν
ὧν φθέγγονται συνίασιν ἔνιοι τῶν Λογικοὺς ἑαυτοὺς
ὀνομαζόντων, ὅταν ἀτονίαν εἶναι φάσκωσιν ἢ κοιλίας,
ἢ ἐντέρων, ἢ ἥπατος, ἢ ὀφθαλμῶν, ἢ ὁτουδηποτοῦν
μέρους. εἰ μὲν γὰρ τὴν περὶ τὸ σύμφυτον ἔργον
ἀσθένειαν οὕτως ὀνομάζουσιν, οὐδὲν ἰδιώτου πλέον
ἴσασιν· ἀκοῦσαι γάρ ἐστι κἀκείνων ἀτονεῖν ἑαυτῶν
τὴν γαστέρα φασκόντων, μὴ πέττειν γοῦν αὐτήν, μὴ
δὲ⁴ τὰ σμικρότατα καὶ κουφότατα τῶν προσφερομένων. εἰ δὲ διάθεσίν τινα λέγουσιν ἐν τῇ γαστρὶ τὴν
ἀτονίαν, ἑρμηνευσάτωσαν ἡμῖν ἥν τινά ποτε ταύτην
εἶναί φασιν, ὡς ἐπὶ φλεγμονῆς ἐποίησαν.

τῷ μὲν γὰρ Ἐμπειρικῷ τὰ συμπτώματα μόνον
ἀρκεῖ τῶν πεπονθότων μορίων ἑρμηνεῦσαι, παρὰ φύσιν
ὄγκον εἰπόντι καὶ ἀντιτυπίαν, ὀδύνην τε σφυγματώδη
461K καὶ τάσιν | ἔρευθός τε καὶ ὅσα τοιαῦτα· καὶ τίθενταί γε
πολλάκις ἓν ὄνομα κατὰ τοῦ σύμπαντος ἀθροίσματος
ἕνεκα διδασκαλίας συντόμου, καθάπερ ἐπὶ τοῦ προειρημένου φλεγμονήν. οἱ Δογματικοὶ δ᾽ οὐχ οὕτως, ἀλλ᾽
αὐτὴν τὴν οὐσίαν ἐπισκέπτονται τοῦ νοσήματος, ᾗ τὸ
προειρημένον ἄθροισμα τῶν συμπτωμάτων ἐξ ἀνάγκης ἕπεται. δοκεῖ γοῦν Ἐρασιστράτῳ τὸ παρεμπεσὸν
εἰς τὰς ἀρτηρίας αἷμα πρὸς τοῦ πνεύματος ὠθούμενον
ἐν τοῖς πέρασιν αὐτῶν σφηνωθῆναι, καὶ τοῦτο εἶναι
τὴν φλεγμονήν. ἀλλ᾽ ἦν, οἶμαι, δίκαιος ἢ Ἐρασίστρα-

____

⁴ B; μήτε K

242

nobody would be able to cure it rationally who is ignorant of the primary elements, and that some of those who call themselves Rationalists understand nothing when they say the stomach, intestines, liver, eyes or any other part whatsoever is weak. If they name the weakness pertaining to the innate action in this way, they know no more than a layman, for it is possible for you to hear the latter say the stomach is weaker than it should be when it does not digest even the smallest and lightest of the things presented to it. If, however, they say the weakness is a certain condition in the stomach, let them explain for us at some time what they are saying this condition is, as they did in the case of inflammation.

For the Empiric it is enough to explain the symptoms of the affected part alone, speaking of it as an abnormal swelling which is resistant, painful, throbbing, tense, red and whatever other such things, and they often assign one name to the whole collection for the sake of concise teaching, as in the case of what was previously mentioned—that is, inflammation. The Rationalists don't do this; instead they consider the actual substance of the disease that the previously mentioned collection of symptoms necessarily follows. So in Erasistratus' opinion, the blood which passes by *paremptosis*[4] into the arteries when it is thrust on by the *pneuma* is obstructed at their ends; and this is inflammation. But either Erasistratus himself or one of his followers

461K

[4] A concept particularly associated with Erasistratus and Asclepiades involving the passage of blood from the veins to the arteries; see section 6 on terminology in the Introduction and J. Vallance (1990), pp. 126–28.

τος αὐτὸς ἢ τῶν ἀπ᾽ αὐτοῦ τις ὁμοίως ἐξηγήσασθαι
καὶ τὴν ἐν ταῖς ἀτονίαις ἑκάστου μορίου διάθεσιν·
ὥσπερ γὰρ ἐπὶ τῆς φλεγμονῆς εἰς ὅ τι μεταπέπτωκεν
ἡ φυσικὴ κατασκευὴ τοῦ μέρους ἐδήλωσεν, οὕτως ὁ
λόγος ἐπιζητεῖ ῥηθῆναι καὶ τὴν εἰς ἀτονίαν ἑκάστου
μορίου μετάπτωσιν. οὐ γὰρ δὴ κατὰ φύσιν γε διακείμενον ἀτονεῖ περὶ τὴν οἰκείαν ἐνέργειαν, ἀλλά τι
πάντως αὐτῷ παρὰ φύσιν αἴτιον ἐγγενόμενον ἐξέλυσέ
τε καὶ κατέβαλε καὶ νεκρῷ παραπλήσιον ἀπέφηνεν· ὃ
οὔτε Ἐρασίστρατος οὔθ᾽ Ἡρόφιλος οὔτ᾽ ἄλλος οὐδεὶς
462K   ἰατρὸς εἶπε | τῶν μὴ τολμησάντων ἀποφήνασθαί τι
περὶ τῆς τῶν πρώτων σωμάτων φύσεως.

3. Ἀλλ᾽ οὐ χρὴ μηκύνειν ἐπὶ πλέον, αὐτάρκως
προαποδεδειχότας ἔμπροσθεν ὡς χρησιμώτατόν τε
τοῦτ᾽ ἐστὶ τὸ γένος τοῦ νοσήματος ἐζητῆσθαι τοῖς
ἰατροῖς καὶ ἀναγκαῖον τὸ περὶ στοιχείων πρότερον
ἐπεσκέφθαι τὸν μέλλοντα καλῶς αὐτὸ μεταχειρίζεσθαι. ὥσθ᾽ ἡμῖν ὁ μέλλων νῦν λεχθήσεσθαι λόγος ἐπὶ
τοῖς προαποδεδειγμένοις περανθήσεται στοιχείοις,
ὑπὲρ ὧν ἕν, ὡς οἶσθα, βιβλίον ἐποιησάμεθα τὸ Περὶ
τῶν καθ᾽ Ἱπποκράτη στοιχείων ἐπιγεγραμμένον. ἔστι
μὲν οὖν καὶ Διοκλεῖ καὶ Μνησιθέῳ καὶ Διευχεῖ καὶ
Ἀθηναίῳ καὶ σχεδὸν πᾶσι τοῖς εὐδοκιμωτάτοις ἰατροῖς, ὥσπερ οὖν καὶ τῶν φιλοσόφων τοῖς ἀρίστοις, ἡ
αὐτὴ δόξα περὶ φύσεως σώματος, ἐκ θερμοῦ καὶ

was, I think, right to explain similarly the condition in the weaknesses of each part for, just as in the case of inflammation the natural state of the part shows into what it has changed, in the same way the argument seeks to explain the change of each part to weakness. What is in a state of accord with nature is certainly not weak in terms of its proper function. Rather, this only happens when some cause totally contrary to nature, having acted on it, has dissipated and put an end to it, has cast it down and made it seem as if dead, which is something neither Erasistratus nor Herophilus said, and neither did any other doctor, among those not bold enough to make a pronouncement about the nature of the primary bodies.

462K

3. But I must not delay any longer since I have previously demonstrated quite adequately that this is a very useful class of disease for doctors to investigate, and that it is essential for someone who intends to manage this class [of disease] properly to have given prior consideration to the elements. As a result, the argument which I am now going to articulate will be realized by means of the previously demonstrated elements, about which I produced one book, as you know, entitled *On the Elements according to Hippocrates*.[5] This is the very same opinion about the nature of the body that was held by Diocles, Mnesitheus, Dieuches, Athenaeus and almost all the doctors of the highest reputation, just as it was also held by the best of the philosophers who thought that all other bodies, no less than the bodies of living creatures in particular, are a mix-

[5] A foundational work for Galen's medical theories, often referred to in the present work. See I.413–508K and the translation by P. H. de Lacy (1996).

ψυχροῦ καὶ ξηροῦ καὶ ὑγροῦ νομίζουσι κεκρᾶσθαι τά
τε ἄλλα σύμπαντα σώματα καὶ τὰ τῶν ζῴων οὐχ
ἥκιστα. τῷ δὲ περὶ τούτων ἁπάντων ἀποφηναμένῳ τε
καὶ ἀποδείξαντι πρώτῳ δίκαιον, οἶμαι, μαρτυρεῖν ἐστι
τὴν εὕρεσιν. καὶ διὰ τοῦθ᾽ ἡμεῖς ὀνομάζομεν αὐτὰ καθ᾽
463K Ἱπποκράτην στοιχεῖα, | κἂν ὅτι μάλιστα Χρύσιππος,
ἢ Ἀριστοτέλης, ἤ τις ἄλλος ἰατρὸς ἢ φιλόσοφος
ὡσαύτως ὑπὲρ αὐτῶν δοξάζῃ.

καὶ τοίνυν ἐπειδὴ τῶν τοῦ ζῴου μορίων ἕκαστον
ἰδίαν ἐνέργειαν ἐνεργεῖ, τῶν ἄλλων ἐνεργειῶν εἰς τοσοῦτον διαφέρουσαν εἰς ὅσον καὶ αὐτὸ διαφέρει τῶν
ἐνεργούντων αὐτάς, διαφέρει δὲ τῷ θερμότερον, ἢ
ψυχρότερον, ἢ ὑγρότερον, ἢ ξηρότερον ὑπάρχειν, ἢ
κατὰ συζυγίαν τι τούτων πεπονθέναι, τὴν κρᾶσιν
αὐτῶν φυλακτέον ἐστὶ τῷ τὴν ἐνέργειαν φυλάττοντι.
φυλαχθήσεται δὲ ψυχόντων μέν, εἰ πρὸς τὸ θερμότερον ἐκτρέποιτο· θερμαινόντων δ᾽ πρὸς τὸ[5] ψυχρότερον, ὑγραινόντων δέ, εἰ πρὸς τὸ ξηρότερον· οὕτω δὲ
καὶ ξηραινόντων,[6] εἰ ὑγραίνοιτο· ξηραινόντων δ᾽ ἅμα
καὶ θερμαινόντων, εἰ ὑγραίνοιτό τε ἅμα καὶ ψύχοιτο,
ξηραινόντων δὲ καὶ ψυχόντων, εἰ πρὸς τὸ ὑγρότερόν τε
καὶ θερμότερον ἐκτρέποιτο, καὶ κατὰ τὰς λοιπὰς δύο
συζυγίας ἀνάλογον. ἀεὶ γὰρ χρὴ τῷ πλεονάζοντι τὸ
ἐναντίον ἀντεισάγειν εἰς τοσοῦτον, ἄχρις ἂν εἰς τὸ
σύμμετρόν τε καὶ κατὰ φύσιν ἀγάγῃς τὸ μόριον.

ἡ μὲν δὴ καθόλου τοῦ γένους ἅπαντος νοσήματος |
464K ἐν ὁμοιομερέσι συνισταμένου μέθοδος ἰάσεως ἤδη μοι
λέλεκται. ὀκτὼ γὰρ ὄντων αὐτῶν, ὡς ἐν τῷ Περὶ τῆς

ture of hot, cold, dry and moist. I think it is right to give credit to the man who first declared and demonstrated the discovery of all these things, and because of this, I name these "elements" following Hippocrates, even if Chrysippus particularly, or Aristotle, or some other doctor or philosopher conceived of them in the same way. 463K

Therefore, when each of the parts of the organism performs a specific function, which differs from other functions to the extent that the part itself differs from those parts that perform those functions—that is, differs by being hotter, colder, moister, or drier, or by being affected in terms of a conjunction of these [qualities]—you must preserve their function by preserving their *krasis*. And [the *krasis*] will be preserved when the parts that are cooled are changed to being hotter; [and similarly] when they are heated, if they are changed to being cooler, when moistened, if they are changed to being drier and when dried, if they are made moist. When they are dried and heated [the *krasis* will be preserved] if they are moistened and cooled at the same time; [similarly] when they are dried and cooled, if they are changed to become moister and hotter, and analogously in respect of the remaining two conjunctions. You must always introduce the opposite to what is in excess to the point where you bring the part to a balance and accord with nature.

I have now stated in general terms the method of cure 464K of every class of disease existing in *homoiomeres*. Since there are eight of these [diseases], as has been demon-

---

5 B (cf. KLat); θερμότερον . . . τό om. K

6 B; post ξηραινόντων add. μέν K

τῶν νοσημάτων ἀποδέδεικται[7] διαφορᾶς, ὀκτὼ καὶ οἱ
τῆς ἰάσεως ἔσονται τρόποι, κοινὸν ἔχοντες σκοπὸν
τὴν ἀλλοίωσιν τοῦ πεπονθότος ὁμοιομεροῦς σώματος·
ἐπειδὴ καὶ τὸ νόσημα αὐτὸ κατὰ δυσκρασίαν καὶ
ἀλλοίωσιν ἐγένετο τῆς κατὰ φύσιν ἑκάστου κράσεως·
ἡ δὲ τῶν κατὰ μέρος ἴασις ἐν δυοῖν τούτοιν προερχο-
μένη γίνεται, τῇ τε τῆς ὕλης εὐπορίᾳ καὶ τῇ ταύτης
ἐπιδεξίᾳ χρήσει. τὴν μὲν δὴ τῶν φαρμάκων εὐπορίαν
ἔκ τε τῆς περὶ τῶν ἁπλῶν φαρμάκων δυνάμεως, ἔτι τε
τῆς περὶ συνθέσεως αὐτῶν πραγματείας ἔξεστί σοι
λαμβάνειν μεθόδῳ· τὴν δὲ ὁδὸν τῶν διαιτημάτων ἐντεῦθεν. ἀλλὰ καὶ ὅπως χρὴ διαγινώσκειν ἑκάστου
μορίου δυσκρασίαν, ἐκ τριῶν πραγματειῶν ἀναμιμνήσκου, πρώτης μὲν τῆς Περὶ κράσεων, δευτέρας δὲ τῆς
Περὶ τῶν πεπονθότων μορίων· καὶ τρίτης, ἣν ἐπι-
γράφομεν Ἰατρικὴν τέχνην. οὔκουν ἔτι δεῖ πολλῶν εἰς
τὰ νῦν ἐνεστῶτα τοῖς φύσει τε συνετοῖς καὶ γε-
465K   γυμνασμένοις τὸν | λογισμὸν ἐν τοῖς πρώτοις· ὅτῳ δὲ
οὐδέτερον ὑπάρχει τούτων, ἔτι ἐνδεῖ τῷ λόγῳ πάμ-
πολυ. εἰ δὲ καὶ μοχθηρᾷ λόγων αἱρέσει συνανετράφη,
διττή γ᾽ οὕτω χρεία τοῦ χρόνου· ἑτέρου μέν, ἵνα
ἀποτρίψηται τὰς μοχθηρὰς δόξας, ἑτέρου δ᾽, ἵν᾽ ἀσκη-
θῇ κατὰ τὰς βελτίους.

4. Σὺ μὲν οὖν εὖ οἶδ᾽ ὅτι καὶ αὐτὸς ἱκανὸς ὑπάρχεις
ἐκ τῶν καθόλου τοῦ γένους εἰρημένων εὑρίσκειν τὰ
κατὰ μέρος· ἀλλὰ καὶ ἡμᾶς ἐθεάσω πολλὰ τῶν τοιούτων νοσημάτων ἰωμένους, ὥστ᾽ ἐξ ἀμφοτέρων ἔχεις

---

[7] K; ἐπιδέδεικται B, recte fort.

strated in the work *On the Differentiae of Diseases*,[6] the types of cure will also be eight, having as their common indicator the change of the affected *homoiomerous* body, and since the disease itself arose in terms of *dyskrasia* and change, the change of the *krasis* of each to an accord with nature. However, the cure of these in the individual case is effected through these two things: the ease of providing material and its proper use. Now it is possible for you to take by method the abundance of the medications from the treatise on the potency of the simple medications, and further, from the treatise on their compounding, and the path of the regimens from here.[7] But also, how you must recognize a *dyskrasia* of each part you will recall from three treatises; the first *On Krasias* (*Mixtures*), the second *On the Affected Parts* and the third, which I entitled *The Medical Art*.[8] There is not much still required in regard to the matters now arising for those who are intelligent by nature, and have trained their reasoning in the fundamentals. For someone who is neither of these things, there is still a great deal wanting in the discussion. If he has also been reared by a sect wretched in their doctrines, there is, on this account at least, a twofold need of time; on the one hand so that he can rid himself of the wretched doctrines, and on the other, so that he may be trained in the better doctrines.

4. I know very well that you yourself are quite able to discover the particular things from what was said regarding the class in general. But you also saw me curing many such diseases, so that on both counts you have no need of

465K

---

[6] See *De differentiis morborum*, 5.3–6 (VI.848–55K), and I. Johnston (2006), pp. 142–44.   [7] For the three works on medications, see Book 3, note 14.   [8] *Ars Medica*, I.305–412K.

τὸ μὴ δεῖσθαι τῶν κατὰ μέρος. ἀλλ' ἐπεί φησιν ὁ
Πλάτων, οὐ γάρ ἐστι τὰ γραφέντα μὴ ἐκπεσεῖν, ἵν'
εἴποτε καὶ εἰς ἄλλον ἀφίκοιτο τὸ βιβλίον ἀγύμναστον
τῷ λογισμῷ, ῥᾷον αὐτὸν διδάξειε, προσθεῖναι χρή
τινα τῶν κατὰ μέρος· ἐξ ὧν εὐθέως ἐνέσται καὶ αὐτὸ
τοῦτο πεισθῆναι σαφῶς, ὃ μικρὸν ἔμπροσθεν εἴρηται,
ὡς δυνατόν ἐστι σαυτῷ τὰ κατὰ μέρος ἐξευρίσκειν.
ἐγὼ γοῦν ἐμαυτῷ πάντα ἐξεῦρον αὐτά, τῷ λόγῳ ποδη-
γούμενος. ὅσον μὲν γὰρ ἐπὶ τοῖς διδασκάλοις, ἐχρῆν
δήπου κἀμὲ τοῖς ἀτόνοις τὴν γαστέρα συμβουλεύειν
466K ἐδέσματα μὲν | τὰ στύφοντα καὶ ὑπόπικρα· καὶ οἶνον
ὡσαύτως τὸν αὐστηρόν· ἀψίνθιόν τε καὶ τοῦ διὰ μήλων κυδωνίων χυλοῦ καὶ ὅσα τοιαῦτα καταπίνεται
φάρμακα. τῶν δ' ἔξωθεν ἐπιτιθεμένων πρώτην μὲν τὴν
δι' ἀψινθίου καὶ ὠμοτριβοῦς ἐλαίου κατάντλησιν· εἶτ'
ἐπίδεσιν ἐρίου πιλήματος, ἐξ αὐτῶν τε τούτων καὶ
προσέτι μύρου μηλίνου καὶ μαστιχίνου καὶ ναρδίνου·
καὶ μετὰ ταῦτα κηρωτὴν διὰ τῶν αὐτῶν ἐσκευασμέ-
νην· εἶτ' ἄλλα φάρμακα κηρωτῶν ἰσχυρότερα, τὰ
πρὸς τῶν ἰατρῶν ἐπιθέματα καλούμενα, διά τε τῶν
εἰρημένων μύρων συγκείμενα καὶ φαρμάκων τῶν
παραπλησίων, ἐν οἷς ἐστιν ἤδη καὶ ἀρωμάτων πλῆθος
οὐκ ὀλίγον, στάχυς νάρδου καὶ ἄμωμον καὶ ὁ ἀρω-
ματικὸς κάλαμος, ἶρίς τε καὶ λάδανον καὶ τὸ τοῦ
μαλαβάθρου φύλλον καὶ στύραξ καὶ βδέλλιον, ὀπο-
βάλσαμόν τε καὶ βάλσαμον καὶ ξυλοβάλσαμον, ὅ τε
λοιπὸς τῶν ἀρωμάτων κατάλογος. εἰ δὲ μηδὲ ταῦτα
μηδὲν ἐνεργοίη, τὸ κοινὸν ἁπάντων τῶν ἀτονούντων

the particulars. But since, as Plato says, it is the case that things written slip away,[9] so that my book, should it ever come into the hands of someone else unpracticed in reasoning, may instruct him the more easily, I must put forward some of the particulars. From these, it will immediately be possible for the very thing stated a little earlier to be clearly believed: that you can discover the particulars for yourself. At any rate, I discovered all these things for myself, being guided by reason. To the extent that I was dependent on my teachers, I suppose I too ought to have recommended for the weaknesses in the stomach, foods that are astringent and somewhat bitter, and likewise, wine   466K that is bitter, and wormwood, and what comes from the juice of quinces and other such medications that are swallowed. Of those things applied externally, make the first perfusion with absinth and omotribes, then [apply] a binding of compressed wool and, apart from those things, also oil of sweet apples, mastich and spikenard, and after these, a salve prepared through those same things; and then other medications stronger than salves, those called "epithemata" by doctors, compounded from the previously mentioned sweet oil and similar medications in which there is already no small abundance of aromatics—spikenard, Nepalese cardamom, aromatic reed, iris, gum ladanum, the leaf of cinnamonum, storax, bdellium, the juice of the balsam tree, balsam, balsam wood and the remaining catalogue of aromatics. If none of these things works, in the end I make up the common remedy of all those

9 See Plato, *Epistles* I, 314c1—not verbatim.

GALEN

ἐπὶ τέλει βοήθημα ποιοῦμεν, καλοῦσι δ' αὐτὸ φοι-
νιγμόν, ἤτοι διὰ θαψίας γιγνόμενον, ἢ διὰ νάπυος, ἤ
467K τινος τῶν τοιούτων· | εἶτ' ἀποπέμψαι πρὸς ὑδάτων
χρῆσιν αὐτοφυῶν.

ουδὲν γὰρ τούτων οἶδε πλέον ὁ Ἐμπειρικός, ὡς καὶ
τὰ συγγράμματα αὐτῶν δηλοῦσι. Κόϊντος μέν γε τοῖς
ἀπεπτεῖν ἢ ἀνορεκτεῖν φάσκουσι πρῶτον μὲν γυμνάζεσθαι συνεβούλευε καὶ ἐσθίειν ὡς εὐπεπτότατά τε
καὶ μὴ πολλά, μηδὲν δὲ ὠφελουμένων ἠναγκάζετο καὶ
αὐτὸς εἰς τὰ τῶν Ἐμπειρικῶν μετιέναι. τίς γὰρ οὐκ
οἶδεν ὡς τὸ θερμὸν καὶ τὸ ψυχρὸν καὶ ξηρὸν καὶ ὑγρὸν
εἰώθει σκώπτειν, βαλανείων ὀνόματα προσαγορεύων,
ὧν χωρὶς ἀδύνατόν ἐστι θεραπεῦσαι μεθόδῳ τὰς ἀτο-
νίας τῶν μορίων; ὥσθ' ὅσον μὲν ἐπὶ τούτοις τὴν αὐτὴν
ὁδὸν ἐβάδιζον ἂν κἀγὼ κατὰ τὴν τῶν τοιούτων διαθέσεων ἴασιν, ἀλλ' ὁ λόγος ἐδίδαξέ με τὰς ὀκτὼ
διαφορὰς τῆς θεραπείας ἀτόνου γαστρός. ἐθεάσω
γοῦν καὶ σύ τινας μὲν ἡμέρᾳ μιᾷ, μᾶλλον δὲ ὥρᾳ,
ψυχροῦ πόσει θεραπευθέντας· ὧν ἐνίοις μὲν οὐ μόνον
τὸ πρόσφατον ἔδωκα πηγαῖον, ἀλλὰ καὶ τὸ διὰ χιόνος
ἐψυγμένον, ὡς ἐν Ῥώμῃ σκευάζειν ἔθος ἔχουσι, προ-
θερμαίνοντες τὴν κατασκευὴν ἣν αὐτοὶ προσαγορεύ-
468K ουσι δηκόκταν· | ἐδέσματά τε τὰ οὕτως ἐψυγμένα πολ-
λάκις ἐθεάσω συγχωροῦντά με λαμβάνειν αὐτοῖς· ἐν
οἷς ἐστι καὶ ἡ μέλκα, τῶν ἐν Ῥώμῃ καὶ τοῦτο ἐν
εὐδοκιμούντων ἐδεσμάτων, ὥσπερ καὶ τὸ ἀφρόγαλα.

10 Made from the date palm; see De simplicium medicamentorum temperamentis et facultatibus, XII.151–52K.

252

[agents] that are weak which they call "phoenix,"[10] or that made with thapsia, or mustard, or another of such things. I   467K then send [the patient] away to use the natural waters.

The Empiric knows nothing more of these things, as their writings show. In fact, for those who say they do not digest or have no appetite, Quintus[11] first recommended exercise and eating those things that are very easily digested (but not much of these), and if these things didn't help, he too was forced to go on to the recommendations of the Empirics. Who does not know that [the Empiric doctor] is accustomed to mock the hot, cold, dry and moist, saying they are just the names of baths? But without these [elemental qualities] it is impossible to treat the weaknesses of the parts by method. As a result, even I, by using these things, would be treading the same path in relation to the cure of such conditions, except that reason taught me the eight differentiae of treatment of a weak stomach. At all events, you also saw some [patients] treated in one day, or rather in an hour, by a drink of cold water. To some of them, I gave not only fresh springwater but also water cooled by snow, as they customarily prepare in Rome when they previously warm the preparation which they themselves call a "decoction." You often saw me allowing them   468K to take foods that had been cooled like this, among which was *melka* (sour milk), this being one of the foods held in high regard in Rome, as is *aphrogala* (frothed milk) also.

11 Quintus (second century AD) was a student of Marinus who practiced in Rome under Hadrian. Elsewhere he is praised by Galen as the best doctor of his time (*Prognosis I*); see also *EANS*, p. 717.

τοῖς δ' αὐτοῖς τούτοις καὶ τὰς ψυχρὰς κατὰ δύναμιν
ὀπώρας ὁμοίως ἀποψύχων ἐδίδουν· καὶ πτισάνην καλῶς ὡσαύτως ἐψυγμένην, ἕτερά τε τοιαῦτα μυρία,
σκοπὸν ἕνα ποιούμενος ἐπ' αὐτῶν τὴν ψύξιν· ἐκώλυον
δ' ἀψινθίου καὶ τῶν στυφόντων ἅψασθαι· καθάπερ γε
καὶ ἄλλους ὁμοίως μὲν ἀπῆγον τῶν ψυχόντων, ἐθέρ-
μαινον δὲ παντοίως οἶνον παλαιὸν τῶν ἱκανῶς θερμῶν
τῇ δυνάμει διδούς, οἷοι μάλιστά εἰσι Φαλερῖνός τε καὶ
Σουρρεντῖνος, καὶ τροφὰς θερμαινούσας μετὰ πε-
πέρεως συχνοῦ. ἐπί τινων δ', ὡς οἶσθα, τὸν σκοπὸν
τῆς θεραπείας ἐποιησάμην ἐν τῷ ξηραίνειν· καὶ ἦν
αὐτοῖς ἐδέσματά τε τὰ φύσει ξηρά, καλῶς ὠπτημένα
καὶ τὸ σύμπαν ὀλίγιστον ποτόν· ἥ τε τῶν στυφόντων
ἁπάντων χρῆσις, ἣν μόνην γινώσκουσιν οἱ χωρὶς
λόγου θεραπεύοντες αὐτούς. ἄλλον δ' οὐ πρὸ πολλοῦ
ξηρότατον ἤδη γεγεννημένων, ὡς ὁμοιότατον εἶναι
469K   τὴν ἰδέαν | τοῖς μαρασμώδεσιν, ἰασάμην, εἰς τὰ ἐναν-
τία πάντα μεταγαγὼν ἢ ὡς οἱ θαυμασιώτατοι ἰατροὶ
συνεβούλευον. οὐδὲ γὰρ οὐδὲ ἐξ ἄλλου τινὸς εἰς τοῦτο
ἧκε κινδύνου, βραχεῖαν τότε κατ' ἀρχὰς ἔχων δυσκρασίαν ἐπὶ ξηρότητι δεομένην ὑγράνσεως· ἀλλ' οἱ
παραλαβόντες αὐτόν, ἀψίνθιόν τε ποτίζοντες καὶ πι-
κροὺς ἀσπαράγους καὶ βολβοὺς ἐσθίειν διδόντες, ἔτι
τε μῆλα κυδώνια καὶ κεστιανὰ καὶ ῥοιάς, ὕστερόν τε,
ὡς οὐδενὸς τούτων ἡ γαστὴρ ἐκράτει καὶ ῥοῦ χυλὸν
ἀναγκάζοντες πιεῖν, ὅσα τε ὀλίγον ἔμπροσθεν εἶπον
ἐπιτιθέντες ἔξωθεν ὀλίγου δεῖν ἀπεφήναντο ἀλίβαντα,
τοῦτον ἡμεῖς ἰασάμεθα, παντοίως ὑγραίνοντες αὐτοὶ

To these same people I also gave fruits that are cool in terms of potency, cooling them in the same way, and ptisan that has been well cooled similarly, and countless other such things, the one objective being to produce cooling. I prevented them from taking absinth or anything astringent. Just as in fact I led others away from cooling agents, I used to heat in all kinds of ways, giving old wines of those sufficiently hot in potency, which are particularly the Falernian and Sorrentine, and also giving heating foods with a large amount of pepper. In some cases, as you know, I placed the goal of therapy in drying, and the foods for them were dry in nature, baked well, and [they had] altogether very little to drink. There was also the use of all the astringents, which is the only thing those without logic know about when treating these cases. Not long ago I cured another [patient], who had already become so dry as to be very similar in appearance to those with marasmus, by changing everything to the opposite of what those most wondrous doctors were advising. It was not that he came to this point from some other danger since there was, at the beginning, a slight *dyskrasia* requiring moistening due to the dryness, but because those who took on his care gave absinth as a drink and bitter asparagus and bulbs[12] to eat, and, in addition, quinces, apples and pomegranates. Later, as the stomach prevailed over none of these things, they also compelled him to drink the juice of pomegranate and, when they were applying externally those things that I mentioned a little earlier, they pronounced him almost dead. I cured this man, making him moist in various ways,

469K

---

12 This is taken to refer to a mixture of bulbs here; see Dioscorides, II.201–2, and Galen, XI.851K.

τὴν ὕλην ἐξευρίσκοντες, ἐκ τῆς γεγραμμένης ἡμῶν μεθόδου κατὰ τὴν περὶ τῶν φαρμάκων πραγματείαν.

5. Ὡς γὰρ κἂν τῷ Περὶ τῆς ἀποδεικτικῆς εὑρέσεως εἴρηται γράμματι, περιαντληθεὶς ὑπὸ τοῦ πλήθους τῆς τῶν ἰατρῶν διαφωνίας, εἶτ᾽ ἐπὶ τὸ κρίνειν αὐτὴν τραπόμενος, ἔγνων χρῆναι πρότερον ἐν ἀποδεικτικαῖς μεθόδοις γυμνάσασθαι. καὶ τοῦτο πράξας ἔτεσιν ἐφε-
470K ξῆς πολλοῖς ὑπέβαλλον | οὕτως ἕκαστον τῶν δογμά- των αὐτῇ καὶ ὡς ἡ τῶν εὑρεθέντων ἔνδειξις ἐποδήγει με, τὰς θεραπείας ἐποιούμην. ἀλλὰ γὰρ οὐ πάντες γε τοιοῦτοι, χρὴ τοίνυν γράφειν αὐτοῖς γυμνάσια τῆς ἐν τοῖς κατὰ μέρος εὐπορίας τῶν ὑλῶν, ἐπὶ τοῖς ὀκτὼ σκοποῖς τῆς θεραπείας, ἀρξαμένους αὖθις ἀπὸ τῆς κοιλίας, ἐπειδὴ πρώτης ταύτης ἔτυχον ἄρτι μνησθείς. εὐιατότατοι μὲν οὖν εἰσιν αἱ κατὰ θερμότητα καὶ ψύξιν ἀλλοιώσεις, ὅτι ταῖς δραστικωτάταις ἐπανορθοῦνται ποιότησι· δυσιατότεραι δὲ αἱ καθ᾽ ὑγρότητα καὶ ξηρό- τητα· ταῖς γὰρ ἀσθενέσι καὶ ὡς ἂν εἴποι τις ὑλι- κωτέραις ποιότησιν ἡ ἴασις αὐτῶν ἐπιτελεῖται, καὶ μάλισθ᾽ ὅταν ὑγραίνειν δέῃ. τὰ μὲν δὴ τοῦ χρόνου τῆς ἐπανορθώσεως ἴσα πώς ἐστι θερμότητί τε καὶ ψυχρό- τητι, τὸ δὲ τῆς ἀσφαλείας οὐκ ἴσον. εἰ μὴ γὰρ ἰσχυρὰ πάντα εἴη τὰ πέριξ μόρια τοῦ θεραπευομένου, κίνδυ- νος αὐτοῖς ὑπὸ τῶν ψυχόντων οὐ σμικρὰν πληγῆναι πληγήν. ἐπὶ δὲ τῶν ὑπολοίπων δυοῖν ποιοτήτων ἡ μὲν ἀσφάλεια παραπλήσιος, ὁ δὲ χρόνος τῆς θεραπείας πολλαπλάσιος ἐπὶ ταῖς ξηραῖς δυσκρασίαις· οἷον γὰρ

finding other relevant material from the method written in the treatises on medications.[13]

5. As I stated in the tract *On Demonstrative Discovery*,[14] when I was completely submerged beneath the discord of the majority of doctors, I then turned to decide in favor of this, knowing I must first become practiced in demonstrative methods. I did this over many years in succession and, in like manner, subjected each of the doctrines to this. As the indication of what was discovered kept leading me on, I fashioned the treatments. But in fact not everyone is like this, so it is necessary to write accounts for them of the advantages of the individual materials in the eight objectives of treatment, beginning again from the stomach, since I happened just now to mention this first. Thus, the changes relating to hot and cold are very easy to cure in that they are restored by the most active qualities. The changes relating to moistness and dryness are more difficult because the cure of these is brought about in those who are weak and, one might say, by more material qualities, especially when there is a need to moisten. Now those [qualities] are somewhat equal to hot and cold in terms of the time of the restoration, but in terms of safety they are not equal because, if all the parts surrounding the part being treated are not strong, there is a danger to them of a significant impact from the cooling agents. In the case of the remaining two qualities, the safety is similar, but the time of treatment for dry *dyskrasias* is very much longer.

470K

---

[13] This is taken to be a general reference to the three pharmacological treatises previously listed in Book 3, note 14.

[14] A lost work; see *On My Own Books*, XIX.44K.

471K τι τὸ γῆράς ἐστιν ἐπὶ τῶν ὑγιαινόντων, | τοιοῦτον ἡ
ξηρὰ δυσκρασία τοῖς νοσοῦσιν, ὥστε καὶ ἀνίατον
ὑπάρχειν, ὅταν ἀκριβῶς συμπληρωθῇ. τὸ δ' ἀκριβὲς
τῆς συμπληρώσεώς ἐστιν ἐν τῷ τὰ τῆς στερεᾶς οὐσίας τῶν ὁμοιομερῶν σωμάτων γεγονέναι ξηρότερα.

6. Ἑτέρα γάρ ἐστιν ἡ ξηρότης τῶν ἐκ τῆς ὑγρο-
παγοῦς οὐσίας συνεστώτων, ὁποῖόν ἐστι πιμελὴ καὶ
σάρξ, ἐκτακέντων. καὶ τρίτη γε πρὸς ταύταις τῆς
οἰκείας ὑγρότητος, ἐξ ἧς τρέφεται τὰ μόρια, τελέως
ἀπολλυμένης. περιέχεται δ' αὐτὴ κατὰ πάντα τοῦ
ζῴου τὰ μόρια δροσοειδῶς ἐν αὐτοῖς παρεσπαρμένη,
καὶ λέλεκται πολλάκις ἤδη περὶ αὐτῆς ἑτέρωθι· ταύτην οὖν ἐνθεῖναι τοῖς μορίοις οὐχ οἷόν τε χωρὶς
τροφῆς· καὶ διὰ τοῦτο χαλεπωτάτη τῶν τοιούτων ἐστὶ
διαθέσεων ἡ ἴασις· ἄλλη δὲ ξηρότης ἐστὶν ἡ κατὰ τὰς
ἀρτηρίας καὶ φλέβας τὰς ἰδίας ἑκάστου τῶν μορίων
συνισταμένη. γίγνεται δὲ καὶ αὕτη δηλονότι κατὰ τὴν
τοῦ αἵματος ἔνδειαν. ἁπάσας ταύτας τὰς[8] ξηρότητας
ἐσχάτως βλάπτουσιν αἱ τῶν αὐστηρῶν ἐδεσμάτων
καὶ πομάτων καὶ φαρμάκων προσφοραί· ἐκδαπανῶσι
γὰρ εἰ καί τι λείποιτο τῆς ἐν αὐτοῖς ὑγρότητος ἐμ-
472K φύτου, | τὸ μὲν ἐκπίνουσαι, τὸ δὲ ἐκθλίβουσαι διὰ τῶν
πόρων εἰς τὴν ἐντὸς εὐρύτητα τῆς κοιλίας, τὸ δ' ἐπὶ
τὰ συνεχῆ μόρια διωθούμεναι. θεραπεύειν οὖν αὐτὰς
προσήκει, τὸ μὲν συνιζηκὸς τῶν πόρων ἀναπεταν-
νύντα, τὸ δ' εἰς τὰ παρακείμενα μόρια διωσθὲν ἐπι-
σπώμενον, ὑγραινούσῃ τε τροφῇ πληροῦντα τῆς οἰκείας ὑγρότητος ἕκαστον τῶν ὁμοιομερῶν, ὥσπερ καὶ

258

Age in those who are healthy is just like a dry *dyskrasia* in   471K
those who are sick, in that it too is incurable whenever it is
completely "consummated." And the strict sense of "con-
summation" lies in the greater dryness existing in the solid
substance of the *homoiomerous* bodies.

6. There is another dryness, when structures formed
from a watery substance like fat and flesh waste away. In
fact, there is also a third dryness in addition to these, when
the proper moisture from which the parts are nourished is
completely lost. This moisture is contained in all the parts
of the animal, dispersed in a dewlike fashion. I have already spoken about this moisture often in other places. It is
not possible to put this into the parts except through nourishment. Because of this, the cure of such conditions is
very difficult. Of another sort, however, is the dryness that
exists specifically in the arteries and veins of each of the
parts. This clearly arises in relation to lack of blood. Administration of astringent foods, drinks and medications
harm all these drynesses to an extreme degree, for they ex-
haust the natural moisture of the parts if any remains,
partly absorbing it, partly squeezing it out through the   472K
channels into the lumen of the stomach, and partly push-
ing it away to the contiguous parts. Therefore, it is appropriate to treat these drynesses in part by opening up collapsed channels, in part by drawing back what has been
pushed away to the contiguous parts, and in part with
moistening nourishment filling each of the *homoiomeres*

---

8 B; τάς *om.* K

ἡμεῖς τὸν ἐξηρασμένον ὑπὸ τῶν ἰατρῶν ἐθεραπεύσαμεν ἄνθρωπον.

ἐν μὲν γὰρ τῇ κατὰ τὸ θερμὸν καὶ ψυχρὸν ἀντιθέσει μηδεμίαν ἐπικρατοῦσαν ἐναργῶς ἔχοντα δυσκρασίαν μήτε καθ᾽ ὅλον τὸ σῶμα μήτε κατὰ τὴν γαστέρα, ξηρὸν δὲ καὶ λεπτὸν ἱκανῶς γεγενημένον, ἐκ τοῦ μὴ πέττειν καλῶς ἀτονούσης τῆς κοιλίας ἐπὶ τῇ κατὰ ξηρότητα δυσκρασίᾳ. τούτου γὰρ ὁ μὲν σκοπὸς τῆς θεραπείας ἦν ὑγρᾶναι τήν τε γαστέρα καὶ σύμπαν τὸ σῶμα. τίσι δ᾽ ἐνεργείαις ἐχρησάμεθα ταῖς κατὰ μέρος, ἐπὶ τίσι τε μάλισθ᾽ ὕλαις ὑπὲρ τοῦ τυχεῖν τοῦ σκοποῦ διελθεῖν ἄμεινον. οἴκημα μὲν αὐτῷ παρεσκευασάμην ἐγγυτάτω τοῦ βαλανείου· τοιαῦτα δ᾽ ἐστίν, ὡς οἶσθα, πολλὰ κατὰ τὰς τῶν πλουσίων οἰκίας· ἐξ |

473K αὐτοῦ δ᾽ εὐθέως ἕωθεν εἰς τὸ βαλανεῖον εἰσεφερόμην ἐπὶ σινδόνων, ὅπως μὴ καταξηραίνοιτό τε δι᾽ ἑαυτοῦ κινούμενος ἐκλύοιτό τε πρὸ τοῦ δέοντος καιροῦ· τούτων γὰρ τὸ μὲν εἰς αὐτὴν τὴν διάθεσιν συντελεῖ, τὸ δὲ συντέμνει τὴν ἐν τῷ βαλανείῳ διατριβήν. ἐπὶ πλεῖστον γὰρ χρὴ τὸν ἄνθρωπον ἐνδιατρίβειν τῷ ὕδατι· καὶ διὰ τοῦτο καὶ αἱ κολυμβῆθραι βελτίους εἰσὶ τῶν μικρῶν πυέλων καὶ μάλισθ᾽ ὅσαι πλησίον ὑπάρχουσι τῆς ἔξωθεν θύρας, ὡς μηδὲ διὰ μακροῦ τὴν ἐκ τοῦ ὕδατος αὐτῷ πρὸς τοὐκτὸς εἶναι φοράν· οὐ γὰρ δὴ τοῦ ἀέρος ἐπὶ τῶν οὕτως ἐχόντων χρῄζομεν. ἔστω δ᾽ ἀκριβῶς εὔκρατον τὸ ὕδωρ· τὸ μὲν γὰρ ψυχρότερον λεληθυῖαν ψύξιν ἐναπεργάζεται τοῖς ἀσθενοῦσι σώμασι, τὸ δὲ θερμότερον συνάγει καὶ σφίγγει τοὺς πόρους

full of the proper moisture. I myself treated a man who had been dried by his doctors in just such a way.

He clearly had no prevailing *dyskrasia* involving the hot and cold opposition, either in the whole body or in the stomach, but he had become excessively dry and thin from the failure of the stomach, weakened by the dry *dyskrasia*, to digest properly. The aim of treatment for this man was to moisten the stomach and the whole body. It is better to go through the measures I used one by one, and particularly the materials used to attain the objective. I prepared a room for him very near the bathhouse—there are, as you know, many such rooms in the houses of the rich. Right away, early in the morning, I brought him from the room to the bathhouse on a linen sheet, so that he might not dry out through his own movements and relax before the required time; for the former contributes to the condition itself, while the latter cuts short the time spent in the bathhouse. It is necessary for the person to spend as long as possible in the water. Because of this, the swimming baths are better than the small bathing tubs, and particularly those that are near the external door, so that for him the movement outside from the water is not long; for in no way do we desire [contact with] the air for those in such a state. The water must be exactly *eukratic*; water which is too cold brings about an imperceptible cooling for weakened bodies, while water that is too hot narrows and closes up the chan-

473K

καὶ πυκνοῖ. ἡμεῖς δ' ἀνεῖναι καὶ χαλάσαι καὶ ἀνευρῦ-
ναι δεόμεθα συνιζηκότας αὐτούς· ὅπερ ἡ τῶν εὐκράτων ὑδάτων ἐργάζεται ποιότης· ἡδίστη γὰρ οὖσα
προκαλεῖται τὴν φύσιν ἐξαπλοῦσθαι καὶ ἀποτείνε-
σθαι πανταχόσε πρὸς τὸ τερπνόν, ἀνάπαλιν τοῖς
ἀηδέσιν· ἀποχωρεῖ γὰρ ἀπὸ τούτων καὶ φεύγει πρὸς
474K   τὸ βάθος. οὐδὲν οὖν θαυμαστὸν | ἐπὶ μὲν τῇ τῶν ἀνι-
ώντων ὁμιλίᾳ πιλεῖσθαι καὶ σφίγγεσθαι καὶ σκληρύ-
νεσθαι τὰ σώματα καὶ τοὺς πόρους αὐτῶν συνάγεσθαι
καὶ πυκνοῦσθαι, τοῖς δ' ἐναντίοις τοῖς ἥδουσιν ἕπεσθαι τὰ ἐναντία· χεῖσθαι μὲν γὰρ καὶ μαλάττεσθαι τὰ
σώματα, τοὺς πόρους δ' εὐρύνεσθαι.

μετὰ δὲ τὸ τοιοῦτον λουτρὸν εὐθέως ἐδίδομεν ὄνειον
γάλα τὴν ὄνον εἰς τὸν οἶκον εἰσάγοντες ἐν ᾧ κατ-
έκειτο. ἐπεπείσμεθα γὰρ ὡς μάλιστα μέν, εἰ αὐτὸν τὸν
ἄνθρωπον οἷόν τε ἦν θηλάζειν τὴν ὄνον, οὕτως ἂν
ἐθεραπεύθη τάχιστα· τούτου δ' ἔχοντος ἀηδίαν ἐλάχιστον χρόνον ὁμιλεῖν αὐτὸ τῷ πέριξ ἀέρι, τάχιστα
μεταβάλλεσθαι πεφυκός, ὁμοίως τῷ σπέρματι· καὶ
γὰρ καὶ τοῦτο χρόνον οὐδένα χρὴ διαμένειν ἔξω τῶν
οἰκείων ὀργάνων, εἰ μέλλει τὴν ἑαυτοῦ φυλάξειν δύνα-
μιν· ἀλλ' ἢ ἐν τοῖς τοῦ ἄρρενος εἶναι μορίοις, ἢ τοῖς
τῆς θηλείας συνῆφθαι. καὶ δὴ καὶ τὸ γάλα κάλλιστον
μὲν εἰ ἐξ αὐτῶν τῶν θηλῶν ἐπισπῷτό τις, ὥσπερ
Εὐρυφῶν καὶ Ἡρόδικος⁹ ἀξιοῦσιν· οἱ τοσοῦτον ἄρα

---

⁹ Ἡρόδικος, lectio in ms quodam quam Nutton monuit (non
vidimus); Εὐρυφῶν καὶ Πρόδικος B; Εὐρυφῶν καὶ Ἡρόδοτος
καὶ Πρόδικος K, KLat

nels and condenses, whereas we need to loosen, relax and dilate those channels that are collapsed, which is what the quality of the *eukratic* waters does. What is most pleasant calls upon Nature to open out and extend in every direction toward what is pleasant, and is opposite to what is unpleasant; for this withdraws from these things and flees toward the depths. It is not, therefore, surprising that from   474K the association with those things that are distressing, bodies are constricted, closed up and made hard, and their channels are narrowed and condensed. Opposite results follow the opposite pleasant things, for bodies are extended and softened, while the channels are dilated.

Immediately after this sort of bathing, I gave [him] ass's milk, after leading the ass to the room in which he lay sick. I was completely convinced that, if it was possible for the man himself to suckle the ass, he would have been cured in this way very quickly indeed. If, however, this is nauseating, [the milk] should be in contact with the ambient air for a very short time as it is very rapidly changed in nature, like sperm, for this too must not remain for any time outside the proper organs, if it is going to preserve its potency, but must either remain in the parts of the male genitalia, or be joined to those of the female. And, certainly, milk is best if someone draws it from the women themselves, as Euryphon and Herodicus did,[15] for they

[15] We are grateful to Vivian Nutton for drawing our attention to a manuscript variation here that we have incorporated. Kühn (and Linacre) have "Euryphon, Herodotus and Prodicus." Both Euryphon (460–400 BC) and Herodicus (440–400 BC) were doctors from Cnidus who advocated the use of women's milk in the treatment of phthisis; see Galen, *De probis pravisque alimentorum succis*, VI.775K.

τεθαρρήκεσαν αὐτῷ πρὸς ἀναθρέψιν σωμάτων ὥστε
475K καὶ τοὺς ὑπὸ φθόης συντετηκότας ἐκέλευον | ἐντιθε-
μένου τοῦ τιτθοῦ τῆς γυναικὸς τὴν θηλὴν βδάλλειν τὸ
γάλα. τοῦτο δὲ οὐχ ὑπομενόντων ποιεῖν τῶν πλείστων
ἄμεινόν ἐστιν ὅτι τάχιστα θερμὸν μεταφέρειν ἐκ τῶν
τιτθῶν εἰς τὴν κοιλίαν τοῦ κάμνοντος αὐτό.

τὸ μὲν οὖν ἀνθρώπειον ὡς ἂν ὁμόφυλον ἄριστον.
ἐπεὶ δ' ὑπομένουσιν οἱ πόλλοι γάλα γυναικὸς προσφέρεσθαι δίκην παιδίων, ὡς ὄνοις αὐτοῖς δοτέον ὄνειον γάλα. τοῦτο γάρ ἐστι πάντων τῶν ἄλλων ἄριστον
εἰς τὴν ἐνεστῶσαν διάθεσιν· λεπτότατον γὰρ ὑπάρχει,
ἥκιστά τε τυροῦται καὶ τάχιστα διαδίδοται πανταχόσε. δεῖται δ' ἀμφοῖν τούτοιν ὁ πεπονθὼς τὴν γαστέ-
ρα· τοῦ μὲν μὴ τυροῦσθαι, διὰ τὴν κοινὴν χρείαν·
ἅπαντας γὰρ βλάπτει τοῦτο· τοῦ διαδίδοσθαι δέ, διά
τε τὸ δεῖσθαι ταχείας θρέψεως καὶ ὅτι μεμύκασιν
αὐτῆς αἱ διέξοδοι. τοῦτό τε οὖν δοτέον αὐτὸ καθ' αὑτὸ
καὶ μέλι χλιαρὸν ὀλίγον ἐπιμιγνύειν αὐτῷ. μεγίστη δ'
ἔστω φροντὶς ἀμφοῖν τῆς ἀρετῆς ἐν ταῖς τοιαύταις
διαθέσεσι. τὸ μὲν οὖν μέλι τὸ ξανθὸν τὴν χροιὰν καὶ
476K ἡδὺ τὴν ὀσμὴν καὶ καθαρὸν οὕτως ὡς ὅλον | διαυ-
γεῖσθαι καὶ γευομένοις ἐπ' ὀλίγον δριμὺ καὶ ἥδιστον·
εἰς τοσοῦτό τε συνεστὼς ὡς ἐπαρθὲν τῷ δακτύλῳ
καταρρεῖν χαμᾶζε, συνεχὲς ἑαυτῷ διαμένον ἁπάντων
ἄριστον. εἰ δ' ἀπορρηγνύοιτο κατά τι καὶ μὴ μένοι
συνεχές, ἐπὶ τὴν γῆν κατατεινόμενον ἤτοι παχύτερόν
ἐστιν ἢ λεπτότερον ἢ ἀνομοιομερές. ὀνομάζω δὲ ἀνομοιομερές, ὥσπερ οὖν καὶ αὐτὸ τοὔνομα ἐνδείκνυται,

had such a degree of confidence in this with regard to the restoration of bodies that they directed those who suffered wasting due to phthisis, when they had applied their 475K mouths to the nipple of the woman's breast, to suck the milk. Since most people cannot bear to do this, it is better to transfer the warm milk from the breasts to the stomach of the patient as quickly as possible.

Human [milk] is best because it is from the same species. However, since many are reluctant to use a woman's milk like infants, you must give them asses' milk, as if they were asses. This is the best of all the others for the present condition because it is very thin, curdles least, and is distributed very quickly everywhere. Someone whose stomach is affected needs both these [qualities]: that it doesn't curdle, for the general benefit (curdling is harmful to all), and that it should be distributed because there is need for quick nourishment and the pathways of this are closed up. You must give the milk either by itself or mix a little lukewarm honey with it. There must be the greatest attention to both components of the honey's goodness in such conditions. Thus, the honey should be yellow in color and sweet in fragrance, so pure as to be completely transparent, and 476K to those who taste it, slightly sharp and sweet. That which is best of all is of such consistency that, when it is lifted with the finger, it flows to the ground, remaining in continuity with itself. If, however, it breaks and does not remain continuous in stretching to the ground, it is either too thick or too thin, or is not homogeneous. What I call "not homogeneous" is just as the name itself indicates—that it is

τὸ συγκείμενον ἐξ ἀνομοίων μερῶν. εὑρήσεις οὖν αὐτὸ
διαθεώμενος ἤτοι παχύτητας ἢ ὑγρότητας ἐμφερο-
μένας ἔχον, οὐχ ὁμοίας ἀλλήλαις τε καὶ τῷ παντί.

κηρωδέστερον οὖν μέλι τό τε παχὺ καὶ ᾧ τοιαῦταί
τινες ἐμφέρονται παχύτητες. περιπτωματικώτερον δὲ
καὶ ἀκατεργαστότερον καὶ δυσπεπτότερον τό θ' ὑγρὸν
ὅλον ᾧ τε πολλαὶ κατὰ τὰ μόρια κατεσπαρμέναι
φαίνονται σταγόνες ὑγρότητος· ᾧ δ' ἐμφέρεταί τις ἢ
κηροῦ ποιότης, ἢ προπόλεως, ἤ τις ἄλλη τοιαύτη
γευομένοις οὐ μόνον οὐκ ἄριστον, ἀλλ' ἤδη καὶ φαῦλον. ὅλως γὰρ οὐδεμίαν ἐξέχουσαν ἑτέρου οὐδενὸς
πράγματος ἐν αὐτῷ χρὴ περιέχεσθαι ποιότητα. δι-
477K όπερ | οὐδὲ τὸ τῶν θύμων ὄζον ἐναργῶς ἐπαινῶ·
ἀκατεργαστότερον γάρ ἐστι τοῦτο καὶ οὐκ ἀκριβῶς
πως μέλι· καίτοι τινὰς οἶδα τὸ τοιοῦτον ἐπαινοῦντας,
ὥστε καὶ τῶν πιπρασκόντων ἔνιοι κόπτοντες τὸν θύ-
μον ἐπιβάλλουσιν, ὅπως αὐτοῦ ὄζοι καὶ δόξειεν εἶναι
κάλλιστον. ἀλλ' ἐγὼ τὸ μὲν οὕτως ἀναιδῶς ὄζον οὐκ
ἐπαινῶ· καὶ πολὺ δὴ μᾶλλον, εἰ ἐπεμβληθείη τι θύμου·
τὸ δ' ἀμυδράν τινα ποιότητα κατ' ὀσμὴν ἢ γεῦσιν
φέρον οὐ μέμφομαι. κατὰ ταὐτὰ δὲ καὶ τὸ γάλα
μηδεμίαν ἔξωθεν ἐπιδεικνύσθω ποιότητα κατ' ὀσμὴν
ἢ γεῦσιν· ἀλλ' ἔστω γλυκὺ μὲν ὡς ἔνι μάλιστα καὶ
συνεχὲς ἑαυτῷ καὶ λαμπρόν, ὡς γάλακτι πρέπει· πρὸ
πάντων δ' ὁμοιομερὲς εἰς ὅσον ἐγχωρεῖ γάλακτι γενέσθαι τοιούτῳ· τελείαν γὰρ οὐκ ἐπιδέχεται τὴν ἀρετήν,
ὥσπερ τὸ μέλι τὸ ἄριστον.

ὅπως δ' εἴη τοιοῦτον, καὶ τροφὰς τῷ ζῴῳ παρα-

made up of dissimilar parts. You will discover this when you describe either the thickness or the moistness as being dissimilar in themselves and to the whole.

Honey, then, is more waxlike when it is thick, and such thickness is carried in it (i.e. is an integral part of it). By being altogether moist, it is more excrementitious, indigestible and difficult to concoct, and many drops of moisture appear dispersed in its parts. But if you taste it when something of a waxy quality, or of propolis, or some other such quality is included in it, not only is it not the best but it is already bad. For it is absolutely necessary that it does not contain any quality that marks it out from anything else. On this account, I don't approve of honey that clearly 477K smells of thyme, for this is more indigestible, and in some way is not exactly honey. And yet, I do know some who praise such honey so that there are some vendors who, having cut thyme, throw it in, so the honey smells of this and seems to be best. But I don't approve of honey that smells intrusively in this way. Much more so do I not approve if some thyme is put in. I do not, however, find fault with honey that carries some indistinct quality, either in terms of smell or taste. By the same token, the milk, too, must not display any foreign quality in terms of smell or taste. It must be sweet, uniform in consistency, and clear, as milk should be. Above all, it must be homogeneous, as far as it is possible for milk to be, since not to be is not consistent with total goodness, just as with the best honey.

In order that the milk might be the best, you must pro-

σκευαστέον ἐπιτηδείους καὶ γυμνάσια γυμναστέον
σύμμετρα, καὶ εἴ τινα θηλάζει πῶλον, ἀφαιρετέον. ὅτι
δὲ καὶ ἡλικίᾳ τὸ ἀκμαιότατον εἶναι προσήκει παντί
που δῆλον. ἐπιμελητέον δὲ καὶ ὅπως εὐπεπτότατον |
478K ἔσται, καταφρονοῦντας τῶν γελασόντων, εἰ καὶ τὰς
ὄνους διαιτήσειν μέλλομεν. εἰ γὰρ οἱ περὶ Βαίνετον
καὶ Πράσινον ἐσπουδακότες ὀσφραίνονται τὰς κό-
πρους τῶν ἵππων ἕνεκα τοῦ γνῶναι πῶς κατεργά-
ζονται τὰς τροφάς, ὡς ἐκ τούτου τὴν ὅλην αὐτῶν
εὐεξίαν κατανοήσωσι, καὶ ἡμᾶς δήπου χρὴ πολὺ δὴ
μᾶλλον εἰς ἀνθρώπου σωτηρίαν ἅπαντα προορᾶσθαι
τὰ τοιαῦτα, καὶ πόας τε παρέχειν τῷ ζῴῳ μὴ λίαν
ὑγρὰς καὶ χόρτου καὶ κριθῆς τὸ σύμμετρον· οὐκ
ἀμελεῖν δὲ οὐδὲν ψύχειν καὶ ἀνατρίβειν καὶ ἀπορρύπτειν καὶ καθαίρειν τὸ τῆς ὄνου σῶμα. εἰ μὲν οὖν
ἀποπατήσειεν ὑγρότερα καὶ δυσωδέστερα καὶ μετὰ
φύσης μεστὰ φαίνοιτο, πρόδηλον δήπουθεν ὡς οὐκ
ἔπεψε καλῶς· ὥστε ἤ τι τῶν τροφῶν ἀφελεῖν, ἢ προσθεῖναι τοῖς γυμνασίοις, ἢ ὑπαλλάξαι χρὴ τὰς ποιότητας, ἢ τὴν περὶ τὰς τρίψεις τε καὶ ψύξεις ἐπιμέλειαν οὐ
τὴν αὐτὴν ποιεῖσθαι, σκληρότερα δ᾽ ἀποπατήσαντος
τοῦ ζῴου πρὸς τἀναντία βλέπειν κἀκείνων τι μετα-
κοσμεῖν. ἐγὼ γὰρ εἰ πάνθ᾽ ὅσα χρὴ σκοπεῖσθαι περὶ
τὸ ζῷον οὗ τῷ γάλακτι χρῆσθαι μέλλοιμεν ἑξῆς ἅπαν-
479K τα λέγοιμι, τῆς ὑγιεινῆς | ὅλης ὑπομνήσομαι πραγμα-
τείας, ἣν ἀναγκαῖον μέν ἐστι γινώσκειν τὸν χρηστὸν

268

vide suitable nutriment for the animal, you must exercise it
in moderation and, if it is suckling its young, you must remove them. It is, somehow, clear to everyone that it is appropriate for the animal to be in its very prime in terms of
age. You must also insure that it will be of very good digestion, paying scant regard to those who will laugh, if we also   478K
intend to prescribe a regimen for asses. If those around
Venetium and Prasinum are being diligent, they smell the
dung of horses for the purpose of knowing how they are
to prepare their nutriments, as from this they may learn
about their good bodily condition as a whole.[16] It also behooves us, charged as we are with the whole preservation
of a person, to look very much more closely at such things,
and to provide grasses that are not overly moist for the animal, and fodder and barley in moderation, and to neglect
nothing in cooling, rubbing down, cleansing and purifying
the body of the ass. Therefore, if she were to excrete what
is more moist, quite foul-smelling and seems full of flatus,
it would, I presume, be clear that she did not digest properly. As a consequence, it is necessary to take away some of
the nutriments, or to add exercise, or to change the qualities, or not to provide the same care for rubbing or cooling.
If, however, when the animal defecates, the feces are quite
hard, we must look to modify those things to their opposites. If I were to say it is necessary to consider individually
all those things pertaining to the animal whose milk we intend to use, I shall be mentioning the whole matter of
health, which it is essential for the person to know who will   479K

16 See A. Cameron, *Porphyrios the Charioteer* (Oxford, 1973),
p. 70.

GALEN

παρασκευάσοντα τὸ γάλα, λέγειν δ' οὐ νῦν καιρός,
ἑτέρωθί γε διειλεγμένον τελεώτατα ὑπὲρ αὐτῆς.

ἀλλ' ἐπὶ τὸ προκείμενον ἐπάνειμι. καί μοι πάλιν
ἀναμιμνήσκου τοῦ λελουμένου τε καὶ τὸ γάλα προσ-
ενεγκαμένου. τοῦτον οὖν τὸν ἄνθρωπον ἡσυχάζειν
ἐάσαντες ἄχρι τοῦ δευτέρου λουτροῦ τρίψομεν τηνικαῦτα μετρίως τε ἅμα καὶ λιπαρῶς, εἰ ἀκριβῶς κατ-
είργασται τὸ δοθὲν γάλα ταῖς ἐρυγαῖς τε καὶ τῷ τῆς
γαστρὸς ὄγκῳ τεκμηρόμενοι. σύμμετρος δὲ ἀπὸ τοῦ
πρώτου λουτροῦ πρὸς τὸ δεύτερον, ὡρῶν ἰσημερινῶν
τεττάρων ἢ πέντε χρόνος, εἴ γε τὸ τρίτον ἔτι μέλλοις
λούειν αὐτόν, εἰ δὲ μή, πλεόνων. λούσεις δὲ τὸ τρίτον,
ἢν εἰθισμένος ᾖ λουτροῖς χρῆσθαι πλείοσιν· οὗτοι
γὰρ καὶ χαίρουσι καὶ ὀνίνανται πολλάκις λουόμενοι.
καὶ μὲν δὴ καὶ ἐπαλείψομεν αὐτὸν ἐλαίῳ πρὶν ἀμφι-
έννυσθαι καθ' ἕκαστον λουτρόν· εἰς ἀνάθρεψιν γὰρ
συντελεῖ καὶ τοῦτο, καθάπερ καὶ αἱ τρίψεις. εἴρηται δὲ
ὑπὲρ ἁπάντων τῶν τοιούτων τῆς δυνάμεως ὁ λογισμὸς
480K ἐν τοῖς Ὑγιεινοῖς· οὐδὲν μὴν χεῖρον ἀναμνῆσαι | καὶ
νῦν ἐπὶ κεφαλαίων αὐτῶν.

εἴτε γὰρ ὕδωρ ἐπιχέοις θερμὸν εὔκρατον ὁτῳδήποτε
μορίῳ τοῦ σώματος, εἴτε τρίβοις, εἴτε λούοις, εἴθ'
ὁπωσοῦν ἄλλως θερμαίνοις, ὄψει γιγνόμενον ὅπερ
Ἱπποκράτης ἐπὶ τῶν ὕδατι θερμῷ καταντλουμένων
εἶπε· τὸ μὲν γὰρ πρῶτον ἀείρεται, ἔπειτα δὲ ἰσχναί-
νεται. διὰ τοῦτο σαρκῶσαι μὲν ἡμῖν βουλομένοις

270

prepare milk so it is good. Now is not the time to speak of this; it has, in fact, been covered completely elsewhere.[17]

But I shall return to the matter before us. Call to mind for me again the man who bathes and takes milk. Thus, after we allow this man to rest quietly until the second bath, we shall, under the circumstances, rub him moderately, and at the same time with oil. We shall judge whether the milk given has been completely worked up for use by his belching and the swelling of his belly. [There should be] a moderate time from the first bath to the second of four or five equinoctial hours, at least if you intend to bathe him a third time. If you don't, the interval should be longer. You will bathe a third time, if the person is accustomed to using a lot of baths, because such people enjoy and benefit from frequent bathing. Furthermore, you will smear him over with oil before dressing him after each bath for this, too, contributes to the restoration, just as the rubbings also do. The thinking about the potency of all such things was spoken of in [my work] *On the Preservation of Health*.[18] There is, however, no harm in also mentioning them now under 480K their chief points.

If you pour warm, *eukratic* water on any part of the body whatsoever, or massage, or bathe, or heat in any other way at all, you will see what Hippocrates spoke of happening in the case of irrigations with hot water—first [the part] is raised and then it is reduced.[19] Because of this, when we

17 Galen also considers the preparation of animals that are to provide milk in *De sanitate tuenda*, Book 5, chapter 7.

18 Massage and exercise are considered in Book 2 of *De sanitate tuenda* and bathing in Book 3.

19 Hippocrates, *The Surgery*, 13.

ότιοῦν σῶμα μέχρι τοσούτου θερμαντέον ὡς εἰς ὄγκον
ἀρθῆναι. διαφορῆσαι δὲ καὶ κενῶσαι μέχρι τοσούτου
χρονιστέον ὡς ἰσχνωθῆναι τὸ πρότερον ὀγκωθέν.
ἀκριβῶς δὲ τὸν νοῦν προσέχειν χρὴ τὸν ἐπιστατοῦντα
ταῖς τοιαύταις ἐνεργείαις καὶ μάλιστα ἐφ' ὧν βού-
λεται σαρκῶσαι· τοῦ μὲν γὰρ διαφορῆσαι πλατὺς ὁ
καιρός, ὥστε οὐδ' ἑκὼν ἁμάρτοις ἂν αὐτοῦ· τοῦ σαρκῶσαι δ' ὀξύς· ὅταν γὰρ πρῶτον ὀγκωθῇ, τότε παύσασθαι χρή· τὸ δ' ὀγκωθῆναι τοῦτο καθ' ἕκαστον τῶν
σωμάτων ἴδιόν ἐστιν. οὐ γὰρ οἷόν τε τὸν τῷ λόγῳ
προκείμενον νῦν τὸν ἰσχνὸν εἰς ὄγκον ἀφικέσθαι τοῖς
ὑγιαίνουσιν ἴσον· ἀλλ' ὅταν ἐπ' ὀλίγον ἀρθῇ, δια-
481K   φορεῖται παραυτίκα. χρὴ τοίνυν ἀκριβῶς | προσέχειν
τὸν νοῦν, ὅπως μὴ λάθῃ σε παρελθὼν ὁ καιρός. ἐὰν
γοῦν ἀνατρίβῃς τὸν ἄκρως ἰσχνόν, ἀρκεῖ τὸ ἐρύθημα
μόνον· ἐάν τε λούῃς, τὸ θερμῆναι μετρίως· ἐπέκεινα δὲ
τοῦδε χρονίζων καταλύσεις μᾶλλον ἢ θρέψεις αὐτόν.
ἐπαλείφειν δὲ ἐλαίῳ μετὰ τὰ λουτρὰ χάριν τοῦ μὴ
διαπνεῖσθαι πλέον τοῦ προσήκοντος, ἀλλ' ἐμπεφρά-
χθαι τοῦ δέρματος τοὺς πόρους. εἴη δ' ἂν εὐθέως
αὐτῷ[10] τοῦτο τῆς μὲν ξηρότητος ἄκος, οἷον πρόβλημα
δέ τι πρὸς τὴν ἐκ τοῦ περιέχοντος βλάβην. καὶ εἰ μὲν
οὖν ἥδοιτο τῷ γάλακτι, καὶ μετὰ τὸ δεύτερον αὐτῷ
δώσομεν λουτρόν· εἰ δὲ μή, πτισάνην ἀκριβῶς καθ-
εψημένην, ἢ χόνδρον ὡς πτισάνην κατεσκευασμένον.
εἶτ' αὖθις ἡσυχάσαντα πάλιν πρὸς τὸ τρίτον ἄξομεν
λουτρὸν ἢ ἄντικρυς ἐπὶ τὸ δεῖπνον. ἄρτος δ' ἔστω
παρεσκευασμένος αὐτῷ κλιβανίτης καθαρὸς ἐπιμελῶς

wish to enflesh any body whatsoever, we must heat to the point of its being raised to a swelling. Then we must spend time dispersing and evacuating to such an extent that we reduce what was previously swollen. It is necessary for the person in charge to pay very careful attention to such actions, and particularly in the case of those [parts] he wishes to enflesh. The time of dispersal is broad, so you should not wittingly err in this. However, the time of enfleshing is short, so you must stop as soon as the part becomes swollen. The production of the swelling is specific to each of the bodies, for it is not possible for the looseness now being put forward in the discussion to come to a swelling equal to those [parts] that are healthy, but whenever it is raised to a slight degree, it is immediately dispersed. It is necessary, therefore, to pay careful attention lest the appropriate time pass you by unnoticed. Anyway, if you rub someone who is extremely lean, redness alone is sufficient, and if you bathe him, then heat him moderately. If you continue beyond this, you will break the person down more than you will build him up. Smear him with oil after the bath, so he is not dissipated by evaporation more than is appropriate but so the pores of the skin are blocked up. Should this immediately be the remedy for his dryness, it is like a barrier against the harm from the ambient air. And if he is pleased with the milk, we shall also give it to him after the second bath. If not, [we shall give] ptisan decocted thoroughly or gruel prepared like the ptisan. Then, when he has again rested, we shall bring him to the third bath, and then directly to a meal. The bread prepared for him

481K

---

10 B; αὐτὸ K

μὲν ὠπτημένος, ἔχων δὲ ζύμης τε καὶ ἁλῶν αὐτάρκων. ὄψον τε τῶν πετραίων ἰχθύων, ἢ ὀνίσκος ἐκ λευκοῦ ζωμοῦ. καὶ μὴν καὶ τὰ πτερὰ καὶ οἱ ὄρχεις τῶν ἐν γάλακτι τρεφομένων ἀλεκτρυόνων ἐπιτήδειοι, μὴ |

482K   παρόντων δὲ τούτων ἄλλοις χρηστέον. οὕτω δὲ καὶ πέρδιξι καὶ στρουθοῖς χρῆσθαι τοῖς ὀρείοις τε καὶ μαλακοσάρκοις, φυλάττεσθαι δὲ τά θ' ἕλεια καὶ σκληρόσαρκα.

συνελόντι δὲ φάναι, τὸ κεφάλαιον τῆς τροφῆς, εὔπεπτος ἔστω καὶ τρόφιμος, ἥκιστα δὲ καὶ γλίσχρος καὶ περιττωματική. τροφιμώτατον οὖν ἐστιν ἁπάντων ὧν ἴσμεν ἐδεσμάτων χοίρειον κρέας, ἀλλ' οὐχ ὁμοίως τοῖς εἰρημένοις εὔπεπτον καὶ χυμοῦ γλίσχρου καὶ παχέος γεννητικόν. οὐδὲ γὰρ οὐδ' ἄλλως οἷόν τε διακεῖσθαι τὸ ἄκρως τρόφιμον. ἴσχεσθαι γὰρ αὐτὸ χρὴ καὶ προσπλάττεσθαι δυσαπολύτως, οὐ διαρρεῖν ὑπὸ λεπτότητος. οὐ μὴν εἴς γε τὰ παρόντα χρηστόν. εἰ μὲν γὰρ ἡ τροφὴ ἑαυτὴν πέττουσα καὶ ἀναδιδοῦσα καὶ τοῖς τρεφομένοις ὁμοιοῦσα, προσεφύετο τοῖς πλείστης ἀναθρέψεως δεομένοις, τροφιμωτάτων ἂν ἔδει σιτίων. ἐπεὶ δ' οὐκ ἄλλο μέν ἐστι τὸ τρέφεσθαι δεόμενον, ἄλλο δὲ τὸ ταῦτα κατεργαζόμενον, ἀλλ' ἑαυτῷ τε τὴν τροφὴν ἐπισπᾶται τὸ τρέφεσθαι μέλλον, ἑαυτῷ τε μεταβάλλει καὶ πέττει καὶ προσφύει καὶ ἐξομοιοῖ,

483K   διττὸν | ἕξει σκοπὸν τῆς τῶν σιτίων ἐπιτηδειότητος· ἕνα μὲν οὖν αὐτοῦ τοῦ σιτίου τὴν φύσιν, ἕτερον δὲ τὴν οἰκείαν τε αὐτῷ καὶ σύμφυτον δύναμιν. ἐξ οὖν τῶν εἰρημένων εὔδηλον ὡς οὔτε τὸ τροφιμώτατον σιτίον

must be pure, carefully baked or roasted, and have enough leaven and salt. Also [give him] fish, one of the rock fish or cod with white juice. Furthermore, the wings and testes of cocks well nourished with milk are suitable although, if 482K these things are not available, we must use other things. In like manner too, use partridges and sparrows that are from the mountains and soft-fleshed, but be on guard against those that are from the marshes and hard-fleshed.

In short, the chief point of nourishment is to let it be easily digested and nutritious, but least viscid and excrementitious. The most nutritious of all the foods we know is the fresh meat of a pig, but it is not easy to digest like those foods mentioned, and is productive of a viscid and thick juice. It is not possible for what is very nutritious to be otherwise. It must be retained and adherent in a way that is difficult to remove and it must not flow through due to thinness. [If it does] it is not, in fact, useful in the present matters; for if the nourishment, when it concocts, distributes and assimilates itself into the parts being nourished, and becomes attached to the parts requiring the greatest restoration, they would be in need of the most nutritious foods. However, what needs to be nourished is not one thing while what brings the nourishment about is another. Rather, if what is going to be nourished draws the nutriment to itself, and changes, digests, attaches and assimilates [the nutriment] to itself, there will be a twofold indi- 483K cator of the suitability of the foods. One is the nature of the food itself, while the other is the specific and innate potency in it. Therefore, it is clear from what was said that it is

ἄριστον εἰς τὰ δεόμενα σώματα πλείονος ἀναθρέψεως·
οὐ γὰρ ὑπάρχει τῷ τοιούτῳ τὸ πέττεσθαι ῥᾳδίως· οὔτε
τὸ εὐπεπτότατον, ὅτι γε μηδὲ τοῦτο δυνατὸν εἶναι
τροφιμώτατον. ἐναντιουμένων οὖν ἀλλήλοις τῶν σκοπῶν οὐ χρὴ τοῦ ἑτέρου τῆς ἀκρότητος ὀριγνώμενον
ἐπιλαθέσθαι θατέρου παντάπασιν, ἀλλ' ἀμφοῖν ἀεὶ
μεμνημένον, εἰς ὅσον οἷόν τέ ἐστι μιγνύειν αὐτούς.
τούτου μὲν δή μοι διαπαντὸς μέμνησο τοῦ παραγ-
γέλματος.

οἴνου δ' ἐφεξῆς ποιότητός τε καὶ ποσότητος μνημο-
νεύσωμεν, μεγίστην ἔχοντος εἰς τὰ τοιαῦτα δύναμιν,
ἐπειδὴ καὶ μόνῳ τούτῳ ποτῷ χρηστέον εἶναί φημι
πᾶσι τοῖς ἀναθρέψεως δεομένοις σώμασιν ἄνευ τοῦ
πυρέττειν. ἔστω τοίνυν ὡς ἐν παραδείγματι μὲν εἰπεῖν,
οἷος ὁ Σαβῖνος μὲν ἐπὶ τῆς Ἰταλίας, Ἀρσύνιος δὲ
κατὰ τὴν Ἀσίαν, ὡς δὲ τύπῳ τε καὶ ὅλῳ τῷ γένει
περιλαβεῖν, ὁ ὑδατώδης μὲν τἆλλα, βραχεῖαν δ' ἔχων
484K τὴν | στύψιν. ὑδατώδη δ' οἶνον ὀνομάζω τὸν λεπτὸν
καὶ λαμπρὸν τὸν λευκόν, ὃς καὶ καθαρὸν καὶ ὀλιγο-
φόρον, ὡς Ἱπποκράτης ἐκάλεσεν. ὀλιγοφόρος δ'
ἐστίν, ὡς ἂν ἐν τῷ κεράννυσθαι, τὴν τοῦ ἥδατος μίξιν
ὀλιγίστην φέρει. τουτέστιν ὁ ἀσθενέστατός τε καὶ
ὑδατωδέστατος ὡς ἐν οἴνοις· ὁ γὰρ ἀνεχόμενος ὕδατος
πλείστου μίξιν ἐν τᾷ κεράννυσθαι σφοδρότατός ἐστι
καὶ ἰσχυρότατος. ὠνόμαζε τὸν τοιοῦτον ὁ Ἱπποκράτης
οἰνώδη. ἀλλὰ τοῦτον μὲν φυλάττεσθαι, πλήττοντα τὰς
ἀσθενεῖς δυνάμεις. ὁ δ' ὑδατώδης μέν, αὐστηρὸς δέ,
ἐπιτήδειος, ὡς ἂν τὴν τοῦ ὕδατος ἐκπεφευγὼς ἀσθένει-

not the most nourishing food that is best for bodies requiring a greater degree of restoration because it is not by being very nutritious that digestion is easy, nor by being very easily digested that it is able to be most nutritious. Thus, when the indicators are in opposition to each other, in striving for the extreme of the one, you must not lose sight of the other altogether, but always bearing both in mind, combine them as far as possible. Remember this directive of mine constantly.

Let me mention wine next, both its quality and quantity, because this has the greatest potency for such things, and so I say that this is the only drink you must use for all those bodies requiring restoration that are without fever. Therefore, let me mention by way of examples [wines] such as the Sabine in Italy and the Arsyine in Asia,[20] so as to encompass in a stroke the whole class, the one being watery in some respects and the other having a slight astringency. I term wine watery that is thin and clear white, which is also what Hippocrates called pure and able to bear little water. Wine that is *oligorophoros*[21] bears the very least admixture [of water] whenever it is mixed. That is to say, it is the weakest and most watery of wines, for what supports a mixture of the greatest amount of water in the mixing is the most powerful and strong. Hippocrates termed such a wine, "vinous."[22] But you must be on guard against this since it overpowers weak capacities. The watery wines, being astringent, are useful because they

484K

---

[20] See Galen's *De sanitate tuenda*, Book 5, chapter 5 (VI.334–39K), on the various wines.   [21] On this term, see Hippocrates, *Acute* 56, and Galen, *De probis pravisque alimentorum succis*, VI.807K.   [22] On this term, see Hippocrates, *Acute* 37, and Galen, *De alimentorum facultatibus*, VI.578K.

αν καὶ μηδέπω τὴν οἴνου βλάβην ἔχων. ἐκ τούτων οὖν
τῶν σκοπῶν εὑρήσεις αὐτοῦ καὶ τὴν ἡλικίαν καὶ τὴν
κρᾶσιν. ὁ μὲν γὰρ νέος ὑδατωδέστερός τε καὶ ἀσθενέ-
στερός πως ἢ ὡς οἴνῳ πρέπει, μετὰ καὶ τοῦ δυσπεπτό-
τερος εἶναι καὶ περιττωματικώτερος. ὁ δὲ πρεσβύτε-
ρος οἰνωδέστερός τε καὶ ἰσχυρότερος ἢ ὡς τοῖς
παροῦσιν ἁρμόττει. διὸ καὶ θᾶττον ἐτῶν ἐξ ὁ εὐγενὴς
Σαβῖνος οὐκ ἐπιτήδειος, ἐμπλέων ἐπὶ πλεῖστον ἔν τε
τῷ στομάχῳ καὶ τῇ γαστρὶ καὶ κλύδωνας ἐργαζό-
485K μενος. | εὐγενῆ δ᾽ ὀνομάζω τὸν αὐστηρόν, ὁμοίως μὲν
Σαβῖνον, ὁμοίως δὲ Ἀδριανόν τε καὶ Ἀλβανόν, ὁμοίως δὲ Ἀρσύνιόν τε καὶ Τιτακαζηνόν, ὅσοι τε ἄλλοι
τοιοῦτοι· λέλεκται γὰρ ἐπὶ πλέον ὑπὲρ ἁπάντων οἴνων
ἰδίᾳ.

πρὸς τούτους οὖν ἀναφέρων τοὺς σκοποὺς καὶ τὴν
πρὸς τὸ ὕδωρ αὐτοῦ ποιεῖσθαι κρᾶσιν· ἔτι δ᾽ ἂν
μᾶλλον ἀκριβῶς στοχάζοιο τῶν εἰρημένων, εἰ τῆς τοῦ
ὕδατος κακίας τὰ κεφάλαια διὰ μνήμης ἔχοις· ἤρτη-
ται δὲ ἐκ τῆς ψυχρότητος αὐτοῦ πάντα, δι᾽ ἣν ἐν τοῖς
ὑποχονδρίοις τε μέχρι πλείστου παραμένει καὶ κλύδωνας ἐργάζεται καὶ πνευματοῦται καὶ διαφθείρεται
καὶ τῆς γαστρὸς ἐκλύει τὸν τόνον, ὡς καὶ τὰς πέψεις
διὰ τοῦτο χείρους γίγνεσθαι· συμπράττει δ᾽ οὐδὲ ταῖς
ἀναδόσεσι τῆς τροφῆς οὐδὲν ὅ τι καὶ ἄξιον λόγου. καὶ
μέν γε καὶ αἱ τῶν προειρημένων οἴνων ἀρεταὶ ταῖς νῦν
εἰρημέναις κακίαις ἐξ ὑπεναντίου τὴν φύσιν ἔχουσιν·
οὔτε γὰρ ἐμφυσῶσι τὸ ὑποχόνδριον, ἀλλ᾽ εἰ καὶ φυ-
σωδέστερόν πως εἴη προστέλλουσιν, οὔτε χρονίζου-

avoid the weakness of water without as yet having the harm
of wine. From these indicators, then, you will discover
both the age and the *krasis*. For what is new is more watery
and weaker, in some way, than is fitting as wine, and along
with this more difficult to assimilate and more excrementitious. That which is older is more vinous and stronger
than is suitable for the present purposes. Wherefore, also,
the good quality Sabine wine greater than six years old is
not suitable, since it floats longer in the esophagus and
stomach and creates "splashing." I call "of good quality"   485K
[wine] that is astringent, like the Sabine, Adrian and
Alban, and like the Arsyine and the Titacazene, and other
such [wines]. More has been said about all the wines individually.

Therefore, when you give consideration to these indicators, make its *krasis* tend toward the watery; still more
might you estimate accurately those things mentioned, if
you keep in mind the chief points of the badness of water.
Everything depends on its coldness, due to which it remains in the hypochondrium for a very long time and creates "splashing," so that it fills [the hypochondrium] with
wind, causes corruption, and breaks down the strength
of the stomach. As a result, the concoctions become worse
because of this. Nor does it contribute anything even
worth mentioning to the distribution of nourishment. Furthermore, the good qualities of the previously mentioned
wines have a nature which is contrary to the evils [of
water] now spoken of, for they do not inflate the hypochondrium but, if it is also somehow more flatulent, they
protect it. Nor do they delay in the one place by virtue of

279

σιν αὐτόθι τῷ τῆς θερμασίας συμμέτρῳ, τάς θ' ὁδοὺς
τῆς ἀναδόσεως ἀνοιγνύντες, αὐτοί τε συνεπωθοῦντες |
486K   ἅμα καὶ συναναφέροντες τὴν τροφὴν εἰς τάχος τῇ
κατὰ τὴν ἀνάδοσιν ἐνεργείᾳ συντελοῦσιν, εὔχυμοι τ'
εἰσὶ καὶ κατακεραστικοὶ καὶ πεπτικοὶ τῶν κατὰ γαστέρα τε καὶ φλέβας. αὐξάνουσι δὲ καὶ τὴν δύναμιν τῶν
ὀργάνων καὶ τὰ περιττώματα πρὸς τὰς ἐκκρίσεις πο-
δηγοῦσι. ταῦτ' ἄρα καὶ οὐρητικοὶ τῶν ἄλλων οἴνων οἱ
τοιοῦτοι μᾶλλόν εἰσιν, ὡς ἂν αὐτοί τε ταχέως διερ-
χόμενοι τὸ σύμπαν σῶμα καὶ τὰ περιττώματα τῇ
ῥύμῃ τῆς φορᾶς ἑαυτοῖς συνεκκρίνοντες. εἴρηται μὲν
οὖν ἑτέρωθι τελεώτερον ὑπὲρ τῆς τῶν οἴνων δυνάμεως,
ὥσπέρ γε καὶ περὶ τῆς τῶν σιτίων·

καὶ χρὴ τὸν μεθόδῳ τὴν τέχνην ἐργάζεσθαι βουλό-
μενον ἰδίας τῶν ὑλῶν ἁπασῶν ἐκμαθόντα τὰς δυνάμεις μηκέτι καθ' ἕκαστον πάθος ἀκούειν ἀξιοῦν, ἀλλ'
αὐτὸ μόνον ἐπιγνόντα τὸ τῆς θεραπείας εἶδος εὑρε-
τικὸν εἶναι τῆς ἁρμοζούσης διαίτης. ἐγὼ δ' οὐκ ὀκνή-
σω, κηδόμενος τῶν τἀληθῆ σπευδόντων ἐκμανθάνειν,
ἐφάψασθαι καὶ τῆς τοιαύτης διδασκαλίας, ἕνεκα τοῦ
τοὺς ἀγυμναστοτέρους τὸν λογισμὸν ἀπὸ τῶν καθόλου μεταβαίνειν ἐπὶ τὰ κατὰ μέρος ὑπὸ τῶν παρα-
487K   δειγμάτων ποδηγουμένους. ὅσον οὖν ὑπόλοιπόν | ἐστι
προσθετέον αὖθις. ὁ σκοπὸς τῆς ποσότητος ἐπὶ μὲν
τοῦ πόματος, ὡς μηδέ ποτ' ἐμπλεῦσαι τῇ γαστρὶ καὶ
κλύδωνος ἐργάζεσθαί τινα αἴσθησιν, ἐπὶ δὲ τῶν
σιτίων, μάλιστα μὲν εἰ οἷόν τε, μηδὲ βαρῦναί ποτε

the moderation of the heat but, by opening up the pathways of distribution, help to push on and, at the same time, 486K
carry along the nutriment quickly. They contribute to function in terms of distribution, they are wholesome and demulcent, and they promote the digestion of what is in the stomach and veins. They also increase the capacity of the organs and lead the superfluities toward the excretions. For this reason, such wines are also more diuretic than the others, as they swiftly pass through the whole body and help to clear out the superfluities by virtue of the flow of their own outward passage. The matter of the potency of wines was spoken of more completely elsewhere, as was that of foods.[23]

And it behooves someone who wishes to practice the craft methodically, after he has thoroughly learned the specific potencies of all the materials, to no longer think it worthwhile to hear about these in relation to each affection but only, when he knows the actual kind of treatment, to be the discoverer of the applicable regimen. However, because my concern is for those who are eager to learn the truth, I shall not hesitate to touch also upon such teachings for the sake of those who are unpracticed in reasoning, so allowing them to move from the general to the particular, being led by examples. I must add in order as much as re- 487K
mains. The indicator of amount in the case of drink is that it does not, at any time, "float" in the stomach and bring about a certain sensation of "splashing." The objective in the case of foods, particularly if it is possible, is that they do

---

[23] See *De sanitate tuenda* in various places on both these subjects. For foods only see *De alimentorum facultatibus*, VI.453–748K.

αὐτήν· χαλεπὸν δὲ τὸ τοιοῦτον ἢ ἴσως ἀδύνατον ἐν ἀτόνῳ γαστρὶ φυλάξασθαι παντάπασιν· δεύτερον δ' ὅτι τάχιστα παύσασθαι τὸ βάρος· ἐφεξῆς δὲ διάτασίν τε καὶ πνευμάτωσιν ὑποχονδρίων φυλάττεσθαι. εἰ μὲν δὴ συμπέσοι τοιοῦτό τι κατὰ τὴν πρώτην ἡμέραν, ἀνάλογον τῷ μεγέθει τοῦ συμπτώματος ἀφελεῖν χρὴ τῶν σιτίων ἐπὶ τῆς ὑστεραίας· εἰ δ' ἀμέμπτως ἅπαντα γίγνοιτο, προσθεῖναι βραχύ τι. κατὰ δὲ τὸν αὐτὸν λόγον ἐπὶ τῆς τρίτης ἡμέρας ἢ ἀφαιρεῖν ἢ προσθεῖναι, τῇ προηγουμένῃ παραβάλλοντας. καὶ οὕτως ἄχρι παντὸς ἀναθρεπτικῶς τε καὶ ἀναληπτικῶς ἅπαντα διαπράττεσθαι κινήσεσιν, αἰωρήσεσι καὶ περιπάτοις ἀνάλογον τῇ τοῦ σώματος ἐπιδόσει, προστιθέντα καὶ τἆλλα πάντα κατὰ τὸν ἀναληπτικὸν ὀνομαζόμενον τρόπον, ὃς οὐ τῷ γένει διενήνοχε τοῦ νῦν ἡμῖν ἐνεστῶτος, ἀλλὰ τῷ σύμπαν ὡσαύτως ἔχειν

488K ἐπ' ἐκείνου τὸ σῶμα, καθάπερ | ἡ γαστὴρ ἐπὶ τοῦδε. συμβαίνει δὲ μάλιστα κατὰ τὰ χρόνια τῶν ἀρρωστημάτων, ὅταν ἡ σύμφυτος ὑγρότης ἑκάστου μέρους, ἐξ ἧς πρώτης τρέφεται, κινδυνεύῃ[11] μηκέτι εἶναι· τὴν μὲν γὰρ αὐτῶν τῶν στερεῶν σωμάτων ξηρότητα τῶν ἀδυνάτων ἐστὶν ἐπανορθώσασθαι, καθάπερ καὶ τὸ γῆρας, ὡς κἂν τῷ Περὶ μαρασμοῦ δέδεικται λόγῳ· τὴν ὑγρότητα δὲ κἂν ἀπόληται τελέως, οἷόν τε γεννῆσαι κατὰ τὴν εἰρημένην ἀρτίως δίαιταν.

ἐν μὲν οὖν ταῖς ἄλλαις ἀναλήψεσιν οὐδὲν ἐξαίρετον ἡ γαστὴρ πέπονθεν· ἐν δὲ τῇ νῦν προκειμένῃ διαθέσει τὸ νόσημα μέν ἐστι τῆς γαστρός, λεπτύνεται δὲ τῷ

not, at any time, weigh the stomach down. If, however, such a thing is difficult, or perhaps impossible, to maintain completely in a weak stomach, the next thing is that the weight comes to an end very quickly and that distension and inflation of the hypochondrium are guarded against. Now if this were to happen during the first day, you should withdraw the foods for the next day by an amount proportional to the magnitude of the symptom. If everything occurs faultlessly, you should add a little [food]. On this basis, for the third day, either take away or add, using the preceding principle. In the same way, continually accomplish everything nutritiously and restoratively with movements, passive exercises, and perambulations in proportion to the progress of the body, also adding all other things according to the so-called restorative way. This does not differ in class from what I now propose, but in that case applies to the whole body in the same way as it does to the stomach in this   488K
case. Particularly during the times of chronic illness what happens is that the innate moisture of each part, from which it is primarily nourished, is at risk of no longer existing. The dryness of the solid bodies themselves cannot be corrected, just as old age cannot, as I showed in my work *On Marasmus*.[24] However, the moisture, even if it is destroyed completely, can be regenerated by the regimen mentioned just now.

Thus, in other restorations, the stomach is not affected egregiously whereas, in the condition now before us, the disease is of the stomach, although over time the whole

[24] *De marasmo*, VII.666–704K.

---

[11] B; κινδυνεύει K

χρόνῳ τὸ σύμπαν σῶμα μὴ καλῶς τρεφόμενον· εὔλογον οὖν ἀκριβεστέρας δεῖσθαι διαίτης τούσδε, διὰ τὸ
περὶ τὰς πέψεις ἄρρωστον, ἀλλ' ὅταν γε βελτίους
γένωνται, κατὰ βραχὺ μετάγειν αὐτοὺς ἐπὶ τὴν ἀναθρεπτικὴν καὶ ἀναληπτικὴν ὀνομαζομένην ἀγωγήν.
γίγνοιτο δ' ἂν τοῦτο τρίψεσί τε πλείοσιν ἢ πρόσθεν
αἰωρήσεσί τε χρωμένων. ὑπαλλάττειν δ' ἐν τούτῳ χρὴ
καὶ τῶν τροφῶν τό τε ποσὸν καὶ τὸ ποιόν, ὡς καὶ
πλείω καὶ ἰσχυρότερα σιτία τῶν ἔμπροσθεν διδόναι. |

489K καὶ τοῦ χρόνου δὲ προϊόντος, ὅταν ἤδη πλησίον ἥκω-
σι τῆς ὑγιεινῆς καταστάσεως, ἀποχωρῆσαι πτισάνης
καὶ γάλακτος· ἔτι δὲ τῶν ἐκ χόνδρου ῥοφημάτων ἐπὶ
τὰ συνήθη προσάγοντας σιτία· εἶτα καὶ τοὺς ἰχθύας
ἀφελεῖν, εἰ μὴ δι' ἔθους εἶεν, ἐπὶ τοῖς πτηνοῖς διαι-
τωμένους· ἐδωδῆς τε χοιρείων κρεῶν, ἄρξασθαι μὲν
ἀπὸ ποδὸς ἐν πτισάνῃ καθεψημένου· μετὰ ταῦτα δὲ
καὶ ὁ κωλὴν ἂν ληφθείη καλῶς, ἔπειτα ἤδη καὶ τἆλλα
μόρια· τὸ μὲν πρῶτον ἱερείου νέου καί, εἰ χειμὼν εἴη,
πρὸ μιᾶς ἡμέρας ἐσφαγμένου· τὸ γὰρ ἔωλον εὐ-
πεπτότερόν ἐστι τοῦ προσφάτου. θέρους δ' ἀρκεῖ τεθύ-
σθαι μὲν ἕωθεν τὸ ἱερεῖον, ἐσθίεσθαι δὲ περὶ δυσμὰς
ἡλίου· πειρατέον γὰρ ἄχρι παντὸς μὲν διαφυλάττειν ὃ
παρῃνέσαμεν ἐν τοῖς ἔμπροσθεν, ἡνίκ' εἰς ἑσπέραν
ἠξιοῦμεν δίδοσθαι τὴν ἰσχυροτέραν τροφήν, οὐ μικρῶν ἀγαθῶν ἀπολαβόντων τῶν οὕτω διαιτωμένων·
γνώσῃ δ' ἐπισκεψάμενος ἀκριβέστερον· ἄμεινον γὰρ
ἐπ' ἀρχὴν ἀναγαγεῖν τὸν λόγον. ἐπειδὴ δέονται μὲν οἱ
οὕτως ἔχοντες ὅτι πλείστου τοῦ θρέψαντος, οὐ δύναν-

body is thinned, if it is not properly nourished. It is reasonable, therefore, to require a more precise regimen in this
case due to the weakness involving the concoctions, but
whenever they are made better, gradually change patients
toward the so-called nutritive and restorative treatment.
This may be through the use of numerous rubbings and
prior passive exercises. In this, it is also necessary to
change the nutriments in terms of both quantity and quality so as to give more and stronger foods than previously.
With the passage of time, when they come near to a state of   489K
health, set aside ptisan and milk, and even more than this,
porridge and gruel, and then go on to their customary
foods. Also take away fish, if these are not customary, and
provide a diet of birds, meat and the flesh of swine, beginning with the feet boiled down in ptisan. After these, leg
ham may be good to take, and then also the other parts.
First, the young hog should be killed one day before it is
eaten, if it is winter, for what is a day old is easier to digest
than what is fresh. In summer, it is enough for the animal
to have been killed in the early morning and eaten around
evening. You must endeavor throughout to observe closely
what I recommended previously. That is, I think it worthwhile to give the stronger nourishment toward evening,
since in this way a lot of the good of what is eaten is received. You will know this when you consider the matter
more precisely because it is better to advance the argument from the beginning. Since those who are so disposed
require much nourishment yet are not able to digest even

ται δὲ πέττειν οὐδὲ τὰ μέτρια, κατὰ βραχὺ καὶ πολ-
490K λάκις ἄμεινον αὐτοῖς | διδόναι. διὰ τοῦτο οὖν οὐ μόνον
ἅπαξ οὐχ / οἷόν τ᾽ ἐστὶ τρέφειν αὐτοὺς ἱκανῶς, ἀλλ᾽
οὐδὲ δίς, ἕως ἂν ὦσιν ἰσχνότατοι· καὶ διὰ ταύτην τὴν
αἰτίαν εἴωθα τρὶς αὐτοὺς τρέφειν.

ὅταν μέντοι βελτίους ἑαυτῶν ἐναργῶς ἤδη φαίνονται γεγονότες, ἀρκεῖ καὶ δὶς αὐτοὺς τρέφειν. τοσαύτην
οὖν εὔλογον εἶναι αὐτοῦ τὴν πρώτην τροφήν, ὡς
ἀκριβῶς ἅπασαν πεπέφθαι πρὶν ἢ τὴν δευτέραν λαμβάνεσθαι. τοῦτο δὲ οὐχ οἷόν τε ταῖς ἰσχυροτέραις
ὑπάρξαι. ληπτέον οὖν ἀσθενῆ τὴν πρώτην, ὡς πεφθῆ-
ναί τε τάχιστα καὶ ὑπελθεῖν αὐτῆς τὸ περίττωμα καὶ
κενῇ καὶ καθαρᾷ τῇ γαστρὶ τὴν δευτέραν δοθῆναι
τροφήν. ἀλλ᾽ ἐπεὶ μὴ μόνον ἡσυχία καὶ ὕπνος, ἀλλὰ
καὶ χρόνος μακρότερος ἐκδέχεται τὴν ταύτης οἰκονο-
μίαν, εὔλογον ὅσα τῶν σιτίων ἰσχυρότερα κατὰ τοῦτον μάλιστα δίδοσθαι τὸν καιρόν. ἀμέλει καὶ οἱ
ἀθλοῦντες οὕτω πράττουσιν ἅπαντες, οὐ λόγου μόνον
ἐξευρόντος αὐτοῖς τινος, ἀλλὰ καὶ τῆς πολυχρονίου
πείρας μαρτυρούσης. καὶ μέντοι καὶ ὅπερ ἐκεῖνοι
πράττουσιν οὐδὲν χεῖρόν ἐστι καὶ τοὺς ἀναληπτικῶς
ἀγομένους ποιεῖν καὶ μὴ πίνειν ἐπὶ τῷ δείπνῳ παρ-
491K αυτίκα, πρὶν πεφθῆναι | τὴν τροφήν· ἐμπλέει γὰρ τὰ
σιτία πινόντων, ὡς μὴ ψαύειν αὐτῶν τὸ τῆς γαστρὸς
σῶμα, μέσης ἱσταμένης τῆς ὑγρότητος. εἰ δὲ διψώδεις
εἶεν, ἀπέχειν μὲν αὐτοὺς τὸ πάμπαν ἀνιαρόν τε ἅμα
καὶ ἀλυσιτελές· ὀλίγον δὲ δοτέον ὡς παραμυθήσα-
σθαι τὴν ἀνίαν καὶ πραῦναι τὸ δίψος. ἐπειδὰν δὲ

moderate [amounts], it is better to give them [the nourishment] little and often. Because of this, it is not only impossible to nourish them adequately at one sitting, but also impossible at two while they are very thin. For this reason, I am accustomed to nourish them three times (a day). 490K

Certainly, when they have already obviously improved, it is sufficient to nourish them twice (a day). Therefore, it is reasonable that the patient's first nourishment should be of such a quantity that it is digested completely before he takes the second nourishment. However, this is not possible with nutriments that are quite strong. You must choose, therefore, a weak first [nourishment], so it is digested very quickly and its superfluities pass. The second nourishment is then given to an empty and purged stomach. Since not only rest and sleep but also a longer time is required for the management of this nourishment, it is reasonable for stronger foods particularly to be given at this time. Of course, all athletes also do this since they discover these things, not only by reason but also on the evidence of experience over a long time. Furthermore, what they do is also acceptable to those who are acting restoratively, like not drinking immediately with the main meal before the 491K nutriment has been digested. If they do drink, the foods flow in such a way that they do not touch the body of the stomach, the fluid being interposed. If, however, the patients are thirsty, keep everything that is both distressing and unfavorable away from them. You must give as little as will comfort the distress and assuage the thirst. When they

πέψωσι τὴν τροφήν, ἐπιτρέπειν αὐτάρκως πιεῖν· ἀνα-
δίδοται γὰρ οὕτω πραξάντων τάχιστα.

μετὰ δὲ τὸν ὕπνον, ἕωθεν μὲν ἀποπατήσαντα καὶ
βραχύ τι περιπατήσαντα τρῖψαι συμμέτρως. ἡ δὲ
σύμμετρος ἐπὶ τούτων τρῖψις ἐστίν, ὡς θερμῆναι τὸ
σῶμα· κἄπειτα αἰώραις χρηστέον· εἶτ᾽ αὖθις λουτέον
τε καὶ τριπτέον ἐντὸς τοῦ μέσου τῆς ἡμέρας, ἢ πάντως
γε περὶ τὸ μέσον ὡς ἱκανὸν γίγνεσθαι τὸ πρὸς τὴν
ἑσπέραν διάστημα. τὸ δ᾽ ὅτι καὶ τῶν οἴκων ἐν οἷς ὁ
ἀναλαμβανόμενος διαιτᾶται φροντιστέον ἐστὶν ἐν
τοῖς μάλιστα, ὅπως μὴ ψυχρότεροι τοῦ προσήκοντος
ἢ θερμότεροι γενηθῶσιν, οὐδὲν δέομαι λέγειν· ὅσα τ᾽
ἄλλα τοιαῦτα δυνατὸν ἑαυτῷ τινα χωρὶς ἡμῶν εὑρίσκειν, οὐ μόνον ἐκ τῶν ἐνταῦθα λελεγμένων ὁρμώ-
492K   μενον, ἀλλὰ καὶ ἐκ τῆς Ὑγιεινῆς | πραγματείας. ἔστι
γὰρ ἡ ἀναληπτικὴ δίαιτα μέση τῆς ὑγιεινῆς τε καὶ
θεραπευτικῆς. ὥστε ἐκ τῶν καθ᾽ ἑκατέραν λεγομένων
οὐδὲν ἔτι χαλεπὸν ἑαυτῷ τινα τὸ μέσον ἐξευρίσκειν.
τά τε γὰρ ἄλλα καὶ οὐχ ἓν εἶδος ἅπασι διαίτης ἐν ἔθει·
τοῖς μὲν γὰρ ἅπαξ, τοῖς δὲ δὶς σιτεῖσθαι, καὶ τοῖς μὲν
θᾶττον, τοῖς δ᾽ εἰς ἑσπέραν ἄρχεσθαι τῆς προσφορᾶς
ἔθος ἐστί. καὶ δὴ καὶ τὰ ἐσθιόμενα τοῖς μὲν ἰχθύες
εἰσὶ τὸ πλεῖστον ἢ ὅλως τὰ ἐκ τῆς θαλάττης, τοῖς δὲ
λάχανα μᾶλλον ἢ ὅλως τὰ ἐκ κήπων τοῖς δὲ ὀπώραι,
τοῖς δὲ ὄσπρια, τοῖς δὲ ἀκρόδρυα, τοῖς δὲ κρέα τετρα-
πόδων ζώων, ἄλλοις ἄλλα, τοῖς δὲ ὀρνίθων, ἄλλων
ἄλλοις καὶ ταῦτα. καὶ δὴ καὶ τυρῷ καὶ γάλακτι χρῶν-

have digested the nourishment, permit them to drink sufficiently because, when they do this, distribution occurs very quickly.

Early in the morning, after sleep, when they have urinated and walked about a little, rub them moderately. The rubbing is moderate in these cases so as to warm the body. Then you must use passive exercises, and next, you must bathe and rub them before the middle of the day, or at least sometime around the middle of the day, so there is a sufficient interval before evening. I need hardly mention that you must give particular thought to the room in which the recovering patient is living so he is not made colder or hotter than is appropriate. It is possible for someone to discover other such things for himself without me, not only being stimulated by what has been said here, but also from my treatise *On the Preservation of Health*.[25] There is a restorative regimen midway between that of health and that of treatment. As a result, it is not difficult for someone to discover for himself a middle course between each of the regimens described. As with other things, there is not one form of regimen habitual for everyone, since it is the custom for some to eat once only, for some to eat twice, for some to eat rather quickly, and for some to begin their intake toward evening. And particularly, those things eaten by some are predominantly fish, or in general, things from the sea, and by some predominantly vegetables, or in general, things from the garden, by some fruits, by some pulses, by some nuts, by some flesh of quadrupeds, and by others, different things, by some birds (different ones for different people), and so on. In fact, some also use cheese

492K

---

[25] See Book 2 of Galen's *De sanitate tuenda* (VI.81–163K).

ταί τινες, οἱ πλεῖστοι μὲν αἰγῶν τοὐπίπαν ἢ βοῶν,
ἔνιοι δὲ καὶ ἑτέρων τινῶν. καὶ πίνουσιν οἱ μὲν θερμόν,
οἱ δὲ ψυχρόν· καὶ οἶνον ἢ πολύν, ἢ ὀλίγον, ἢ οὐδόλως,
ἢ τοῖον, ἢ τοῖον. εἰς ταῦτ᾽ οὖν ἐπανάγειν ἕκαστον εἰς
ἅπερ εἴθισται.

γέγραπται δὲ καὶ περὶ ἔθους ἀκριβέστερον ἰδίᾳ.
493K καὶ νῦν ἀρκεῖ τό | γε τοσοῦτον εἰπεῖν, ὡς οὐ μόνον ἐπὶ
τῆς ἀναληπτικῆς διαίτης, ἀλλὰ καὶ κατὰ τὰς νόσους
παμπόλλην μοῖραν εἰς ὕλης τροφῆς καὶ φαρμάκου
ἔκλεξιν ἐκ τῶν ἐθῶν ἐστι λαβεῖν. ὁ μὲν γὰρ λόγος
ὅλου τοῦ γένους ὑπαγορευτικώτερός ἐστι· τὰς δὲ ὕλας
ἔκ τε τῶν ἄλλων ὧν εἶπον, ἀτὰρ οὐχ ἥκιστα καὶ ἐκ τῶν
ἐθῶν ληπτέον. ἐπιβλέπειν δὲ καὶ τὰς τῶν φύσεων
ἰδιότητας. οἶδα γοῦν τινας, εἰ διὰ τῆς νυκτὸς ὁτιοῦν
ἀναγκασθεῖεν ἐγρηγορικὸν διαπράξασθαι, μηκέτι κοι-
μηθῆναι δυναμένους. ἐκείνοις οὖν οὐ χρὴ συμβουλεύειν ἐν τῇ νυκτὶ πίνειν, μή πως δι᾽ αὐτὸ τοῦτο περι-
πεσόντες ἀγρυπνίᾳ βλαβῶσι τὰ μέγιστα· ξηραίνει
γὰρ ὅλου τοῦ σώματος τὴν ἕξιν ἡ ἀγρυπνία καὶ διὰ
τοῦτό ἐστι βλαβερωτάτη τοῖς κατὰ ξηρότητα νοσοῦσιν. ἔνιοι δὲ εἰ γεύσαιντο πτισάνης, αὐτίκα ναυτιῶσιν·
ἐνίοις δ᾽ ὀξύνεται ῥᾳδίως. ὅτι δὲ καὶ τὰ βόεια κρέα
πέπτουσί τινες ῥᾳδίως, ἔν τι τῶν τεθρυλημένων ἐστί·
καὶ ὡς ἄλλοι πρὸς ἄλλα βρώματά τε καὶ πόματα
διάκεινται εὖ τε καὶ κακῶς. οὐ μόνον δ᾽ ἐκ τούτων,

---

26 This is presumably a reference to Galen's De consuetudini-
bus, which is not in Kühn but can be found in volume II of Galeni
Scriptora Minora (ed. I. Müller, 1891), and in a French translation
in vol. 1 of Daremberg's Galen translations.

and milk, the majority from goats or cows entirely, but some also from certain other [animals]. And some prefer hot drinks and others cold, and either a lot of wine or a little, or none at all, or of such and such a kind. Therefore, introduce each person to those things to which he is accustomed.

I have also written specifically and more comprehensively about customs.[26] For now, it is enough to say just 493K this—that not only with regard to the restorative regimen, but also in relation to diseases in great part, make the choice of both nutritive material and medication from those things that are customary, as the discussion is more prescriptive of the whole class. You must choose the materials from the other things I spoke of, but no less also from those things that are customary. However, also look closely at the specific aspects of patients' natures. Anyway, I know some who, if they are compelled to do anything whatsoever in a wakeful state during the night, are not able to sleep any longer. Therefore, it is necessary to advise those people not to drink during the night lest, in some way due to this very thing, when they encounter sleeplessness, they harm themselves greatly, for sleeplessness dries the state of the whole body and, because of this, is most harmful to those who are sick with dryness. However, some, if they taste ptisan, are immediately nauseated, whereas in others it readily turns sour. That some digest the flesh of the ox easily is one of the things that has become public knowledge, as is the fact that some are well disposed toward other meats and drinks and some are not. But it is not just from those factors: advantage derived from the mate-

ἀλλὰ κἀκ τῶν ὡρῶν τε καὶ χωρῶν εὐπορία τῆς ὕλης
τῶν διαιτημάτων γίγνεται. καὶ ἤδη πολλοῖς τῶν πρὸ
494K   ἐμοῦ πρεσβυτέρων γέγραπται | ταῦτα· διὸ κἀγὼ
παρατρέχω μὲν τὰ τοιαῦτα, μόνον ἀναμιμνήσκων αὐτῶν, ὑπὲρ τοῦ μηδὲν παραλείπεσθαι.

   τὸ παραλελειμμένον δὲ τοῖς νεωτέροις ἰατροῖς ἅπασιν ἐπέξειμι, τὸ τὰς διαθέσεις τῶν σωμάτων θεραπεύειν μεθόδῳ τῶν παλαιῶν, ὡς εἶπον καὶ πρόσθεν,
ἀρξαμένων μὲν καλῶς, οὐ συντελεσάντων δὲ τὴν μέθοδον. ἀλλ' ἐπὶ τὸ προκείμενον ἰτέον αὖθις. αἱ ξηραὶ
δυσκρασίαι τῶν σωμάτων αὐτῶν ἁψάμεναι τῶν στερεῶν, εἰ μὲν περί τι μόριον συσταῖεν, ὥσπερ ὑπόκειται
νῦν κατὰ τὴν γαστέρα, τελέως οὐκ ἄν ποτε θερα-
πευθεῖεν, οὐδ' ἐπανέλθοιεν εἰς τὴν πρὸ τοῦ νοσῆσαι
διάθεσιν. ἀλλ' ὅπερ εἴωθε λέγεσθαι συνήθως ὑπ'
αὐτοῦ τοῦ πράγματος διδαχθεῖσι τοῖς πολλοῖς, ὡς ἡ
νόσος ἐκάκωσε τὴν γαστέρα καὶ ἀσθενῆ παρεσκεύ-
ασεν εἰς τὸν ἔπειτα βίον, ἐπὶ ταῖς τοιαύταις μάλιστα
συμπίπτει δυσκρασίαις. γέροντος γὰρ οὗτοι ἴσχουσι
γαστέρα πρὶν γηρᾶσαι· διὸ καὶ βλάπτονται ῥᾳδίως
ὑπὸ σμικρῶν αἰτιῶν, ὥσπερ οἱ γέροντες, καὶ πέττειν
οὐ δύνανται καλῶς· ἐξ οὗ συμβαίνει καὶ τὸ σύμπαν
495K   αὐτοῖς σῶμα | χεῖρον ἴσχειν· οὕτως δὲ κἂν ἄλλο τι
πάθωσι μόριον, ἂν δὲ καὶ τὴν καρδίαν αὐτήν, ἐν τάχει
μαραίνονται· καὶ καλεῖ τὸν τοιοῦτον μαρασμὸν ὁ Φί-
λιππος ἐκ νόσου γῆρας.

292

rial of food and drink also arises from the seasons and places. These things have already been written about by many older than myself. Accordingly, let me run through 494K such things only to call them to mind so that nothing is left out.

I shall go through in detail what has been left out by all the younger doctors; that is, treating the conditions of bodies by a method of the ancients who, as I also said previously, although beginning well, did not bring the method to completion. But I must go back again to what was proposed. When the dry *dyskrasias* involve bodies that are themselves solid, if they should exist in relation to some part, as is now postulated in relation to the stomach, they would never be treated completely and would not return to the condition prior to the disease. What is usually stated on a regular basis by the majority who were taught by the matter itself is that the disease afflicts the stomach and renders it weak for the rest [of the patient's] life, and this particularly happens due to such *dyskrasias*. These people have the stomach of an old man before they are old, and this is why they are easily harmed by minor causes just as the aged are, and are not able to digest properly. From this, what also happens to them is that the whole body is worse. 495K And so, if they are affected in some other part, and if it should be the heart itself, they quickly become wasted. Philip calls such a marasmus "old age from a disease."[27]

27 This is taken to be a reference to Philip of Egypt (AD 100–170), who wrote on marasmus and on a regimen for retaining eternal youth. He is to be distinguished from Philip of Rome (AD 45–95), whom Galen refers to in a later book on the subject of bathing.

293

οὗτος μὲν οὖν ὁ μαρασμὸς ἐπὶ θάνατον ἄγει συν-
τόμως· ἐφεξῆς δὲ καὶ ὁ ἀπὸ τοῦ ἥπατος ἀρξάμενος·
εἶθ᾽ ὁ ἀπὸ τῆς γαστρός· οἱ δ᾽ ἐπ᾽ ἄλλοις μορίοις εἰς
τοσοῦτον χρονιώτεροι εἰς ὅσον ἕκαστον αὐτῶν ἀκυρώ-
τερόν ἐστι τῶν εἰρημένων. εἰ δὲ βραχύ τι τὸ τῆς
καρδίας σῶμα ξηρανθείη, ταχύγηροι μὲν γίγνονται,
διαρκοῦσι δ᾽ εἰς ἔτη πλείω τῶν ἰσχυρῶς βλαβέντων.
μετὰ ταῦτα δ᾽ ἐφ᾽ ἥπατι καὶ γαστρί, καθάπερ εἴρηται,
κἀπὶ τοῖς ἄλλοις ἅπασιν ἀνάλογον. ἀλλ᾽ ἥ γε δίαιτα
κατὰ γένος ἡ αὐτὴ καὶ ταύτῃ τῇ ξηρότητι καὶ τῇ τὴν
ὑγρότητα μόνην ἐκδαπανώσῃ, τὴν τρέφουσαν τὰ στε-
ρεά· καὶ πρὸς ταύταις γέ τοι τῇ τρίτῃ ῥηθείσῃ ξηρότητι κατὰ τὸν ἔμπροσθεν λόγον. εὐιατοτάτη δὲ πασῶν
ἐστιν ἡ τετάρτη, καθ᾽ ἣν αἱ φλέβες ἐκκενοῦνται τῶν
χυμῶν. καὶ μᾶλλον ὡς ψυχρότης ἕξεως ἡ τοιαύτη
θεραπεύεται διάθεσις ἤπερ ὡς ξηρότης· καὶ ἵνα
σαφέστερον[12] εἴπω, τὸ μὲν τῆς ψύξεως ἐπικρατεῖ καὶ ὁ
496K κίνδυνος | κατὰ τοῦτο. δευτέρα δ᾽ ἐστὶν ἡ ξηρότης,
ὅθεν καὶ ἡ ἴασις ἑτοίμη τε καὶ ταχίστη. δύο γὰρ
ἡμέραις εἴ τις αὐτοὺς διαιτήσειε πεφυλαγμένως θερ-
μαίνων τε ἅμα τὰ μέτρια καὶ ἀνατρέφων ἀσφαλῶς,
εὐθέως τῇ τρίτῃ προσίενται δίαιταν ἁδροτέραν ἀβλα-
βῶς· καὶ πολὺ δὴ μᾶλλον τῇ τετάρτῃ καὶ τῇ πέμπτῃ.
καὶ τοῦτο ἄρα ἦν τὸ ὑφ᾽ Ἱπποκράτους λεγόμενον, Τὰ
ἐν πολλῷ χρόνῳ λεπτυνόμενα νωθρῶς ἐπανατρέφειν,
τὰ ἐν ὀλίγῳ ὀλίγως.

7. Ἰστέον δ᾽ ὅτι ταῖς ξηρότησι τῶν στερεῶν

This marasmus leads to death in a short time. Next, there is that which begins from the liver, then that which begins from the stomach. Those due to other parts are more chronic to the extent that each of them is more unimportant than those spoken of. If the body of the heart is made dry to a slight extent, people soon become decrepit, but they hold out for many years compared to those who are harmed severely. After these, come those made dry in relation to the liver and stomach, just as I said, and analogously in relation to all the other parts. But the regimen in terms of class is in fact the same, both in this dryness and in that which only exhausts the moistness that nourishes the solid parts, and in addition to these, in the third dryness spoken of in the previous discussion. The fourth dryness is the easiest of all to remedy—that in which the veins are emptied of humors. Such a condition is treated more as a cold state than as a dry state. To be more explicit, it is the coldness that prevails, as does the danger associated with   496K
this. The dryness is secondary, which is why the cure is both easy and very quick. Thus, in two days, if someone regulates these things carefully, heating moderately and at the same time nourishing safely, on the third day they progress immediately to a thicker diet without harm, and much more particularly, on the fourth and fifth days. And this was what Hippocrates said: "Restore slowly those who are made thin over a long time, and restore quickly those who are made thin over a short time."[28]

7. You must be aware that cooling inevitably follows the

[28] Hippocrates, *Aphorisms* II.7.

---

[12] K; ἀσφαλέστερον B

σωμάτων ψύξις ἐξ ἀνάγκης ἀκολουθεῖ χρονιζούσαις.
διαμένειν γὰρ αὐτὴ καθ᾽ ἑαυτὴν ἡ ξηρότης, ἀμέμπτου
τῆς κατὰ τὸ θερμὸν καὶ ψυχρὸν ἀντιθέσεως ὑπαρ-
χούσης, οὐ δύναται. τάχιστα γὰρ ἐπὶ ταῖς ἀτροφίαις
ἀποψύχεται τὰ μόρια, διότι καὶ ἡ θρέψις αὐτοῖς ἐστιν
ἐκ θερμοῦ χυμοῦ τοῦ αἵματος. ἀλλ᾽ ὅπερ ἐν ἀρχῇ
λέλεκται, ξηρότητος ἴασιν ἐποιησάμεθα νῦν ἐξ ἀρχῆς
τε συστάσης μόνης ἄχρι πλείστου τε μηδεμίαν ἀξιόλογον ἀκολουθοῦσαν ἐχούσης ψύξιν. ἑξῆς ἂν οὖν εἴη
μιγνύειν αὐτῇ ψύξιν ἐναργῆ μὲν ἔχουσαν τὰ γνω-
497K ρίσματα, μὴ μέντοι μεγάλα.[13] | ἔσται τοίνυν ὁ σκοπὸς
τῆς θεραπείας οὐκέθ᾽ ἁπλοῦς, ὡς ὁ καὶ πρόσθεν· οὐ
γὰρ ὑγραίνειν μόνον, ἀλλὰ καὶ θερμαίνειν δεήσει.

προσέχειν οὖν ἀκριβῶς ταῖς ὕλαις τῶν βοηθημάτων μή τι λάθωμεν ἡμᾶς αὐτούς, ἐπιπλέκοντές τι τῶν
ἄγαν θερμαινόντων τοῖς ὑγραίνουσιν· ὑπάρχει γὰρ
τούτοις τὸ ξηραῖνον ἰσχυρῶς. ἀλλὰ καὶ τὸ ποσὸν τῆς
βλάβης ἐν ἑκατέρᾳ δυσκρασίᾳ πρὸς τὴν τῆς ὕλης
εὕρεσιν ἱκανὸν ἔσται ποδηγεῖν, οἷον αὐτίκα μεγίστης
μὲν ξηρότητος ἴασιν ἄρτι πέπαυμαι γράφων· αἱ μέ-
τριαι δὲ τῆς οὕτως ἀκριβοῦς οὐ δέονται διαίτης, ὅτι
μηδὲ τὰ τῆς δυνάμεως ἐπ᾽ αὐτῶν ἐσχάτως καταπέπτω-
κεν. οὐδὲ γὰρ οὐδὲ δι᾽ ἄλλο τι χαλεπὸν γίγνεται
θεραπεῦσαι καλῶς ἰσχυρὰν ξηρότητα σώματος ἢ ὅτι
χρῄζει μὲν ἀναθρέψεως, ἡ τροφὴ δ᾽ οὐ πέφυκεν ἑαυτὴν
προσφύειν, ἀλλ᾽ αὐτοῦ δεῖται τοῦ τρεφομένου δημι-
ουργοῦντος αὐτήν, ὅπερ ἐστὶν ἄτονον. ὅταν οὖν ᾖ
μετρίως τοῦτο ξηρόν, ἄτε μηδὲ τῆς δυνάμεως ἐσχάτως

dryness of a solid body that is long-lasting. It is impossible for dryness to remain pure in itself and to be uncontaminated by the opposition between hot and cold. Due to the atrophies, parts become cooled very quickly because their nourishing is from the hot humor of the blood. But what I said at the start is that we effect a cure of dryness from the beginning when it exists alone to a great extent; there is no significant cooling that follows. Next would be to mix a distinct cold having visible manifestations with the dryness—not major ones, however. Accordingly, the indicator 497K of treatment will no longer be simple as it was before, because not only will this require moistening, but it will also require heating.

Therefore, we must pay careful attention to the materials of the remedies so that we don't fail to notice something when we combine one of the extreme warming agents with those that are moistening, because the drying is strong with these. But also the degree of harm in each *dyskrasia* will be sufficient to lead us to the discovery of the material, as in the case of the cure of severe dryness which I shall finish writing about just now. The moderate drynesses do not require a regimen that is precise in this way in that the components of the capacities in these cases have not fallen away to an extreme degree. There is no difficulty in properly treating strong dryness of a body other than the fact that the nourishment it requires restoratively is not of a nature to assimilate itself, but needs the effector of what is actually nourishing it, which is weak. Therefore, whenever this is moderately dry, inasmuch as the capacity has not

---

καταπεπτωκυίας, ἁδρότερόν τε διαιτᾶν ἐγχωρεῖ καὶ κίνδυνος οὐδεὶς ἁμαρτεῖν τοῦ μέτρου. διὰ τοῦτο καὶ

498K ἡμεῖς ἐπὶ τὴν χαλεπωτάτην δυσκρασίαν | τὸν λόγον ἀγαγόντες ἐγυμνάσαμεν τοὺς φιλομαθεῖς ἐπ᾽ αὐτῆς ὡς ἐπὶ παραδείγματος· ἐντεῦθεν γὰρ ὁρμηθέντες αὐτοὶ ἐξευρήσουσιν ὕλας ἐπιτηδείας εἰς ἐπανόρθωσιν ξηροτήτων μετρίων.

ὑποκείσθω δὲ καὶ νῦν ἔτι παραπλησία μὲν ἡ ξηρότης τῇ πρόσθεν, ἐζεύχθω δ᾽ αὐτῇ ψυχρότης. εἰς τοσοῦτον οὖν ἐπιμίξομεν τοῖς ἔμπροσθεν εἰρημένοις τὴν τῶν θερμαινόντων ὕλην εἰς ὅσον ἔψυκται τὸ μόριον. ὑποκείσθω δὴ πρῶτον ἐψῦχθαι μετρίως αὐτό· προσθήσομεν οὖν τοῖς εἰρημένοις ὀλίγον ἔμπροσθεν, ἐν μὲν τῇ τοῦ γάλακτος χρήσει τὸ μέλι πλέον, ἐν δὲ τῇ τοῦ οἴνου τὸν αὐτὸν μὲν τῷ γένει δώσομεν, ἧττον δ᾽ ὑδαρῆ τοῦ πρόσθεν, ἐτῶν δὲ πλειόνων. ἀλλὰ καὶ αὐτὰ τὰ ἐδέσματα σύμπαντα θερμότερα δοτέον οὐ ταῖς φυσικαῖς κράσεσι μόνον, ἀλλὰ καὶ ταῖς προσφάτοις ποιότησι. καὶ ναρδίνῳ μύρῳ συνεχῶς ἐπαλειπτέον τὴν κοιλίαν ὡς μηδέποτε γενέσθαι ξηράν. εἰ δὲ τοῦτο μὴ παρείη, τὸ μαστίχινον ἱκανόν· ἀλλὰ καὶ ὀποβαλσάμῳ χρίσομεν αὐτήν, αὐτῷ τε καθ᾽ ἑαυτὸ καὶ μιγνύντες τοῖς προειρημένοις. καὶ εἰ μᾶλλον ἐθέλοιμεν ἔχεσθαι

499K τοῦ χρωτὸς αὐτά, καὶ κηροῦ τι | προσμίξομεν. εἰ δὲ καὶ τὸ περιέχον εἴη ψυχρόν, ἔριον βρέξαντες ἐπιθήσομεν. ἀλλὰ καὶ τὴν Χίαν μαστίχην ἐν ὀποβαλσάμῳ λειώσαντες, ἢ μύρῳ ναρδίνῳ, κἄπειτα ἔριον δεύσαντες ἐπιθήσομεν τῇ γαστρί. κάλλιον δὲ τὸ δευόμενον ἀρι-

298

fallen away to an extreme degree, it is possible to feed more abundantly without any danger of erring in moderation. And because of this, when I brought the argument to 498K the most difficult *dyskrasia*, in this case I trained those who are fond of learning by way of an example. Henceforth, being encouraged, they will discover suitable materials for the correction of moderate dryness.

Now let us assume dryness similar to that previously [spoken of], but let coldness be joined with it. We shall, then, mix with the aforementioned things the material of those agents that are heating to the extent that the part was cooled. Let us suppose first that it was cooled moderately. We shall apply those things spoken of a little earlier in the use of milk and more honey, while in the use of wine, we shall give that which in class is less watery than before and of greater age. But you must also give all the foods that are themselves warmer, not only in their natural *krasis*, but also in their fresh qualities. And you must continually smear the belly with oil of spikenard so that it never becomes dry. If this is not available, mastich is adequate, but we shall also rub the abdomen with the oil of the balsam tree, either by itself or mixed with those things previously mentioned. If we wish these same things to adhere to the skin more, we shall mix in some wax. If the ambient air is 499K cold, we shall apply wool soaked [in these things], but we shall also apply Chian mastich triturated with the oil of the balsam tree or oil of spikenard and then, after moistening the wool, we shall apply it to the abdomen. Better is the

στην εἶναι πορφύραν οἷα πέρ ἐστιν ἡ Τυρία· στύφειν
τε γὰρ χρὴ μετρίως αὐτήν, ἐμπλάττειν τε τοῦ δέρματος τοὺς πόρους.

τὰ μὲν γὰρ ἀραιοῦντα καὶ χαλῶντα διαφορεῖ τε
ἅμα καὶ τοῖς ἔξωθεν προσπίπτουσι ψυχροῖς εὐάλωτα
παρασκευάζει τὰ μόρια· τὰ δ' ἐπὶ πλέον στύφοντα
ξηραίνει. ὅθεν οὔτε διαφορητικῆς εἶναι προσήκει δυνάμεως τὸ θερμαῖνον φάρμακον τὰ οὕτως διακείμενα
σώματα καὶ μὴ πολλὴν ἔχειν ἐπιμεμιγμένην στύψιν.
ἀναμνήσθητι γὰρ ὧν ἐπί τε τῶν ἀρρώστων ἐθεάσω
κἀν ταῖς περὶ τῶν φαρμάκων πραγματείαις ἐδιδά-
χθης, ὡς κατὰ μὲν τὴν ἑαυτῆς δύναμιν ἡ στρυφνὴ
ποιότης ψύχει τε καὶ ξηραίνει· ξυμμιγνυμένη δέ τινι
τῶν ἰσχυρῶς θερμαινόντων τὸ συγκείμενον ἐξ ἀμφοῖν
ἧττον μὲν ἐργάζεται θατέρου θερμαῖνον, οὐχ ἧττον δ'
500K ἑκατέρου ξηραῖνον. ἐναντιώτατα | δὲ τούτων οἱ νῦν
ἰατροὶ διαπράττονται, ξηρῷ καὶ ἀτρόφῳ μορίῳ πολλάκις ἐπιτιθέντες ἐκ θερμῆς καὶ στρυφνῆς ὕλης σύνθετον φάρμακον. ἔστι δὲ οὐδὲν τῶν στρυφνῶν μετρίως
στῦφον, ὅθεν οὐδὲ ξηραίνει μετρίως· ἀφεκτέον οὖν
ἐστιν αὐτῶν ἐπὶ τῶν προκειμένων διαθέσεων. εἰ δὲ καὶ
ἡ ψύξις εἴη πολλὴ μετὰ τῆς κατὰ ξηρότητα δυσκρασίας, ὡς ἀναγκάζειν ἐπὶ πλέον θερμαίνειν, πρῶτον
μὲν ἀναμιμνήσκου τοῦ τῆς δυσκρασίας εἴδους· ἐδείχθη γὰρ ἁπάντων χαλεπώτατον· ὥστε δυσεπανόρθω-
τον αὐτὸ γίνωσκε καὶ πολλῆς ἀκριβείας δεόμενον. ἀπ-
έχεσθαι δὲ ὁμοίως τῶν ἰσχυρῶς θερμῶν καὶ στρυφνῶν

moistening of the best purpura such as the Tyrian is,[29] and this must be moderately astringent to stop up the pores of the skin.

For those things that are of loose texture and relaxing disperse and, at the same time, prepare the easily affected parts against cold things that befall them externally. Those things that are more astringent dry, hence it is inappropriate for the medication heating the bodies so disposed to have a discutient potency or to have much intermixed astringency. You must bear in mind what you saw in those who are sick, and even what you learned in the treatises on medications: that the astringent quality, by virtue of its own potency, cools and dries, whereas, when it is mixed with one of those things that is strongly heating, what is compounded from both brings about less heating than the latter, but is not less drying than either. Doctors now accomplish the most opposite effects to these since they often apply a medication compounded from hot and astringent material to a dry and atrophic part. None of the astringents is moderately astringent, which is why none dries moderately. You must, therefore, avoid these in the conditions under consideration. If there is severe cooling in conjunction with a dry *dyskrasia,* so it is necessary to heat still more, you must first bear in mind the kind of *dyskrasia*, for it was shown to be the most difficult of all to deal with. Consequently, you must know it is difficult to correct and requires great precision. On this basis, you must try to avoid those things that heat strongly and are as-

500K

[29] The brief entry in Dioscorides, II.4 states (in Goodyer's translation): "The Purpura being burnt hath a facultie of drying and cleansing ye teeth, of repressing excrescent flesh, of drawing boyles and healing them" (p. 93).

πειρῶ, ἐλπίζων ἐν χρόνῳ πλείονι διὰ τῶν ἀσφαλεστέρων ἰᾶσθαι τὴν διάθεσιν.

ἀσφαλέστεραι δέ εἰσιν αὗται, αὖθις γὰρ εἰπεῖν ὑπὲρ τῶν ὄντως ἀναγκαίων ἄμεινον, ἐν ναρδίνῳ μύρῳ λειώσας μαστίχην Χίαν ὡς λιπαρωτάτην, ἀναλαμβάνων πορφύρᾳ χρῶ. μιγνύναι δ' ἄμεινον εἰ παρείη καὶ τὸν ὀπὸν τοῦ βαλσάμου. διδόναι δὲ καὶ τὸ μέλι πλέον 501K ἅμα τῷ γάλακτι προαπηφρισμένον, ὅπως ἧττόν | τε εἴη περιττωματικὸν καὶ μᾶλλον τρόφιμον. ἀλλὰ καὶ αὐτὸ καθ' ἑαυτὸ ἀφηψημένον τὸ μέλι τροφὴ καλλίστη ψυχρᾷ γαστρί, πολεμιώτατον δὲ τῇ θερμῇ. καὶ χρὴ μεμνῆσθαι τούτων ἀμφοτέρων ἐν τοῖς μάλιστα, μὴ γινωσκομένων τοῖς πλείστοις ἰατροῖς, καὶ μήθ' αἱρεῖσθαί τι μέλιτος μᾶλλον ἐν ψυχροῖς σώμασι μήτε φεύγειν ἐν θερμοῖς. ἡ μὲν δὴ πλείστη τροφὴ τῆς τοιαύτης διαθέσεως ἔστω μὲν μέλι τὸ κάλλιστον ἀφηψημένον ἐπ' ἀνθράκων δρυΐνων ἢ ἀμπελίνων εἰς τέλος διακαῶν, ἀφῃρημένου παντὸς τοῦ ἀφροῦ. τὸν δ' οἶνον ἀεὶ καὶ μᾶλλον αἱρεῖσθαι παλαιότερον ὅσῳ περ ἂν ἡ ψύξις ἐπὶ μᾶλλον κρατῇ. καὶ εἰ δεήσειεν ὅλον τὸ γένος ὑπαλλάττειν, Ἀδριανὸν μὲν τὸ πρῶτον, εἶτ' αὖθις Τιβουρτῖνον, ἤ τινα τῶν ὁμοίων αἱρούμενον ἅπαντας παλαιοὺς μέν, ἀλλ' οὐχ οὕτως ὥστε πικροὺς ὑπάρχειν ἤδη· ξηραίνουσι γὰρ οἱ πικροί, περαιτέρω τοῦ δέοντος.

ἄριστον δὲ φάρμακον ἐπὶ τῶν τοιούτων ἁπάντων ᾧ καθ' ἑκάστην ἡμέραν οἱ πιττωταὶ χρῶνται. καταχρίειν οὖν αὐτῷ προσήκει τὴν γαστέρα σύμπασαν· εἶτ'

tringent, hoping to cure the condition over a longer time through those agents that are safer.

The ones that are safer are as follows (it is better to state again those that are truly necessary): Chian mastich triturated in oil of spikenard so as to be very oily, bearing in mind that you must use purpura. It is better also to mix the sap of the balsam tree, if it is available, and to give more honey along with the milk, after removing the froth, so it is less excrementitious and more nutritious. But also, honey   501K
boiled down by itself is the best nutriment for a cold stomach but is very hostile to a hot stomach. It is especially necessary to remember both these things in such situations, things not known to the majority of doctors, i.e. neither to choose honey, especially in cold bodies, nor to avoid it in hot bodies. Let the major nourishment of such a condition be honey, which is best when boiled down in coals of oak, or vine cuttings that are very hot indeed, when all the froth is taken away. Always choose wine for preference that is older because the cooling predominates still more. If there is a need to change the whole class, first the Adrian and then again the Tiburtine, or one of those that is similar is chosen, all being old but not in such a way that they are already bitter—those that are bitter, dry more than is required.

The best medication for all such conditions, and one which [doctors] use every day is pitch. It is appropriate to smear the whole abdomen with this, then to remove it be-

ἀποσπᾶν πρὶν ψυχθῆναι· καὶ τοῦτο ἀρκεῖ ποιῆσαι δὶς |
502K ἐφεξῆς ἐν ἡμέρᾳ μιᾷ· πλεονάκις δ' οὐ χρή· διαφορή-
σεις γὰρ οὕτως, οὐ πληρώσεις αἵματος χρηστοῦ τὸ
μόριον· ἡμεῖς δὲ οὐ διαφορεῖν τὸ περιεχόμενον, ἀλλ'
ἐκ τῶν πλησίον ἐπισπᾶσθαι βουλόμεθα. τοῦτο καὶ ἐπὶ
τῶν ἠτροφηκότων ἁπάντων μορίων θαυμαστὸν φάρμακον, ἤν τις ἐπίστηται μετρίως χρῆσθαι. καὶ μέντοι
καὶ τὸ θερμαίνειν τὴν γαστέρα, μὴ ποιότητι παρ-
αύξοντα τὸ θερμόν, ἀλλ' ὅτι μάλιστα φυλάττοντα μὲν
αὐτοῦ τὴν κατὰ φύσιν εὐκρασίαν, αὐξάνοντα δὲ τὴν
οὐσίαν, οὐχ ὁ φαυλότατός ἐστι τῶν ἐν ταῖς τοιαύταις
διαθέσεσι σκοπῶν. ἐργάζεται δὲ τοῦτο κατὰ μὲν τὴν
δίαιταν ἅμα τοῖς εἰρημένοις ἔμπροσθεν οἶνος μάλι-
στα· τῶν δ' ἔξωθεν τῇ γαστρὶ προσφερομένων εὔσαρ-
κον παιδίον συγκοιμώμενον, ὡς ψαύειν ἀεὶ τῶν κατ'
ἐπιγάστριον. ἔνιοι δὲ καὶ κυνίδια λιπαρὰ τῆς αὐτῆς
ἕνεκα χρείας ἔχουσιν, οὐκ ἐν τῷ νοσηλεύεσθαι μόνον,
ἀλλὰ καὶ ὑγιαίνοντες. τὰ μὲν οὖν τοιαῦτα καὶ τοῖς διὰ
ξηρότητα μόνην ἀτονοῦσι τὴν κοιλίαν ἐπιτήδεια. καὶ
χρὴ περὶ παντὸς ποιεῖσθαι τὸ παιδίον ἄνικμον ἔχειν
τὸν χρῶτα· τὰ γὰρ ἐφιδροῦντα διὰ νυκτὸς ψύχει
503K μᾶλλον | ἢ θερμαίνει. βλάπτουσι δὲ καὶ αἱ θερμαὶ
πυρίαι τὰς τοιαύτας διαθέσεις, αἱ μὲν οὖν ἅμα ξηρότητι συνιστάμεναι, διότι τῶν ὁμοιομερῶν σωμάτων
ἐκβόσκονται τὰς νοτίδας· αἱ δὲ σὺν ὑγρότητι τῷ διαφορεῖν τε ταύτας, καὶ μάλισθ' ὅταν ἐπὶ πλέον χρησώ-
μεθα, καὶ τῷ σφόδρα μανὸν ἐργάζεσθαι καὶ τὸ σῶμα
καὶ εὔψυκτον. περὶ μὲν οὖν ταύτης τῆς συζυγίας τῶν
δυσκρασιῶν ἱκανὰ καὶ ταῦτα.

fore it is cooled. It is enough to do this twice in succession   502K
in one day. More frequently is not necessary because, if
you disperse like this, you do not fill the part with useful
blood. Our wish is not to disperse [the blood] contained [in
the part], but to draw it in from the parts nearby. Pitch is a
remarkable medication in the case of all parts that have
wasted away, if a person knows how to use it moderately.
Indeed, to heat the stomach so as not to increase the heat
in quality, but particularly to preserve its natural *eukrasia*
while increasing the substance is by no means the least important objective in such conditions. Pitch does this in relation to the regimen alone along with those things previously mentioned, particularly wine. Among those things
applied externally to the stomach, there is a young slave,
well-fleshed, lying beside [the patient] so as to be in constant contact with the epigastrium. Some also have little
puppies in good condition for this same use, not only in attending to a sick person, but also for those who are healthy.
Such things are also suitable for those who are weak in the
stomach due to dryness alone. And above all, you ought to
insure that the young slave is without moisture on his skin;
for to perspire throughout the night cools more than it   503K
heats. The hot vapor baths also harm such conditions
because, when these conditions coexist with drying, they
consume the moisture of the *homoiomerous* bodies,
whereas when they coexist with moisture, by dispersing
this [moisture], and particularly whenever we use them
still more, by bringing about excessive looseness, they also
make the body cool. This is enough about this conjunction
of the *dyskrasias*.

8. Μιγνύσθω δ' ἐφεξῆς δαψιλεῖ ξηρότητι, καθάπερ ὑπόκειται, μὴ πάνυ πολλὴ θερμότης. ἐπὶ μὲν τῆς τοιαύτης ὑπαλλάξεως τὴν πρώτην ἀγωγὴν φυλάξομεν, οὐχ ὅπως αὐξάνοντες ἢ τὴν τοῦ οἴνου δύναμιν, ἢ τὸ πλῆθος τοῦ μέλιτος, ἀλλὰ καὶ καθαιροῦντες ὡς μέλιτος μὲν μηδόλως γεύεσθαι, τὸν δ' οἶνον ἥκιστα παλαιὸν προσφέρεσθαι. χλιαρὰ δὲ τὰ ἐδέσματα προσοίσομεν, ἃ καλεῖν ἔθος ἐστὶ τοῖς ἰατροῖς γαλακτώδη· θέρους δ' ὄντος ἃ καλοῦσι κρηναῖα· καὶ τὴν κοιλίαν ἐπαλείψομεν ὀμφακίνῳ τε καὶ μηλίνῳ· πολὺ δὲ δὴ μᾶλλον ἐὰν πλείων ἡ θερμότης ᾖ, τὸν οἶνον ὑδαρέστερόν τε δώσομεν καὶ οὕτως ἔχοντα ψύξεως ὡς τὸ

504K κρηναῖον ὕδωρ | ἐν ἦρι μέσῳ, γινώσκοντες ὡς ἡ τοιαύτη διάθεσις ἀνάλογός ἐστι πυρετῷ· ὅπερ γὰρ ἐκεῖνος ἢ κατὰ σύμπαν τὸ ζῷον, ἢ κατὰ τὴν καρδίαν ἐστί, τοῦθ' ἡ νῦν προκειμένη διάθεσις ἐν τῇ γαστρί. καὶ ἐγὼ πρῶτον μὲν ἁπάντων οἶδά τινα θεασάμενος ἅμα τοῖς διδασκάλοις ἄνδρα τῆς καθεστώσης ἡλικίας, ἐνοχλούμενον ἤδη μηνῶν οὐκ ὀλίγων· ἀλλ' οὔτ' ἐκείνων τις ἐγίνωσκε τὴν διάθεσιν οὔτ' ἐγώ· μετὰ ταῦτα δ' ἀνεμνήσθην εὑρηκὼς ἤδη τὴν θεραπευτικὴν μέθοδον, ὡς τοῦτ' ἄρ' ἦν ἐκεῖνο τὸ θεωρηθέν μοι πάλαι. κάλλιον δ' αὐτὸ καὶ διηγήσασθαι, πάντως γὰρ δή που καὶ τοὺς ἀκούσαντας ὀνήσει, καθάπερ κἀμέ.

τετταρακοντούτης μὲν ἦν ὁ ἄνθρωπος, ἕξεως δὲ συμμέτρου κατὰ πάχος καὶ λεπτότητα κατὰ τὸν τῆς ὑγείας χρόνον. ἐδίψα δὲ σφόδρα καὶ μισεῖν ἔφασκε τὸ θερμόν, ἐδίδου δ' αὐτῷ οὐδεὶς ψυχρὸν ἱκανῶς λιπα-

8. Next, suppose there is heat mixed with abundant dryness, but not in great amount. In the face of such an alteration, I shall maintain the primary method [of treatment], not going so far as to increase the potency of the wine or the amount of the honey, but purifying it so that the honey is not tasted at all, while the wine to be given is least aged. I shall introduce lukewarm foods, which doctors customarily call "milk-warm." In summer [I shall introduce] those they call "fountain-warm." Also, I shall smear the abdomen with [a medication] made from unripe olives and quinces, and much more, if the heat is great, I shall give wine that is more watery, and in this way cold like spring water in the middle of spring, knowing that such a condition is analogous to a fever because that [heat] which is in the whole organism or in the heart is now the proposed condition in the stomach. I know this first of all, having seen with my teachers a man of mature age who had already been troubled for quite a few months, although none of those [teachers] recognized the condition, and nor did I. Subsequently, however, having already discovered the therapeutic method, I recalled that this was something I had seen previously. It is better that I set this out in detail, for it will certainly be very beneficial to those hearing it, just as it was to me.

504K

The man was forty years old, of moderate condition in terms of fatness and thinness when he was healthy. He was very thirsty and said that he was averse to what was hot, but that nobody was giving him a cold drink despite his per-

ρούντι· πυρέττειν μέντοι τοῖς ἰατροῖς οὐκ ἐδόκει· καὶ ἡ
γαστὴρ ἐξέκρινε τὰ ληφθέντα τριῶν ἢ τεττάρων ὡρῶν
ὕστερον ἅμα τῷ ποτῷ. ταῦτ᾽ ἄρα καὶ ἰσχνὸς ἦν ἤδη
505K καὶ πλησίον ἀφῖκτο κινδύνου, | μηδὲν ὀνινάμενος ὑπὸ
τῶν αὐστηρῶν καὶ στρυφνῶν ἐδεσμάτων τε καὶ φαρμάκων. ἐλάμβανε δὲ καὶ οἶνον αὐστηρὸν ἐπὶ τῇ τροφῇ
οὗτος ὁ ἄνθρωπος, ἅμα μὲν οὐκέτι φέρων τὸ δίψος,
ἅμα δὲ καί, ὡς ἔφασκεν, ἑλόμενος ἀποθανεῖν μᾶλλον
ἢ ζῆν ἀνιώμενος. ὕδατος ψυχροῦ δαψιλὲς ἀθρόως ἐπὶ
τῇ τροφῇ προσενεγκάμενος αὐτίκα μὲν ἐπαύσατο
διψῶν, ἐξήμεσε δ᾽ ὀλίγον ὕστερον τὸ πλεῖστον. φρι-
κῶδες δὲ τοὐντεῦθεν γίνεται τὸ σύμπαν σῶμα καὶ
δεῖται σκεπασμάτων πλειόνων ἀπορρίπτων ἔμπροσθεν ἅπαντα· τό τε οὖν λοιπὸν ἅπαν τῆς ἡμέρας ἐφεξῆς
τοῦτο ἔπραξε καὶ δι᾽ ὅλης νυκτὸς ἐφ᾽ ἡσυχίας ἔμεινεν,
ἐνθάλπων ἑαυτὸν ἐπιβλήμασιν. ἐξέκρινε δ᾽ ἅπαξ ἡ
γαστὴρ αὐτῷ μετρίως συνεστῶτα διὰ μέσης νυκτός·
ὥστε οὐδὲ διψώδης ἦν ἔτι κατὰ τὴν ὑστεραίαν, εὐ-
χρούστερός τε μακρῷ καὶ ἰσχυρότερος ἀπείργαστο.
καὶ σκέψις μὲν ἐγένετο τοῖς ἰατροῖς εἰ λουστέον αὐτόν,
ἐνίων μὲν κελευόντων, ἐνίων δ᾽ ἀπαγορευόντων· ἐκράτει δὲ ἡ τοῦ λούειν δόξα. καὶ τοίνυν λουσάμενος
εὐφόρως, μετρίως τε διῃτήθη καὶ κρεῖττον ἢ κατὰ τὴν
προτεραίαν ἔπεψεν ἡ γαστήρ. ἐμέμψατο δὲ τὴν |
506K κατάποσιν ὡς δυσχερῆ καὶ πᾶσιν ἐδόκει διὰ τὸν
ἔμετον ἀήθως γενόμενον ἐσπαράχθαι τε καὶ κάμνειν
τὸν στόμαχον, ὡς δὲ καὶ τῶν ἑξῆς ἡμερῶν ἔμενε τὸ
σύμπτωμα, δῆλον ἐγίγνετο πᾶσιν ἡμῖν ὡς ἡ μὲν

sistent entreaty. And indeed, he did not seem to the doctors to be febrile, and the stomach expelled those things taken three or four hours later with a drink. He had already become thin to the point of danger and had not benefited 505K from harsh and astringent foods and medications. This man was also taking astringent wine for nourishment, partly because he could no longer bear the thirst and partly also, as he said, because he chose to die rather than to live in [such] distress. When I gave him cold water in abundance all at once by way of nourishment, he immediately ceased to be thirsty and a little later vomited copiously. Thereupon his whole body started shivering and he begged for the many coverings, all of which he had previously cast off. He did this continually all the rest of the day and remained the whole night in a quiet state, warming himself with bedcovers. During the middle of the night the stomach expelled all at once those things moderately compacted by it, so that he was not still thirsty on the next day, was of good color, and was made much stronger. The opinion of the doctors on the question of whether to bathe him was that some ordered it and some forbade it, but the notion of bathing prevailed. And further, when he bathed without distress, and was nourished moderately, the stomach digested better than on the day before. He complained of his swallowing being difficult, and it seemed to everyone 506K that, due to the extraordinary vomiting, the esophagus was torn and was causing him distress, as the symptom also remained over the following days. It became clear to all of us

γαστὴρ ἐθεραπεύθη τὴν ἀτονίαν, ὁ στόμαχος δὲ
ἐψύχθη. καὶ οὐδὲν ἄρα θαυμαστὸν ἦν εἰς συμμετρίαν
μὲν ἐπανελθεῖν τῇ πόσει τὸ ὑπερτεθερμασμένον,
ψυχθῆναι δὲ τὸ μετρίως θερμόν. οὐ μὴν οὐδὲ ἠδυνήθη
τις αὐτοῦ θεραπεῦσαι τὸν στόμαχον, ἀλλ᾽ ἕτερον ἀνθ᾽
ἑτέρου κακὸν ἀνταλλαξάμενος ἐτελεύτα τῷ χρόνῳ.

ἄλλον δὲ τοιοῦτον ἔτεσιν ὕστερον οὐκ ὀλίγοις ἐθεασάμην, ἤδη διαγινώσκειν εἰδὼς ἁπάσας τῆς γαστρὸς
τὰς δυσκρασίας, ἐδόκει μοι δὴ ψύχειν ἀγωνιστικώτε-
ρον εὐθέως πρὶν ἐπὶ πλέον λεπτυνθέντα παραπλήσιόν
τι τῷ πρόσθεν ἐπὶ τῇ ψύξει παθεῖν. ἀσφαλέστερον οὖν
ἐφαίνετό μοι τὴν πρώτην ἀποπειραθῆναι τῶν κατὰ τὸ
ὑποχόνδριον ἐπιτιθεμένων ψυκτηρίων φαρμάκων. καὶ
οὕτω πράξαντος ἠλαττώθη μὲν τὸ καῦμα τῆς κοιλίας,
ἀνέπνει δὲ τοῖς στενάζουσιν ὁμοίως ὁ ἄνθρωπος, ὡς
μόλις κινῶν ὅλον τὸν θώρακα. καὶ μέντοι καὶ αὐτὸς
507K ἀνερωτώμενος | ὡμολόγει[14] τοιούτου τινὸς αἰσθάνε-
σθαι παθήματος. ἔγνων οὖν ἐψῦχθαι τὰς φρένας
αὐτῷ· καὶ διὰ τοῦτο ἀπορρίψας τὰ ψυκτήρια, κατ-
ήντλουν ἐλαίῳ θερμῷ. τάχιστα δὲ τῆς ἀναπνοῆς ἀπο-
λαβούσης τὸν κατὰ φύσιν ῥυθμόν, ἐπαυσάμην μὲν
τῆς αἰονήσεως, ἔγνων δὲ μετρίως αὐτὸν ψύχειν ἐν
χρόνῳ πλείονι. τά τε οὖν ἔξωθεν ἐπιτιθέμενα κάτω
μᾶλλον ἐπετίθην ἀποχωρῶν τοῦ διαφράγματος, ὡς
ἐπὶ τὸν ὀμφαλόν· ἅπαντά τε τὰ ἐσθιόμενα καὶ τὰ
πινόμενα πλὴν τοῦ γάλακτος ἐδίδων ψυχρά, παραπλησίως ὕδατι κρηναίῳ. καὶ οὕτως ἐν χρόνῳ πλείονι
κατέστη, μηδὲν τῶν ἄλλων βλαβείς. ἰστέον δέ σοι καὶ

that, while the stomach was being treated for weakness, the esophagus was being cooled. It was not surprising that what had been overheated returned to a balance with the drink and that what was moderately hot was cooled. But it was not possible for anyone to treat his esophagus so, having changed one evil for another, in time he died.

I saw another such patient quite a few years later and, since I now knew how to recognize every *dyskrasia* of the stomach, it seemed to me more fitting for the contest to cool him immediately before he became even thinner, like the man previously who suffered due to cooling. It seemed to me safer in the first instance to make trial of the applications of cooling medications to the hypochondrium. Having done this, the burning heat of the abdomen was reduced, while the man breathed with deep sighs as if he were moving the whole chest with difficulty. And indeed, when questioned, he did allow that he sensed some 507K such affection. Therefore, I realized that his midriff was cooled and, because of this, withdrew the cooling agents and bathed the abdomen with warm oil. Very quickly the breathing took on a natural rhythm and I stopped the irrigation, as I knew it would moderately cool him over a longer time. Therefore, I made the external applications lower, moving away from the midriff and toward the umbilicus. Everything eaten and drunk apart from milk, I gave cold, like water from a spring. In this way, he settled down over a longer time and was not harmed by other things. You

---

14 K; ὁμολόγει (*sic*) B

τοῦτο πρὸ πάντων, ὡς ἐπειδὰν μὲν ἰσχυρῶς ἀλλοιωθῇ
κατ᾿ ἀμφοτέρας τὰς ποιότητας ὁτιοῦν μόριον, ἀπόλλυται τὸ ἔργον αὐτοῦ σύμπαν. οὐ ῥᾴδιον δὲ οὐδὲ τὸ
οὕτω διατεθὲν ἐπανελθεῖν εἰς τὸ κατὰ φύσιν. ἐπειδὰν
δὲ ἡ ἑτέρα μόνη ποιότης ἐπὶ πλέον ὑπαλλαχθῇ,
καθάπερ νῦν ὑπεθέμεθα τὴν ξηρότητα, τῶν δ᾿ ἄλλων
τις ᾖ μετρία, δυνατὸν ἰάσασθαι τὸν οὕτω διακείμενον
ἄνθρωπον. |

508K   9. Ὑποκείσθω δὴ πάλιν ἐπικρατεῖν μὲν τὴν θερμὴν
δυσκρασίαν, μίγνυσθαι δ᾿ αὐτῇ ποτὲ μὲν ὑγρότητα,
ποτὲ δὲ ξηρότητα, μετρίαν ἑκατέραν, καὶ προτέραν γε
τὴν ὑγρότητα. τὴν τοιαύτην δυσκρασίαν ἀδεέστερον
ὕδατι θεραπεύσομεν ψυχρῷ διὰ τὸ μὴ βλάπτεσθαι
πρὸς αὐτοῦ τὰ γειτνιῶντα μετρίως διακείμενα. κατὰ
γὰρ τὰς ξηρὰς διαθέσεις ἀναγκαῖόν ἐστιν οὐ τὰ
πλησιάζοντα μόνον, ἀλλὰ τὸ σύμπαν ἰσχνότερον
γενέσθαι τὸ σῶμα· τῆς κοιλίας δ᾿ αὐτῆς μηδέπω
κατεξηρασμένης, ὡς ὑπόκειται νῦν, οὐδὲ τὸ σύμπαν
σῶμα λεπτύνεσθαι δυνατόν· ὥστε οὐδὲ βλάπτεσθαι
τῇ πόσει τοῦ ψυχροῦ. εἰ δ᾿ οὕτω ποτὲ ἰσχυρὰ δυσκρασία θερμότητος εἴη κατὰ τὴν κοιλίαν ὡς μέχρι τῆς
καρδίας ἐξικνεῖσθαι, πυρέττειν ἀνάγκη τὸν ἄνθρωπον·
ὥστε καὶ τὸν κίνδυνον ὀξύτερον εἰκὸς ἕπεσθαι ταῖς
τοιαύταις δυσκρασίαις· ἡ δ᾿ ἴασις ἡ αὐτὴ μὲν κατὰ
γένος, εἰρήσεται δὲ αὖθις ἐν ταῖς θεραπείαις τῶν
πυρετῶν. ἡ δὲ μετὰ ξηρότητος θερμότης τοῖς αὐτοῖς
μὲν ὑπάγεται κατὰ γένος, οὐ μὴν ὁμοίως γε ἀδεὴς τῶν
ψυχόντων ἡ χρῆσίς ἐστι, δι᾿ ἃς εἶπον αἰτίας. ἡ δὲ

312

must realize this above all else: whenever any part whatsoever is changed strongly in two directions in respect of qualities, its whole action is destroyed. Nor is it easy for what is disposed in this way to return to normal. Whenever one quality alone is changed even more, as I postulated just now with respect to dryness, while one of the others is moderate, it is possible to treat the person in such a state.

9. Now let us assume, in turn, that a hot *dyskrasia* prevails, but that moistness is mixed with it in one instance, and dryness in another, each in moderation. First, in the case of moisture, we shall treat such a *dyskrasia* more confidently with cold water because no harm is done by this to the parts that are adjacent to it (i.e. to the stomach), which are in a balanced state. This is because, in dry conditions, it is inevitable that not only those parts that are near but also the whole body becomes more emaciated, whereas when the stomach itself has not as yet been thoroughly dried up, as is now supposed, it is impossible for the whole body to become thin, so there is no harm from the cold drink. If, in this way, there is at some time a severe hot *dyskrasia* in the stomach, such as to reach as far as the heart, the person is, of necessity, febrile. As a result, the danger likely to follow such *dyskrasias* is more acute. But the cure is the same in terms of class, and I shall speak of it again in my treatments of fevers. The hot *dyskrasia* conjoined with dryness is subject to these same remedies in terms of class, but cooling agents cannot be used confidently in the same way for the

508K

GALEN

509K ὑγρὰ δυσκρασία καὶ | μόνη συστᾶσα πασῶν ἐστιν
εὐιατοτάτη καὶ μετὰ θερμότητος ἢ ψυχρότητος ἐπι-
πλεκομένη.

καὶ διότι γε συνεχῶς αἱ τρεῖς αὗται δυσκρασίαι
καταλαμβάνουσι τὴν γαστέρα καί εἰσιν ἰαθῆναι
ῥᾷσται, τὴν θεραπείαν δὲ ἀεὶ τούτων ἔχοντες ἐν μνή-
μῃ, πάντες οἱ χωρὶς μεθόδου θεραπεύοντες τὰς νόσους
ἐπὶ τὰς ἄλλας, ὡς εἴρηται, μεταφέρουσιν ἀγνοοῦντες
ὅτι πλείους εἰσίν. ἔστι δὲ δήπου τῆς μὲν ὑγρᾶς δυσκρασίας μόνης συνισταμένης βοηθήματα τὰ ξηραίνοντα τῶν ἐδεσμάτων, ἄνευ τοῦ θερμαίνειν ἢ ψύχειν
ἰσχυρῶς· ἔτι τε πρὸς τούτοις ἔνδεια τῶν συνηθῶν
ποτῶν· τῆς[15] δὲ μετὰ θερμότητος ἡ τῶν[16] στυφόντων
ἐδεσμάτων καὶ ποτῶν χρῆσις· ἔστω δὲ καὶ ταῦτα
χωρὶς τοῦ θερμαίνειν αὐστηρά, καὶ μὲν δὴ καὶ τὸ
ψυχρὸν ποτὸν ἐπιτήδειον αὐτοῖς. τῆς δὲ μετὰ ψύξεως
ὑγρᾶς δυσκρασίας ἄριστα μὲν ἰάματα τὰ δριμέα
σύμπαντα· μιγνύσθω δ' αὐτοῖς καὶ τὰ στρυφνὰ χωρὶς
τοῦ ψύχειν σαφῶς. ἄριστον δ' ἴαμα καὶ τούτοις ὀλίγιστον πόμα καὶ τοῦτό τις τῶν θερμαινόντων ἰσχυρῶς
οἴνων· εὔδηλον δὲ καὶ ὡς ὁ τοιοῦτος οὐδὲ νέος ἐστίν.
510K ἀνάλογον δὲ τοῖς ἐσθιομένοις | καὶ πινομένοις καὶ
τἆλλα ἔστω σύμπαντα τὰ ἔξωθεν προσαγόμενα.

10. Ἐπεὶ δὲ καὶ περὶ τούτων αὐτάρκως εἴρηται,
πάλιν ἐπ' ἀρχὴν τὸν λόγον ἀναγαγόντες, ἀθροίσωμεν
αὐτοῦ τὰ κεφάλαια. τὴν μὲν οὖν θερμὴν δυσκρασίαν
ψυκτέον ἐστί, τὴν δὲ ψυχρὰν θερμαντέον· οὕτω δὲ καὶ
τὴν μὲν ὑγρὰν ξηραντέον, τὴν δὲ ξηρὰν ὑγραντέον. εἰ
δὲ κατὰ συζυγίαν τινὰ εἴη γεγενημένη, μιγνύειν

314

reasons I spoke of. However, the moist *dyskrasia*, either existing alone or intermingled with hot or cold, is the most easily curable of all. 509K

And because these three *dyskrasias* continually befall the stomach and are the most easily cured, and because we always have the treatment of these in mind, all those who treat the diseases without method transfer the treatment to the other *dyskrasias*, as I said, unaware that there are more. Of course, the remedies of the moist *dyskrasias* existing alone are foods that are drying without heating and cooling strongly, and in addition to these, there is need of the customary drinks. For when the *dyskrasia* is conjoined with heat, there is the use of astringent foods and drinks. However, these must be astringent without being heating. Furthermore, cold water is suitable for them. The best cures of the moist *dyskrasia* conjoined with cold are all those things that are acrid. Also mix the astringents with them, obviously without what is cooling. And the best cure for these is a very small drink of any of the strongly heating wines. It is clear, too, that such [a wine] should not be new. All the other things applied externally must be analogous to the foods and drinks. 510K

10. Since enough has been said about these matters, let me return again to the beginning of the discussion, and summarize its chief points. These are that you must cool a hot *dyskrasia*, but you must heat a cold *dyskrasia*. In like manner, you must dry a moist *dyskrasia* and moisten a dry *dyskrasia*. If the *dyskrasia* has occurred by way of some

---

15 B; τῶν K
16 B; τῶν om. K

ἀμφοτέρους τοὺς σκοπούς, τὴν μὲν ὑγρὰν καὶ θερμὴν
ξηραίνοντας καὶ ψύχοντας, τὴν δ' ὑγρὰν καὶ ψυχρὰν
ξηραίνοντας καὶ θερμαίνοντας· οὕτως δὲ καὶ ξηρὰν
καὶ θερμὴν ὑγραίνοντας καὶ ψύχοντας, τὴν δὲ ψυχρὰν
καὶ ξηρὰν ὑγραίνοντας καὶ θερμαίνοντας. ἁπασῶν δὲ
χειρίστην ἰστέον εἶναι τῶν μὲν ἁπλῶν τὴν ξηράν, τῶν
δὲ συνθέτων τὴν ξηράν τε ἅμα καὶ ψυχράν. ἐν τούτῳ
μὲν ἡμῖν τέλος ἐχέτω τὰ περὶ τῆς δυσκρασίας τῆς
γαστρός, ἄνευ περιττῆς τινος ἔξωθεν ὑγρότητος.

11. Ἐξῆς δ' ἐπὶ τὰς διά τι τοιοῦτο γιγνομένας ἴωμεν
τῷ λόγῳ. πολλάκις μὲν οὖν ἐν αὐτῷ τῷ κύτει τῆς |
511K κοιλίας τι[17] περιεχόμενον ὑγρὸν ἤτοι θερμαίνειν, ἢ
ψύχειν, ἢ ὑγραίνειν, ἢ ξηραίνειν αὐτὴν πεφυκός, ἢ
κατὰ συζυγίαν τινὰ τούτων ποιεῖν· ἐνίοτε δ' εἰς αὐτοὺς
τοὺς χιτῶνας, ὡς ἂν εἴποι τις, ἀναπεπωμένον, ἢ ἐμπε-
πλασμένον ἐργάζεται τὴν δυσκρασίαν. ἡ μὲν οὖν
προτέρα διάθεσις εἰ μὲν ἅπαξ συσταίη, δι' ἐμέτων
ἐκκαθαρθεῖσα καθίσταται ῥᾳδίως· εἰ δ' αὖθις καὶ
αὖθις ἐξ ἑτέρου τινὸς ἢ ἑτέρων ἐπιρρέοι μορίων, ἀκριβοῦς δεῖται τῆς διαγνώσεως. ἡ θεραπεία δ' εὐθέως
ἐπακολουθήσει τῇ γνώσει τῆς ποιούσης αἰτίας.
ἐλέχθη γὰρ ἡμῖν ἐν ἄλλοις τέ τισι κἀν Ταῖς τῶν
πυρετῶν διαφοραῖς ὡς ἐπιρρεῖ πολλάκις ἐξ ἑτέρων
μορίων ἑτέροις ἡ περιουσία· καθάπερ, εἰ οὕτως ἔτυχε,
τῶν κατὰ τὴν κεφαλὴν τοῖς ὀφθαλμοῖς. ἔνθα χρὴ τὰ
πέμποντα θεραπεύοντα μόρια μόνου τοῦ τόνου προ-

17 K; τι om. B

316

conjunction, mix both indicators, drying and cooling the moist and hot, but drying and heating the moist and cold. In the same way too, you must moisten and cool the dry and hot, but moisten and heat the cold and dry. You must be aware that the dry *dyskrasia* is the worst of all the simple *dyskrasias*, and the dry and cold *dyskrasia* the worst of the compound *dyskrasias*. With this, let me end the matters concerning *dyskrasias* of the stomach occurring apart from some excess moisture from without.

11. Next in the discussion, let me proceed to the *dyskrasias* that do occur due to this. Often some fluid contained naturally in the actual lumen of the stomach either heats, cools, moistens or dries it in nature, or produces one of these [*dyskrasias*] involving a conjunction. Sometimes what is absorbed by or adheres to the walls themselves, as one might say, brings about the *dyskrasia*. The former condition, if it occurs once only, is easily settled when purged by vomiting. However, if there is flow repeatedly from one or several parts, there is need of an accurate diagnosis. The treatment will follow directly from the knowledge of the effecting cause. As I said in some other works, and even in *On the Differentiae of Fevers*, the surplus often flows from some parts to others, as might happen for example from those parts in the head to the eyes.[30] Therefore, when treating the parts sending forth [the flux], it is necessary to

511K

---

[30] *De differentiis febrium, libri II*, VII.273–405K. The differentiae of fluxes from the head are considered in Galen's *In Hippocrates librum de acutorum victu commentarii*, XV.788K.

GALEN

νοεῖσθαι τῶν ὀφθαλμῶν, ὡς μὴ ῥᾳδίως δέχοιντο μηδὲν τῶν ἐπιρρεόντων αὐτοῖς. οὕτως οὖν χρὴ καὶ ἐπὶ τῆς κοιλίας εὑρόντα τὸ πέμπον τὰ περιττὰ σύμπασαν μὲν ἐκείνῳ προσάγειν τὴν θεραπείαν, τῆς γαστρὸς δὲ

512K προνοεῖσθαι τοσοῦτον[18] | μόνον, ὡς μὴ ῥᾳδίως δέχοιτο τὰ ἐπιρρέοντα. γένοιτο δ᾽ ἄν, οἶμαι, τοῦτο κατὰ διττὸν τρόπον· ἕνα μὲν τὸν κοινὸν ἁπάντων ῥευμάτων, ἀναστέλλεται γὰρ ὑπὸ τῶν στυφόντων· ἕτερον δὲ καὶ τοῦτον κοινὸν ἁπάντων τῶν εἰς εὐεξίαν ἀγομένων μορίων ὑπὸ τῆς ἐν τῇ κράσει συμμετρίας.

ἐπισκέπτεσθαι τοιγαροῦν ὅλον μὲν τὸ σῶμα πρότερον χρή, εἰ περιττωματικὸν ἢ πληθωρικὸν ὂν ἐπιπέμπει ῥεῦμα τῇ γαστρί· δεύτερον δ᾽, εἰ μὴ τοῦτο φαίνοιτο, τὸ καθ᾽ ἓν ἕκαστον τῶν μορίων σκοπεῖσθαι. κατὰ μὲν δὴ τὴν τοῦ παντὸς σώματος ἐπίσκεψιν ἔμαθες μὲν καὶ τὰ τοῦ πλήθους καὶ τὰ τῶν περιττωμάτων γνωρίσματα διά τε τῆς Ὑγιεινῆς πραγματείας καὶ τοῦ Περὶ πλήθους βιβλίου. προσεπισκέπτου δ᾽ αὐτοῖς καὶ τάδε· πρῶτον μὲν ἐπὶ γυναικῶν, εἰ μὴ καθαίρονται κατὰ φύσιν· ἑξῆς δ᾽ ἐπ᾽ ἀνδρῶν, εἰ ἡ συνήθης ἔκκρισις ἐπέσχηται. πολλοῖς μὲν γὰρ αἱμορροῖς εἴθισται τὸ περιττὸν ἐκκενοῦν· ἐνίοις δ᾽ αἱμορραγίαι διὰ ῥινῶν, ἢ ἔμετος, ἤ τι κατ᾽ ἄλλον τινὰ τόπον ἐν ὡρισμέναις περιόδοις ἀποστομούμενον ἀγγεῖον· οἶδα γοῦν ἐν

513K στόματι καὶ φάρυγγι πολλοῖς τοῦθ᾽ ὑπάρχον. | ἔστι δ᾽ οἷς καὶ διὰ γαστρὸς αἵματος ἔκκρισις ἀνὰ χρόνον

---

[18] K; τοσοῦτο B

318

give prior consideration only to the strength of the eyes so that they do not readily receive any of the things flowing to them. Similarly, it is also necessary, in the case of the stomach, when you discover what is sending the superfluities, to apply the whole treatment to that, and to give forethought to the stomach only to the extent that it does not readily receive those things flowing in. This, I believe, may happen in a twofold manner. One, which is common for all flows, is repulsion by astringents; the other, and this is also common, is to bring all the parts to a good condition by a balance of their *krasis*.

512K

So it is necessary to give consideration to the whole body first, if it is excrementitious or plethoric and sends a flux to the stomach. Second, if this does not seem to be the case, it is necessary to look at each one of the parts. Certainly, in relation to the investigation of the whole body, you learned the signs of both abundance and superfluities by way of the treatise *On the Preservation of Health* and the book *On Plethora*.[31] Besides these, also consider the following: first in women, if the menses are normal; next in men, if some habitual excretion is retained, for it is customary to evacuate the superfluity through many forms of bloody discharge. In some, the hemorrhages are via the nostrils or through vomiting, or in some other place in distinct cycles when a vessel is blocked. At all events, I know this to exist in the mouth and pharynx in many instances. In some, the expulsion of blood occurs through the stomach

513K

[31] *De sanitate tuenda*, VI.1–452K (particularly VI.238 and 442K and VI.63 and 249K), and *De plenitudine*, VII.513–83K (particularly VII.514K).

γίγνεται, καὶ μάλισθ᾽ ὅσοι γυμνασίων ἰσχυρῶν ἀποστάντες οὐκ ἀπέστησαν τῆς ἔμπροσθεν διαίτης, ἤ τι
κῶλον ὅλον ἀφῃρέθησαν, ὡς ἐδήλωσε καὶ ὁ Ἱπποκράτης. ἐνίοις δὲ καὶ διάρροιαι καὶ χολέραι διὰ χρόνων τινῶν γιγνόμεναι κενοῦσιν ἅπαντος τοῦ σώματος
τὴν περιουσίαν. ἐπὶ δὲ τοῖσδε σκεπτέον ὁποίᾳ τινὶ
κέχρηται διαίτῃ· πότερον ἀπεψίαις πλείοσιν, ἢ πλη-
σμοναῖς, ἐν ἀργῷ βίῳ καὶ μάλιστα παρὰ τὸ ἔθος, ἢ
τοὐναντίον.

εἶθ᾽ ἑξῆς ἐπισκεπτέον εἰ μέλος ἄλλο τοῦ σώματος
ἐνοχλούμενον ἐν περιόδοις τισὶ χρόνων οὐκ ἐνοχλεῖται
νῦν. ἐνίοις μὲν γὰρ ἀρθρῖτις, ἢ ἰσχιάς, ἢ ποδάγρα·
κεφαλαία δ᾽ ἄλλοις, ἢ εἰς ὀφθαλμοὺς ἢ εἰς ὦτα κατα-
σκῆπτον ῥεῦμα μετέστη νῦν εἰς τὴν γαστέρα. πολλοῖς
δὲ καὶ κατάρροι καὶ κόρυζαι συνεχῶς γινόμεναι μεθ-
ίστανται κατά τινα χρόνον εἰς τὰ κατὰ τὴν γαστέρα
χωρία. ταῦτ᾽ οὖν ἅπαντα κατασκεψάμενος, εἰ μὲν ἐξ
ἀκυρωτέρων ἡ μετάστασις εἴη γεγενημένη, πρὸς ἐκεῖ-
514K να πάλιν ἀντισπάσεις τὸ περιττεῦον· εἰ δ᾽ | ἐκ κυρι-
ωτέρων, ἀμφοῖν ὁμοίως προνοήσεις τὴν ἐργαζομένην
τὸ ῥεῦμα διάθεσιν ἐκκόπτων ἀεί. γίγνοιτο δ᾽ ἂν τοῦτο
πάντα μὲν εἰς εὐεξίαν ἀγόντων ἡμῶν τὰ τοῦ σώματος
μόρια, πάσας δὲ τὰς φυσικὰς ἐκκρίσεις εὔρους ἀεὶ
παρασκευαζόντων. ὅπως δ᾽ ἄν τις ἐργάζοιτο ταῦτα διὰ
τῆς Ὑγιεινῆς πραγματείας δεδήλωται. πρόδηλον δ᾽
ὅτι κἂν εἰ διὰ πληθώραν ὅλου τοῦ σώματος ἡ γαστὴρ

320

over a period of time, and particularly in those who, when they gave up vigorous exercise, did not desist from their prior regimen, or have a whole limb removed, as Hippocrates also showed.[32] In some too, diarrhea and cholera occurring over a certain time empty out the surplus of the whole body. In regard to these, you must consider what sort of regimen has been used, whether [it is associated with] more frequent indigestion or repletions, in an idle life and contrary to custom in particular, or the opposite.

Then next, you must consider if another member of the body is distressed at certain periods of time but is not distressed at the present time. In some, there is arthritis, sciatica or gout, and in others, a cranial flux which, having come down to the eyes or ears, now changes to the stomach. In many cases both catarrh and coryza continuously occur and change over a certain time to the places in the stomach. Therefore, when you have considered all these, if the transference is one that has occurred from less important [parts], draw the surplus back again toward those. If, however, it has occurred from the more important parts, 514K give prior consideration to both in like manner, always eradicating the condition which creates the flux. This would occur if we were to bring all parts of the body to a good state and always cause all the natural excretions to be free-flowing. How someone might bring these things about has been shown by way of the work *On the Preservation of Health*.[33] It is clear too that, if the stomach receives a flux due to *plethora* of the whole body, it is the whole

[32] See Hippocrates, *On Joints*, LXIX, and Galen, *De symptomatum causis*, VII.243K. [33] Presumably a reference to Book 1 of Galen's *De sanitate tuenda*, chapters 3, 13, and 14.

δέχηται ῥεῦμα, τὸ σύμπαν σῶμα κενωτέον ἐστίν. ὥσπερ γε καὶ εἰ διὰ κακοχυμίαν, ἐκείνην ἐκκαθαίρειν προσήκει, κἄπειθ᾽ οὕτως ἐπ᾽ αὐτὴν ἰέναι τὴν γαστέρα. πάντως γὰρ δήπου πλείοσιν ἡμέραις εἰς αὐτὴν ἐνηνεγμένων τῶν περιττῶν ἀπήλαυσέ τι τῆς μοχθηρίας αὐτῶν. ὥστε ἐν καιρῷ χρήσαιτ᾽ ἂν ὁ οὕτω κάμνων ἀψινθίῳ ποτῷ. δήλη δ᾽ οἶμαί σοι καὶ ἡ μετὰ ταῦτα πρόνοια μέχρι τοῦ τὴν ἔμπροσθεν ἕξιν ἀνακτήσασθαι τῇ γαστρί. κατὰ γάρ τοι τὴν κρᾶσιν τοῦ λυπήσαντος αὐτὴν ῥεύματος ἔστιν ὅτε δύσκρατος γενομένη τῶν ἐναντίων εἰς ἐπανόρθωσιν ἰαμάτων δεῖται. ῥᾷστον δὲ

515K δήπου τὴν τοιαύτην ἰάσασθαι βλάβην, | ὅταν ὀλίγων ἡμερῶν ᾖ. ὡς εἴ γε δι᾽ ἀπορίαν τῶν ἰασομένων διαμείνῃ χρόνῳ πλείονι, ποτὲ μὲν οἱ χιτῶνες αὐτῆς ἐμπίπλανται τῆς κακοχυμίας, ἔστι δ᾽ ὅτε καὶ αὐτὸ τὸ ὁμοιομερὲς αὐτῶν σῶμα δύσκρατον ἱκανῶς γενόμενον ὁμοίας χρῄζει θεραπείας τῇ λελεγμένῃ πρόσθεν ἐν τῷδε τῷ λόγῳ τῶν δυσκρασιῶν.

ὑπόλοιπον οὖν ἐστι διελθεῖν ὅπως ἄν τις ἰῷτο τὰς ἐν τοῖς χιτῶσι τῆς γαστρὸς κακοχυμίας. ἔστι δὲ ἐπιτήδεια πρὸς τὰς τοιαύτας διαθέσεις φάρμακα τῷ γένει μὲν ἐκ τῶν καθαιρόντων μετρίως, ὡς μὴ προϊέναι τὴν δύναμιν αὐτῶν ἀνωτέρω τῶν κατὰ γαστέρα τε καὶ τὰ ἔντερα χωρίων· ἢ εἴπερ ἄρα πρὸς τούτοις, ἅπτεσθαι τῶν κατὰ τὸ μεσεντέριον ἀγγείων. ἐν εἴδει δὲ τὰ διὰ τῆς ἀλόης ἐστὶ κάλλιστα καὶ αὐτὴ καθ᾽ αὑτὴν ἡ ἀλόη, μὴ παρόντων ἐκείνων. ἄπλυτος μὲν οὖν ἱκανώτερον ἐκκενοῖ, πλυθεῖσα δὲ ἧττον μὲν καθαίρει,

body you must purge, just as if it is due to a *kakochymia*, it is appropriate to purge that, and then in like manner to go to the stomach itself. At all events, presumably if the superfluities have been carried to it over many days, it will have the "benefit" of their bad [quality]. As a result, at the appropriate time, the patient might use a wormwood potion in this way. Your clear intention, after these things, is I believe the restoration of the stomach to its former state. Surely, in terms of *krasis*, at such times as it becomes *dyskratic* after the flux has distressed it, it requires the opposite cures for restoration. It is easier, of course, to cure this kind of damage when it is over a few days. If, because of difficulty for those curing it, it remains over a longer time, the walls of the stomach are sometimes filled with bad humors, and sometimes, when the actual *homoiomerous* body of these walls becomes sufficiently *dyskratic*, there is need of a treatment similar to that spoken of previously in the discussion of the *dyskrasias*.

515K

What remains is [for me] to go over how someone might cure *kakochymia* in the walls of the stomach. There are useful medications for such conditions that are moderately purging in class, so they don't extend their potency beyond the spaces in the stomach and intestines, or if in fact they do, in addition come into contact with the vessels in the mesentery. In kind, the best are those made from aloes, or aloes by itself, if the former are not available. Unwashed aloes purges more adequately; washed aloes

ῥώννυσι δὲ μᾶλλον τὴν γαστέρα. χρὴ τοίνυν ἔχειν τὴν καλουμένην ἤδη πρὸς ἁπάντων τῶν ἰατρῶν πικράν, ἐσκευασμένην διττῶς, ἐξ ἀπλύτου τε καὶ πεπλυμένης ἀλόης. μίγνυται δὲ αὐτῇ κινναμώμου τε καὶ

516K ξυλοβαλσάμου καὶ ἀσάρου | καὶ νάρδου στάχυος καὶ κρόκου καὶ Χίας μαστίχης. ἔστωσαν δὲ ἐξ μὲν τούτων ἑκάστου δραχμαί, μόνης δὲ τῆς ἀλόης ἑκατόν.

ἄριστον δὲ τοῦτο τὸ φάρμακον εἰς τὰς τῶν χιτώνων τῆς γαστρὸς κακοχυμίας· ἄμεινον δὲ χρῆσθαι ξηρῷ δι᾽ ὕδατος ὅσον δυοῖν κοχλιαρίων μικρῶν τὸ πλῆθος, ὅταν γε μετρίως καὶ μέσως αὐτῷ χρῆσθαι προαιρώμεθα· τὸ γὰρ δὴ πλεῖστον καὶ τελεώτατον οὐ σμικρῶν δυοῖν, ἀλλὰ μεγάλων ἡ πόσις, ἐν ὕδατος εὐκράτου τρισὶ κυάθοις· ἐλαχίστη δὲ δόσις ἑνὸς κοχλιαρίου μικροῦ. χρῆσθαι δ᾽ ἐν τῷ καιρῷ τῷδε καὶ πτισάνης χυλῷ, πρῶτον αὐτὸν ἀπὸ τοῦ βαλανείου λαμβάνοντα. τὸ μὲν γὰρ φάρμακον αὐτὸ κατὰ τὸν συνήθη τοῖς καθαίρουσι διδόσθω καιρόν. ἐπ᾽ αὐτῷ δὲ περιπατήσαντα σύμμετρα καὶ κινηθέντα μέτρια τῶν ἄλλων τῶν κατὰ τὴν δίαιταν ὑπαλλάττειν μηδέν. ἔνιοι δ᾽ ἀποτίθενται τὸ φάρμακον, ἀναλαμβάνοντες ἀπηφρισμένῳ μέλιτι, καὶ γίνεται μὲν οὕτω μονιμώτερον, οὐ μὴν δὲ ὁμοίως γε πρὸς τὰς ὑποκειμένας ἐν τῷ παρόντι λόγῳ διαθέσεις ἀγαθόν· ἧττον γὰρ ῥώννυσι τὴν κοιλίαν καὶ

517K μᾶλλον ὑπάγει. γλίσχρου μέντοι φλέγματος | ἐμπεπλασμένου τῇ γαστρὶ δοτέον αὐτοῖς πρότερον ὅσα τέμνει τοῦτο, κἄπειτα οὕτω καθαρτέον. εἰ δὲ καὶ πρὸς ἔμετον ἐπιτηδείως ἔχοιεν, οὐδὲν ἂν εἴη χεῖρον ἐμεῖν

324

purges less but strengthens the stomach more. Accordingly, it is necessary to have what is now called by all doctors "bitter" [aloes], prepared in a twofold manner; from unwashed and washed aloes. Mix with it cinnamon, balsam wood, hazelwort, spikenard, saffron and Chian mastich. 516K Let there be six drachms of each of these and a hundred of aloes alone.

This is the best medication for the *kakochymias* of the walls of the stomach. It is better to use the dry [medication] with water in the amount of two small spoonfuls whenever we choose to use it in due measure and moderately. But the greatest and most powerful [dose] is not two small spoonfuls but two large spoonfuls in three cups of *eukratic* water, while the least dose is one small spoonful. At this time, also use the juice of ptisan when the person is first taken from the bath. The medication itself must be given at the time customary for purgings. In addition to this, when the person has walked around to a moderate extent and moved in moderation, change none of the other things relating to the regimen. Some employ the medication, taking it up with skimmed honey; in this way, it becomes more stable, but is not, after making it up, equally good for the conditions considered in the present discussion because it strengthens the stomach less and reduces it more. Nevertheless, when viscid phlegm has plugged up 517K the stomach, you must first give the patients things that cut this, and then you must purge. If they also have a tendency to vomit, you would not be amiss to bring about vomiting

ἀπὸ ῥαφανίδων δι᾽ ὀξυμέλιτος. εἰ δὲ μήτε γλίσχρος ὁ
χυμὸς εἴη μήτε παχύς, ἀρκεῖ καὶ ὁ ἀπὸ χυλοῦ τῆς
πτισάνης ἔμετος μόνον καὶ ὁ ἀπὸ τοῦ μελικράτου·
λαμβάνειν δ᾽ ἑκατέρου πλέον ἢ ὡς ἄν τις ἔλαβεν, ἤτοι
τροφῆς ἕνεκεν ἢ ὑπαγωγῆς γαστρός. ἐπιτήδειον δὲ
καὶ τὸ μελίκρατόν ἐστιν, ἀφεψημένου καθ᾽ αὑτὸ τοῦ
ἀψινθίου πίνεσθαι· καὶ γὰρ καὶ τοῦτο καλῶς ὑπάγει
κάτω τοὺς λεπτοὺς χυμούς, ὅσοι περ ἂν ἐν τῷ στόματι[19] τῆς γαστρὸς αὐτῆς περιέχονται.

κοινωνεῖ δὲ ὁ λόγος ὅδε σύμπας ὁ περὶ τῆς δυσκρασίας τοῖς Ὑγιεινοῖς παραγγέλμασι, καθ᾽ ἕν τι
μέρος ἑαυτοῦ τὸ τῶν ἀσθενῶν ἐνεργειῶν, οὐ γὰρ δὴ
τῶν γε ἀπολωλυιῶν, ἢ πλημμελῶς γιγνομένων· ἐκπέπτωκε γὰρ ἐκεῖνο τελέως τῶν ὑγιεινῶν· ἡ δ᾽ ἀσθενὴς
ἐνέργεια κατὰ τὸ μᾶλλόν τε καὶ ἧττον τῆς ὑγιεινῆς
ἐστι πραγματείας ἢ τῆς θεραπευτικῆς, ὀλίγον μέν τι
παραποδιζομένη τῆς ὑγιεινῆς, ἐπὶ πλέον δὲ τῆς θερα-
πευτικῆς. ὅρος δ᾽ ἑκατέρας τὸ | τῶν συνήθων πράξεων
ἤτοι γε ἀφίστασθαι διὰ τὸ μέγεθος τῆς βλάβης ἢ μή.

12. Λέλεκται δὲ καὶ περὶ τούτων ἐν τοῖς Ὑγιεινοῖς·
καὶ χρὴ καὶ διὰ τοῦτο θᾶττον ἀφίστασθαι τοῦ λόγου,
γεγραμμένης γε δὴ συμπάσης ἤδη τῆς μεθόδου, καθ᾽
ἣν χρὴ τὰς δυσκρασίας ἰᾶσθαι τῆς γαστρός, εἴτε καθ᾽
ἕξιν, εἴτε κατὰ σχέσιν ἐν αὐτῇ γένοιντο. δῆλον δὲ ὡς
καὶ μικτή ποτ᾽ ἐστὶ διάθεσις ἐν τῇ γαστρὶ τῶν εἰρη-

518K

---

[19] K; σώματι B, recte fort. (cf. 520.4K)

by radish root with oxymel. If the humor is neither viscid nor thick, vomiting induced by the juice of ptisan or melikraton is sufficient alone. [The patient should] take more of each than someone might take of either for the sake of nourishment or for downward purging of the stomach, and it is suitable to drink melikraton provided the absinth is boiled on its own. This also leads the thin humors downward well; [that is] those that might be contained in the opening of the stomach itself.

This whole discussion shares a common ground with that about the *dyskrasias* in one part of *On the Preservation of Health*, which is about weakened functions, but not about those that are destroyed or deficient, since these represent a complete departure from health.[34] Weak function is a matter that pertains more or less to health or to therapy. If it is slight, it is an impediment to health; if it is greater, it is a matter of therapy. The demarcation between the two depends on whether there is departure from the customary actions due to the magnitude of the damage or not.

518K

12. I have also spoken about these things in my work *On the Preservation of Health*;[35] and so, because of this, I ought to move away from the discussion more quickly, having already written the whole method by which we must cure the *dyskrasias* of the stomach, whether they arise in it as a result of a permanent or changeable state. It is clear also that there is sometimes a mixed condition in the stom-

[34] The *dyskrasias* are considered in Book 6, chapters 2–5 and 9–10 of *De sanitate tuenda*.

[35] The *dyskrasias* of the stomach are considered specifically in Book 6, chapter 10 of *De sanitate tuenda*.

μένων τριῶν τῶν ἁπλῶν· ὥστε καὶ αὐτὰ τὰ στερεὰ
μόρια δυσκράτως ἔχειν καὶ χυμοὺς μοχθηροὺς ἐμπε-
πλάσθαι δυσεκνίπτως τῇ γαστρὶ καὶ κατὰ τὴν ἔνδον
εὐρυχωρίαν ἑτέρους περιέχεσθαι. δύναται μὲν γάρ
ποτε καὶ τὸ πρῶτον ἅμα τῷ δευτέρῳ ῥηθέντι συστῆ-
ναι· δύναται δὲ καὶ τὸ δεύτερον ἅμα τῷ τρίτῳ· καὶ
αὖθίς γε τὸ πρῶτον ἅμα τῷ τρίτῳ· καὶ πάνθ' ἅμα
πολλάκις. ἐφ' ὧν ἀναμνησθήσῃ μὲν δήπου καὶ τῆς
ἔμπροσθεν εἰρημένης μεθόδου ἐν τοῖς περὶ τῶν ἑλκῶν
λόγοις ἁπασῶν τῶν ἐπιπλεκομένων ἀλλήλαις διαθέσεων, οὐδὲν δ' ἂν εἴη χεῖρον καὶ νῦν αὐτὴν διελθεῖν.

ἐπίσκεψαι γὰρ ἐν ταῖς τοιαύταις ὑπαλλάξεσι |
519K πρῶτον μὲν ἀφ' οὗ μάλιστα κινδυνεύειν εἰκός ἐστι τὸν
κάμνοντα, δεύτερον δέ, τί μὲν ἐξ αὐτῶν ἢ τίνα λόγον
αἰτίας ἔχει, τίνα δὲ ἀποτελεῖται πρὸς αὐτῶν, καὶ
τρίτον οἷς τ' ἀδύνατόν ἐστι ἰαθῆναι πρὸ τῶν ἄλλων
καὶ ὅσοις δυνατόν, ὥσπερ ἐπὶ τῶν ἑλκῶν ἐδείκνυμεν
τῶν ἅμα φλεγμοναῖς συνισταμένων. ἔνθα μὲν γὰρ
ἀπό τινος τῶν διαθέσεων οὐ μικρὸς ὁ κίνδυνός ἐστιν, ὁ
πρὸς τὸ κατεπεῖγον σκοπὸς αἱρετέος· ἔνθα δὲ τὸ μὲν
ποιοῦν ἐστι, τὸ δὲ γιγνόμενον, ὁ πρὸς τὸ αἴτιον· ἔνθα
δὲ οὐχ οἷόν τε θεραπευθῆναι τόδε πρὸ τοῦδε, ὁ ἀπὸ τῆς
τάξεως. ἐν μὲν οὖν τῷ πρὸς τὸ κατεπεῖγον σκοπῷ τὸ
μέγεθος τῆς διαθέσεως σκεπτέον. ἔστι δὲ τρία μεγέθη, τὸ μέν τι κατὰ τὸ τῆς βεβλαμμένης ἐνεργείας
ἀξίωμα, τὸ δέ τι κατὰ τὴν οἰκείαν οὐσίαν τῆς διαθέσεως, καὶ τρίτον ἐπὶ τοῖσδε τὸ πρὸς τὴν διοικοῦσαν
τὸ βεβλαμμένον σῶμα δύναμιν. ἐν δὲ τῷ πρὸς τὸ

328

ach involving the three simple [conditions] spoken of—
that is, the solid parts themselves are *dyskratic*, the bad
humors are blocked up in the stomach in a way that is difficult to cleanse, and other [humors] are contained in the
open space within (the lumen). It is also possible, on some
occasions, for the first of these three to exist along with the
second spoken of, and it is possible for the second to exist
along with the third, and again the first along with the
third, and frequently all [these] at the same time. In these
cases, you will of course recall the method previously spoken of in the discussions about wounds when all the conditions combine with each other. It would be good to also go
over this now.

In such combinations, consider first from which one   519K
the patient is most likely to be in danger; second, consider
which of them has either a certain ground or cause, and
what things are brought about by them; and third, consider
which among them cannot be healed before the others,
just as I showed in the case of wounds coexisting with inflammation. Where the danger from any one of the conditions is not slight, the indicator must be chosen on the
grounds of urgency, whereas where there is what is acting
and what is occurring, the indicator must be chosen on the
grounds of cause. Where one cannot be treated before another, [the indicator] is based on the order. Thus, in the indicator that pertains to urgency, you must consider the
magnitude of the condition. There are three magnitudes:
one relates to the importance of the function that has been
damaged, one relates to the specific substance of the condition, and the third after these relates to the capacity
governing the damaged body. In the indicator pertaining

αἴτιον σκοπῷ θεωρητέον τί μὲν ἐκ τῶν ἐπιπεπλεγμένων ἀλλήλοις αὐξάνειν ἢ γεννᾶν πέφυκε τὰ λοιπά,
τίνα δ᾽ ἤτοι τὴν γένεσιν, ἢ τὴν αὔξησιν ὑπ᾽ ἐκείνου
λαμβάνει. ἐν δὲ τῷ κατὰ τὴν τάξιν, τί πρὸ τίνος, ἢ τί
520K σὺν τίνι, ἢ τί μετὰ | τί δυνατὸν ἰᾶσθαι.

ὑποκείσθω γοῦν εἰς τὴν γαστέρα καταρρέουσά τις
ἐξ ἐγκεφάλου περιουσία ψυχρὰ καὶ διὰ ταύτην ἤδη
μέν τις οὖσα δυσκρασία κατὰ τὴν κοιλίαν· ἔτι δὲ
μᾶλλον καὶ χυμοὶ μοχθηροὶ καθ᾽ ὅλον αὐτῆς τὸ σῶμα
τὰς μεταξὺ χώρας ἁπάσας τῶν ὁμοιομερῶν κατειλη-
φότες. ἐν ταῖς τοιαύταις ἐπιπλοκαῖς ἀξιώματι μέν ἐστι
μείζων ἡ δυσκρασία, διότι καὶ ἡ εὐκρασία· κατὰ δὲ
τὴν οἰκείαν οὐσίαν ὅ τι περ ἂν αὐτῶν τύχῃ περὶ τὸν
ὑποκείμενον ἄρρωστον ἐν τῷ λόγῳ. καὶ κείσθω πρός
γε τὸ παρὸν ἐν τοῖς εἰρημένοις τρισὶ μείζων εἶναι
διάθεσις ἡ τῶν ἐν τοῖς χιτῶσι τῆς γαστρὸς χυμῶν.
ἔστω δ᾽, εἰ βούλει, καὶ τοῦτο αὐτὸ πάθημα δῆξιν
ἐργαζόμενον, καὶ διὰ τοῦτο αὐτὸ λειποψυχίαν τε καί
τινας ἱδρῶτας ἅμα συγκοπτικαῖς ἐκλύσεσιν, ὡς εἶναι
μεῖζον αὐτὸ τῶν ἄλλων, καθ᾽ ὅσον ὑπὲρ τὴν δύναμίν
ἐστιν. εὔδηλον οὖν ὅτι τὴν μὲν πρώτην ἄν τις ἐπιμέλειαν ἀντιτάξαιτο πρὸς τὸ καταλῦον τὴν δύναμιν.
ἑξῆς δ᾽ ἐπειδὴ μέγιστον τῶν ἄλλων ὑπέκειτο τοῦτο
κατὰ τὴν οἰκείαν οὐσίαν, ἐπὶ πρῶτον ἄν τις αὐτὸ
521K παραγίγνοιτο | καὶ πρῶτον ἰῷτο. εἰ δ᾽ ἡμῖν ἤδη μετρί-
ως ἔχει τοῦτο καὶ μήθ᾽ ἡ δύναμις ἔτι καταπίπτει μήτε
ἡ κατὰ τὴν γαστέρα κακοχυμία παμπόλλη τις εἴη,
τότε ἐπὶ τὴν αἰτίαν ἀφιξόμεθα, γιγνώσκοντες ὡς οὐχ

330

to the cause, you must consider which of the [three] magnitudes intermingled with each other either increases or generates by nature the others, and which takes either its genesis or increase from that one. In terms of sequence, you must consider which condition can be cured before the others, which conditions can be cured together, and which conditions can be cured after others.   520K

Anyway, let us assume that some cold superfluity flows from the brain to the stomach and there is now a *dyskrasia* involving the stomach due to this. Further, let us also assume that the bad humors involve the whole body of the stomach, occupying all the spaces between the *homoiomeres*. In such combinations, the *dyskrasia* is of greater importance because the *eukrasia* is also [of greater importance]. However, in terms of the specific substance, whatever any of them (i.e. the combinations) may be, in relation to the weakness presumed in the discussion, [the *dyskrasia* is more important]. Let us suppose, for the present, that of those three things described, the greatest is the condition of the humors in the walls of the stomach. If you wish, let it also be the same affection which brings about a biting, and because of this, a swooning, and also sweats along with syncopal episodes, as this is more important than the other components to the extent that it is in excess of the capacity. It is quite clear, then, that you direct the primary treatment against what dissipates the capacity. Next in order, in respect of the specific substance, since this is supposed to be the most important among the other factors, one should attend to this first and cure it first. If [it   521K
seems] to me that this is already in balance and that the capacity neither falls further nor the *kakochymia* in the stomach is very marked, I will at that time proceed to the

οἷόν τ᾽ ἐστὶ τελέως ἰαθῆναι διάθεσιν οὐδεμίαν, ἔτι
μενούσης τῆς ἐργαζομένης αὐτὴν αἰτίας. ἡ δέ γε τάξις
τῆς θεραπείας ἐνίοτε μὲν ὡς ἐπὶ φλεγμονῆς καὶ ἕλκους
ἐναντίων ἐστὶν ἐνδεικτικὴ βοηθημάτων· ἐδείκνυτο γὰρ
ὡς τὰ τὴν φλεγμονὴν θεραπεύοντα μεῖζον ἀποφαίνει
τὸ ἕλκος, ἐνίοτε δ᾽ οὐδὲν βλάπτει τὴν ἑτέραν διάθεσιν,
ὥσπερ ἐπὶ τῆς ἐνεστώσης ὑποθέσεως. ὁ γὰρ ἐκκενῶν
τὸ περιεχόμενον ὑγρὸν ἐν τῷ τῆς γαστρὸς κύτει τῶν ἐν
τοῖς χιτῶσι περιεχομένων ὑγρῶν ὑποτέμνεται τὴν
τροφήν· ὥσπερ γε καὶ ὁ τὸν ἐγκέφαλον ἰώμενος, ὡς
μηδεμίαν ἐν ἑαυτῷ γεννᾶν αἰσθητὴν περιουσίαν, ἐκ-
κόπτει τὴν οἷον πηγὴν τῶν εἰς τὴν γαστέρα ῥευμάτων.

ἐπὶ μὲν δὴ τούτων ἡ τάξις τῆς θεραπείας τῇ τάξει
τῶν αἰτίων ὁμολογεῖ, κἂν ἐν ἄλλοις διαφέρηται. τὸ δ᾽
ἀπὸ τοῦ μεγέθους χρὴ διορίζεσθαι, καθάπερ εἴρηται.
522K καὶ πρῶτον μὲν πειρᾶσθαι τῷ μεγίστῳ τὴν | οἰκείαν
ὕλην ἐφαρμόζειν τῶν βοηθημάτων· ἑξῆς δὲ κατὰ τὸν
τῶν αἰτίων στοῖχον ἰᾶσθαι τὸ σύμπαν. ὅταν γὰρ μήτε
πλῆθος ἀξιόλογον ᾖ τῶν κατὰ τὴν γαστέρα περιττωμάτων μήτ᾽ ἐν τῇ κοιλότητι περιεχομένων αὐτῆς μήτ᾽
ἐν τοῖς χιτῶσι, μήτε δῆξις ἀδικοῦσα τὴν δύναμιν,
ἁπάντων μὲν χρὴ πρῶτον ἐκνοσηλεῦσαι τὸν ἐγκέφαλον, ἐφεξῆς δὲ τὸ κύτος τῆς γαστρὸς ἐκκενῶσαί τε καὶ
διαρρύψαι, καὶ μετὰ ταῦτα ἐκκαθῆραι τοὺς χιτῶνας,
εἶτα τὴν δυσκρασίαν ἰᾶσθαι. εἴρηται δὲ ἑκάστου τὰ
ἰάματα πρόσθεν, ὥστε οὐ χρὴ μηκύνειν ἔτι τὰ κατὰ
τὸν λόγον, ἀλλ᾽ ἤδη τέλος ἐπιθεῖναι τῷ γράμματι. τὰς
γὰρ τῶν ἄλλων ἁπάντων μορίων δυσκρασίας ἀνάλο-

cause, since I know that it is not possible for any condition to be cured completely when the cause bringing it about still remains. In fact, the order of the treatment is indicative of contradictory remedies, as in the case of inflammation and a wound, for it was shown that when treating the inflammation, the wound sometimes becomes more apparent, although sometimes this does not harm the other condition, as in the case of what is presently before us. For when someone evacuates the fluid contained in the lumen of the stomach, the fluid contained in the walls cuts off the nourishment of the surrounding fluids in the walls, just as also someone curing the brain does, when he cuts off the fount, as it were, of the fluxes to the stomach.

Certainly, in these cases, the order of the treatment agrees with the order of the causes, even if it differs in other respects. However, it is necessary to make a distinction based on magnitude, as I said. The first thing is to attempt to adapt the specific material of the remedies on the basis of the magnitude; and next, in relation to the 522K sequence of causes, attempt to cure the whole because, whenever there is neither a very significant amount of the superfluities in the stomach, nor of those things contained in its lumen or walls, nor something biting which injures the capacity, it is necessary first of all to cure the brain completely, next to evacuate and thoroughly cleanse the lumen of the stomach, and after this to purge the walls, and then to cure the *dyskrasia*. As I spoke of the cures of each of these before, it is not necessary to extend the discussion further but to put an end to the book now. Someone may cure the *dyskrasias* of all the other parts analogously to

γον ἄν τις ἰῷτο τοῖς ἐπὶ τῆς γαστρὸς εἰρημένοις. ἀεὶ γὰρ χρὴ τό τ᾽ ἐναντίον τῇ διαθέσει πορίζεσθαι βοήθημα καὶ τὴν ἀπὸ τῆς φύσεως τοῦ πεπονθότος ἔνδειξιν τῶν ὠφελησόντων προστιθέναι, περὶ ἧς εἴρηται μὲν ἤδη καὶ διὰ τῶν ἔμπροσθεν, εἰρήσεται δὲ καὶ διὰ τῶν ἑξῆς ἐπὶ πλέον. ἀναγκαῖον δ᾽ ἂν εἴη καὶ νῦν ἐπ᾽ ὀλίγον αὐτῆς μνημονεῦσαι. |

523K   13. Πρώτη μὲν ἔνδειξις ἀπὸ τῆς τοῦ πεπονθότος τόπου κράσεώς ἐστι, τὸ μέτρον ὁρίζουσα τοῦ θερμαίνειν, ἢ ψύχειν, ἢ ξηραίνειν, ἢ ὑγραίνειν, ἢ κατὰ συζυγίαν τι πράττειν αὐτῶν. δευτέρα δὲ ἀπὸ τοῦ κοινὸν εἶναι τὸ ἔργον ἅπασι τοῖς τοῦ ζῴου μορίοις, ἢ κοινὴν τὴν δύναμιν χορηγεῖν. καὶ τρίτη παρὰ τῆς διαπλάσεως αὐτοῦ. καὶ τετάρτη παρὰ τῆς θέσεως ἧς μέρος ἐστὶν ἡ πρὸς τὰ πλησιάζοντα τοῦ ζῴου μόρια κοινωνία· καὶ πρὸς τούτοις ἅπασι τὸ τῆς αἰσθήσεως ποσόν. λεχθήσεται μὲν οὖν κἀν τοῖς ἑξῆς ὑπὲρ ἑκάστου τῶν εἰρημένων ἐπὶ πλέον· ἐπέλθωμεν δὲ καὶ νῦν αὐτὰ διὰ βραχέων.

    ὅσα μὲν θερμότερα φύσει μόρια, κατὰ τὴν ψύξιν νοσεῖ, μειζόνως τε ταῦτα καὶ μέχρι πλείονος χρόνου θερμαίνεσθαι δεῖται, καθάπερ γε καὶ τὰ ψυχρότερα κατὰ θερμότητα νοσοῦντα ψύχεσθαι, καὶ τὰ ξηρότερα καθ᾽ ὑγρότητα ξηραίνεσθαι· καὶ δὴ κατὰ τὸν αὐτὸν λόγον ὑγραίνεσθαι τὰ φύσει μὲν ὑγρότερα, κατὰ ξηρότητα δὲ νοσοῦντα, μειζόνως τε καὶ μέχρι πλείονος χρῄζει τοῦ χρόνου. τοσοῦτον γὰρ ἐπανελθεῖν εἰς
524K   τὸ κατὰ φύσιν ἕκαστον αὐτῶν ἀναγκαῖόν | ἐστιν,

those spoken of in relation to the stomach. It is always necessary to provide the opposite remedy to the condition, and to add the indication of those things that are useful from the nature of what has been affected—something which I also spoke about previously and will speak of further in what is to follow. It is also necessary to make mention of this briefly now.

13. The first indication is from the *krasis* of the affected place, which determines the measure to effect for heating, cooling, drying, moistening, or whatever conjunction there is of these. The second indication is that which is the action common to all the parts of the organism, or which furnishes the common capacity. The third indication is from the conformation of the part. The fourth indication is from the position which the part has in association with the adjacent parts of the organism. And besides all these, there is the "amount" of sensation. I shall say more about each of the aforementioned [indications] in what follows. However, let me also now go over these things in brief. 523K

Those parts that are hotter by nature become diseased in relation to cooling and need to be warmed quite significantly over a fairly long time period. In the same way, those that are colder in nature, and become diseased in relation to heating, need to be cooled. Likewise, those that are drier in nature and become diseased in relation to moistening, need to be dried. Further, on the same grounds, those that are more moist in nature and become diseased in relation to drying, need to be moistened more significantly and over a longer time period. It is necessary for each of these parts to return to an accord with nature to 524K

ὁπόσον εἰς τὰ παρὰ φύσιν ἐξετράπετο· καὶ καθάπερ
ὁδόν τινα παλινδρομοῦσαν εἰς τοὐπίσω τῆς γενομένης
αὐτῷ μεταστάσεως. ἔμπαλιν δὲ τῷ μὲν θερμοτέρῳ
φύσει μορίῳ θερμὸν νόσημα νοσοῦντι βραχείας ἢ
ὀλιγοχρονίου δεῖται τῆς ψύξεως, τῷ δὲ ψυχροτέρῳ τῆς
θερμάνσεως· ὡσαύτως δὲ κἀπὶ τῶν ὑγρῶν τε καὶ
ξηρῶν. ὀλίγη γὰρ ἡ εἰς τὸ παρὰ φύσιν ἐκτροπὴ τοῖς
τοιούτοις ἐστὶ καὶ ἡ εἰς τὸ κατὰ φύσιν ἐπάνοδος. ὅθεν
καὶ κινδυνεύουσιν ἧττον οἷς ἂν οἰκεία τῆς φύσεως ἡ
νόσος ὑπάρχει. δῆλον δ' ὅτι καὶ κατὰ συζυγίαν αἱ
δυσκρασίαι γιγνόμεναι τηλικαύτην ἕξουσι τὴν εἰς τὸ
κατὰ φύσιν ἐπάνοδον ἡλίκην ἐποιήσαντο τὴν εἰς τὸ
παρὰ φύσιν ἐκτροπήν. οὕτω μὲν οὖν ἀπὸ τῆς κράσεως
τῶν πεπονθότων μορίων ἡ τῶν βοηθημάτων ἔνδειξις
γίγνεται, ἀπὸ δὲ τοῦ κατὰ τὴν ἐνέργειαν ἤτοι χρήσιμον ἅπασι τοῖς τοῦ ζῴου μορίοις ὑπάρχειν, ἢ ὀλίγοις τισὶν, ἢ ἑαυτῷ μόνῳ κατὰ μὲν τὰς ἄνευ χυμῶν
δυσκρασίας οὐδὲν ἐκ τούτου τοῦ σκοποῦ λαμβάνεται.

525K   καὶ γὰρ ψύχεσθαι καὶ θερμαίνεσθαι καὶ | ξηραίνε-
σθαι καὶ ὑγραίνεσθαι καὶ κατὰ συζυγίαν πάσχειν τι
τούτων, ἕκαστον τῶν μορίων ἀνάλογον δεῖται τοῦ
νοσήματος. κατὰ μέντοι τὰς ἐπὶ τοῖς χυμοῖς, εἰ μὲν
ἑαυτῷ μόνῳ τὸ μόριον ἐνεργοίη, θαρρῶν ποιήσεις τὰς
κενώσεις, ὡς ἂν ἡ διάθεσις ὑπαγορεύῃ, συνεπιβλέπων
τὴν δύναμιν· εἰ δ' ἀναγκαῖον εἴη τὸ ἔργον αὐτοῦ πᾶσι
τοῖς τοῦ ζῴου μορίοις, ὥσπερ τὸ τῆς γαστρός τε καὶ
τοῦ ἥπατος, οὐ μικρὰν χρὴ φροντίδα ποιεῖσθαι τοῦ
τόνου τῆς δυνάμεως, μή πως καταλύσωμεν αὐτὸν

the extent that there was a turning aside to a contrariety to nature, just as it is also necessary, when change has occurred in the part, to follow the same path back. Contrariwise, for the part that is hotter in nature when it suffers a hot disease, there is need of a small amount or a short time of cooling, [and the same applies to] the colder part in respect of heating. It is the same in the cases of the moist and dry parts. Where the turning aside to what is abnormal for such parts is slight, so also is the return to what is normal. Wherefore, they are also less in danger when the disease is of a similar nature to them. However, it is clear too that the *dyskrasias* occurring in relation to a conjunction that is substantial will have a return to normal that is as substantial as the turning aside to abnormality was. In this way, then, the indication of the remedies arises from the *krasis* of the affected parts, whereas in the *dyskrasias* arising apart from humors, nothing is taken from that indicator which relates to function or use for all the parts of the organism, or some few, or one by itself, in respect of the *dyskrasias* arising without humors.

When cooled, heated, dried, or moistened, or when any 525K conjunction of these is suffered, each of the parts needs what is in proportion to the disease. Moreover, in the *dyskrasias* due to the humors, if the part acts by itself alone, carry out the purgings confidently, as the condition might direct, taking into account the capacity at the same time. If, however, the action of the part is necessary to all the [other] parts of the organism, as is the case with the stomach and liver, we should pay great attention to the strength of the capacity, lest in some way we destroy this by

ἀθροωτέρᾳ κενώσει χρησάμενοι. κοινὸν δὲ τὸ ἔργον
ἐστὶ καὶ φλεψὶ καὶ ἀρτηρίαις καὶ καρδίᾳ καὶ θώρακι
καὶ νεφροῖς καὶ κύστει τῇ τε τὸ οὖρον ὑποδεχομένῃ
καὶ τῇ χοληδόχῳ, σὺν τοῖς ἀπ᾽ αὐτῆς ἀγγείοις. ἐγκεφάλου δὲ τὸ ἔργον ὑγιαινόντων μὲν ἡμῶν εἰς τὰς καθ᾽
ὁρμὴν ἐνεργείας ἀναγκαῖόν ἐστι τοῖς τὰς τοιαύτας
ἐνεργείας διαπραττομένοις ὀργάνοις, οὐκ ἀναγκαῖον
δὲ νοσούντων, ὅτι μὴ μόνον τοῖς ἀναπνευστικοῖς. ἄλλων δ᾽ ὀργάνων ἐνέργεια μὲν οὐκ ἔστιν, ὑπηρεσία δὲ
ἀναγκαία πρὸς τὸ ζῆν, ὡς πνεύμονος καὶ τραχείας ἀρ-
τηρίας καὶ φάρυγγος. ἐκ μὲν δὴ τῶν τοιούτων μορίων |
526K ὅτι τάχιστα τὰς περιουσίας ἐκκενοῦν ἔξεστιν ὡς ἂν
ἐθέλῃς· ἐκ γαστρὸς δὲ καὶ ἥπατος οὐχ ὡς ἔτυχεν,
ἀλλὰ σὺν τῷ φροντίζειν μηδὲν βλάψαι τὴν δύναμιν.

ἐπεὶ δ᾽ οὐκ ἴσον ἐστὶ τὸ τῶν ἔργων ἀξίωμα πρὸς τὸ
διασῴζεσθαι τοὺς νοσοῦντας, ἀλλὰ τὸ μὲν μᾶλλον
αὐτοῖς ἐστι, τὸ δὲ ἧττον χρηστόν, ἀνάλογον χρὴ τῶν
ἀξιωμάτων ἑκάστου μορίου προνοεῖσθαι τοῦ τόνου.
μέγιστον μὲν οὖν ἀξίωμα τὸ τῆς καρδίας ἐστὶν ἔργον,
καὶ πάντων ἀναγκαιότατον τοῖς νοσοῦσι· τοῦ δ᾽ ἐγκεφάλου πρὸς μὲν τὴν ζωὴν ὁμότιμον, οὐ μὴν τῆς
ἴσης ῥώμης ἐπὶ νοσούντων αὐτοῦ χρῄζομεν· οὐ γὰρ
ἀναγκαῖον πᾶσιν ἐνεργεῖν νεύροις καὶ μυσίν. ὥστ᾽
ἀρκεῖ τοσοῦτον αὐτοῦ διασῳζόμενον ᾧ μέλλει μόνον
ἐπιτελεῖσθαι τὸ τῆς ἀναπνοῆς ἔργον. ἡ δ᾽ ἥπατος
ἐνέργεια πᾶσι μὲν ἀναγκαία τοῖς μορίοις ἐστίν, οὐ
μὴν οὕτω γε διηνεκῶς ἔχει τὸ ἀναγκαῖον ὡς ἡ τῆς
καρδίας. ἀνάλογον δ᾽ αὐτῷ καὶ ἡ τῆς γαστρός. ἐπι-

using too concentrated a purging. The actions of veins, arteries, heart, thorax, kidneys and the two bladders (both that which receives urine and that which contains bile), along with the vessels from them, are general. The action of the brain, when we are healthy, is essential to the voluntary functions and to the organs that accomplish such functions, but is not essential when we are sick, except for the organs of respiration alone. However, the function of other organs is not essential, although there are services essential to life such as those of the lungs, trachea and pharynx. Now it is possible to purge the superfluities from such parts as quickly as you might wish, whereas from the stomach and liver this is not the case. Instead, [it must be] with due care so as not to harm the capacity. 526K

Since the importance of the actions is not equal for the recovery of those who are diseased, but their usefulness is greater or less for them, it is necessary to give forethought to the strength of each part in proportion to its importance. Of the greatest importance, then, is the action of the heart—this is the most essential part of all for those who are diseased. The brain has equal rank as regards life, although we do not require equal strength of it in those who are sick because it is not necessary for all nerves and muscles to be in operation. As a result, it is enough for there to be such preservation of the brain that will enable it simply to accomplish the action of respiration. The function of the liver is esssential to all parts, although it does not have a similar necessity to function continuously in the same way as the heart. The [function] of the stomach is analogous to

στάμενος οὖν ἁπάντων τῶν μορίων τὰς ἐνεργείας καὶ
τὰς χρείας ἐν ἑτέροις ὑπομνήμασιν ἀποδεδειγμένας
527K οὐ χαλεπῶς ἂν εὑρίσκοις εἰς ὅσον | ἑκάστου φυλάττειν
χρὴ τὸν τόνον. οὕτω δὲ καὶ ὅσα τῶν μορίων ἑτέροις
ἀρχαὶ δυνάμεών εἰσι, διασῴζειν αὐτῶν χρὴ τὸν τόνον
ἀνάλογον τῇ χρείᾳ τῶν δυνάμεων, ἵν᾽ ὑπηρετῇ τοῖς
τοῦ ζῴου μέρεσιν. ἔμαθες δὲ καὶ περὶ τῶν τοιούτων
μορίων ὡς ἐγκέφαλός ἐστι καὶ καρδία καὶ ἧπαρ· ὁ μὲν
τοῖς νεύροις τε καὶ μυσίν, ἡ δὲ ταῖς ἀρτηρίαις, τὸ δ᾽
ἧπαρ ταῖς φλεψὶ χορηγὰ τῆς δυνάμεως.

ἀπὸ δὲ τῆς διαπλάσεως ἡ ἔνδειξις γίγνεται τοῦ
τρόπου τῆς κενώσεως τῶν περιττῶν ὡδέ πως. ἐμέτοις
μὲν καὶ διαχωρήμασιν ἡ γαστὴρ ἐκκενοῦται, τὰ δ᾽
ἔντερα μόνοις τοῖς κάτω διερχομένοις, ὥσπερ γε καὶ
τοῦ ἥπατος τὰ σιμά. νεφροὶ δὲ καὶ κύστις καὶ τοῦ
ἥπατος τὰ κυρτὰ πολλῆς μὲν ἐμπεπλησμένα κακοχυμίας διὰ τῶν ὑπηλάτων τε καὶ κατωτερικῶν ὀνομα-
ζομένων ἐκκαθαίρεται φαρμάκων, μετρίας δὲ διὰ τῶν
οὐρητικῶν πόρων, ἐγκέφαλος δὲ δι᾽ ὑπερῴας καὶ ῥινῶν
καὶ ὤτων, θώραξ δὲ καὶ πνεύμων διὰ τραχείας ἀρτηρίας καὶ φάρυγγος. ἀπὸ δὲ τῆς θέσεως ἔνδειξις ἔσται
εἴς τε τὰς δυσκρασίας τῶν στερεῶν, οὐχ ἥκιστα δὲ καὶ
528K τὰς | κακοχυμίας, οὐ μικρά. τοῖς μὲν γὰρ ἐπιπολῆς
δυσκράτοις, οἵων δεῖται φαρμάκων, τοιαῦτα προσοί-
σομεν· τοῖς δ᾽ ἐν τῷ βάθει προσλογιούμεθα τὸ διά-
στημα· εἰ γὰρ οἵων δεῖται, τοιούτοις χρῶτο, πολὺ ἂν

this. Therefore, since you know the functions and uses of all the parts, which have been demonstrated in other treatises,[36] you will find no difficulty [in understanding] the extent to which it is necessary to preserve the strength of 527K each part. In the same way too, in respect of those parts that are origins of capacities for others, it is necessary to preserve their strength in proportion to the use of their capacities, so they may provide a service to the parts of the organism. You learned about parts like the brain, heart and liver. The brain is the provider of capacity for nerves and muscles, the heart for the arteries, and the liver for the veins.

The indication of the manner of the evacuation of the superfluities arises in some way from the conformation [of the part] as follows. The stomach is evacuated by vomiting and defecation, while the intestines [are evacuated] by defecation alone, just as the concavity of the liver also is. The kidneys, bladder and convexity of the liver, when they are filled with much bad humor (*kakochymia*), are purified through the so-called downward purging and purgative medications, but if [the *kakochymia*] is moderate, through the urinary channels. The brain is purified through the palate, nostrils and ears, and the chest and lungs through the trachea and pharynx. From the position [of the part] there will be an indication in regard to the *dyskrasias* of the solid bodies, and not least also in regard to the *kako-* 528K *chymias*—and not an insignificant one. For those with *dyskrasias* of the surface, we shall apply such medications as are required, while to those in the depths, we shall also take into account the interval. If he (i.e. the doctor) re-

36 Presumably, *De usu partium*, III.1–939K and IV.1–366K, but also perhaps *De locis affectis*, VIII.1–451K.

ἀσθενέστερα τῆς χρείας γιγνόμενα πρὸς τὸν πεπονθότα τόπον ἀφίκοιτο. καὶ πρὸς τὰς κενώσεις δὲ ἡ
θέσις συνενδείκνυταί τι διδάσκουσα τὴν κοινωνίαν
τοῦ πεπονθότος μορίου πρὸς τὰ ἄλλα. εἰ μὴ γὰρ
εἰδείημεν ὅτι τοῖς σιμοῖς τοῦ ἥπατος ὑπόκειται τὰ
κατὰ τὴν γαστέρα καὶ τὴν νῆστιν, ὅτι τε τὴν μὲν
νῆστιν ἐκδέχεται τὸ λεπτὸν ἔντερον, ἐκεῖνο δὲ τὸ
κῶλον, εἶτ' αὖθις ἐκεῖνο τὸ ἀπευθυσμένον, οὐκ ἂν οἷοί
τε εἴημεν εὑρεῖν, ἐκκαθαίρεσθαι μὲν δύνασθαι καὶ διὰ
τῆς κάτω γαστρὸς καὶ δι' ἐμέτων τὸ ἦπαρ, ἀμείνω δ'
εἶναι τὴν προτέραν κένωσιν, ἀνέχεσθαι μεμελετη-
κότων τῶν ἐντέρων ἀεὶ τῆς ἐξ ἥπατος περιουσίας.

 οὐ σμικρὰ δ' ἔνδειξις οὐδὲ παρὰ τῆς εὐαισθησίας
τε καὶ δυσαισθησίας τῶν μορίων ἐστί· καταφρονεῖ
μὲν γὰρ πάντων τῶν φαρμάκων τὰ δυσαίσθητα, κἂν
ἱκανῶς ᾖ δακνώδη· τὰ δ' εὐαίσθητα μόρια καταλύεται
529K τὴν | δύναμιν ἀνιώμενα. χρὴ τοίνυν ταῦτα μὲν οὐκ
ἀθρόως οὐδ' ἅπαξ ἰσχυροῖς ἐπιχειρεῖν ἰᾶσθαι φαρμάκοις, ἀλλ' ἐν χρόνῳ μᾶλλον ἀσφαλῶς. ἐπὶ δὲ τῶν
δυσαισθήτων οἷόν τ' ἐστὶ καὶ διὰ ταχέων ἀκοῦσαι τὸ
δέον, εἰσάπαξ χρησαμένους τῷ προσήκοντι βοηθήματι. ἀλλ' ὥσπερ ἔφην εἰρήσεται καὶ αὖθις ἐπὶ πλέον
ὑπὲρ τῆς ἀπὸ τῶν τόπων ἐνδείξεως, καὶ μάλισθ' ὅταν
ἡμῖν ὁ λόγος γίγνηται περὶ τῆς τῶν παρὰ φύσιν
ὄγκων ἰάσεως.

quires such [medications], let him use them, although they become much weaker than is needed as they approach the affected place. Regarding the evacuations, the position [of the part] jointly indicates something, teaching the association of the affected part with others. Thus, if we did not know that the stomach and the jejunum lie under the concavities of the liver, and that the jejunum receives the thin intestine, and the latter the colon, and the latter again the rectum, we would not have been able to discover that the liver can be purged both through the stomach downward and through vomiting. The former evacuation is better, since the intestines are practiced in always receiving the superfluity from the liver.

There is no small indication from the normal or disturbed sensation of the parts, since parts with impaired sensation "disregard" all medications, even if they are exceedingly stinging, whereas parts with normal sensation break down the capacity when they are distressed. Accordingly, we must not attempt to cure these with strong medications given in a concentrated fashion or all at once, but more safely over a period of time. However, in the case of the parts with impaired sensation, it is possible [for us] to understand what is required quickly, using the appropriate remedy as a single administration. But, as I mentioned, more will be said about the indications from the places, and particularly when my discussion about the cure of swellings contrary to nature occurs.[37]

529K

37 In the final two books (13 and 14) of the *MM*.

ΒΙΒΛΙΟΝ Θ

530K 1. Ὡς δ' ἄν τις καὶ τοὺς πυρετοὺς ἰῷτο μεθόδῳ, τοῖς ἄλλοις ὡσαύτως νοσήμασι τοῖς κατὰ δυσκρασίαν συνισταμένοις ἐν ἡμῖν, ἃ διῆλθον ἐν τῷ πρὸ τοῦδε γράμματι, καιρὸς ἂν εἴη λέγειν. ὅτι μὲν οὖν, ὦ Εὐγενιανέ, ἐκ θερμοῦ καὶ ψυχροῦ καὶ ὑγροῦ καὶ ξηροῦ κεραννυμένων ἀλλήλοις ἡ τῶν ἁπλῶν καὶ πρώτων μορίων συνέστηκε φύσις, ἅπερ Ἀριστοτέλης προσαγορεύει ὁμοιομερῆ δι' ἑνὸς ἀποδέδεικται γράμματος, ἐν ᾧ Περὶ τῶν καθ' Ἱπποκράτην στοιχείων ἐσκοπού-
531K μεθα. ὅτι δ' ἐν τῷ τὸ μὲν θερμότερον | εἶναι, τὸ δὲ ψυχρότερον, ἢ ὑγρότερον, ἢ ξηρότερον, ἢ κατὰ συζυγίαν τι τούτων πεπονθέναι, τῶν ἁπλῶν καὶ πρώτων μορίων ἐστὶν ἡ πρὸς ἄλληλα διαφορά, διὰ τῆς Περὶ κράσεων ἐδείχθη πραγματείας. ἐν δέ γε τῷ Περὶ τῆς τῶν νοσημάτων διαφορᾶς βιβλίῳ τῶν μὲν ἁπλῶν τοῦ ζῴου μορίων δύο τὰ πάντ' εἶναι γένη νοσημάτων ἐδείκνυτο· κοινὸν μὲν τὸ ἕτερον αὐτοῖς, πρὸς τὰ σύνθετά τε καὶ δεύτερα καὶ ὀργανικὰ μόρια, καλοῦμεν δὲ αὐτὸ λύσιν ἑνώσεως· ἕτερον δ' ἐξαίρετον ἴδιον ἐν

---

[1] See Aristotle, *Parts of Animals* II, 648a6–55b27, and *Meteorologica* X–XIII, 388a10–90b20.

344

BOOK VIII

1. Now would be an appropriate time to state that some-   530K
one might also cure fevers by method in a similar way to
the other diseases that arise in us due to *dyskrasia*, which I
went over in the preceding book. Thus, Eugenianus, that
the nature of the simple and primary parts, which Aristotle
calls *homoiomeres*,[1] arises from the mixing of hot, cold,
moist, and dry with each other, was shown through one
treatise in which I gave them consideration: *On the Elements according to Hippocrates*.[2] That the differentiation
of the simple and primary parts from each other lies in
their being hotter, colder, more moist, or drier, or their   531K
having been affected by some conjunction of these [quali-
ties] was shown in the treatise *On Krasias (On Mixtures)*.
In the book *On the Differentiae of Diseases* it was shown
that there are in all two classes of diseases of the simple
parts of the animal. One, which is common to simple and
compound, secondary and organic parts, I call "dissolution
of union (continuity)." The other, which is selective and

[2] The works referred to in this opening statement are several
of those which form the theoretical foundation of Galen's method
of treatment. These are, in order, *De elementis secundum Hip-
pocratem* (I.413–508K), *De temperamentis* (I.509–694K), *De
differentiis morborum* (VI.831–80K), and *De morborum causis*
(VII.1–41K).

δυσκρασίᾳ θερμοῦ καὶ ψυχροῦ καὶ ξηροῦ καὶ ὑγροῦ
συνιστάμενον. εἶναι δὲ τῶν δυσκρασιῶν τὰς μὲν κατὰ
μίαν ἀντίθεσιν ἁπλᾶς, τὰς δὲ κατ᾽ ἄμφω συνθέτους. ἐν
μὲν οὖν τῇ κατὰ τὸ θερμὸν καὶ ψυχρὸν ἀντιθέσει δύο
συνίστασθαι δυσκρασίας ἁπλᾶς· ἑτέραν μέν, ὅταν
ἑαυτοῦ γένηταί τι θερμότερον εἰς τοσοῦτον ὡς ἤδη
βλάπτεσθαι τὴν ἐνέργειαν αὐτοῦ· δευτέραν δέ, ὅταν
ὁμοίως ἐπικρατήσῃ τὸ ψυχρόν. ἐν δὲ τῇ κατὰ τὸ ξηρὸν
καὶ ὑγρὸν ἑτέρας αὖ δύο καὶ κατὰ τάσδε δυσκρασίας,
532K   ὑγρὰν καὶ ξηράν. ἐπιπλεκομένων δ᾽ ἀλλήλαις | τῶν
ἁπλῶν ἄλλας συνίστασθαι τέτταρας, ὑγρὰν καὶ ψυχράν, ὑγράν τε καὶ θερμήν, καὶ ξηρὰν καὶ θερμήν,
ξηράν τε καὶ ψυχράν. εἴρηται δὲ καὶ γεννῶντα τὰς
δυσκρασίας αἴτια δι᾽ ἑτέρου γράμματος, ἐν ᾧ τὰς τῶν
νοσημάτων αἰτίας ἐδηλοῦμεν.

ἐν δέ τι τῶν κατὰ δυσκρασίαν νοσημάτων ἐστὶ καὶ
ὁ πυρετός, εἰς τοσαύτην ἀμετρίαν αἱρομένης τῆς θερμασίας ὡς ἀνιᾶν τε τὸν ἄνθρωπον ἐνέργειάν τε βλά-
πτειν, ὡς εἴ γε μήπω μηδέτερον ἔχει, κἂν ὅτι μάλιστα
θερμότερος ἑαυτοῦ τύχῃ γεγονώς, οὐδέπω πυρέττειν
αὐτὸν ἐδείκνυμεν. ἐπεὶ δ᾽ ἕκαστον τῶν παρὰ φύσιν ἐν
ἡμῖν συνισταμένων, εἴτε εἰδικωτέρας τυγχάνῃ προσηγορίας, εἴτε γενικωτέρας, οὐκ ἔστιν ἐκτὸς τοῦ σώματος, ἀλλὰ καὶ ἡ φλεγμονὴ καὶ ἡ πλευρῖτις καὶ ἡ
νόσος ἅμα τὰ τρία καθ᾽ ἓν γίνεται σῶμα τὸ Δίωνος, εἰ
οὕτως ἔτυχεν, ἐνδείκνυται δὲ καὶ τούτων μὲν ἕκαστον
ἰδίαν ἔνδειξιν, οὐδὲν δ᾽ ἧττον αἱ διαφοραὶ τοῦ πάντων
εἰδικωτάτου, διὰ τοῦτ᾽ οὖν ἄμεινον ἔνδοξεν ἰδίᾳ πρα-

specific, lies in a *dyskrasia* of hot and cold, and moist and dry. However, among the *dyskrasias*, there are those that are simple, involving a single opposition, and those that are compound, involving a double opposition. Thus, in the opposition of hot and cold, two simple *dyskrasias* exist: one, when something becomes hotter than it was to the extent that is now damaging to its function, and a second, when cold prevails in a similar manner. In the opposition of dry and moist there are again two *dyskrasias*; namely, moist and dry. When the simples combine with each other, four 532K other *dyskrasias* arise: moist and cold, moist and hot, dry and hot, and dry and cold. And the causes which generate the *dyskrasias* were spoken of by way of another treatise in which I set out the causes of diseases.

Fever is also one of the diseases involving *dyskrasia* when the heat is raised to such a disproportionate degree that it distresses the person and harms function. I demonstrated that someone was not yet febrile if he does not yet have either of these [consequences], even if he has by chance become much hotter than he was. However, since each of the abnormal things which arise in us, whether they acquire a specific or generic name, do not exist outside the body, but the three things, inflammation, pleurisy and disease also occur simultaneously in one body—that of Dion should this be the case—each [body] demonstrates a specific indication of these and the differentiae are nothing less than what is most specific of all of them, it seemed better, on account of this, to deal separately with the dif-

γματεύεσθαι περὶ τῆς διαφορᾶς τῶν πυρετῶν, ἵν᾽ ἐφ᾽
ἑτοίμοις ἅπασιν ὁ λόγος ἡμῖν ἐνθάδε περαίνοιτο,
533K χωρὶς τῆς πρὸς τοὺς κακῶς ὑπὲρ αὐτῶν | γράψαντας
ἀντιλογίας.

ἀναμνησθῶμεν οὖν τὰ κεφάλαια τῆς διαφορᾶς
αὐτῶν, ἵν᾽ ἐξ ἑκάστου τὴν οἰκείαν ἔνδειξιν ἀναλάβω-
μεν. ἔστιν οὖν μία μὲν αὐτῶν διαφορὰ καθ᾽ ἣν ἤτοι
πάρεστιν ἔτι τὸ ποιοῦν αἴτιον τὸν πυρετόν, ἢ πέπαυται
ποιῆσαν· ἑτέρα δὲ καθ᾽ ἣν ἤτοι γ᾽ ἑκτικός ἐστιν ὁ
γεγονὼς πυρετός, ἢ ὡς ἂν εἴποι τις σχετικός. ἐπειδὴ
γὰρ ὀνομάζουσι τὰς μὲν εὐλύτους διαθέσεις ἐν σχέ-
σει, τὰς δὲ μὴ τοιαύτας, ἐν ἕξει, συγχωρητέον ἐστὶν
ἕνεκα σαφοῦς διδασκαλίας καὶ αὐτῶν τῶν πυρετῶν
τοὺς μὲν δυσλύτους ἑκτικούς, τοὺς δ᾽ εὐλύτους σχετι-
κοὺς ὀνομάζεσθαι. καὶ λεκτέον ἐνίους μὲν τῶν πυρετῶν εὔλυτον ἔχειν τὴν διάθεσιν οὕτως ὡς τὸ θερμαν-
θὲν ἐν ἡλίῳ ξύλον, ἐνίους δὲ δύσλυτον, ὡς τὸ μέχρι
τοῦ τύφεσθαί τε καὶ καπνὸν ἀποβάλλειν ὑπὸ πυρὸς
θαλφθέν. αὗται μὲν αἱ πρῶται διαφοραὶ τῶν πυρετῶν,
ἃς συνάψαντες πρότερον ταῖς γενικωτέραις ἐνδείξεσιν, αὖθις ἐπὶ τὰς αὐτῶν τῶν εἰρημένων εἰδικωτέρας
ἀφιξόμεθα. πρώτη μὲν γὰρ ἔνδειξις ὡς ἀπὸ νοσήματός ἐστιν, ἅπαντος πυρετοῦ τὴν ἀναίρεσιν ἐνδεικνυ-
534K μένη· δευτέρα δὲ ὡς τοιοῦδε νοσήματος, | οὐκέθ᾽
ἁπλῶς ἀναίρεσιν, ἀλλὰ τοιάνδε τινά. κατὰ γάρ τοι

---

3 Galen essentially defines his use of these two terms here.

ferentiae of fevers so that, through everything being prepared, my discussion may be brought to completion here, apart from the refutation of those who have written badly on them. 533K

Let us call to mind, then, the chief points of the differentiation of these so that we might take the proper indication from each [Figure 9]. Thus, there is one differentia of fevers depending on whether the cause bringing about the fever is still present or has ceased to act. There is another differentia depending on whether the fever that has arisen is established, or is, one might say, temporary. Since people term the conditions that are readily resolved "temporary" and those that are not readily resolved "established," we must agree for the sake of clear exposition to term those actual fevers that are difficult of resolution "hectic" and those that are easy of resolution, "schetic."[3] It must be said that some of the fevers have a condition that is easy to resolve, as is the case with wood heated in the sun, whereas some have a condition that is difficult to resolve, as is the case with wood heated by fire to the point of smoldering and giving off smoke. These are the primary differentiae of the fevers. When we have joined them first with the more generic indications, we will come in turn to the more specific indications of those things mentioned. For the primary indication, as it comes from the disease, indicates the eradication of fever as a whole. The second, as it comes from this particular disease, no longer indicates simply 534K eradication, but a particular eradication. Certainly, when

Both Linacre and Peter English use the Greek terms—the latter has "hecticall" and "schiticall." The former is still in use, albeit with a more specific meaning, whereas the latter is not.

Figure 9. A modern classification of the various kinds of fever. From *Stedman's Medical Dictionary*, 27th ed. (Philadelphia: Lippincott Williams & Wilkins, 2003), p. 659; with permission.

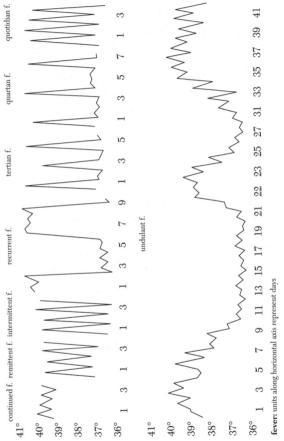

fever: units along horizontal axis represent days

δυσκρασίαν νοσούντων τῶν πυρεττόντων ἀναγκαῖον
καὶ τὴν ἴασιν αὐτῶν ἀναίρεσιν εἶναι τῆς δυσκρασίας.
ἅπασα δ᾿ ἀμετρία διὰ τῆς ἐναντίας ἀμετρίας ἰᾶται,
καὶ γὰρ καὶ τοῦτ᾿ ἔμπροσθεν ἐδείχθη. καὶ τοίνυν καὶ ἡ
τῆς τοῦ πυρετοῦ δυσκρασίας ἀμετρία διὰ τῆς ἐναντίας
ἀμετρίας ἰαθήσεται. ἔστι δ᾿ ἡ ἀμετρία τῆς τῶν πυρεττόντων δυσκρασίας ἐν πλεονεξίᾳ θερμότητος· ἡ τοίνυν
εἰς εὐκρασίαν ὁδὸς αὐτοῖς διὰ ψύξεως ἔσται. εἰ μὲν
οὖν ἤδη γεγονότες εἶεν οἱ πυρετοί, τὸ δὲ ποιῆσαν
αὐτοὺς αἴτιον οἴχοιτο, μόνος ἂν οὗτος εἴη τῆς θεραπείας αὐτῶν ὁ σκοπὸς ἡ ψύξις. ὥστε ζητητέον ἡμῖν
τὰς ὕλας ὅσαι ψύξιν ἰσόρροπον ἔχουσι τῇ πλεο-
νεκτούσῃ θερμότητι. γιγνομένων δ᾿ ἔτι τῶν πυρετῶν
τὴν ποιοῦσαν αὐτοὺς αἰτίαν ἀναιρετέον· εἰ δὲ τὸ μὲν
ἤδη γεγονὸς αὐτῶν εἴη, τὸ δ᾿ ἔτι γίγνοιτο, τοῦ συναμφοτέρου στοχαστέον, ἀναιρούντων πρότερον μὲν
τὴν ἀνάπτουσαν αἰτίαν αὐτούς, ἐφεξῆς δὲ σβεννύν-
των, τὸ φθάσαν ἀπ᾿ αὐτῆς ἀνῆφθαι. τίνες οὖν αἰτίαι
πυρετοὺς γεννῶσαι καὶ κατὰ τίνα σώματος ἑκάστου
535K διάθεσιν, | εἰ μή τις εἴη μεμνημένος ἀκριβῶς, οὐκ ἄν
ποτε μεθόδῳ προνοήσαιτο πυρετῶν.

2. Ἐλέχθη γοῦν ἐν ἐκείνοις τοῖς σώμασιν ἡ
στέγνωσις ἐργαζομένη πυρετόν, ἐν οἷς ἐστι τὸ διαπνεόμενον οὐκ ἀτμῶδες οἷόν περ ἐν τοῖς εὐχύμοις,
ἀλλὰ δακνῶδες καὶ δριμὺ καὶ καπνῷ παραπλήσιον ἢ
λιγνύϊ. τούτοις οὖν τοῖς σώμασιν ὑγιεινότατα μὲν
λουτρὰ γλυκέων ὑδάτων εὐκράτων καὶ τρίψις ἀραι-
ωτικὴ καὶ γυμνάσια σύμμετρα καὶ δίαιτα γλυκύχυ-

febrile people become sick in relation to a *dyskrasia*, the cure of these is also necessarily an eradication of the *dyskrasia*. Every excess is cured by the opposite excess—this was also shown earlier. Therefore, the excess of the *dyskrasia* of the fever will also be cured by the opposite excess. Where the excess of the *dyskrasia* in those who are febrile lies in a preponderance of heat, the path to *eukrasia* for them will accordingly be through cooling. If there are fevers that have already occurred, while the cause bringing them about is gone, the sole objective of treatment for these would be cooling. As a result, what we must seek are those materials that are cooling in equal measure to the preponderant heat. If fevers are still in the process of occurring, what must be eradicated is the cause bringing them about. If part of the fever has already occurred and part is still occurring, we must aim at both together, first removing the cause that kindled them and next quenching what has already been kindled by this cause. Therefore, unless someone were to call to mind accurately what the causes which generate fevers are, and in relation to what condition of each body, he would never give forethought to fevers on the basis of method.

535K

2. Anyway, what I said was that stoppage of the pores creates a fever in those bodies in which what is being dissipated is not vaporous, as it is for example in those which are *euchymous*, but biting and sharp, and like smoke or soot. Therefore, what are most healthy for these bodies are baths of sweet water that is *eukratic*, massage which rarefies, moderate exercise and food with sweet juices.[4] What

---

[4] On this see Galen, *De simplicium medicamentorum temperamentis et facultatibus*, XI.494K.

μος, ἐναντιώτατα δὲ λουτρὰ ψυχρὰ καὶ στυπτηριώδη
καὶ ἀλουσία καὶ γυμνάσιον ὀξὺ καὶ τρίψις ἤτοι μηδ᾽
ὅλως ἢ σκληρὰ γινομένη καὶ δίαιτα κακόχυμος, ἀγρυπνία τε καὶ θυμὸς καὶ λύπη καὶ φροντίς, ἔγκαυσίς τε
καὶ κόπος. ἀλλὰ περὶ μὲν τῶν ἄλλων αὖθις ἐπάνειμι·
περὶ δὲ τῆς στεγνώσεως νῦν ὅθεν ὁ λόγος ὡρμήθη
γινώσκειν χρὴ σαφῶς ὅτι καὶ μόνη χωρὶς τῶν ἄλλων
ἁπάντων ἱκανὴ γεννῆσαι πυρετούς. ἐὰν οὖν εἴρξῃς τὰς
τοιαύτας φύσεις τῶν οἰκείων λουτρῶν, αὐτίκα πυρέτ-
536K τουσιν. ἐθεάσω δὲ δήπου καθ᾽ ὃν ἐν | Ῥώμῃ σὺν ἡμῖν
διέτριψας χρόνον ἐνίους οὕτω νοσήσαντας· ἐφ᾽ ὧν
οἱ τὴν θαυμαστὴν διάτριτον ἀσίτους ὑπερβάλλειν
ἀξιοῦντες αὐτοὶ κακοηθεστάτους πυρετοὺς κατασκευ-
άζουσι, δέον αὐτίκα τοῦ πρώτου παροξυσμοῦ παρ-
ακμάζοντος ἀπάγειν εἰς τὸ βαλανεῖον, ἐπιτρέπειν τε
λούεσθαι πολλάκις εἰ βούλοιντο, μὴ μόνον ἅπαξ,
ἀλλὰ καὶ δίς. ἐγὼ γοῦν ἐάσας ἅπαντας τοὺς ἄλλους
ἀναμνήσω σε τοῦ λουσαμένου μὲν ἐν τοῖς στυπτη-
ριώδεσιν ὕδασιν, ἃ καλοῦσιν Ἄλβουλα, πυκνωθέντος
δ᾽ ἐκ τούτου τὸ δέρμα, κἀντεῦθεν ἀρξαμένου πυρέττειν, ἀρκέσει γὰρ ἕνεκα σαφηνείας οὗτος οἷον παρά-
δειγμά τι τοῦ λόγου γενέσθαι.

παρῆσαν μὲν ἐπισκοπούμενοι τῶν οὐκ ἀφανῶν
τινες ἰατρῶν ὁ μὲν Ἐρασιστράτειος, ὁ δὲ Μεθοδικός·
ἔδοξε δ᾽ ἀμφοτέροις ἀσιτῆσαι τὸν ἄνθρωπον. οὐ μὴν
εἰάσαμέν γε ἡμεῖς, χωρισθέντων αὐτῶν ἐλθόντες, ἀλλ᾽
εἰς βαλανεῖον εἰσαγαγόντες εὐθέως καὶ χλιαρὸν ἔλαιον ἐπὶ πλεῖστον αὐτῷ περιχέαντες, ἀνατρίψαντές τε

are most inimical [to such bodies] are cold baths that are astringent, remaining unwashed, vigorous exercise, either no massage at all or that which is hard, *kakochymous* foods, sleeplessness, anger, grief, anxiety, heatstroke and fatigue. But I shall return later to the other things. Now, concerning stoppage of the pores, which is where the discussion started, you should know clearly that this alone, apart from all the other things, is enough to generate fevers. Thus, if you keep natures of this sort away from their normal baths, they immediately become febrile. You saw, I am sure, dur- 536K ing the time you spent with me in Rome, some people who became ill in this way. In their case, those who think it right that they should go beyond the wondrous three-day fast, themselves create the most *kakoethical* fevers, making it necessary, immediately the first paroxysm abates, to lead the patient to the bathhouse, and allow frequent washing, if they wish, not only once but twice. Anyway, leaving aside all the other things, I shall remind you of the man bathing in astringent waters, which they call Albula.[5] After the skin was thickened by this, he began to be febrile. This man will be a satisfactory kind of example for the purpose of bringing clarity to the discussion.

Certain quite well-known doctors were present observing, one an Erasistratean and the other a Methodic. Both decided to fast the man. I, arriving after they departed, did not allow this but at once led him to the bathhouse, and after pouring a copious amount of lukewarm oil over him

5 Albula was an ancient name for the Tibur. Albulae Aquae were medicinal waters near Rome; Pliny, *HN*, XXXI.2 (6).

πραότατα, τὸ πλεῖστον τοῦ χρόνου μέρος ἐν τῷ τῆς
θερμῆς δεξαμένης ὕδατι διατρίβειν ἐκελεύσαμεν. εἶτα
537K ἐξελθόντα καὶ | χρησάμενον ὕδατι ψυχρῷ, κατὰ τὰ
εἰωθότα σκεπάσαντες σινδόνι καὶ βραχὺ καθῖσαι
κελεύσαντες, ὡς ἀνακτήσασθαι τὴν δύναμιν, αὖθις
εἰσαγαγόντες εἰς τὸ βαλανεῖον ὁμοίως τε πάλιν ἀλεί-
ψαντές τε καὶ τρίψαντες καὶ κατὰ τὸ θερμὸν ὕδωρ
χρονίσαι κελεύσαντες· εἶτ' αὖθις ἐξαγαγόντες καὶ τῷ
ψυχρῷ βάψαντες, ἀπομάξαντές τε τροφὴν ἐδώκαμεν
αὐτίκα μὲν ἐξελθόντι μετὰ τὸ πιεῖν ὕδατος πτισάνης
χυλόν, εἶτα βραχὺ διαλιπόντες θριδακίνην, καὶ μετ'
αὐτὴν ἐξ ἁπλοῦ λευκοῦ ζωμοῦ τῶν ἁπαλοσάρκων
ἰχθύων, οἷοί περ οἱ πετραῖοι πάντες εἰσὶ καὶ οἱ ὀνίσκοι
καλούμενοι. κάλλιον δ' ἕνεκα τοῦ παραλελεῖφθαι μηδὲν ἅπαντα προσθεῖναι τῇ διηγήσει.

τοῦ μὲν ἔτους ἦν ὁ καιρὸς ἐκεῖνος ἐν ᾧ ταῦτ'
ἐπράττετο βραχύ τι μετὰ τὰς ὑπὸ κύνα προσαγο-
ρευομένας ἡμέρας, ὁ δὲ ἄνθρωπος ὡς πέντε καὶ τριάκοντα ἐτῶν, μελάντερος τὴν χροιὰν καὶ λεπτὸς τὴν
ἕξιν καὶ δασύς, ἁπτομένοις τε σαφῶς δακνώδη τὴν
θερμασίαν ἔχων, ὁπόθ' ὑγίαινεν, οὖρα κατακορῆ
ξανθὰ καὶ εἰ ἐπὶ πλέον ἀσιτήσειε δάκνοντα, γαστὴρ
ἐξηραίνετο συνεχῶς καὶ ἦν τὰ διαχωρήματα βραχέα
538K καὶ δριμέα καὶ ξηρά· τὸ | δὲ τῆς ψυχῆς ἦθος ὀξύθυμόν
τε καὶ φροντιστικὸν ὑπῆρχεν, ὀλιγόϋπνός τε τὰ πάντα
καὶ συνεχῶς ἀγρυπνίαν μεμφόμενος. οὗτος ἐν χωρίῳ
τινὶ πράξεων ἕνεκα γενόμενος ἐχρῆτο τοῖς Ἀλβούλοις
πλησίον οὖσιν, ὥρας τε ἑβδόμης, ὡς ἔφασκε, καὶ τρίς

and massaging him very gently, directed him to spend the greater part of the time in the water of the hot tank. Then, when he came out and used cold water, as was customary, I 537K covered him with muslin and directed him to sit for a short time so as to recover his strength before again leading him to the bathhouse, and in like manner anointing and massaging him again, and directing him to spend time in the warm water. Then, having led him out once more and plunged him into cold water, I dried him off and gave him nourishment immediately after he emerged, and next a drink of water and the juice of ptisan. After a short interval, I gave him lettuce, and after that, some simple white soup made from soft-fleshed fish—fish like all those that live among the rocks, and the so-called cod. It is better to detail all the things to give so that nothing is left out.

The time of year in which this was being done was a little after what they call the dog days.[6] The man himself was about thirty-five years old, rather dark in complexion, thin in build, and hirsute. To those who touched him when he was healthy, he clearly had a mordant heat, and he had strongly yellow urine which, if he fasted still more, was mordant. His stomach was continuously drying and his excretions scanty, sharp and dry. The disposition of his soul 538K was choleric and anxious, he slept very little overall, and continually complained of insomnia. Being in a certain place for business purposes, he began using the Albula which was to hand and, at the seventh hour of the day, as he

[6] The dog days are the period when the Dogstar rises and sets with the sun, usually reckoned as July 3 to August 11. It was regarded as a time when dogs were prone to hydrophobia.

GALEN

γε καὶ τετράκις ἐλούετο, χρηστόν τι δὴ τοῦτο νομίζων
ἐργάζεσθαι. ἐκεῖθέν τε πάλιν εἰς τὸν ἀγρὸν ἀφικόμενος καὶ τροφὴν προσαράμενος, εἶθ᾽ ὑπνώσας βραχέα
παρεγένετο μὲν εἰς τὴν πόλιν ἑσπέρας βαθείας, ἀπο-
ρήσας δ᾽ ἐπιτηδείου λουτροῦ διὰ τῆς νυκτὸς ἐπύρεξε.
θεασάμενοι δὲ αὐτὸν οἱ ἰατροὶ σχεδὸν ὥρας τρίτης
τῆς ἡμέρας ἐκέλευσαν οὐ μόνον ἐκείνην, ἀλλὰ καὶ τὴν
ἑξῆς ἀσιτεῖν ὅλην, ἵνα τὴν διὰ τρίτης νύκτα φυ-
λάξαιντο, καὶ ταῦτ᾽ εἰπόντες ἀπηλλάττοντο. λούσαντες οὖν ἡμεῖς αὐτὸν καὶ διαιτήσαντες, ὡς εἴρηται, τοῖς
οἰκέταις ἐκελεύσαμεν, ἐὰν εἰς ἑσπέραν οἱ ἰατροὶ παρα-
γενηθῶσιν ἡσυχάζειν τε φάναι τὸν ἄνθρωπον, ἀπο-
πέμπειν τ᾽ ἐκείνους παραχρῆμα, τῆς ἐπιούσης ἡμέρας
ἥκειν ἀξιώσαντας. ὡς δ᾽ ἐχωρίσθησαν, αὖθις ὁμοίως
λούσαντές τε καὶ διαιτήσαντες αὐτόν, ἐξ αὐτῶν τού-
των ὧν ἐπράξαμεν ὑπνῶσαι καλῶς ἐποιήσαμεν. |

539K παραγενόμενοι δὲ οἱ ἰατροὶ κατὰ τὴν ὑστεραίαν ἕωθεν,
ἠξίουν αὐτὸν ἔτι κἀκείνην ἀσιτῆσαι τὴν ἡμέραν, εἰ
καὶ ὅτι μάλιστα τελέως ἀπύρετος εἴη. τοῦ δ᾽ ὑποσχο-
μένου πράξειν ὡς κελεύουσιν, οἶσθα μὲν δή που τὸν
γέλωτα τὸν γενόμενον ἀπελθόντων αὐτῶν, ὃν ἡμεῖς
ὀλίγον ὕστερον ἀφικόμενοι κατελάβομεν. ἐπίστασαι
δὲ καὶ ὅπως ἐπράχθη τά κατὰ τὸν ἄρρωστον.

Ἐπειδὴ γάρ, ἔφην, ὑγιαίνων ἐλούου δίς, οἱ δ᾽ ἰατροὶ τελέως ἀσιτῆσαί σε κελεύουσιν, οὔτ᾽ ἐκείνοις τὸ
σύμπαν πεισθῆναι δίκαιον, διαλλάξαι τέ τι τῶν ἐπὶ
τῆς ὑγείας προσήκει.

τοῦ δ᾽ οἰομένου με συμβουλεύειν αὐτῷ λούσασθαι

358

said, he would bathe three or even four times, thinking this would surely be a beneficial thing to do. When he again arrived at his estate from that place and took some nourishment, he slept for a short time and came to the city at a late hour, but being unable to avail himself of a suitable bath, he was febrile during the night. When the doctors saw him, almost at the third hour of the next day, they directed him to fast, not only that day but also the whole of the day following, so that they might watch him during the night of the third day. Having said this, they departed. So when I had bathed and fed him, as I described, I instructed his household slaves that, if the doctors came to attend him toward evening, to say the man was quiet and to send them away immediately, suggesting it would be appropriate for them to come the following day. When they departed, I bathed and fed him again in the same way, and by the very things I did, enabled him to sleep well. When the doctors attended early in the morning of the next day, they required him to continue to fast on that day, particularly if he was completely afebrile. After he promised to do as they ordered, you know, of course, of the laughter that occurred following their departure, which I detected when I arrived a little later. You are also aware of what was done for the patient.

"For," I said, "since you were bathing twice when healthy, and the doctors order you to fast completely, it is not right to be entirely persuaded by them, and it is appropriate to change some of those things [you do] when you are healthy."

On the assumption that I was advising him to bathe, but

539K

μέν, ἀλλ' οὐ δίς, ἑτοίμου τε πράττειν εἶναι φάσκοντος
ὅ τί περ ἂν ἐγὼ κελεύσαιμι·

Μὴ τοίνυν, ἔφην, μήτ' ἀσιτήσῃς μήτε λούσῃς δίς,
καὶ γὰρ καὶ καμεῖν μοι δοκεῖς τρίτην ἡμέραν καί που
καὶ θερμανθῆναι σφοδρότερον ὑπὸ τοῦ ἡλίου· διά τε
οὖν τὸν κάματον, ἔφην, καὶ τὴν ἔγκαυσιν ἕν σε χρὴ
προσθεῖναι λουτρὸν τοῖς δύο καὶ τρὶς λούσασθαι
πειθόμενον ἐμοί, τοῦτο γὰρ ὑπέσχου ποιήσειν.

ὁ δὲ μειδιάσας, Χαλεπὰ μέν, ἔφη, προστάττεις, |
540K ἀλλ' ἐπεὶ συνθήκας φυλάττειν δίκαιον, ἃ κελεύεις
ποιήσω.

τὰ δὲ πραχθέντα μετὰ ταῦτα τοῖς ἰατροῖς οἶσθα
δήπου σαφῶς ὁπόσου καταγέλωτος ἦν ἄξια. δὶς μὲν
ἐλέλουτο καὶ ἠριστηκὼς ἐκεκοίμητο, περὶ δὲ δυσμὰς
ἡλίου τῶν οἰκετῶν τις ἀγγέλλει παρεῖναι τοὺς ἰατρούς.
ὁ δὲ προσποιεῖται πυρέττειν καὶ περιβαλλόμενος ἱμά-
τιον, εἴσω τε στραφείς, ὅπως μὴ καταφανὴς αὐτοῖς
γένοιτο πεπωκὼς οἶνον, τῶν φίλων τινὶ κελεύει τοῖς
ἰατροῖς ἀνθ' αὑτοῦ ποιεῖσθαι τὰς ἀποκρίσεις, ἤν τι
πυνθάνωνται. ἔμελλον δὲ δήπου τὸ συνηθέστατον αὐτοῖς πρῶτον ἐρήσεσθαι, θεώμενοι περιβεβλημένον
ἐπιμελῶς ἅμα τῇ κεφαλῇ τὸν ἄνθρωπον, ἥτις ἦν ὥρα
καθ' ἣν ὁ παροξυσμὸς εἰσέβαλεν. ἀποκρινομένου δὲ
τοῦ φίλου σχεδὸν οὐδεμίαν ὁλόκληρον ὥραν γεγονέναι μεταξύ, πότερον μετὰ φρίκης, ἢ μετὰ περι-
ψύξεως ὑπήρξατο πυνθάνονται. τοῦ δὲ μετὰ φρίκης
εἰπόντος ἅπτονται τοῦ ἀνθρώπου, διά τε τὸ γελᾶν καὶ
περικεκαλύφθαι νοτιζομένου. ἐπαινέσαντες οὖν αὐτὸν

not twice, he said that he was ready to do whatever I might direct.

"Accordingly," I said, "you are neither to fast nor bathe twice [a day], for in fact you seem to me sick for a third day and, somehow, also to be heated too strongly by the sun. Therefore, because of the suffering and the heatstroke, it is necessary for you to add one bath to the two and bathe three times, and for you to comply with my directions and promise to do this."

Smiling, he said, "You give difficult orders, but since it is right to observe our agreement, I shall do as you direct." 540K

You know clearly, I am sure, how much the things that were done by the doctors after this were worthy of derision. He had bathed twice and, having eaten, had fallen asleep. Around sunset, one of the household slaves announced that the doctors were present. The man pretended to be febrile, and throwing his cloak over his shoulders and wrapping himself in it, went back inside. So it would not seem to them that he had been drinking, he directed one of his friends to answer the doctors on his behalf, if they were asking about anything. They were intending, I presume, to ask first what was most usual for them, when they saw the man carefully wrapped up including his head: i.e. what was the hour at which the paroxysm struck. When the friend answered that scarcely a whole hour had passed, they asked whether it began with shivering or with chilling. When he said with shivering, they touched the man who was sweating due to laughing and having covered himself completely. Therefore, they commended the man

ἐπὶ τῷ πεισθῆναί σφισι καὶ μηδὲν ἁμαρτεῖν, πρὸ τοῦ
διὰ τρίτης παροξυσμοῦ,

Τοιγάρ τοι διὰ τοῦτο, ἔφασαν, ἤδη μέν σοι πέπαυ-
541K ται | τὰ τῆς φρίκης, ἱδρῶτος δ᾽ ἐστὶν ὑπόφασις καὶ
νοτὶς πολλὴ περὶ τὸ δέρμα, καὶ ταῦτ᾽ οὐκ ἂν ἐγένετο,
μὴ ἀσιτήσαντός σου καὶ τὴν διάτριτον ὑπερβάλ-
λοντος.

κελεύσαντες οὖν τοῖς οἰκείοις, ἐὰν γένωνταί τινες
αὐτῷ νοτίδες, ἀπομάττειν ἐπιμελῶς, ὅπως μὴ ψυχθείη,
παραγενήσεσθαί τε φάντες ἕωθεν ἀπαλλάττονται,
μηδὲ τότε κελεύοντες ἐπὶ τοῖς ἱδρῶσι τραφῆναι τὸν
ἄνθρωπον· ἐδόκουν γὰρ αὐτοῖς αἱ κατ᾽ ἐκεῖνον τὸν
καιρὸν νύκτες εἶναι μικραὶ καὶ κάλλιον ἐφαίνετο κατὰ
τὴν ὑστεραίαν ἕωθεν τρέφειν. ἀπαλλαγέντων οὖν αὐτῶν ἐν ἱδρῶτι ῥεόμενος ὁ νεανίσκος, ὡς ἐνετετύλικτο
τοῖς ἱματίοις, καταδραμὼν εἰς τὸ βαλανεῖον, ἐλούσατο
τὸ τρίτον, ὁμοίως τε διῃτήθη. κἄπειτα κατὰ τὴν ὑστεραίαν, πρὶν ἀφικέσθαι τοὺς ἰατρούς, προῆλθε τῆς
οἰκίας ἐπίτηδες, ὅπως μὴ καταλάβοιεν αὐτόν. οἱ δ᾽
ἄρα μικρὸν ὕστερον ἥκοντες καὶ μαθόντες αὐτὸν προ-
εληλυθέναι καθ᾽ ἑαυτοὺς ἐθαύμαζον, ὅ τι ποτ᾽ εἴη τὸ
προελθεῖν ἀναγκάσαν ἠσιτηκότα δυοῖν ἡμέραιν τὸν
ἄνθρωπον. εἰ μὲν οὖν ἓν τοῦτο μόνον ὑπὸ τῶν τοιούτων
ἰατρῶν οἱ κάμνοντες ἠδικοῦντο, τὸ μέχρι πλειόνων
ἡμερῶν ἐπὶ τῆς κλίνης κατέχεσθαι, δυνάμενοι καὶ
542K χωρὶς ἀσιτίας ἀκαίρου | πολὺ θᾶττον ἐπὶ τὰς συνήθεις
ἀφικνεῖσθαι πράξεις, ἦν μὲν ἂν δήπου καὶ οὕτω δεινὸν

for putting his trust in them and doing nothing wrong during the three days before the paroxysm.

"It is certainly because of this," they said, "that your shivering has already stopped while much moisture of the 541K skin is a sign of sweating, and this would not have happened if you had not fasted and gone through the three day period."

They then directed the household slaves to wipe the man down carefully if any moisture was present on him, so he would not become cooled, and saying they would return at dawn the following morning, they departed, giving directions at that time that the man was not be nourished after his sweats. For the nights seemed short to them at that time and it seemed to be better to give nourishment early in the morning of the following day. When they were gone, the young man was covered in sweat so, when he had wrapped himself in his cloak, he ran to the bathhouse, bathed for the third time, and fed in like manner. Then, on the next day, before the doctors arrived, he left the house deliberately so they would not find him. When they arrived a little later and learned that he had left beforehand, they were amazed among themselves at whatever it was that had compelled the man to go out after fasting for two days. Therefore, if patients are harmed through this one thing alone by such doctors—that they are confined to bed for many days when they are able, without untimely fasting, to 542K return much more quickly to their customary activities— the occurrence would have been strange too, of course,

τὸ γιγνόμενον, ἀλλ᾽ ἧττον μακρῷ τῶν καταλαμβανόντων αὐτούς.

ἐπειδὴ δὲ χαλεπωτάτοις ἁλίσκονται πυρετοῖς αἱ εἰρημέναι φύσεις, ὅταν οὕτω διαιτηθῶσιν, οὐδὲν ἀποδεῖν μοι δοκοῦσι δημίων οἱ τὰ τοιαῦτα διαπραττόμενοι τῶν ἰατρῶν. οὐ γὰρ ὥσπερ οἱ ὑγροὶ καὶ εὔχυμοι κατὰ τοὺς ἐφημέρους πυρετοὺς ἀναγκασθέντες ὑπερβάλλειν τὴν δαιμονίαν διάτριτον ἐν τοῦτο ἀδικοῦνται μόνον, τὸ κατατρίβεσθαι μάτην, οὕτω καὶ αἱ προειρημέναι φύσεις, ἀλλ᾽ ἐπὶ ταῖς μακροτέραις ἀσιτίαις ἁλίσκονται πυρετοῖς δριμυτάτοις τε καὶ ὀξυτάτοις, ἐξ ὧν ἡ μετάπτωσις εἰς τοὺς ἑκτικοὺς γίνεται ῥᾳδίως, κἀξ ἐκείνων αὖθις εἰς τὸν περιφρυγῆ μαρασμόν, ἢ ἐὰν εὐαδίκητον εἴη τὸ στόμα τῆς γαστρός, εἰς τὸν συγκοπώδη. πολλάκις γοῦν ἤκουσας ἡμῶν λεγόντων ἐνίοις τῶν τοιούτων ἰατρῶν ὡς ἔξεστιν αὐτοῖς ἐναργέστατα μαθεῖν ἡλίκον ἐργάζονται κακὸν ἐν ἀσιτίᾳ φυλάττοντες τὰς προειρημένας φύσεις, ἢν ἐθελήσωσιν ὑγιαίνοντας ἀμέμπτως αὐτοὺς ἀσιτῆσαι κελεῦσαι
543Κ δυοῖν ἡμέραιν· | ὄψονται γὰρ αὐτίκα πυρέττοντας δι᾽ οὐδὲν ἄλλο δήπουθεν, ἢ τὸν λιμόν. ᾧ γὰρ οὔτ᾽ ἦν ἔμπροσθεν οὐδὲν ἄλλο περὶ τὸ σῶμα μέμψεως ἄξιον, οὔτ᾽ ἐν τῷ μεταξὺ προσεγένετό τι νεώτερον ἔξωθεν, οὗτος ἐναργῶς ὑπὸ τῆς ἀσιτίας ἐπύρεξεν.

ὅταν δ᾽, ὡς εἴρηται, μὴ μόνον εἷς ἢ δύο τῶν οὕτω δυσκράτων, ἀλλ᾽ ἐφεξῆς ἅπαντες ἀσιτήσαντές τε καὶ ἀλουτήσαντες ἁλίσκωνται πυρετοῖς, οὐδ᾽ ἀμυδρὰν ὑπόνοιαν ἔτι δυνατὸν ἡμῖν γίγνεσθαι τοῦ δι᾽ ἄλλο τι

but much less strange than when [those doctors] get hold of them.

Since the natures mentioned are seized by the most severe fevers when they are treated in such a way, those doctors who do such things seem to me to be nothing less than executioners. For the previously mentioned natures are not harmed like this in this one respect alone, to be worn out to no purpose, like those who are moist and *euchymous* when they are compelled to go through the marvelous "three day period" but also, due to the overlong fasting, they are seized by fevers that are very sharp and acute. From these, the change to fevers that are hectic readily occurs, and from those again to the dry marasmus, or, if the opening of the stomach is readily susceptible to injury, to syncope. Anyway, you have often heard me say to doctors of this sort that it is possible for them to learn very clearly how great a harm they bring about by fasting when they are keeping watch over the aforementioned natures, if they want to direct them to fast for two days when they are faultlessly healthy; for they will see them immediately become febrile due, of course, to nothing else than their hunger. In the case of that man, because he clearly became febrile due to the fasting, nothing else previously was deserving of blame concerning the body, nor did anything new emerge externally in the intervening period.

543K

Whenever, as I said, not only one or two of those who are *dyskratic* in this way, but all of them in turn, are seized by fevers if they fast and do not bathe, it is no longer possible for us to harbor some vague suspicion that their being

καὶ μὴ διὰ τὴν ἀσιτίαν πυρέττειν αὐτούς. οἷς γὰρ ἐν τῇ κράσει τοῦ μὲν ὑγροῦ τὸ ξηρόν, τοῦ δὲ ψυχροῦ τὸ θερμὸν πλεονεκτεῖ, τούτοις ἡ μὲν ἕξις τοῦ σώματος ἰσχνὴ καὶ δασεῖα καὶ μελαντέρα, καὶ εἰ ἄψαιο, θερμοτέρα τῶν ἄλλων τριῶν τῶν δυσκράτων, τῶν ἧττον θερμῶν, ἐντεύξῃ· παμπόλλη δ᾽ ἡ ξανθὴ χολὴ καὶ οὖρα καὶ διαχωρήματα κατακορῆ καὶ οἱ σφυγμοὶ μεγάλοι καὶ ὕπνοι λεπτοὶ καὶ ὀλίγοι καὶ ὁ θυμὸς σφοδρός.

εἴρηται τοιγαροῦν ἐπ᾽ αὐτῶν ὀρθῶς Ἱπποκράτει τό τε καθόλου τῆς διαίτης εἶδος ὁποῖόν τι χρὴ ποιεῖσθαι καὶ ὅσα καὶ οἷα καταλαμβάνει συμπτώματα τοὺς τοιούτους ἀνθρώπους, οὐ μόνον ἐπειδὰν δι᾽ ὅλης ἡμέ-
544K ρας | ἀσιτήσωσιν, ἀλλὰ κἂν τῆς ἑτέρας τροφῆς ἀπόσχωνται τῆς κατὰ τὸ ἄριστον. ἐν μὲν γὰρ τῷ ἕκτῳ τῶν Ἐπιδημιῶν φησίν, Ἐν θερμῷ φύσει ψύξις, ποτὸν ὕδωρ, ἐλιννύειν. ἐν δὲ τῷ Περὶ διαίτης ὀξέων, ὅ τινες ἐπιγράφουσι Πρὸς τὰς Κνιδείας γνώμας, ὑπὲρ τῶν ἐκλειπόντων τὸ εἰθισμένον ἄριστον διαλεγόμενος, ἐμνημόνευσε καὶ τούτων τῶν φύσεων ὀνομάζων αὐτὰς πικροχόλους ἀπὸ τοῦ κρατοῦντος χυμοῦ. ἔχει δ᾽ ἡ ῥῆσις ὧδε· Ἀλλὰ μὴν καὶ οἱ μεμαθηκότες δὶς σιτέεσθαι τῆς ἡμέρης, ἢν μὴ ἀριστήσωσιν, ἀσθενέες καὶ ἄρρωστοί εἰσι καὶ δειλοὶ εἰς πᾶν ἔργον καὶ καρδιαλγέες, ἐκκρεμᾶσθαι γὰρ δοκέει αὐτοῖς τὰ σπλάγχνα καὶ οὐρέουσι θερμὸν καὶ χλωρὸν καὶ ἡ ἄφοδος συγκαίεται· ἔστι δ᾽ οἷσι καὶ πικραίνεται τὸ στόμα καὶ οἱ ὀφθαλμοὶ κοιλαίνονται καὶ οἱ κρόταφοι πάλλονται καὶ τὰ ἄκρα διαψύχονται. ταῦτα προειπών, εἶθ᾽ ἑξῆς ἕτερά

febrile is due to something else and not the fasting. For in those in whom the dry is predominant over the moist and the hot over the cold in their *krasis*, the state of the body is thin, hirsute and quite dark, and if you were to touch it, you would find it hotter than the other three *dyskrasias* which are less in terms of heat, while the yellow bile is of great amount, saturating the urine and feces, and the pulse is large, sleep light and brief, and the spirit violent.

Accordingly, Hippocrates was right in stating in these cases what kind of diet in general we should fashion, and how many and what kind of symptoms constrain such people, not only whenever they fast through the whole day, but 544K also if they abstain from food other than breakfast. For in *Epidemics VI*, he said: "Cold in a hot nature, water to drink, and to take rest."[7] In the work *Regimen in Acute Diseases*, which some subtitle *Against the Cnidian Sentences*,[8] when he takes issue with those who leave out the customary breakfast, he made mention also of these natures, calling them "picrocholic" on the basis of the predominant humor. This is his statement: "But if those who have learned to eat twice a day do not take breakfast, they are weak and sickly, and are useless for every task, and have heartburn, for the viscera seem to hang in them, the urine is hot and pale, and its passage burning. In them, the mouth has a bitterness, the eyes are hollow, the temples pulsate and the extremities become chilled."[9] Having said

---

[7] See Hippocrates, *Epidemics*, VI.4 (13).

[8] The view is that the Hippocratic work was composed as a rebuttal of the *Cnidian Sentences*; see J. Jouanna, *Hippocrates* (Baltimore, 1999), pp. 409–10.   [9] See Hippocrates, *Regimen in Acute Diseases* 30, LCL, *Hippocrates*, vol. II, pp. 86–87.

τινα περί τε τοῦ δείπνου καὶ τῆς νυκτός, ἐπὶ τελευτῇ
τοῦ λόγου φησὶ τοὺς πικροχόλους χαλεπώτερόν τε
φέρειν τήν τε μονοσιτίαν καὶ τὴν ἀσιτίαν τῶν ἄλλων
ἀνθρώπων, ἕν τι μέρος ἐξηγούμενος ἐπ᾽ αὐτῶν ἐν
545K ἐκείνῳ τῷ βιβλίῳ, τὸ κατὰ τὴν | δίαιταν, ὧν ἐν τῷ
καθόλου παρήνεσεν ἐν τῷ ἕκτῳ τῶν Ἐπιδημιῶν.

ἡ γάρ τοι ψῦξις ἣν ἐν ταῖς θερμαῖς φύσεσι συνεβούλευε, γίγνεται μὲν καὶ ἐξ ἄλλων τινῶν, οὐχ ἥκιστα δὲ καὶ διὰ τῆς τῶν σιτίων προσφορᾶς. ὅσοις
γὰρ αὐχμῶδές τέ ἐστι καὶ δριμὺ τὸ ἔμφυτον θερμὸν
ἀμετρότερον ἐξ ἀρχῆς κραθεῖσι,[1] τούτοις αἱ τροφαὶ
πραότερον αὐτὸ καὶ ἧττον ἀποφαίνουσι δακνῶδες.
εὔδηλον δὲ δήπουθεν ὡς εὐχύμους τε καὶ ἀκριβεῖς
εἶναι χρὴ τροφὰς τὰς διδομένας αὐτοῖς, οὐ τῷ
μετέχειν τῆς τοιαύτης ὀλίγης οὐσίας ὀνομαζομένας τε
καὶ τρέφειν μὴ δυναμένας, ἀλλὰ τῷ τὸ πλεῖστον ἐν
ἑαυταῖς ἔχειν χρηστόν. εἰσὶ μὲν γάρ πως καὶ αἱ
μαλάχαι καὶ αἱ κράμβαι καὶ τὰ τεῦτλα καὶ αἱ θρι-
δακίναι καὶ ἁπλῶς εἰπεῖν τὰ λάχανα σύμπαντα καὶ αἱ
ὀπῶραι τροφαί, περιέχεται γάρ τις ἐν αὐταῖς οὐσία
τρόφιμος. ἀλλ᾽ ὥσπερ τούτων ὀλίγον μέν ἐστι τὸ
τρέφον, οὐκ ὀλίγον δὲ τὸ ἄχρηστον, οὕτως ἐπ᾽ ἄρτου
καθαροῦ καὶ ᾠῶν καὶ κρεῶν καὶ χόνδρου πλεῖστον μὲν
τὸ τρέφον, ἐλάχιστον δὲ τὸ μὴ τοιοῦτον. ἐὰν δὲ δὴ
πρὸς τῷ μὴ τρέφειν τὸ μεμιγμένον ἄχρηστον τῷ
546K χρησίμῳ συμβαίνῃ δριμεῖαν, ἢ ὀξεῖαν, ἢ | στρυφνήν,
ἢ πικράν, ἢ ὅλως μοχθηράν τινα δύναμιν ἔχειν, οὐ
μόνον ἐκπέπτωκε τὰ τοιαῦτα τοῦ μὴ κυρίως ὑπάρχειν

these things first, he next says certain other things in turn about the midday and evening meals, and at the end of the discussion, that those who are picrocholic have greater difficulty than other people in tolerating one meal a day or fasting, devoting one part in that book to these matters, that is, with regard to diet which he advised in general in *Epidemics VI*.

545K

The cold which he recommended for hot natures also occurs from certain other things, and not least due to the taking of foods. The innate heat is dry and sharp for those whose *krasis* was rather out of balance from the start; for them, the nourishments render it more mild and less biting. It is clear, of course, that it is necessary to give them *euchymous* and "frugal" nutriments, so called not by their partaking of such a small substance and being unable to nourish, but because the benefit in them is very great. They include mallows, cabbages, beets, lettuces and, in a word, all nutritious vegetables and fruits, since some nutritious substance is contained in them. Yet just as what is nutritious in them is slight but what is useless is not slight, so in pure wheat bread, eggs, meats and gruel what is nutritious is great, while what is not nutritious is slight. So then, if besides being nonnutritious, what is useless is mixed with what is useful, and it should happen to have some sharp, acidic, astringent, bitter, or in general some bad quality, not only are such things not properly nutritious,

546K

---

[1] B; κρατηθεῖσι K

369

τροφαί, ἀλλὰ καὶ τῆς τῶν φαρμάκων δυνάμεως οὐ σμικρόν τι μετείληφε. τρέφειν οὖν χρὴ τοὺς τοιούτους ἀνθρώπους ἐκείναις ταῖς τροφαῖς ὅσαι μήτε τινὰ φαρμακώδη δύναμιν ἔχουσιν αἰσθητήν, ἀλλὰ μηδ᾽ αὐτὸ τὸ τρόφιμον πάμπολυ· πρόσεστι γάρ τις καὶ τούτῳ κακία πλεονασθέντι.

πρῶτον μὲν γὰρ οὐδ᾽ ἡ κατὰ τὴν γαστέρα πέψις ἐπ᾽ αὐτῶν ἀκριβοῦται· εἰ δ᾽ ἄρα καὶ κρατήσειεν αὐτῶν ἡ γαστήρ, ἀλλ᾽ ἥ γε κατὰ τὰς φλέβας οὐκ ἀκριβὴς γίγνεται. εἰ δὲ κἂν ταύτῃ μηδέν τι μέγα πλημμεληθείη, τὴν γοῦν τρίτην πέψιν, ἥτις ἐστὶν ἑκάστου τῶν μορίων οἰκεία, χαλεπὸν ἐν ἅπασιν αὐτοῖς κατορθωθῆναι. πρόσεστι δὲ δήπου τῷδε καὶ τὸ χρῆναι τὰς τροφάς, ἐπειδὰν θερμότερόν τε καὶ ξηρότερον ἑαυτοῦ γένηται τὸ σῶμα, ψυχούσας πως εἶναι καὶ ὑγραινούσας. ὅθεν ἄριστος μὲν ὁ τῆς πτισάνης χυλὸς ἀκριβῶς καθεψημένης, οὐδὲν ἐχούσης ἐν ἑαυτῇ περίεργον ἥδυσμα. καὶ γὰρ ἐμψῦξαι καὶ ὑγρᾶναι καὶ ἄδιψον φυλάξαι τὸν ἄνθρωπον ἱκανὴ πρὸς τῷ μήτ᾽ ἀτμῶδές |
547K τι μήθ᾽ ὑγρὸν μήτε ξηρὸν ἐπέχειν περίττωμα. καλῶ δ᾽ ἀτμώδη μὲν ὅσα διὰ τοῦ δέρματος εἰς τοὖκτὸς ἀπορρεῖν πέφυκεν, ὑγρὰ δ᾽ ὅσα διὰ τῆς κύστεως, ξηρὰ δ᾽ ὅσα διὰ τῆς κάτω γαστρός· ἀπὸ γὰρ τῆς ἐπικρατούσης ἐν ἑκάστῳ τῶν εἰρημένων οὐσίας ἐθέμην αὐτοῖς τὰς προσηγορίας. ὅτι δ᾽ οὐδέν τι τούτων ἐφέξει πτισάνης χυλὸς εὔδηλον. αἱ μὲν γὰρ διὰ τῆς κάτω γαστρὸς ὑποχωρήσεις τῶν περιττῶν ὑπὸ τῶν στυφόντων ἴσχεσθαι πεφύκασιν, αἱ δὲ δι᾽ οὔρων τε καὶ σύμπαντος τοῦ

but they also partake of the potency of medications to no small extent. It is, then, necessary to nourish such people with those nutriments that neither have any detectable medicinal potency, nor are themselves very nourishing, for there is also some damage from that which is in excess.

Firstly, the digestion of these things in the stomach is not perfect and, if the stomach does prevail over them, the digestion in the veins is not complete. And even if nothing major is at fault in this, it is at least difficult for the third digestion, which is specific to each of the parts, to be accomplished successfully in them all. Over and above this, there is of course the need for the nutriments to be in some way cooling and moistening whenever the body is hotter and dryer than it should be. For this reason, the juice of ptisan is best when it has been boiled down thoroughly and has no superfluous sauce in it, because it is enough both to cool and moisten, and to guard the person against thirst, in addition to which it does not hold back any vaporous, moist or dry superfluity. I call "vaporous" [superfluities] those that are of a nature to flow off through the skin to the exterior; I call "moist," those that flow off through the bladder; and I call "dry," those that flow off through the stomach downward, for it is from the prevailing substance in each of those mentioned that I applied the terms to them. It is clear that the juice of ptisan will hold back none of these because the excretions of the superfluities through the stomach downward are of a nature to be held back by those things that are astringent, while the excretions through the

547K

371

δέρματος, ὑπό τε τῶν ἐμπλαττομένων τοῖς πόροις καὶ σφοδρῶς στυφόντων, ἅπερ ὀνομάζουσι στρυφνά· πτισάνη δ' οὔτε ἐμπλαστικῆς οὔτε στυπτικῆς μετέχει δυνάμεως. ὅτι μὲν οὖν οὐ στύφει δηλοῖ καὶ ἡ γεῦσις, ἧς μόνης ἴδιόν ἐστιν αἰσθητὸν τὸ στῦφον. ὅτι δ' οὐδ' ἐμπλάττει λογίσασθαι πάρεστιν ἐκ τῶν ἄλλων αὐτῆς ἔργων. εἴτε γὰρ ἔξωθεν ἀνατρίβοις τῇ πτισάνῃ ῥυπαρὸν σῶμα, ῥύπτει τοῖς ἀφρονίτροις τε καὶ τῷ μέλιτι παραπλησίως, εἴτ' ἐν τῇ γαστρὶ περιεχόμενον μέτριον φλέγμα δι' ἐμέτων ἐκκενῶσαι βουληθείης, οὐδὲν ἧττον μελικράτου πτισάνης χυλὸς χρήσιμος. |

548K   πρόσκειται δ' ἐν τῷ λόγῳ τὸ μέτριον, ὅτι, εἰ πολὺ καὶ γλίσχρον ἰσχυρῶς καὶ παχύ, ῥαφανῖδες δι' ὀξυμέλιτος, οὐ μελίκρατον καὶ πτισάνη πεφύκασιν ἀπορρύπτειν. ἔστι μὲν δὴ καὶ μελίκρατον εἰς ἁπάσας τὰς εἰρημένας ἐκκρίσεις ἐπιτήδειον, ἀλλ' οὔτε ἄδιψον ὥσπερ ἡ πτισάνη κἂν ταῖς πικροχόλοις φύσεσιν ἐκχολοῦται ῥαδίως. ἄριστον μὲν οὖν, ὡς εἴρηται, σιτίον ἐπὶ τῶν τοιούτων κράσεων ἡ πτισάνη. δεύτερον δ' ἐπ' αὐτῇ χόνδρος ὄξος χάριν ἀναδόσεως ὀλίγον εἰληφώς. τρίτον δ' ὁ χωρὶς ὄξους ὑγρός, οὐ παχύς, ὁποῖον ἔνιοι σκευάζουσιν ἔτνει παραπλήσιον· ἐμβαλεῖν δὲ καὶ τῷδε βέλτιον ἐψημένῳ πράσου· δῆλον δ' ὅτι καὶ ἀνήθου καὶ ἁλῶν ἐξ ἀνάγκης μετέχειν. καὶ μὲν δὴ καὶ οἱ κλιβανῖται[2] τῶν ἄρτων ἀγαθὸν ἔδεσμα καὶ

---

[2] B; κριβανῖται K

bladder and the whole skin are of a nature to be held back by those things that are emplastic and strongly astringent, which they term "astringents." Ptisan, however, partakes of neither emplastic nor astringent potency. That it is not astringent, the taste also makes apparent, by which alone what is astringent is specifically detectable. That it is not emplastic may be reckoned from its other actions because, if you rub a dirty body clean externally with ptisan, it cleans in a similar way to aphronitum and honey. Or, if you wish to evacuate a moderate amount of phlegm contained in the stomach by vomiting, the juice of ptisan is no less useful than melikraton.

But measure is to be added to the discussion, in that, if [the phlegm] is great in amount, strongly viscid and thick, radishes with oxymel,[10] and not melikraton or ptisan is of a nature to cleanse thoroughly. Now melikraton is also useful for all the aforementioned excretions, but does not quench thirst, just as ptisan in picrocholic natures easily changes to bile. The best food, then, in the case of such *krasias* is, as I said, ptisan. Second to this is pottage with a little added vinegar for purposes of distribution. Third is the fluid without vinegar, but not thick like the thick soup some prepare. It is better to throw some leek into this while it is boiling, and it is clear that it must partake of both dill and salt. Furthermore, bread baked in an oven is a

548K

---

[10] In the entry on radish (*Raphanus sativus*) in Dioscorides (II.137), there is the following: "Being so, and soe taken, it is good for such as have had the cough for a long tyme, and who breed thick matter in their breasts, but the skinne of it being taken with oxymel is stronger to make one vomit" (Goodyer's translation, p. 148).

τῶν ἰχθύων οἱ πετραῖοι πάντες. οὐδὲν δ' ἧττον αὐτῶν
οἱ ὀνίσκοι, πλὴν εἴ που λίμνης ἢ ποταμῶν μεγάλων
ἐμβαλλόντων εἰς θάλατταν νεμόμενοι πλησίον ὑγρο-
τέραν τε καὶ γλισχροτέραν ἔχοιεν τὴν σάρκα. ἐφεξῆς
δὲ τῶν ὀνίσκων τά τε βούγλωσσά ἐστι καὶ αἱ νάρκαι
549K καὶ τῶν ἰχθύων ὁ λάβραξ | καὶ ἡ τρίγλα κἄπειτα
πελάγιος κέφαλος. ἅπαντες δὲ χείρους ἰχθύες ὅσοι
λίμνης θαλάττῃ συναπτούσης ἢ στόματος ποταμοῦ
μεγάλου νέμονται πλησίον. ἐκ δὲ τῶν πτηνῶν ἄριστοι
μὲν οἱ πέρδικες ἅμα τοῖς ὀρείοις ἅπασι στρουθοῖς,
ἐφεξῆς δὲ ἀλεκτορίδες καὶ φασιανοὶ καὶ νέαι περι-
στεραί· μὴ καθείρχθω δ' ἔνδον μηδὲ ταῦτα, περιττωματικὰ γὰρ ἅπαντα τὰ ἀγύμναστα, περιττωματικὰ δὲ
καὶ τὰ λιμναῖα σύμπαντα καὶ ἑλώδη· διὸ καὶ τῶν
χοιρείων αὐτῶν ἀμείνω τὰ ἐκ τῶν ὀρῶν· ἐφεξῆς δὲ
αὐτῶν οἱ ἔριφοι. πάντων δὲ τῶν εἰρημένων φευκτέα τά
τε παλαιὰ ζῷα καὶ τὰ νεωστὶ γεγενημένα· τὰ μὲν γὰρ
σκληρότερά ἐστι καὶ δυσπεπτότερα, τὰ δὲ πέρα τοῦ
δέοντος ὑγρὰ καὶ περιττωματικά, καὶ μάλισθ' ὅσα
φύσει τῶν ζῴων ἐστὶν ὑγρά, καθάπερ ὄϊές τε καὶ σύες.
ἀλλὰ περὶ μὲν τῆς ἐν ἅπασι τοῖς ἐδέσμασι δυνάμεως
ἑτέρωθι διῄρηται·

νυνὶ δ' ἕνεκα παραδείγματος ὥσπερ τἆλλα τὰ κατὰ
τήνδε τὴν πραγματείαν, οὕτω λελέχθω καὶ ταῦτα·
πρόκειται γὰρ οὐ τὰς ὕλας ἡμᾶς ἐξευρεῖν ἐν αὐτῇ,
καθάπερ ἔν τε τοῖς περὶ τῶν ἐδεσμάτων καὶ φαρ-
550K μάκων, ἀλλὰ μόνας | τὰς καθόλου δυνάμεις· οὐδὲν δὲ
δήπου διοίσει δυνάμεις γενικὰς ἢ καθόλου προσαγο-

374

good food, as are all the fish caught among the rocks. No less good than these are cod unless, in some way, they have fed near a lake or large river draining into the sea, when they will have flesh that is moister and more glutinous. Next in order to cod are sole and rays, and among fish, bass, 549K red mullet, and then sea mullet. However, all fish that feed near a lake adjacent to the sea or the mouth of a large river are not as good. Of birds, partridges are best, along with all the mountain sparrows, and next to these, poultry, pheasants and young pigeons; but don't let these be kept caged, for all those without exercise are excrementitious, as are those from lakes and marshes. On this account also, among pigs themselves, those from the mountains are better, and next to these, young goats. Of all the things mentioned, those to be avoided are old animals and those that are newly born, for the former are harder and more difficult to digest while the latter are more moist than is required and excrementitious, especially those animals that are naturally moist like sheep and pigs. But matters pertaining to the potency in all the foods are discussed elsewhere.[11]

Now let me speak of these things for the purposes of exemplification, just as I did with other things in this treatise. It is not my intention to look into materials in this treatise, as in those on foods and medications, but only to look into 550K potencies in general—it will make no difference, I am sure, to name them generic potencies or potencies in a

[11] Presumably in *De alimentorum facultatibus libri III*, VI.453–748K.

ρεύειν αὐτάς. ἐπανέλθωμεν οὖν αὖθις ἐπὶ τὸ προκείμενον. οἷς δ᾽ ἀπορρεῖ τι δριμὺ καθ᾽ ἑκάστην ἡμέραν,
ἕτοιμον τούτοις ἁλῶναι πυρετῷ στεγνωθεῖσι τὸ δέρμα. στεγνοῖ δ᾽ αὐτὸ τά τε στύφοντα πάντα, καθάπερ
ὀλίγον ἔμπροσθεν ἐμνημονεύσαμεν τῶν Ἀλβούλων
ὑδάτων, ὅσα τ᾽ ἐμπλάττεται τοῖς πόροις. ἔστι δ᾽ ὅτε
καὶ ἡ ξηρότης αὐτοῦ τοῦ δέρματος ἀμετροτέρα γινομένη πυκνοῖ τοὺς πόρους αὐτοῦ. διὰ τοῦτο οὖν ἡμᾶς
ἐθεάσω καὶ τὸν οἰνάνθῃ χρώμενον ἐν τῷ βαλανείῳ
κωλύσαντας χρῆσθαι, καθ᾽ ὃν ἔφησε χρόνον, ἀνωμα-
λίας αἰσθάνεσθαι. καὶ τὸν ὡσαύτως ἐκείνῳ μετ᾽ ἐλαίου
βραχέος οἴνῳ πολλῷ, καὶ τὸν τῷ σχινίνῳ δὲ χρώμενον, ἡνίκα πυκνώσεως ᾔσθετο, καὶ τοῦτον, ὡς
οἶσθα, τοῦ λοιποῦ χρῆσθαι διεκωλύσαμεν. εἴρξαμεν
δὲ καὶ τὸν εἰληθεροῦντα καθ᾽ ἑκάστην ἡμέραν καὶ τὸν
τῇ κόνει χρώμενον. ἑτέρῳ δ᾽ εἰς ἀνωμαλίαν ἀφικομένῳ
πυρετώδη τὸ σφοδρὸν τῆς τρίψεως ἐπανεῖναι προσε-
τάξαμεν. ἄλλον δ᾽, ὡς οἶσθα, ἐθεράπευσα θαυμάσαν-
551K   τα πῶς ἑνὶ λουτρῷ μετὰ τρίψεως | τῆς προσηκούσης
ἡμερῶν παμπόλλων ἀνωμαλίαν αὐτοῦ φρικώδη διελύ-
σαμεν. ἐχρῆτο δὲ κἀκεῖνος ἐλαίῳ στύφοντι, τούτῳ δὴ[3]
τῷ δικαίως ἐνδόξῳ διὰ τὴν εἰς τἆλλα χρείαν, ὃ
καλοῦσιν Σπανόν. ἀφελόντες οὖν τοῦτο καὶ τῷ Σαβίνῳ
λιπαρῶς τε καὶ μαλακῶς ἐπὶ πλέον ἀνατρίψαντες καὶ
λούσαντες δίς, εὐθέως ἀπεφήναμεν ὑγιῆ.

   ταῦτά τε οὖν ἅπαντα φυλακτέα τοῖς τὰ καπνώδη τε
καὶ λιγνυώδη διαπνεομένοις, ὅσα τε γλίσχρα καὶ
παχύχυμα· καὶ γὰρ καὶ ταῦτ᾽ ἐμπλάττοντα τοὺς πό-

general sense. Therefore, let me return once more to the matter before us. If the skin is blocked up in those in whom something sharp flows away every day, they are readily seized by a fever. All astringents stop up the skin, just as I mentioned a little earlier in regard to the Albulian waters that block up the pores. Sometimes, also, when too excessive a dryness of the skin itself occurs, this thickens its pores. Because of this, you saw me forbid the man who was using a salve in the bathhouse from using it at the time when he said he felt something irregular. It was the same with the man using a lot of wine with a little oil, and with the man using mastich when he felt thickening, and as you know, I forbade him from using this in the future. I also prevented the man who basked in the sun every day and the one who used powder [from doing these things]. Another, who came to an irregular fever of some severity, I ordered to abandon massage. And another man, as you know, was surprised how I treated him by means of a single bath in conjunction with the appropriate massage, and put an end to his irregular shivering of very many days duration. That man was also using astringent oil, the one which is justly held in high regard due to its use for other [purposes], and which they call "Spanos." When I took this away and massaged him still more with the greasy and soft Sabine oil, and bathed him twice, I immediately restored him to health.

These are all things to be guarded against in those who transpire what is smoky and sooty, as also are those things that are viscid and thick-humored, for these things too,

551K

---

3 B; δή *om.* K

ρους ἐπέχει τὰς διαπνοάς. ἐπέχει δ', ὡς εἴρηται, καὶ τὰ
ξηραίνοντα τὸ δέρμα πέρα τοῦ μετρίου. πολλοὺς γοῦν
ἐν ἡλίῳ χρωμένους τοῖς τοιούτοις, ἄλλους δὲ νίτρῳ
πλέονι, τινὰς δ' ἁλσίν, ἢ γυμνασίοις πολλοῖς, ἢ τρίψει
σκληρᾷ, ξηροτριβίᾳ τε καὶ κόνει νοσοῦντας ἐκ τούτων
καθ' ἕκαστον ἐνιαυτόν, οἶσθ' ὡς ἐπισχόντες αὐτῶν
ἀνόσους ἤδη παμπόλλων ἐτῶν ἐφυλάξαμεν. ἔστι δ'
εἰκότως τὰ τοιαῦτα παραγγέλματα κοινὰ τῶν θ' ὑγιαινόντων ἔτι καὶ τῶν ἤδη δυσαρεστουμένων. οὐ γὰρ τῶν
552K γεγονότων, ἀλλὰ τῶν γιγνομένων | ἔτι πυρετῶν ἐστι
προφυλακτικὰ κατὰ τὴν τῶν ποιούντων αὐτοὺς αἰτιῶν
ἄρσιν· ὁ γὰρ ἔτι γιγνόμενος πυρετὸς ἀρθέντος τοῦ
αἰτίου συναναιρεῖται καὶ αὐτός. ὥσθ' ὅσαπερ ἐν τῇ
Τῶν ὑγιεινῶν πραγματείᾳ περὶ τῶν τοιούτων κράσεων
εἴρηται, μεταφέρειν ἅπαντα προσῆκεν ἐπὶ τὴν τῶν
ἀρχομένων πυρέττειν, εἴτε ἴασιν χρὴ λέγειν, εἴτε προ-
φυλακήν. ἰώμεθα γὰρ τὸ ἤδη γεγονὸς αὐτῶν, κἂν
ὀλίγον ᾖ, προφυλαττόμεθα δὲ μὴ γενηθῆναί τι πλέον
ἀναιροῦντες τὸ αἴτιον.

3. Ὥσπερ δὲ εὔρουν τε καὶ εὔπνουν[4] εἶναι τὸ σῶμα
πάντῃ κοινὸς ἐπὶ τῶν εἰρημένων μάλιστα φύσεων
ὑγιαινόντων τε καὶ δυσαρεστουμένων σκοπός, οὕτω
καὶ τὸ φεύγειν ἀγρυπνίας καὶ θυμοὺς καὶ λύπας καὶ
φροντίδας, ἐγκαύσεις τε καὶ ψύξεις καὶ κόπους· ἑτοιμότατα γὰρ ἐφ' ἑκάστῳ τῶν εἰρημένων αἱ τοιαῦται
κράσεις ἀναλίσκονται πυρετοῖς. ἰᾶσθαι δὲ αὐτάς,

---

[4] B (cf. transpirabile KLat); ἔμπνουν K

when they block up the pores, hold back transpirations. And those things that dry the skin beyond moderation also hold back transpirations, as I said. At all events, you know that by prohibiting these things, I have by now, over very many years, kept disease-free many who used such things in the sun, others who used too much nitrate, some who used salts and numerous exercises, or hard massage, dry rubbing, or powder, when they became ill from these every year. Such instructions are, in all likelihood, generally applicable, both to those who are still healthy and to those already suffering sickness. For these are prophylactic, not for fevers that have occurred but for those still occurring by virtue of the removal of the effecting causes; for when the cause of a fever that is still in evolution is taken away, so too is the fever itself. As a result, what was said in the treatise *On the Preservation of Health* regarding such *krasias* is that it is appropriate to change all of them in those who are beginning to be febrile, whether you speak of cure or prophylaxis.[12] For if we take away the cause, we cure the fever that has already occurred although, if it is slight, we act prophylactically so that they do not become severe.

3. Just as in the case of the natures already described, both those that are healthy and those suffering malaise, the common objective is for the body to have its pores open in every way and to transpire, so too is it to avoid insomnia, anger, grief and anxiety as well as heatstroke, cooling and fatigue; for such *krasias* are most readily consumed by fevers due to each of those things mentioned.

552K

[12] Prophylaxis is the subject of Book 6 of Galen's *De sanitate tuenda*; i.e. VI.381–452K.

ὅταν ἐπί τινι τῶν λελεγμένων προφάσεων πυρέξωσιν,
ἔστιν οὖν ᾗ μὲν ὁμοίως τοῖς ἐπὶ στεγνώσει ῥηθεῖσιν, ᾗ
553K δὲ καὶ διαφερόντως. ἄγειν μὲν | γὰρ⁵ ἐπὶ τὰ λουτρὰ
χρὴ πάντας ἐν τῇ παρακμῇ τοῦ πρώτου παροξυσμοῦ·
τρίβειν τε λιπαρῶς τε ἅμα καὶ μαλακῶς τοὺς μὲν ἐπί
τινι τῶν στυψάντων τὸ δέρμα, καθάπερ εἴρηται, τοὺς
δ' ἐπὶ ξηρότητι τρίβειν μέν, ἀλλ'⁶ ἐλάττω τούτοις,
λούειν δὲ πλείω. στοχάζεσθαι δ' ἀεὶ τοῦ μέτρου τῆς
δυνάμεως ἐν ἅπαντι τῷ τοιούτῳ βοηθήματι. θυμοὶ μὲν
οὖν καὶ λῦπαι καὶ ἀγρυπνίαι καὶ φροντίδες οὔτε
τρίψεων οὔτε λουτρῶν δέονται πολλῶν, ἀλλὰ πολὺ μὲν
ἐφ'⁷ ἁπάντων ἐκ τοιᾶσδε πυρεξάντων αἰτίας, ἔλαιον
περιχεῖν χλιαρὸν ἥκιστα μετέχον στύψεως· ἀνατρίψαντες δὲ βραχέα, λούειν ὡς ἔθος. αἱ δ' ἐγκαύσεις
εὐθέως μὲν ἐξ ἀρχῆς τῶν ψυχόντων δέονται καὶ πλειόνων λουτρῶν, ἥκιστα δὲ ἐλαίου δαψιλοῦς, ἢ συχνῆς
τρίψεως. ἔστω δὲ τὰ ψύχοντα ῥόδινόν τε καὶ τὸ καλούμενον ὀμφάκινον ἔλαιον, ὅπερ, οἶμαι, καὶ ὠμοτριβὲς
ὀνομάζουσιν. ἀμείνω δ' ἐξ αὐτῶν ὅσα μηδ' ὅλως ἐν
ἑαυτοῖς⁸ ἔχει τῶν ἁλῶν. ἐργάζεσθαι δὲ ψυχρὸν τὸ
ἀγγεῖον ἐν ᾧ μᾶλλον περιέχεται, κρεμῶντας⁹ ἐν φρέ-
554K ατι, | ψαῦον τοῦ ὕδατος. ἔτι δὲ μᾶλλον εἰ κρουνὸς
ὕδατος ψυχροῦ καταράττει τοῦ ἀγγείου, ψύχει τὸ
ὑγρόν. εἰ δ' ἐπὶ πλέον ἐψῦχθαι βούλοιο, χιόνι περι-
πλάττειν αὐτό.

καταχεῖν δὲ τοῦ βρέγματος μετέωρον ἐξαίροντα
τὴν χεῖρα δι' ἐρίου συμμέτρου τῷ μεγέθει. ταῦτα

However, to cure these *krasias* whenever [patients] are febrile due to one of the stated causes is partly similar to curing those spoken of due to stoppage of the pores, and
partly different. For it is necessary to bring everyone to   553K
bathing at the abatement of the first paroxysm, and to massage smoothly and gently at the same time those whose
skin was affected by one of the astringents, as I said. For
those whose skin is affected by drying, it is necessary to
massage, but less, yet to bathe them more. Always estimate
the measure of the potency in every such remedy. Anger,
grief, insomnia and anxiety require neither much massage
nor bathing; but in the case of all those who are febrile
from such causes, pour on a large amount of lukewarm oil
that partakes least of astringency. And after rubbing a little, bathe them as is customary. Those with heatstroke require cooling agents and rather more frequent baths right
from the start, but little in the way of an abundance of oil or
frequent massage. Let the cooling agents be oil of roses
and oil of the so-called unripe olives, which, I believe, they
also call "omotribes."[13] Of these, those are better that do
not have any salt in them at all. Make the vessel in which
they are contained colder, hanging it in a well, touching the   554K
water. Still more, if a stream of cold water falls down onto
the vessel, it cools the fluid. If you wish it to be cooled even
more, cover the vessel with snow.

Raising your hand in the air, pour the cooling agent
over the patient's forehead through a piece of wool of mod-

13 On this term, see Dioscorides, I.29.

---

5 K; γάρ *om.* B   6 K; ἀλλ᾽ *om.* B   7 B; ἐK
8 K; ἑαυτοῖς *om.* B   9 K; κρεμῶντας *om.* B

ποιήσαντα μέχρι τῆς παρακμῆς ἄγειν ἐπὶ τὸ λουτρόν. εἰ δὲ καὶ ψυχθεὶς πυρέξειεν ὁ τὴν προειρημένην ἔχων κρᾶσιν, ὑπὲρ οὗ ταῦτα πάντα λέγεται, καὶ τοῦτον ἐπὶ τὸ βαλανεῖον ἀκτέον ἐν ταῖς παρακμαῖς. εἰ δὲ σὺν κορύζῃ καὶ κατάρρῳ πυρέττοι, πρὶν πεφθῆναι ταῦτα, λούειν οὐ χρή. τοὺς δ᾽ ἐπ᾽ ἐγκαύσει καὶ τούτων παρόντων λουστέον, ὡς ἐν τοῖς ὑγιεινοῖς ἐπεδείκνυμεν. ἐπὶ δὲ τῷ λουτρῷ τὴν κεφαλὴν τοῖς αὐτοῖς ἑκατέρων βρέχειν οἷς πρὸ τοῦ, ψύχουσι μὲν τῶν ἐπ᾽ ἐγκαύσει πυρεξάντων, θερμαίνουσι δὲ τῶν ἐπὶ ψύξει. μέτρια δ᾽ ἔστω καὶ ταῦτα μετὰ τὸ λουτρόν, οἷον τό τε καλούμενον ἴρινόν ἐστι καὶ τὸ νάρδινον μύρον.

ἀλλ᾽ οὐ χρὴ μηκύνειν ἔτι περὶ τῶνδε λελεγμένων ἐν τοῖς ὑγιεινοῖς. ἀρκέσει δ᾽ εἰπεῖν ὡς τὰ μὲν ἄλλα
555K πρακτέον ὁμοίως ἐπὶ τῶν πυρεξάντων, | ἅπερ ἐπὶ τῶν ἐγκαυθέντων τε καὶ ψυχθέντων ἐλέγομεν χρῆναι πράττειν, ἄνευ τοῦ πυρέξαι· οὐ μὴν ἅπαντί γε καιρῷ τοὺς πυρέξαντας λουστέον, ἀλλ᾽ ἐν τῇ παρακμῇ τοῦ παροξυσμοῦ. καὶ μέντοι καὶ ὡς ἐπιτείνειν μὲν χρὴ τῶν πυρεξάντων ἐπ᾽ ἐγκαύσει τὴν ἐμψυκτικὴν ὀνομαζομένην ἀγωγήν, ἐκλύειν δὲ τῶν ψυχθέντων τὴν θερμαντικήν.

ὑπόλοιπον δ᾽ ἂν εἴη τὴν ἐν τῇ διαίτῃ κοινότητά τε καὶ διαφορὰν ἐπελθεῖν ἁπάντων τῶν εἰρημένων ἐπὶ τῆς ὑποκειμένης κράσεως, ἀλούσης ἐφημέρῳ πυρετῷ.

---

14 See *De sanitate tuenda*, Book 5 (VI.308K) and Book 6 (VI.398K). Bathing in general is discussed in Book 3 of that work.

erate size. When you have done this to the point of abatement, bring him to the bath. And if the person who has the aforementioned *krasis* (that is, the person whom all this is about) is febrile, once he is cooled, you must bring him to the bathhouse during the abatements. If, however, he is febrile with coryza and catarrh, you must not bathe him before these come to a crisis. You must, however, bathe those who are febrile due to heatstroke, even when these things are present, as I showed in the work *On the Preservation of Health*.[14] After the bath, moisten the head in both groups with the same cooling agents which you used before for those who are febrile due to heatstroke, but with heating agents for those who are febrile due to cooling. And let the things subsequent to the bath be moderate, like, for example, the so-called iris and oil of spikenard.

But there is no need to go on at length about things that have been said in the work *On the Preservation of Health*. It will be enough to say that you must do the other things in the same way in those who are febrile which I said it was 555K necessary to do them in those who have suffered heatstroke and those cooled without becoming febrile. You must not bathe those who are febrile at any time, but only in the abatement of the paroxysm. And yet it is also necessary to increase in intensity the so-called cooling treatments of those who are febrile due to heatstroke, and conversely, to reduce them in those who are febrile due to cooling.

What remains is to go over what is common and what is different in the diet of all those spoken of in respect to the underlying *krasis* when it has been seized by an ephemeral

κοινὸν μὲν δὴ πάντων εἶδος ἔστω σοι διαίτης, εὔ-
χυμον, εὔπεπτον, οὐδαμόθι κατὰ τοὺς πόρους ἰσχόμε-
νον· ἀλλὰ τοῖς μὲν ἐγκαυθεῖσι μετὰ τοῦ ψύχειν καὶ
ὑγραίνειν. οὕτως δὲ καὶ τοῖς θυμωθεῖσι. τοῖς δὲ
ψυχθεῖσι, μετὰ τοῦ θερμαίνειν μετρίως. τοῖς δ' ἀγρυ-
πνήσασι καὶ λυπηθεῖσι καὶ πλεῖον φροντίσασιν,
ὑγρὸν καὶ ὑπνῶδες. οὕτως δὲ καὶ τοῖς μὲν κοπωθεῖσι
τροφιμώτερον, τοῖς δὲ ψυχθεῖσιν ἀτροφώτερον, τοῖς δ'
ἄλλοις τὸ μέσον ἀμφοῖν. οἶνον δὲ πίνειν[10] γινώσκεις
556K δή που[11] καὶ σὺ διδόντα με τοῖς | τοιούτοις ἅπασι, τὸν
ὑδατώδη καὶ ὄψει καὶ δυνάμει· βελτίων γὰρ ὕδατος
αὐτὸς τὰ πάντα, τῇ πέψει συναιρόμενος καὶ οὖρα καὶ
ἱδρῶτας προτρέπων. δῆλος δ' ἐστὶ καὶ ὁ Ἱπποκράτης
οὐ μόνον ἐν τοῖς ἐφημέροις πυρετοῖς, ἀλλὰ κἂν τοῖς
ὀξέσι διδοὺς οἶνον, ἐξ ὧν ἐν τῷ Περὶ διαίτης ὀξέων
γράφει. κατὰ δὲ τὸ ἕκτον τῶν Ἐπιδημιῶν εἴτε οὖν
αὐτὸς εἴτε καὶ Θεσσαλὸς ὁ υἱὸς αὐτοῦ παραιτεῖται τὸν
οἶνον οὐ μόνον εἰ πυρέξειαν οἱ θερμοὶ ταῖς κράσεσιν,
ἀλλὰ κἂν τῷ τῆς ὑγείας χρόνῳ· ἐν θερμῇ γοῦν φύσει
παραινεῖ ψύξιν τε καὶ ποτὸν ὕδωρ καὶ ἐλιννύειν· εἰ μὴ
ἄρα τὸ πλείονι χρῆσθαι τῷ ὕδατι τοῦ οἴνου κατὰ τὰς
τοιαύτας φύσεις, ᾗπερ ἀρέσκει τισί, φαίημεν ὑπ'
αὐτοῦ παραινεῖσθαι.

τοῦτο μὲν οὖν αὐτὸ καθ' αὑτὸ ζητείσθω, περὶ δὲ
ποτοῦ ψυχροῦ διοριστέον. ἐγὼ μὲν δή φημι τὰς τοι-

[10] K; πίνειν om. B, recte fort.
[11] K; γινώσκεις μὲν δή που B

384

fever. You must make the general kind of diet in all cases *euchymous*, easily digested, and in no way obstructive to the pores, but in those who have suffered heatstroke, with what cools and moistens. The same applies in those who are angry. In those who are cooled, it should be with what heats moderately. To those who are sleepless, grieving or overanxious, [give something that] moistens and induces sleep. Similarly, to those who are fatigued, [give] what is more nourishing, to those who have been chilled, what is less nourishing, and to the others, what is intermediate between both. You are aware, I presume, that I give wine to drink to all such cases, but wine that is watery in both appearance and strength; for this is better than water in all respects, combining to increase digestion and promote urination and sweating. It is also clear, from what he writes in the work *Regimen in Acute Diseases*, that Hippocrates gave wine, not only in these ephemeral fevers, but also in the acute fevers. However, in *Epidemics VI*, either he himself or his son Thessalus prohibited wine, not only if those who were hot in terms of *krasis* were febrile, but also in a time of health. In a hot nature, at any rate, he recommends cooling, water as a drink, and rest—unless we were to say he recommended the use of water more than wine in such natures to satisfy some.[15]

556K

You must, then, investigate this in its own right and come to a decision about cold water. I say that natures of this kind are very greatly benefited by such a drink, and

15 See Hippocrates, *Regimen in Acute Diseases*, L–LII on wine. We could not find any mention of the prohibition of wine in *Epidemics VI*.

αὐτὰς φύσεις ὑγιαινούσας τε μᾶλλον πρὸς τοῦ τοιούτου πόματος ὀνίνασθαι μεγάλα καὶ κατὰ τὸν ἐφήμερον πυρετὸν οὐκ ἀφαιροῦμαι τοὺς εἰθικότας· ἀήθει
δὲ οὐκ ἀρξαίμην τηνικαῦτα διδόναι, ἀλλὰ πρότερον
ἀκριβῶς ἀποδοὺς[12] αὐτῷ τὴν ἀρχαίαν ὑγείαν. εἰ δέ[13]
557K δὶα δειλίαν, ἢ | ἔθος ἐκ παίδων, ἢ ὑπόληψιν ψευδῆ
φοβοῖτο τὸ ψυχρόν, ἔργῳ τε πείσαιμ᾽ ἂν αὐτόν, ἐπιτη-
ρήσας ποτὲ αὐτὸν καυσούμενον ἐν θέρει καὶ λόγῳ
μετὰ τὸ ἔργον. ὅταν γὰρ ἐπὶ τῇ χρήσει τοῦ τοιούτου
πόματος ἀπαλλαχθῇ μὲν αὐτίκα τοῦ καυσοῦσθαί τε
καὶ ἀλύειν, ἑαυτοῦ δὲ εὐτονώτερον ἐπὶ τῷδε τὸν στόμαχον ἔχειν φαίνεται, καὶ ἡ γαστὴρ καταρρήξῃ χο-
λῶδες καὶ πέψῃ καλῶς, αὖθίς τε καὶ αὖθις χρησαμένῳ
ἀνατρέφηται τὸ σῶμα, τότε ἂν ἤδη καὶ τοὺς ἄλλους
αὐτῷ δείξας οὓς ὁμοίως ὤνησα, διέλθοιμι πρῶτον μὲν
ὡς παντὸς μᾶλλον ἀληθές ἐστι τὸ τὰ ἐναντία τῶν
ἐναντίων ὑπάρχειν ἰάματα· γέγραπται δέ μοι περὶ
τούτου πρόσθεν.

εἶθ᾽ ὡς ἡ κρᾶσις αὐτῷ τὴν ὑγιεινὴν ἐκβεβηκυῖα
συμμετρίαν πλησίον ἥκει νόσου, καὶ ὡς τοῖς τοιούτοις
σώμασιν ἡ ὑγίεια σφαλερὰ καὶ κατὰ σχέσιν μᾶλλον
ἢ καθ᾽ ἕξιν ὑπάρχει, καὶ ὡς εἰ μὲν ἀγαθῆς ὢν ἔτυχε
κράσεως, ἐφύλαττον ἂν ἐκείνην αὐτῷ τρόπῳ παντί,
μοχθηρᾶς δ᾽ ὑπάρχοντι φύσεως ἐπὶ τὸ κρεῖττον ἐξαλ-
λάττω. λέλεκται δὲ καὶ περὶ τούτων ἐπὶ πλέον ἐν τοῖς
558K ὑγιεινοῖς· ὅθεν οὐ χρὴ μηκύνειν ἐνταῦθα | μιμούμενον

12 K; ἀποδιδούς B   13 K; δέ om. B

more so when healthy, and in an ephemeral fever I do not deprive them of what they are used to. But to someone un-used [to cold water], I would not begin to give this under these circumstances, before I had restored him thoroughly to his original health. If, however, due to timidity, custom   557K
from childhood or misplaced suspicion, he is afraid of cold water, I would persuade him by its action, after observing this at some time when he is intensely hot in summer, and by argument after the action. For whenever, by the use of such a drink, he is immediately freed from being intensely hot and distraught, and due to this appears to have the opening of his stomach more vigorous than it was, and the stomach discharges bile and digests well, and by repeated use the body is restored, at that time I would have shown him other people also for whom I provided benefit in like manner. And I would first go through in detail what is truer than anything—that "opposites are the cures of opposites." I have written about this before.[16]

[I would say], then, that as the *krasis* in him has devi-ated from a healthy balance and come closer to disease, and that as health in such bodies is precarious and more in respect of disposition than habitus, and that if the *krasis* in him happens to be good, I would be protecting that for him in every way and, if he were of a bad nature, I would be ef-fecting a change to what is better. More has been said about these matters in *On the Preservation of Health*,[17] on which account it is not necessary to speak at length here, imitating those writers who, when they know a very small   558K

[16] See Hippocrates, *Airs, Waters, Places*, 1.5, and *MM*, Book 7, chapter 3 (463–64K).
[17] See Book 6 of *De sanitate tuenda*.

ἐκείνους τῶν συγγραφέων, ὅσοι παντάπασιν ὀλίγα
γινώσκοντες, ἐν ἅπασι τοῖς βιβλίοις αὐτὰ διέρχονται.
τουτὶ γὰρ οὐχὶ πολλὰ διδάσκειν, ἀλλὰ γράφειν ἐστὶ
πολλὰ καὶ πολλοὺς χάρτας ἀναλίσκειν, ἐνὸν ὀλίγους,
καὶ πολὺν κατατρίβειν χρόνον, ἐνὸν μή.

οὐκοῦν μηδὲ τὰς διαγνώσεις τῶν ἐφημέρων πυρετῶν ἐνταυθοῖ γράφω, προειρηκὼς μὲν ἐν τῷ δευτέρῳ
Περὶ κρίσεων ὀλίγου δεῖν ἁπάσας, εἰρηκὼς δὲ τὰ
λείποντα κἂν τῷ πρώτῳ Περὶ τῆς διαφορᾶς τῶν πυρετῶν. οὐδὲν δ᾿ ἧττον κἂν τοῖς περὶ σφυγμῶν ὅσον ἐξ
αὐτῶν χρηστὸν ἑρμηνεύσας, ὡς οἶσθα, μόνος ἁπάντων τῶν ἔμπροσθεν οὕτω σαφῶς ὡς μηδ᾿ ἂν παῖδα
λαθεῖν τὸ δηλούμενον, Ἀρχιγένους ὡς ἐπεδείξαμεν, οὐ
διδάξαντος, ἀλλὰ περιλαλήσαντος τὰ πλεῖστα κατά
τε τὸ περὶ σφυγμῶν σύγγραμμα καὶ τὴν πραγματείαν
ἣν αὐτὸς ἐπιγράφει περὶ τῆς τοῦ πυρετοῦ σημειώσεως.
ὥσπερ δὲ τὸν ἐν τῷ στυπτηριώδει λουσάμενον ὕδατι
τὴν τρίτην ἡμέραν ἀσιτεῖν ἐκέλευον οἱ ἰατροί, κατὰ
τὸν αὐτὸν τρόπον οἶσθα καὶ τῶν ἐγκαυθέντων τε καὶ
κοπωθέντων ἐπί τε τοῖς ἄλλοις τοῖς εἰρημένοις αἰτίοις |
559K   πυρεξάντων, οὐδένα λούοντας, οὐδὲ τρέφοντας αὐτοὺς
πρὶν ὑπερβαλεῖν τὴν διάτριτον. ἀνάγκη δήπου κἀμὲ
λέγειν διάτριτον, ὅπως μὴ καὶ περὶ τοῦδε τῆς φλυα-
ρίας αὐτῶν ἐνέχωμαι, φασκόντων ἕτερον μὲν εἶναι τὸν
διὰ τρίτης παροξυσμόν, ἑτέραν δὲ τὴν πρώτην διάτρι-
τον· ἤκουσας γάρ ποτε καὶ σὺ μέχρι τοσούτου τῆς
ἀδολεσχίας αὐτῶν ἐκταθείσης.

number of things, go over them in all their books. This particular approach is not to teach many things but to write many things, and to use many sheets of papyrus when it is possible to use a few, and to waste much time when it is not necessary.

Therefore, I am not writing here about the diagnoses of the ephemeral fevers, having previously spoken about almost all these in the second book of *On Crises* and about those things left out in the first book of *On the Differentiae of Fevers*.[18] No less, in the works on the pulses, did I explain as much of them as was useful, as you know, and alone compared to all my predecessors, so clearly that what is set forth would not be obscure even to a child, whereas Archigenes, as I showed, did not teach but waffled on about most things in his work on the pulse and in the treatise he wrote about the symptomatology of fever.[19] Just like the person the doctors directed to fast for the third day after bathing in astringent water, in the same way you know that they directed them to bathe none of those with heatstroke, fatigue or fevers due to the other causes mentioned, nor to nourish them before they went through the   559K
three day period. It is, I suppose, necessary for me to say "three day period" so that I am not caught up in their nonsense as well when they assert that the paroxysm during the third day is one thing and the first three day period is another, for you too have heard at some time their prattle drawn out to such an extent.

[18] For these two works, *De crisibus* and *De differentiis febrium*, see respectively IX.550–768K and VII.273–405K.

[19] The writings of Archigenes of Apameia (fl. AD 95–115), which include a work on fevers, survive only in fragments.

ἀλλ᾽ εἴτε πρώτην διάτριτον, εἴτ᾽ ἄλλο τι καλεῖν
ἐθέλουσιν, οὐχ ἅπαξ ἢ δὶς ἢ τρίς, ἀλλ᾽ ὁσάκις οὐδὲ
μεμνῆσθαι δυνατὸν ἔν τε τῇ πρώτῃ τῶν ἡμερῶν,
εὐθέως εἰς ἑσπέραν ἐλούσαμεν τοὺς τὸν ἐφήμερον
πυρέξαντας, ἔν τε τῇ δευτέρᾳ πολὺ μᾶλλον, ἀλλὰ καὶ
κατ᾽ αὐτὴν τρίτην, ὡς ὀλίγον ἔμπροσθεν ἔφαμεν ἐπὶ
τοῦ στεγνωθέντος τὸ δέρμα. τουτὶ μὲν οὖν τὸ αἴτιον
ὀλιγάκις ἐθεασάμην ἀνάψαν πυρετόν, ὅθεν αὐτὸ μάλιστα προὐχειρισάμην, ἔγκαυσιν δὲ καὶ κόπον ἕκαστόν
τε τῶν ἄλλων πάνυ πολλάκις. ἐθεάσω δὲ καὶ σὺ
πάντας αὐτοὺς τῇ τετάρτῃ τῶν ἡμερῶν ἕωθέν τε προσ-
ελθόντας ἐπὶ τὰς συνήθεις πράξεις, ἀπαντῶντάς τε
τοῖς ἰατροῖς ὅσοι καὶ τὴν τρίτην ἡμέραν ᾤοντο τελέως
αὐτοὺς ἠσιτηκέναι. καὶ τοίνυν θαυμάζοντί σοι καὶ
ζητοῦντι, τί ποτ᾽ ἐστὶ τὸ αἴτιον τοῦ μὴ κἂν νῦν γοῦν |
560K ἀποστῆναι τοὺς τοιούτους ἰατροὺς τῶν κακῶς αὐτοῖς
ὑπειλημμένων, ὡς νομίζω, προσηκόντως εἶπον ὅτι
μήτε γινώσκουσι τὴν ἀληθῆ θεωρίαν, αἰδοῦνταί τε νῦν
ἄρξασθαι μανθάνειν αὐτήν. ἔνιοι δ᾽, ὡς ἤκουσάς ποτέ
τινος ἐξ αὐτῶν ὁμολογήσαντος ἡμῖν ἰδίᾳ τἀληθῆ μετὰ
τοῦ δακρύειν, οὐδὲ τῶν ἐπιτηδείων εὐπορῆσαί φασιν,
ἐὰν οἱ νομίζοντες αὐτοὺς ἐπίστασθαί τι θεάσωνται
μανθάνοντας. ἀλλὰ σὺ μὲν ἅπαξ ἤκουσας τοῦτο,
θεοὺς δ᾽ ἐγὼ σύμπαντας ἐπόμνυμι πολλοὺς ἐμοὶ πολλάκις ὡμολογηκέναι μόνῳ ταῦτα μετὰ τοῦ δακρύειν τε
καὶ τὴν ἑαυτῶν ὀδύρεσθαι δυστυχίαν ἐπὶ τῷ τὸν χρόνον ἀπολέσαι μοχθηροῖς διδασκάλοις χρησαμένους.

ἐπόμνυμι δ᾽ αὖθίς σοι τοὺς θεοὺς καὶ τοῦτό μοι

But during the first three day period, or whatever else they might want to call it, I bathed those with an ephemeral fever immediately toward evening on the first day, not once, twice or three times, but more times than I can recall, and much more on the second day, but also on the third day itself, as I said a little earlier, when there is blockage of the skin pores. Seldom did I see this particular cause kindle a fever, which is why I especially chose it very frequently in respect of heatstroke, fatigue and each of the other [conditions]. You also saw them all going about their customary activities early on the morning of the fourth day, meeting the doctors who thought they had fasted completely for the third day. So to you, wondering and inquiring what the reason is for them not sending away such doctors even now, when they have been treated badly by them in my view, it is appropriate for me to say that these doctors do not know the true theory, and are ashamed to start learning it now. But some, as you heard one of them tearfully admitting the truth to me privately on one occasion, say they don't have much going for them if those who think they know something see them learning. Now you have heard this once only, whereas I swear by all the gods that many have admitted these things to me privately along with weeping and lamenting their own misfortune for the time frittered away using unsound teachings.

I also swear to you by the gods that this was said to me

560K

λεχθῆναι πάνυ πολλάκις ὑπὸ τῶν συμφοιτησάντων
ἔτι μειρακίῳ τὴν ἡλικίαν ὄντι. ἐπειδὴ γὰρ εἴωθα,
ὁπότε τις ἔροιτο περί τινος δόγματος εἰ ἀληθὲς εἴη,
μηδὲν ἀποκρίνασθαι δύνασθαι φάσκειν, εἰ μὴ καὶ
τὰ τῶν ἄλλων ἁπάντων μάθοιμι καί τινα μεθόδον
κτησαίμην ᾗ διαγινώσκοιμι τὸ βέλτιον, οὐκ ὀλίγοι
τῶν ταῦτα ἀκουόντων συμφοιτητῶν ἔφασάν μοι, Σὺ
561K   μὲν καὶ φύσει διαφερούσῃ | κέχρησαι καὶ παιδείᾳ
θαυμαστῇ διὰ τὴν τοῦ πατρός σου φιλοτιμίαν, καὶ
ἡλικίαν δυναμένην μανθάνειν ἔχεις ὅθεν τε δαπανᾶν
χρὴ σχολάζοντα μαθήμασι κέκτησαι· τὸ δ' ἡμέτερον
οὐχ ὧδ' ἔχει· καὶ γὰρ ἀπαίδευτοι τὴν πρώτην παιδείαν
ἐσμὲν καὶ οὐκ ὀξεῖς τὴν διάνοιαν ὥσπερ σὺ καὶ
ἀναλίσκειν οὐκ ἔχομεν· ἀγαπητέον οὖν ἡμῖν ἐστιν
ὁποῖά ποτ' ἂν εἴη ταῦτα τὰ νῦν γινωσκόμενα.

μὴ τοίνυν ἔτι θαύμαζε διὰ τί πολλοὶ τῶν ἰατρῶν
οὐκ αἰσχύνονται φθεγγόμενοι τὴν διάτριτον, καίτοι
μυριάκις ἡμᾶς ἑωρακότες ἐπὶ τῶν νοσούντων προ-
γινώσκοντας εἴτ' ἐφήμερός ἐστιν ὁ πυρετὸς εἴτε καὶ
κατὰ τὴν τρίτην ἢ τετάρτην ἡμέραν οἴσει τινὰ παροξυσμόν. ἀλλ' ἐκεῖνο σκόπει τὸ ὑφ' Ἱπποκράτους
εἰρημένον, ἀνθρώπου φιλοσοφωτάτου τὸ ἦθος. Ὅκου
μὲν οὖν κάτοξυ τὸ νόσημα, αὐτίκα καὶ τοὺς ἐσχάτους
πόνους ἔχει καὶ τῇ ἐσχάτως λεπτῇ διαίτῃ ἀναγκαῖον
χρέεσθαι. ὅκου δὲ μή, ἀλλ' ἐνδέχεται ἁδροτέρως διαιτᾶν, τοσοῦτον ὑποκαταβαίνειν, ὁκόσον ἂν ἡ νοῦσος
μαλθακωτέρα τῶν ἐσχάτων εἴη. πόνους μὲν γὰρ λέγει
πάντα τὰ λυποῦντα τὸν ἄνθρωπον, ἐσχάτους δὲ ὀνο-

very often by my fellow students when I was still a young man. For whenever someone asked me about a certain opinion, whether it was true, my practice was to say I could not answer unless I had also learned the opinions of everyone else, and acquired some method by which I might distinguish what is better. Not a few of my fellow students, when they heard these things, said to me: "You have made use of a discriminating nature and a remarkable education 561K by reason of your father's ambition, and are of an age when you are able to learn, which is why you must have had time to spend acquiring teachings. For us it is not like this; we are unschooled in our initial childhood education and are not of keen intellect as you are, and do not have the time to spend. We must be happy, then, with those things we do now know, such as they are."

Therefore, do not go on being surprised as to why many doctors are not embarrassed when they speak of the "three day period," even though on countless occasions they have seen me prognosticating on those who are sick, whether it is an ephemeral fever or will bring some crisis on the third or fourth day. Rather consider what was said by Hippocrates, a man whose distinguishing trait was a great love of wisdom: "Whenever a disease is acute and there is extreme suffering immediately, it necessarily requires an extremely thin diet. Whenever it is not, but allows a thicker diet, it necessarily diminishes to the degree that the disease is milder than the extremes."[20] For, he says, "sufferings" are all those things that distress the person, whereas he terms

[20] Hippocrates, *Aphorisms*, I.7.

562K μάζει τοὺς μεγίστους, | οἷός περ[14] ἐν ταῖς ἀκμαῖς
γίνονται. κελεύει δ' εἰς τοῦτον τὸν χρόνον ἀποτίθεσθαι
τῆς διαίτης τὸ λεπτότατον, οὐδέν τι διδάσκων ἀλλοι-
ότερον ἢ ὁπότ' ἔφασκεν· Ὁκόταν ἀκμάζῃ τὸ νόσημα,
τότε καὶ τῇ λεπτοτάτῃ διαίτῃ ἀναγκαῖον χρέεσθαι. καὶ
μὴν οὐχ οἷόν τε τὸ λεπτότατον τῆς διαίτης εἰς τὴν
ἀκμὴν ἀναβάλλεσθαι, χωρὶς τοῦ κατὰ τὴν ἀρχὴν
ἐστοχάσθαι περὶ πόσην ἡμέραν ἀκμάσει τὸ νόσημα.

ἀλλὰ καὶ Ἱπποκράτη φασὶ ληρεῖν ταῦτα γράψαντα
καὶ ἡμῶν ἔργῳ διαδεικνυμένων αὐτά, πάντα μᾶλλον
νομίζουσιν ἢ δυνατὸν ἀνθρώπῳ κατὰ τὸν πρῶτόν ποτε
παροξυσμόν, ἀκριβῆ γνῶσιν ἔχειν τῆς τοῦ κάμνοντος
διαθέσεως. ἐγὼ δ' οὔτ' ἐφ' ἁπάντων τῶν νοσούντων
ἀκριβῶς φημι γινώσκειν τὴν διάθεσιν ἐν τῇ πρώτῃ
τῶν ἡμερῶν οὔτε τῶν ἐφημέρων τινὰ πυρετὸν λαθεῖν
ἄν με· τοῦτο δ' ἀρκεῖ κατά γε τὰς πρώτας ἡμέρας εἰς
τὸ μηδὲν ἁμαρτεῖν ἐν τῇ διαίτῃ, τὸ διακρῖναι τὸν
ἐφήμερον πυρετὸν τῶν ἄλλων. διὰ τοῦτό γέ τοι τῇ
τετάρτῃ τῶν ἡμερῶν, ἐν ᾗ πρῶτον ἐκεῖνοι προταρι-
χεύσαντες τὸν ἄνθρωπον ἄρχονται τρέφειν, ἡμεῖς ἤδη
τὰ συνήθη τῶν ἔργων παρεχόμεθα πράττειν, κωλύ-
563K ομεν δὲ | κἀκ τῶν ἐφημέρων πυρετῶν τὴν μετάστασιν
εἰς τοὺς ὀξεῖς γενέσθαι τῷ τρόπῳ τούτῳ τῆς διαίτης·
ὡς εἴ γέ τις ἑτέρως διαιτήσειεν, ἐξ ἀνάγκης ἀκολου-
θήσουσι κατά γε τὰς προειρημένας κράσεις ὅσοι
διακαέστατοι τῶν πυρετῶν. ἕτεραι γάρ εἰσι φύσεις ἐφ'

[14] B; οἵπερ K

394

"extreme" those things that are greatest and of the kind   562K
that occur in the peaks. He orders the thinnest diet to be
set aside for this time, which is to teach nothing too different from when he said: "Whenever the disease comes to a
peak, it is also necessary, at that time, to use the thinnest
diet."[21] And yet it is not possible to put off the thinnest diet
until the peak without estimating at the start the number of days for the disease to peak.

But they say that Hippocrates too was talking nonsense
in writing these things, and so am I when I demonstrate
them in practice; they think anything else rather than that
it is possible for a person to have a precise knowledge of
the condition of the patient at some time during the first
paroxysm. However, I don't say that I have a precise knowledge of the condition on the first day in all those who are
sick, or that no ephemeral fever could escape my notice. In
distinguishing the ephemeral fever from the others it is
enough not to make a mistake in the diet during the first
days. Now because of this, on the fourth day, which is when
for the first time those who have previously fasted the man
to a reduced state begin to nourish him, I already allow
him to carry out his customary activities, and prevent the   563K
change from the ephemeral fevers to those that are acute
by this form of diet, since if someone were to treat otherwise, the most burning fevers would necessarily follow, at
least in the aforementioned *krasias*. There are other na-

21 Hippocrates, *Aphorisms*, I.8.

ὧν ἢ οὐδὲν ἢ οὐκ ἀξιόλογόν γέ τι τοσούτων ἡμερῶν
βλάπτει λιμός· ὥσπερ ἕτεραί γε πάλιν ἐφ' ὧν οὐ
μόνον ἀβλαβής, ἀλλὰ καὶ πρὸς ὠφέλειάν ἐστιν ἡ
ἀσιτία τῶν τριῶν ἡμερῶν. ἀλλὰ περὶ μὲν τῶν ἄλλων
φύσεων ἑξῆς διορίσω. τὴν ὑποκειμένην δ' ἐξ ἀρχῆς
κρᾶσιν εἰς τοὔσχατον ἄγουσι κίνδυνον αἱ ἀσιτίαι
κατὰ πάντας τοὺς πυρετούς, οὐ μόνους ὧν νῦν ἐμνημό-
νευσα, τοὺς ἐφημέρους ὀνομαζομένους.

4. Εἰ γὰρ καὶ δι' ἔμφραξιν, ὡς ἐλέχθη, στεγνω-
θείη ποθ' ἡ διαπνοή, μικρὰν μὲν ὑπάρχουσαν αὐτήν,
εἴ τις ὀρθῶς διαρρύψειε, κωλύσει γενέσθαι δεύτερον
παροξυσμόν· ἀξιολόγου δὲ γενομένης, οὐκ ἔθ' ὁμοίως
ἐπαγγείλασθαι δυνατόν. εἰ δὲ καὶ πολλοὶ καὶ παχεῖς
καὶ γλίσχροι σφηνωθεῖεν χυμοί, τὸν ἐπὶ τούτοις πυρετὸν ἀδύνατόν ἐστιν ἐν τοῖς ὅροις διαμεῖναι τῶν ἐφημέ-
564K ρων. ἐξ ἀνάγκης γὰρ αὐτὸν ὁ ἐπὶ σήψει | χυμῶν δια-
δέξεται κατὰ τὰς κακοχύμους φύσεις. ὡς ἐπ' ἄλλων γε
κράσεων ἃς ἐν τοῖς ἑξῆς ἑρμηνεύσω γένοιτ' ἄν τις
ἰδέα πυρετοῦ συνόχου διὰ τὴν τοιαύτην αἰτίαν ἄνευ
σήψεως χυμῶν. ἕπεσθαι δὲ τοῖς τοιούτοις λόγοις ἀδύνατον ὅσον ἄν τις μὴ πρότερον ἐν ταῖς εἰρημέναις
πραγματείαις γυμνάσηται, καὶ μάλιστα τῇ Περὶ τῆς
διαφορᾶς τῶν πυρετῶν. ἐπειδὰν γὰρ ἔμφραξίς τις ᾖ
μετρία δι' ἣν ἐπύρεξαν οἱ κακόχυμοι κατὰ μὲν τὴν
πρώτην καὶ δευτέραν ἡμέραν ἄχρι τινός, οὐκ ἔστιν ἐν
τῷ σφυγμῷ τὸ τῆς σήψεως σημεῖον, ὥσπερ οὐδ' ἐν
τοῖς οὔροις τὸ τῆς ἀπεψίας. μηκυνομένης μέντοι τῆς
παρακμῆς καὶ μηκέτ' ἀνάλογον ἐνδιδούσης ὥσπερ

tures which starvation for this many days either does not harm at all, or at least to no great extent, just as there are others again in whom a three-day fast is not only not harmful but is even beneficial. But I shall distinguish the other natures in due course. Fasting brings the *krasis* underlying from the beginning to extreme danger in all the fevers, not only in those I have now mentioned which are termed ephemeral.

4. As I said, when transpiration is at some point stopped up due to blockage, if someone were to cleanse it thoroughly while it is slight, he would prevent a second paroxysm from occurring whereas, if it has become substantial, he cannot still make the same promise. If, however, many thick and viscid humors are impacted, the fever due to these cannot remain within the limits of the ephemeral, for the fever which occurs due to putrefaction of humors will necessarily succeed this in *kakochymous* natures. Even so, in other *krasias*, which I shall explain subsequently, some kind of continuous fever may occur due to such a cause without putrefaction of humors. It is impossible for someone to follow such arguments unless he is previously practiced in the aforementioned treatises, and particularly *On the Differentiae of Fevers*.[22] For whenever those who are *kakochymous* are made febrile on the first or some part of the second day due to a blockage which is moderate, the sign of the putrefaction is not in the pulse, just as the sign of the failure of digestion is not in the urine. Nevertheless, when the abatement is delayed and does not

564K

22 *De differentis febrium libri II*, VII.273–405K.

ὑπήρξατο μετὰ τὴν ἀκμὴν ὕποπτός ἐστιν ἡ διάθεσις,
ὥσπερ ἐφ᾽ ὧν ἐπείγεται πρὸς τὴν λύσιν, ἀτμοί γέ
τινες, ἢ νοτίδες, ἢ ἱδρῶτες ἐξεκρίθησαν, ἢ οὖρα πλείω
χρηστὰ μετὰ τῶν κατὰ τὸν σφυγμὸν σημείων ἀγαθῶν
ὑπαρχόντων, ἐφήμερός ἐστι βεβαίως ὁ πυρετός. ἀλλ᾽
ὅ γε διὰ μικρὰν ἔμφραξιν ὅμοιος μὲν τοῖς ἐφημέροις
ἐστίν, ὡς εἴρηται, κατά γε τὰ πρῶτα, μηκυνομένης δὲ
τῆς παρακμῆς ὕποπτος γίγνεται, κἂν μηδέπω τὸ τῆς
565K σήψεως | σημεῖον ἐμφαίνηται τοῖς σφυγμοῖς· ἐὰν γὰρ
μὴ διαπνευσθῇ τελέως, ἀναγκαῖόν ἐστι σαπῆναι τὴν
κακοχυμίαν. ἵν᾽ οὖν διαπνεύσῃ, μείζονος ἡμῖν ἤδη
βοηθείας ἐστὶ χρεία.

εἰ μὲν οὖν ἤτοι παιδίον ἢ γέρων εἴη, φλεβοτομεῖν
οὐκ ἔτ᾽ ἐγχωρεῖ· μεταξὺ δὲ τῶν ἡλικιῶν τούτων ῥώμης
παρούσης τῷ κάμνοντι φλεβοτομητέον, εἰ καὶ τὰ τοῦ
πλήθους ἀπείη σημεῖα. βέλτιον γὰρ ἐκκενώσαντα τὸ
πλέον τῆς κακοχυμίας, ἐπὶ τὸ διαρρύπτειν τὰς ἐμφράξεις ἰέναι· κίνδυνος γὰρ ἐκφραττόντων, πρὶν ἐκκενῶσαι, σφηνωθῆναι μᾶλλον αὐτάς. ἐκφράττουσι μὲν
οὖν αἱ ῥυπτικαὶ δυνάμεις, ἔξωθέν τε προσπίπτουσαι
τῷ δέρματι καὶ εἴσω τοῦ σώματος λαμβανόμεναι.
κίνδυνος δὲ καὶ τὰς ἐκτὸς ἑλκούσας ἅμα τοῖς ἐμ-
φράττουσι συνεπισπᾶσθαί τι καὶ ἄλλο, καὶ τὰς ἔνδοθεν ἀναφερομένας ἑαυταῖς συνεπισύρεσθαί τι τῶν ἐν
τοῖς ἀγγείοις. ὅπερ εἰ μὲν γλίσχρον ἢ παχὺ τύχοι,
διπλασιάσει τὴν ἔμφραξιν· εἰ δὲ μή, ἀλλὰ τῷ πλήθει
γε, πάντως οὐ σμικρὰ λυπήσει. φαίνεται γὰρ ἐνίοτε
καὶ τὸ πολὺ ῥεῦσαν ἀθρόως οὐδὲν ἧττον ἐμφράττε-

still go forward in proportion to how it began after the peak, the condition is suspected, just as the fever is certainly ephemeral in those in whom it hastens toward lysis when certain vapors, moistures or sweats are expelled, or there is much useful urine along with good signs in the pulse. But the fever which is due to a small blockage is similar to those that are ephemeral, at least at the start, as was said, and when the abatement is delayed suspicion arises, even if the sign of putrefaction is not yet apparent in the pulse because, if it is not dispersed completely, of necessity the *kakochymia* is putrefied. Therefore, in order that it may be dispersed, we already have need of a better remedy. 565K

If [the patient] is either a child or an old person, phlebotomy is ruled out. But between these ages, when bodily strength is present in the patient, you must carry out phlebotomy, if the signs of abundance are also absent. It is better, when you have evacuated the bulk of the *kakochymia*, to proceed to the thorough cleansing of the blockages; for when blockages are removed prior to purging, there is a danger of them becoming more impacted. The cleansing potencies remove blockages, whether applied externally to the skin or taken internally into the body. There is also a danger that those cleansing agents which are external draw with them something else along with the blocking agents, or that those which carry inwardly draw with them something of what is in the vessels. If this happens to be viscid or thick, it will double the blockage. If it does not, but by virtue of the abundance it is not altogether insignificant, it will cause distress. It sometimes seems that when there is a great amount flowing all at once it ob-

566K σθαί τε | καὶ σφηνοῦσθαι τῶν γλίσχρων καὶ παχέων·
εἰ δὲ καὶ διὰ πόρων στενῶν ἡ διέξοδος εἴη, πολὺ δὴ
μᾶλλον ἴσχεσθαι. ὅπως οὖν ἀσφαλὴς καὶ τούτοις ἡ
φορὰ καὶ τοῖς ἔξωθεν ἡ ὁλκὴ γίγνοιτο, χρησιμώτατόν
τέ ἐστι κενῶσαι τὸ σῶμα. καὶ μέντοι καὶ τὰ διαπνεόμενα τῶν χυμῶν καπνώδη καὶ λιγνυώδη περιττώματα μειωθέντων αὐτῶν ἐλάττω γίγνοιτο ἄν. ὥστε
καὶ διὰ τοῦτο βραχύτερον ἔσται τὸ πυρετῶδες θερμόν,
ὡς ἐγχωρεῖν αὐτὸ διαπνεῖν εἰς τοὐκτός, εἴπερ εἶεν αἱ
ἐμφράξεις βραχεῖαι. ὅταν οὖν ποθ' ὑπονοήσῃς ἐστε-
γνῶσθαι τὸ δέρμα δι' ἔμφραξιν, ἔκ τε τοῦ[15] μὴ γίγνεσθαι κατὰ τὴν παρακμὴν τοῦ πρώτου παροξυσμοῦ
κενώσεις ἀξιολόγους καὶ διαμένειν ὁμοίαν ἑαυτῇ τὴν
παρακμήν, εἰ μὲν ἀπὸ μηδενὸς τῶν προφανῶν αἰτίων
ἐπύρεξεν ὁ ἄνθρωπος, ἔτι καὶ μᾶλλον ἂν εἴη σοι
βεβαιοτέρα τῆς ἐμφράξεως ἡ διάγνωσις· εἰ δὲ σύν
τινι τῶν προφανῶν, οὐκ ἀδύνατον ἅμα συνεληλυθέναι
διττὰς αἰτίας. εἰ γὰρ ἡ ἑτέρα μόνη τῶν αἰτιῶν ἡ ἐκ τοῦ
προκατάρξαντος αἰτίου ἦν, ἐμειοῦτο ἂν ἡ παρακμὴ
σὺν αἰσθηταῖς ἐκκρίσεσιν.

οὔκουν ἀναβάλλεσθαι τῶν τοιούτων ἀρρώστων |
567K ἀφαιρεῖν αἵματος, ἀνάλογον τῇ τε δυνάμει καὶ τῷ
μεγέθει τῆς ἐμφράξεως. ἔνδειξις δὲ τοῦ ποσοῦ τῆς
ἐμφράξεως ἐκ τοῦ μεγέθους ἔσται τοῦ κατὰ τὸν πυρε-
τόν· μείζων μὲν γὰρ ἐπὶ ταῖς σφοδροτέραις ἐμφράξε-
σιν, ἐλάττων δ' ἐπὶ ταῖς μικροτέραις συμπίπτει. κενώσαντα δὲ τὸ σύμμετρον οὐ μετὰ πολὺ διδόναι τῶν
ῥυπτικῶν τινα τροφῶν ἢ φαρμάκων. ἔστι δ' οὐ πολλὰ

structs and impacts no less than the viscid and thick humors. And if the passage outward is through narrowed pores as well, it is certainly held up much more. Therefore, so that the transfer to these by one group of agents and the drawing out externally occurs safely, it is most useful to evacuate the body. Furthermore, the smoky and sooty superfluities of the humors which are transpired would become less because the humors themselves are reduced. Consequently, because of this, the feverish heat will also be less, thus permitting it to transpire externally if the blockages are slight. Therefore, whenever at any time you suspect the skin is plugged up due to blockage, because significant evacuations do not occur during the abatement of the first paroxysm and the abatement persists as it was, if the person is febrile from none of the obvious causes, your diagnosis of blockage would be even more certain. If, however, it is with one of the obvious causes, it is not impossible for two causes to have come together at the same time. For if there was only the one cause, and this was *prokatarktic* in type, the abatement would be diminishing along with perceptible excretions.

Do not, then, delay the removal of blood from those who are sick in proportion to the capacity and magnitude of the blockage. There will be an indication of the amount of the blockage from the magnitude of the fever because what happens is that it (the fever) is greater with the more severe blockages and less with those that are smaller. When you have purged moderately, give one of the cleansing nutriments or medications quite soon. There are not

---

15 B; τῶν K

GALEN

τὰ τοῦτο δυνάμενα ποιεῖν ἐν πυρετοῖς, ἀλλ' ὡς εἴρηται
καὶ πρόσθεν, ἐν τῷ τροφῆς μὲν λόγῳ ἡ πτισάνη τ'
ἐστὶ καὶ μελίκρατον, ἐν φαρμάκοις δὲ τό τε ὀξύμελι
καὶ ὅσα τῷ μελικράτῳ συνεψεῖν οἷόν τε, ὡς καλαμίνθη
καὶ ὕσσωπον καὶ ὀρίγανον καὶ ἕρπυλον καὶ ἴρις καὶ
σέλινον. ἀλλὰ ταῦτά γε πάντα πλέον ἢ δεῖ θερμὰ καὶ
διὰ τοῦτο ἐξάπτει τοὺς πυρετοὺς ὥσπερ οἶνος. ὀξύμελι
δὲ μόνον οὔτε τοὺς πυρετοὺς ἐξάπτει καὶ ῥύπτει γεν-
ναίως, ὡς διαλύειν μὲν ὅσα γλίσχρα καὶ παχέα, τοὺς
πόρους δ' ἐκφράττειν. ἐν δὲ τῷ μεταξὺ μελικράτου τε
καὶ ὀξυμέλιτος εἴη ἂν τὸ καλούμενον ἀπόμελι, χρῶνται δ' αὐτῷ κατὰ τὴν Πελοπόννησον πλεῖστον. θείη δ'
568K ἄν τις εὐλόγως, οἶμαι, | καὶ τὸ σάκχαρ ἐν τοῖς τοι-
ούτοις. ἐπὶ δὲ τῇ τοιαύτῃ τροφῇ παραφύλαττε πόσον
ἀφαιρεῖται τῆς πυρετώδους θερμασίας.

εἰ γάρ σοι φαίνοιτο κατὰ τὴν τρίτην ἡμέραν ἕωθεν
ἐλάχιστόν τι λείπεσθαι τοῦ πυρετοῦ, μήτε τοῦ τῆς
σήψεως τῶν χυμῶν σημείου παρόντος ἐν τοῖς σφυγμοῖς μήτ' ἐν τοῖς οὔροις τοῦ τῆς ἀπεψίας, ἡ δ'
ὕποπτος ὥρα καθ' ἣν εἰσέβαλεν ὁ πυρετὸς ἐν τῇ
πρώτῃ τῶν ἡμερῶν, ἐξωτέρω τῆς μεσημβρίας εἴη,
θαρρῶν λοῦε τὸν ἄνθρωπον ὅσον οἷόν τε τάχιστα πρὸ
πολλοῦ τῆς ἕκτης ὥρας. ἱκανὸν δὲ κἂν ἐντὸς τρίτης
ὥρας[16] ξυμπληρωθῇ τὸ λουτρόν. οὕτω δὲ καὶ εἰ δεκά-
την ὥραν ὑποπτεύοις, ἄχρι τῆς ἑβδόμης λούειν ἐγχωρεῖ, τριῶν ἢ τεττάρων ὡρῶν ἰσημερινῶν αὐταρκεστά-
των οὐσῶν εἰς διάστημα τοῦ τε βαλανείου καὶ τοῦ
παροξυντικοῦ καιροῦ. χρὴ δ' ἐν τῷ βαλανείῳ προ-

402

many things that can do this in fevers, but as I also said before, in the consideration of nutriment, there is ptisan and melikraton, and among the medications, oxymel and those that can be boiled down with melikraton, like catmint, hyssop, oregano, tufted thyme, iris and celery. But all these things are hot beyond what is required, and because of this, kindle fevers, just as wine does. However, oxymel alone does not kindle fevers, and washes effectively so as to disperse those things that are viscid and thick, and unblocks the pores. Intermediate between melikraton and oxymel is the so-called apomel which they use a lot in the Peloponnese. I think someone might reasonably also put sugar among such things. Watch closely how much of the feverish heat is taken away due to such nutriment.   568K

If, on the morning of the third day, very little of the fever seems to you to remain, and if no sign of putrefaction of the humors is present in the pulse, or of failure of digestion in the urine, and if the suspected hour at which the fever struck on the first day should be outside the middle of the day, confidently bathe the person as much and as quickly as possible, well before the sixth hour. It should be sufficient if the bath is completed by the third hour. On the same basis, if you suspect [the onset to have been in] the tenth hour, it is possible to bathe up to the seventh hour, since it is most satisfactory if there is an interval of three or four standard hours between bathing and the time of the paroxysm. In the bathhouse, it is necessary, after anointing

---

16 K; ὥρας om. B, recte fort.

ἀλείψαντα καὶ προανατρίψαντα μετριώτατα διαρρύπτειν τε καὶ ἀποπλύνειν ἔξωθεν τὸ σῶμα. τὴν δ' ὕλην τῶν ῥυπτικῶν ἐν τοῖς περὶ φαρμάκων ὑπομνήμασιν ἔχεις. ἐκλέγειν οὖν ἐκεῖθεν δεῖ τὰ πρὸς τὸ παρὸν ἁρμόττοντα· διῄρηται γὰρ ἐν αὐτοῖς ὅσα τε μᾶλλον αὐτῶν ὅσα θ' ἧττον ῥύπτει.

569K   λεχθήσεται | δὲ καὶ νῦν ἕνεκα παραδείγματος ὀλίγα. μετριώτατα μὲν οὖν ῥύπτει τό τε τῶν ὀρόβων ἄλευρον καὶ τὸ τῶν κριθῶν καὶ τὸ τῶν κυάμων, ἔτι καὶ τὸ μελίκρατον τὸ ὑδαρές· μᾶλλον δὲ τούτων ἶρις καὶ πάνακος ῥίζα καὶ ἀριστολοχία καὶ τὸ μέτριον ἐν τῇ κράσει μελίκρατόν ἐστιν, ὥσπερ γε καὶ τοῦδε μᾶλλον ὅταν ἀκρατέστατον ᾖ. γίνεται δὲ τοιοῦτο τῷ μέλιτι μιχθέντος ὕδατος βραχέως, ὡς χυθὲν τοῖς μικροῖς τοῦ δέρματος εὐκόλως ἐνδῦναι πόροις. ἔτι δὲ τοῦδε μᾶλλον ὁ ἀφρὸς τοῦ νίτρου καὶ τὸ νίτρον αὐτὸ καὶ τὸ ἀφρόνιτρον. εἶναι δὲ χρὴ τὸ μὲν νίτρον λεπτομερές, οἷόν πέρ ἐστι τὸ Βερενίκιον· εἰ δὲ τὸ λιθῶδες εἴη, καίειν τε καὶ λειοῦν ἀκριβῶς αὐτό. μέλι δὲ τὸ καθαρώτατον καὶ δριμύτατον, οἷόν πέρ ἐστι τὸ Ὑμήττιον. ἔστι δὲ δήπου καὶ ὁ σάπων ὀνομαζόμενος ἐν τοῖς μάλιστα ῥύπτειν δυναμένοις. ἅλις ἔστω σοι παραδειγμάτων τῆς τῶν ῥυπτόντων ὕλης.

μετιέναι γὰρ ἤδη καιρὸς ἐπὶ τὰ μετὰ τὸ λουτρόν. ἔστι δ' οὐ πολλά· πλὴν γὰρ ὕδατος ἔχοντος ἐναφη-

---

²³ See *De simplicium medicamentorum temperamentis et facultatibus*, XI.743K, and *De compositione medicamentorum per genera*, XIII.499K.

404

and massaging very moderately beforehand, to clean thoroughly and wash the body externally. You have the material of the cleansing agents in the treatises on medications.[23] You must choose from that place those that are suitable for the purpose to hand, for a distinction is made among them between those that cleanse more and those that cleanse less.

I shall also say a little now by way of example. The 569K things that cleanse most moderately are the meal of bitter vetch, of barley, and of beans; and further, watery melikraton. More cleansing than these are iris, root of panax, aristolochia and melikraton that is moderate in *krasis*, the last being more [cleansing] when it is very pure. Such a thing occurs when water is mixed to a slight extent with honey in that, when it is dissolved, it enters the small pores of the skin easily. Still more [cleansing] than this are foam of niter, niter itself, and aphronitrum. And it is necessary for the niter to be fine-particled as, for example, that from Berenice is.[24] If, however, it is stony, burn and thin it thoroughly. Honey that is the purest and sharpest is, for example, that from [Mount] Hymettus.[25] There is, of course, also what is called soap among those things that have a particular capacity to cleanse. Let these be enough for you by way of examples of the material of things that cleanse.

It is now time to proceed to the things that come after bathing. There are not many of these for, apart from water having a little celery boiled up in it, you must give them

[24] See *De compositione medicamentorum per genera*, XIII.568K.

[25] According to Dioscorides, II.101: "Attick Honey is the best, and of this that which is called Hymettian" (Goodyer).

570K ψημένον ἑαυτῷ[17] σέλινον ὀλίγον, οὐδὲν αὐτοῖς | δο
τέον, κἂν[18] τὸ μεταξὺ τοῦ τε λουτροῦ καὶ τῶν ὑπό
πτων διάστημα τριῶν ὡρῶν ἔσεσθαι στοχαζώμεθα.
περὶ μέντοι τὴν ἑσπέραν καὶ θᾶττον ὥραις δυοῖν
τοῦ κατὰ τὴν πρώτην ἡμέραν εἰσβεβληκότος παρ
οξυσμοῦ καὶ λοῦσαι δυνατὸν ἔωθεν καὶ θρέψαι μόνον
πτισάνης χυλῷ. εἶτα εἰ μὲν μηδόλως ἐνέγκοιεν αἱ
ὧραί τι, λούειν αὖθις εἰ βούλοιτο καὶ τρέφειν, ὧν
εἴρηκα σκοπῶν ἐχόμενον. εἰ δέ τι γένοιτο, παρα
βάλλειν τε τοῦτο τῷ κατὰ τὴν πρώτην ἡμέραν παρ
οξυσμῷ, συνεπισκοπεῖσθαι τά τε οὖρα καὶ τοὺς
σφυγμούς· ἅπαντα δὲ δήπου φαίνεταί σοι μέτρια,
βραχείας ὑπολειπομένης ἐμφράξεως, ὥστε καὶ λού
σεις τῇ τετάρτῃ τῶν ἡμερῶν αὐτὸν καὶ θρέψεις τῶν
εἰρημένων σκοπῶν ἐχόμενος· ἐλπίσεις τε μηδὲν ἔτι τῇ
πέμπτῃ συμπεσεῖσθαι. μεγάλης δὲ τῆς ἐμφράξεως
γεγενημένης οὐκέτ᾽ ἂν ὁ τοιοῦτος ἐκ τῶν ἐφημέρων εἴη
πυρετῶν. ὥστε εἰς τὸν ἑξῆς ἀναβεβλήσθω λόγον, ἐν
γάρ τοι τῷδε περὶ τῶν ἐφημέρων μόνων πρόκειται
διελθεῖν.

5. Ἐπεὶ τοίνυν ἐφήμεροι πυρετοὶ ταῖς ὑποκειμέ
ναις κράσεσι τῶν σωμάτων ἐπιγίγνονται πολλάκις ἐπ᾽ |
571K ἀπεψίᾳ[19] σιτίων, ἑξῆς ἂν εἴη περὶ αὐτῶν διελθεῖν.
χειρίστη μὲν οὖν ἐστιν ἐν ταῖς τοιαύταις κράσεσιν
ἀπεψία καὶ τάχιστ᾽ ἀνάπτει πυρετόν, ἐν ᾗ διαφθεί
ρεται τὰ σιτία πρὸς τὸ κνισῶδές τε καὶ καπνῶδες· ὡς
ἥ γε εἰς ὀξύτατον τρέπουσα σπανιάκις τε γίγνεται
ταῖς τοιαύταις φύσεσι καὶ ἧττον βλάπτει. λεχθήσεται

406

nothing, although I estimate that [the period] between the 570K
bathing and those things suspected will be an interval of
three hours. However, if it is around evening, or the paroxysm attacked more quickly by two hours than on the first
day, it is possible to bathe [the patient] in the early morning and to nourish with the juice of ptisan alone. Then, if
the hours bring nothing else, bathe him again if he should
wish it, and nourish him, bearing in mind the objectives
I have mentioned. If, however, something should occur,
compare this with the paroxysm during the first day, and
examine together the urine and the pulse. If everything
seems to you to be within measure, since there is only a
slight blockage remaining, you will bathe the patient on
the fourth day and nourish him, bearing in mind the stated
aims, and you will hope that nothing further happens on
the fifth day. But when the blockage which has occurred is
major, such a fever would no longer still be one of the
ephemeral kind. Consequently, let me defer that to the following discussion, for in this one I propose to go through
the ephemeral fevers only.

5. Therefore, since ephemeral fevers frequently super-
vene on the underlying *krasias* of bodies due to failure to 571K
digest foods, it would be appropriate to go over these in order. Worst, then, in such *krasias* is the apepsia that very
quickly kindles a fever in which the foods are corrupted toward the steamy and smoky, since the *krasis* turning to a
very sharp quality occurs infrequently in such natures and
is less harmful. I shall say something about each of these

17 B; αὐτῷ K   18 K; ἄν B
19 B; καὶ ἀπεψίας K

407

δὲ ὑπὲρ ἑκατέρας ἰδίᾳ ἡμῖν τὴν ἀρχὴν ἀπὸ τῆς πλει-
στάκις τε γιγνομένης αὐτῆς καὶ μᾶλλον βλαπτούσης
ποιησαμένοις. ἔστι δ᾽, ὡς εἴρηται, κνισώδης τέ τις
αὕτη καὶ καπνώδης, ἔλαττον μὲν ὑπάρχουσα κακόν, εἰ
καταρρήξειεν ἡ γαστήρ, οὐ μικρὸν δὲ εἰ ἐπισχεθείη.
πυρέττουσι γὰρ ἐπὶ ταῖς ἐπισχέσεσιν αὐτῆς ἑτοίμως
αἱ κακόχυμοι φύσεις καὶ μάλισθ᾽ ὅταν ἐπὶ τὰς συνήθεις τράπωνται πράξεις, ὑπεριδόντες τοῦ μένειν ἔνδον ἐφ᾽ ἡσυχίας ἐν τῇ στρωμνῇ. πυρέττουσι δ᾽ ἔνιοι
καὶ τῶν ἁλόντων διαρροίαις ἐπ᾽ ἀπεψίᾳ, συναυξανο-
μένης αὐτοῖς τῆς πυρετώδους διαθέσεως, οὐχ ἥκιστα
κἀκ τοῦ πλήθους τῶν ἐξαναστάσεων. εἰ δὲ καὶ δῆξίς
τις, ἢ πόνος, ἢ θερμότης ἄμετρος ἐν τοῖς κατὰ τὴν
γαστέρα καὶ τὰ ἔντερα γένοιτο, πολὺ δὴ μᾶλλον
572K   ἐντεῦθεν ἡ πυρετώδης αὐξάνεται | διάθεσις. ἴασις δ᾽
οὐχ ἡ αὐτὴ τῶν ἐπισχεθέντων τοῖς ἐκκρίνουσι· λεκτέον οὖν ὑπὲρ ἑκατέρων ἐν μέρει· φανεῖται γὰρ οὕτως
οὐ μόνον ὅπῃ διαλλάττουσιν, ἀλλὰ καὶ ὅσον ἐν ἀμφοτέροις κοινόν.

εἰ μὲν δὴ φαίνοιτο μόνα τὰ διεφθαρμένα κεκενῶ-
σθαι, λουστέον τ᾽ ἐστὶ καὶ θρεπτέον ἐν τῇ παρακμῇ
τοῦ πυρετοῦ,[20] πρόνοιάν τινα τῶν κατὰ τὴν γαστέρα
ποιησαμένοις πρότερον. εἰ δ᾽ οὕτως εἴη πολλὴ κένωσις ἤτοι γεγονυῖα πρόσθεν ἢ καὶ νῦν ἔτι γιγνομένη
κατὰ τὸν ἄνθρωπον ὡς κεκμηκέναι τὴν δύναμιν, ἄμεινον θρέψαι χωρὶς τοῦ λοῦσαι προνοησαμένους τῶν
κατὰ τὴν γαστέρα. πρόνοια δ᾽ ἐπὶ τῶν οὕτως ἐχόντων,
εἰ μὲν μηκέτι γίγνοιτο κένωσις, ἐμβροχὴ δι᾽ ἐλαίου τε

separately, making a start from what occurs very frequently and is more harmful. As I said, this steamy and smoky quality itself is less harmful if the stomach is discharging, but more harmful if it is retaining. For the *kakochymous* natures readily become febrile due to the retention of it, and particularly when patients return to their customary activities after being supported while they remained indoors in quietude and confined to bed. There are also some who become febrile when seized by diarrhea due to apepsia, since the febrile condition is exacerbated in them, not least due to the number of times they must get up to defecate. And if some gnawing, distress or excessive heat occurs in the stomach and intestines among those patients, the febrile condition is certain to be markedly increased due to that. The cure for those who are retaining is   572K
not the same as for those who are expelling. I must, then, speak about each of these separately, as it will be apparent in this way not only how they differ but also how much is common in both.

Now if it does appear that only those things that have been corrupted have been evacuated, you must bathe and nourish in the abatement of the fever after previously giving forethought to those things in the stomach. In this way, if a major evacuation has either occurred earlier or is still now occurring in the person such that the capacity has suffered, it is better to nourish without bathing, giving prior thought to those things in the stomach. When consideration [has been given] to the things so affected, if evacuation has not yet occurred, perfuse with oil and wormwood,

---

20 K; τοῦ πυρετοῦ *om.* B, *recte fort.*

καὶ ἀψινθίου· χρὴ δὲ προδιαβρέχειν ὕδατι ζέοντι τὸ
ἀψίνθιον, ὅπως μὴ κνισωθείη. καὶ εἴ γέ τις αἴσθησις
ὑπολείποιτο κατὰ τὴν γαστέρα δήξεως ἢ πόνου, πίλημα μετὰ τὴν ἐπιβροχὴν ἤτοι θερμὸν καὶ ξηρὸν ἢ
βεβρεγμένον μὲν ἐλαίῳ ἐν ᾧ τὸ ἀψίνθιον ἀπεζέσθη, τὸ
πλεῖστον δ᾽ ἐκπεπιεσμένον ἐπιβάλλειν χρή. κάλλιον
573K δ᾽ εἰ παρείη μύρον νάρδινον ἐπιμελῶς | ἐσκευασμένον,
ἐν ἐκείνῳ δεύσαντα κατὰ τὸν αὐτὸν τρόπον ἐπιτιθέναι
τὸ πίλημα. καὶ πολύ γε κάλλιον, εἰ πορφύρα τοῦτ᾽ εἴη
θαλαττία· καὶ κάλλιστόν γε εἰ Τυρία ἀρίστη τυγ-
χάνοι· λεπτομερεστέραν γὰρ ἔχει τὴν στύψιν ἥ τε
τοιαύτη πορφύρα καὶ ἡ ἀρίστη νάρδος, ὥστε διὰ τοῦ
βάθους διεξέρχηται[21] τῶν κατὰ τὴν γαστέρα σωμάτων ἡ δύναμις αὐτῶν ξηραινόντων καὶ θερμαινόντων
καὶ τονούντων. εἰ δ᾽ ἀτονώτερον ὑπάρχει τὸ στόμα τῆς
γαστρός, ὅπερ καὶ στόμαχον ὀνομάζειν εἰθίσμεθα
καταχρώμενοι τῇ προσηγορίᾳ, καὶ τῆς Χίας μαστί-
χης ἐν τῇ νάρδῳ λειώσαντες, ὡς γλοιῶδες γενέσθαι,
δεύσαντες ἐξ αὐτοῦ, τὴν πορφύραν ἐπιθήσομεν.

ἔστω δὲ θερμὸν ἱκανῶς ἕκαστον τῶν τοιούτων κατὰ
τὴν πρώτην ἐπιβολήν· ἐκλύει γὰρ ἅπαντα τὰ χλιαρὰ
τὸν τόνον τῆς γαστρός. οἶσθα δὲ καὶ ὡς ἐπ᾽ ἀγγείου
διπλοῦ προϋποβεβλημένου κατὰ τὸ ἕτερον αὐτῶν τὸ
μεῖζον ὕδατος ζέοντος, ἔθος ἡμῖν ἐστι θερμαίνειν τὰ
μύρα· διαφθείρεται γὰρ ἡ δύναμις αὐτῶν ἑτέρως θερ-

---

[21] B; διεξέρχεσθαι K

although you must soak the wormwood beforehand in boiling water so it is not greasy. And if some sensation of gnawing or pain remains in the stomach, it is necessary to lay on compressed wool after the perfusion, either hot and dry, or soaked with oil in which the wormwood was boiled, having squeezed out the excess. Better, if it is available, is sweet oil of nard carefully prepared; apply the compressed wool after moistening with that in the same way. And in fact, it is much better if this is the marine purple, and best if it happens to be the finest Tyrian,[26] for either such purpura or the finest nard has a quite fine-particled astringency, so that the potency of those things that are drying, heating, and strengthening passes deeply through the bodies in the stomach. If the opening of the stomach (which we are also accustomed to term the "stomachus"—i.e. the cardiac orifice of the stomach—when we use the name)[27] is rather weak, and when I have triturated the Chian mastich in spikenard, so that it becomes glutinous, and after I have moistened it with this, I will apply the purpura.

573K

Each one of such things must be sufficiently hot at the first application; for all those things that are lukewarm relax the tone of the stomach. And you know it is my custom to heat the unguent in a double vessel, having beforehand boiling water in the larger one of them. Otherwise

[26] This is presumably *Murex brandaris* (purple murex), briefly mentioned in Dioscorides II.4. C. has under murex: "a gastropod mollusc of the Murex genus, yielding Tyrian purple dye."

[27] The nomenclature applied to the stomach is somewhat confusing. *Stomachos* seems sometimes to mean the cardiac orifice of the stomach, sometimes the esophagus, and sometimes the stomach itself.

μανθέντων. ἔτι δὲ ὑπιούσης τῆς γαστρὸς ἄμεινον χρῆσθαι μηλίνῳ καλλίστῳ τε καὶ προσφάτῳ, τὸν

574K αὐτὸν τρόπον | τῇ νάρδῳ· χρηστὸν δ' εἰς τὰ τοιαῦτα σύμπαντα καὶ τὸ μαστίχινον. οὐχ ἥκιστα δὲ καὶ τὰ τῆς τρυφῆς ἕνεκα τῶν διατεθρυμμένων γυναικῶν σκευαζόμενα μύρα ταυτὶ τὰ πολυτελῆ, χρήσιμα πρὸς τὰς τοιαύτας διαθέσεις τῆς γαστρός, ἅπερ ἔοικεν ὑπὸ τῆς ἐν Ῥώμῃ τρυφῆς εὑρεθέντα καὶ τὰς προσηγορίας ἔχειν Ῥωμαϊκάς· ὀνομάζεταί γέ τοι σπικάτα γε καὶ φουλιάτα. μὴ φερόντων δὲ τῶν καμνόντων τὴν ζώνην, ἤτοι γ' ἐξ ἔθους ἢ τρυφῆς, ἐγχωρεῖ διὰ κηρωτῆς ἀνύειν ταὐτόν. ἔστι δ' εἰς τὰ τοιαῦτα καλλίστη κηροῦ Τυρρηνικοῦ τηχθέντος ἐν μύρῳ ναρδίνῳ, ψυχθείσῃ καὶ ξυσθείσῃ τῇ κηρωτῇ μιχθείσης λείας ἀκριβῶς ἀλόης τε καὶ μαστίχης. ἑκάστου δὲ αὐτῶν τὸ πλῆθος εἶναι χρὴ τοσόνδε· κηροῦ μὲν καὶ νάρδου τὸ ἴσον, ἀλόης δὲ τὸ ὄγδοον· ὥσπερ οὖν καὶ τῆς μαστίχης, ἢ εἰ βούλει βραχύ τι ταύτης πλέον. εἰ δὲ καὶ τῶν εἰρημένων τι μύρων τῶν πολυτελῶν ἀναμίξαις τῇ κηρωτῇ, βέλτιον ἔσται σοι τὸ φάρμακον.

ἐγκαιομένης δὲ τῆς γαστρός, ὡς καὶ φλεγμονώδη διάθεσιν ἐν αὐτῇ ξυνίστασθαι δοκεῖν, ἀμείνων ἡ κη-

575K ρωτὴ ἡ διὰ τοῦ μηλίνου γιγνομένη. | πολλὰ δὲ καὶ ἄλλα φάρμακα τὰ μὲν δι' οἰάνθης ἐστί, τὰ δὲ δι' ὑποκυστίδος καὶ βαλαυστίου καὶ φοινίκων σαρκός, ἐπιτήδεια καὶ ψύχειν τὴν γαστέρα καὶ ῥώμην ἐντιθέναι. τούτων οὖν ἐκλέγου τὸ κάλλιστον εἰς τὰ παρόντα πρὸς τοὺς εἰρημένους σκοποὺς ἀποβλέπων.

their potency is destroyed when they are heated. If the stomach is still relaxing, it is better to use oil of apples of the best quality, and fresh in the same way as the nard. 574K Mastich is also useful for all such things. No less useful are the expensive oils prepared for fastidious women; they are useful for such conditions of the stomach. These, it seems, were discovered by the luxuriousness of Rome and have the name "Roman," being called "spicata" and "foliata."[28] If the patient cannot tolerate the binding, whether through custom or delicacy, it is possible to accomplish the same result with a salve. What is best for such things is Tyrrhenian beeswax dissolved in sweet oil of spikenard, when fine aloes in precise quantities and mastich have been mixed with the cooled and scraped salve. The required amount of each of these ingredients is as follows: equal parts of beeswax and spikenard, an eighth part of aloes, and the same also of mastich, or, if you wish, a little more of this. If you also mix one of the aforementioned expensive sweet oils with the salve, your medication will be better.

If the stomach is burning so that it seems as if there is an inflammatory condition involved in it, the salve made with apple is better. And there are many other medications— 575K those from oenanthe, hypocystis, the flower of wild pomegranate, and the flesh of the date palm—which are suitable for cooling the stomach and giving it strength. Choose the best of these for the prevailing situation, keeping an eye on the stated objectives.

[28] Both are preparations for external application made from nard. Spicata is also mentioned by Galen at VI.427K and VIII.292K. Foliatum is mentioned by Pliny at *HN*, 13,1,2, #15.

GALEN

τρέφειν δ᾽ ἑξῆς τοῖσδε χρή· ῥεούσης μὲν ἔτι τῆς
γαστρός, ἀλφίτοις τε καὶ τοῖς καλουμένοις ὀξυλι-
πέσιν ἄρτοις, ὀλίγιστον ὄξους ἔχουσιν, οὐχ ὡς ἐπὶ
δυσεντερίας ἢ χρονίας διαρροίας εἰώθαμεν σκευάζειν.
ἐπιπάττειν δὲ τὸ ἄλφιτον, ἐνίοτε μὲν ὕδατι θερμῷ
δαψιλές, ἐνίοτε δ᾽, ὅταν ἡ γαστὴρ ἔτι ἐκκρίνῃ πλέον
τοῦ δέοντος ἤτοι ῥοιᾶς, ἢ ἀπίων, ἢ μήλων χυλῷ καὶ
μάλιστα τῶν κυδωνίων. εἰ δ᾽ ἥδιον τοῖς κάμνουσι καὶ
ἀφέψημά τι παρασκευάζοντας ἀπίων, ἢ μήλων, ἢ
μύρτων, ἐπιπάττειν αὐτοῖς τὸ ἄλφιτον. εἰ δὲ μηκέτ᾽
ἐκκρίνει ἡ γαστήρ, ὅ τε χόνδρος ἱκανὸς ὁμοίως ἠρτυ-
μένος πτισάνῃ, τουτέστιν ὄξους ἔχων, τό τε δι᾽ αὐτοῦ
ῥόφημα χωρὶς ὄξους οἵ τε ὄρχεις τῶν ἀλεκτρυόνων
576K καὶ οἱ πετραῖοι τῶν ἰχθύων | ἐκ τοῦ λευκοῦ ζωμοῦ καί
τις τῶν στυφουσῶν ὀπωρῶν καὶ μόνη καὶ μετ᾽ ἄρτου.
ταύτας δὲ πολὺ μᾶλλον ἐπιδώσεις, ἔτι ῥεούσης τῆς
γαστρός.

εἰ δ᾽ ἀνορέκτως ἔχοιεν, ὡς ἀπεστράφθαι τὰ σιτία,
γίνεται γὰρ καὶ τοῦτο πολλοῖς τῶν ἁλόντων διαρροίαις ἐπ᾽ ἀπεψίᾳ, τοῦ διὰ τῶν κυδωνίων μήλων χυμοῦ
προσδοτέον αὐτοῖς ὅσον κοχλιάριον ἓν ἢ δύο· εἰ δὲ μὴ
παρείη τοῦτο, τὸ διὰ τῆς σαρκὸς αὐτῶν. εἴρηται δ᾽ ἡ
σύνθεσις τῶν τοιούτων φαρμάκων ἐν τοῖς Ὑγιεινοῖς.
οὕτω μὲν ἐν τῇ πρώτῃ παρακμῇ τοὺς ἐπ᾽ ἀπεψίᾳ
πυρέξαντας ἰᾶσθαι προσήκει, τῆς γαστρὸς ἐκκρινού-
σης. εἰ δ᾽ ἐπέχοιτο τελείως, ἁψάμενος πρότερον τῶν
ὑποχονδρίων, εἶθ᾽ ὅλης τῆς γαστρός, ἐπίσκεψαι σα-

414

You should nourish with the following things in order: when the stomach is still flowing, use barley groats and the so-called bread loaf dressed in vinegar and fat since these have very little acid and are not as we are accustomed to prepare for the dysenteries and chronic diarrheas. Sprinkle the barley groats on some occasions with abundant warm water, and on others, when the stomach is still excreting more than it should, with the juice of pomegranate, pears, apples or, particularly, of quinces. If, however, it is more pleasant for the patients, prepare a decoction from pears, apples, or myrtle berries, and sprinkle the barley groats on them. If the stomach is not still excreting, gruel is similarly sufficient when prepared with ptisan—that is to say, when it is acidic—and the broth from this that is not acidic, and the testes of roosters, and from the white broth of the fish caught among the rocks, and one of the astringent fruits, either alone or with bread. Much more will you give these when the stomach is still flowing.

576K

If there are those who so lack appetite that they have been turned away from foods—for this also occurs in many who are seized with diarrhea in addition to apepsia— you must give them in addition as much as one or two spoonfuls of the juice made from quinces; if this is not available, give the flesh of quinces. The preparation of such medications was detailed in *On the Preservation of Health*.[29] It is appropriate to cure those who are febrile due to the apepsia in the same way at the first abatement, if the stomach is excreting. If, however, it is retaining completely, when you first palpate the hypochondrium and

[29] See the final chapter of Book 6 of Galen's *De sanitate tuenda*.

φῶς εἰ ὑπελήλυθεν εἰς τὸ λεπτὸν ἔντερον ἢ εἰς τὸ κῶλον ἡ τροφή. κἄπειτ᾽ ἐρώτησον ἐξῆς τὸν κάμνοντα κατὰ τί μὲν αἰσθάνεται μάλιστα μέρος, ἤτοι δήξεως ἢ βάρους, ὁποῖαι δέ τινες αἱ ἐρυγαί. διαγνοὺς δ᾽ ἐκ τούτων ἐν ᾧ μάλιστα μέρει τῆς συμπάσης γαστρός

577K ἐστιν ἡ τροφή, μετεώρου μὲν οὔσης ἔτι τοῦ διὰ | τῶν τριῶν πεπέρεων διδόναι μὴ τοῦ φαρμακώδους, ἀλλ᾽ ὅπερ ἐπηνέσαμεν ἐν τοῖς Ὑγιεινοῖς ὡς ἁπλούστατόν τε καὶ τοῖς ἠπεπτηκόσιν ἐπιτήδειον· εἶθ᾽ ἑξῆς αἰονᾶν ὑποχόνδριά τε καὶ σύμπασαν τὴν γαστέρα. τοῦτο δὲ κἂν ἤδη κατωτέρω προήκῃ τὰ διεφθαρμένα, τῷ μᾶλλόν τε καὶ ἧττον ἐξαλλάττοντα· πλέονος γὰρ δεῖται καὶ χρόνου καὶ καταντλήσεως εἰς τὴν παρασκευὴν τῶν ἐφεξῆς πρακτέων οἷς ἔτι μετεώρός ἐστιν ἡ διεφθαρμένη τροφὴ κατὰ τὴν γαστέρα.

κινηθείσης δ᾽ αὐτῆς ἀξιολόγως ἐπὶ τὰ κάτω, συμπράττειν ἤτοι διὰ τῶν προσθέτων ἢ διὰ κλυσμάτων πράεων, εἰ μὲν δῆξις συνείη, ταύτην πραΰνοντας· εἰ δ᾽ ἐμπνευμάτωσις ἐκείνη, καθιστῶντας· εἰ δὲ μηδέτερον, ἐκ μέλιτος καὶ ὕδατος ἐλαίου τε βραχέως συντιθεμένου τοῦ κλύσματος. ἔνθα δὲ εἴη σφοδροτέρα δῆξις, ἔλαιον Σαβῖνον ἐπιτηδειότερον ἐνιέναι τηχθέντος στέατος ἐν αὐτῷ χηνός· εἰ δὲ μὴ παρείη τοῦτο, τοῦ τῆς ἀλεκτορίδος· εἰ δὲ μηδὲ τοῦτο, τοῦ τῆς αἰγός· ἀπορούντων δὲ καὶ τοῦδε, κηροῦ βραχύ τι προσεπεμβάλλειν τῷ ἐλαίῳ, καὶ μᾶλλον εἰ πεπλυμένος εἴη. τὰς

---

[30] Linacre has this term, which is used directly by Peter En-

then the whole abdomen, clearly detect whether the nutriment has advanced to the small intestine or to the colon. Next, ask the patient in what part he feels most sensitive, whether the sensation is gnawing or deep, and what the eructations are like. When you have diagnosed from these things in what part of the whole gut, in particular, the nutriment is, if it is still high up, give some of the *diatrion* 577K *piperion*, not of the medicinal kind but that which I praised in *On the Preservation of Health* as very pure and suitable for those who are not digesting.[30] Then, next, moisten the hypochondrium and the whole abdomen. [Do] this even if the things that are corrupted have already advanced lower, changing it in terms of more or less; for there is need for a longer time and for irrigations in the preparation of the things to be done in succession in those in whom the corrupted nutriment is still high in the abdomen.

If, however, the nutriment has been moved to the parts below to a significant extent, help this either with suppositories or with gentle clysters. If gnawing is also present, soothe it; if there is flatulence, suppress it. If there is neither, the clyster is put together from honey, water and a little oil. When the gnawing is quite severe, it is more useful to insert Sabine oil with goosefat dissolved in it. If this is not available, insert the fat of a rooster; if this too is not available, insert the fat of a goat. And if there is not even enough of this, add in besides a little beeswax to the oil, preferably when it has been washed. Cure flatulence by

glish (p. 170). It presumably refers to a mixture of peppers. In the *De sanitate tuenda*, white pepper is specifically mentioned for its strengthening effect on the stomach (VI.265K) and for obstruction due to foods (VI.341K).

417

GALEN

578K δὲ ἐμπνευματώσεις ἰᾶσθαι | συνέψοντας τῷ ἐλαίῳ
πηγάνου τέ τι καὶ τῶν ἀφύσων σπερμάτων σελίνου
καὶ κυμίνου καὶ μαράθου καὶ σίνωνος, ὅσα τ’ ἄλλα
τοιαῦτα. κενωθείσης δὲ τῆς γαστρός, αὐτίκα τρέφειν
τὰ μὲν ἄλλα παραπλησίως τοῖς ἔμπροσθεν εἰρημένοις, ἀφαιρεῖν δὲ τὰ στύφοντα.

εἰ μὲν οὖν ἐν τῇ πρώτῃ τῶν ἡμερῶν εἰς ἑσπέραν
πραχθείη ταῦτα, ἢ νυκτὸς ὥρας ἡστινοσοῦν, προνοεῖσθαι χρὴ κατὰ τὰ παραπλήσια καὶ τῇ δευτέρᾳ τῶν
ἡμερῶν, καὶ λούεσθαί γ’ ἐν αὐτῇ συγχωρεῖν, ἐὰν
ἀκριβῶς ἀπύρετος ᾖ. κἀπειδὰν τῆς ἐπιούσης νυκτὸς
ἀλύπως ὑπνώσῃ, τελέως ἤδη νομίζειν ὑγιαίνειν αὐτόν.
εἰ δ’ ἡ μετὰ τὴν ἀπεψίαν ἡμέρα μετὰ τῆς ἐπιούσης
νυκτὸς ἐνέγκοι τὸν πυρετόν, ὡς ἐν τῇ δευτέρᾳ τῶν
ἡμερῶν, προνοηθῆναί τε καὶ τραφῆναι τὸν ἄνθρωπον.
ἐνέγκοι δέ τι εἰ καὶ ἡ διὰ τρίτης νὺξ βραχὺ πυρετῶδες,
οὐδ’ οὕτως χρὴ δεδιέναι, ἀλλὰ καὶ τούτους ἐπὶ τῆς
ἐρχομένης ἡμέρας καὶ λούειν καὶ τρέφειν, ἅπαντά τε
τἄλλα ποιεῖν ἀνάλογον ἐπ’ αὐτῶν, ὡς ἔμπροσθεν
διῄρηται. τὰς δ’ εἰς ὀξεῖαν ποιότητα μεταβολὰς τῆς
ἀπεπτηθείσης τροφῆς οὔτε γιγνομένας ἐστὶν ἰδεῖν ἐν
579K ταῖς τοιαύταις φύσεσιν, | ὅτι μὴ σπανίως, ἐπὶ τροφαῖς
ἑτοίμως ὀξυνομέναις οὔτε πυρετὸν ἀναπτούσας ἐφήμερον, ὥσπερ οὐδὲ τὰς βραδυπεψίας· ὡς εἴ γέ ποτε διὰ
τοιαύτην ἀπεψίαν πυρέξειαν αἱ πικρόχολοι κράσεις,
ἔμφραξίν τε καὶ σῆψιν χυμῶν ὑποπτεύειν προσήκει,
περὶ ἧς ἀκριβέστερον ἑξῆς διοριῶ κατὰ τὸν περὶ
σήψεων λόγον.

418

boiling up with oil, rue and the seeds of flatulence suppres-   578K
sants—celery, cumin, fennel, stone parsley, and other such
things. Once the stomach has been evacuated, immediately nourish with other things similar to those previously
mentioned, avoiding those that are astringent.

If these things are done toward evening on the first day,
or during whatever hour of the night, it is necessary to give
prior consideration to similar things also on the second
day, and to allow bathing on that day, if the patient is absolutely afebrile. And if, when night comes, he goes to sleep
without pain, you may now finally consider him healthy. If
the day after the apepsia, the coming of night brings fever,
the person is to be provided for and nourished as on the
second day. If the night of the third day brings on a slight
feverishness, you should not be anxious on this account,
but should also bathe and nourish these people in the coming day and do all the other things in proportion in the case
of these patients, as previously defined. In such natures it
is not possible to see the changes of undigested nutriment
to a sharp quality occurring because they are not frequent.   579K
Nor do the changes due the nutriments that have become
acidic readily provoke an ephemeral fever, just as the
bradypepsias do not. If, at some time, the picrocholic
krasias become febrile due to such a digestive failure, it is
appropriate to suspect a blockage and putrefaction of humors, which is something I shall define more precisely in
due course in the discussion on the putrefactions.

νυνὶ δὲ τῇ τῶν ἐφημέρων πυρετῶν διδασκαλίᾳ τοσοῦτον ἐπιπροσθεὶς ἀπαλλάξομαι. καπνώδης ἀναθυμίασις οὐκ ἄν ποτε γένοιτο διὰ βραδυπεψίαν, ἢ ἀπεψίαν ὀξυρεγμιώδη. ψυχροὶ γὰρ ἀτμοὶ ἐκ τῶν τοιούτων χυμῶν ἀπορρέουσιν ὀλίγιστοι παντάπασιν, οὐκ ἀναθυμιάσεις πολλαὶ καπνώδεις τε καὶ λιγνυώδεις· οὐδὲ γὰρ οὐδ᾽ ἐπὶ τῶν ἐκτὸς ἑτέρως ἔστιν ἰδεῖν καπνὸν ἢ λιγνύν, ἀλλὰ καπνὸν μὲν ἐπὶ ταῖς ἡμικαύστοις ὡς ἂν εἴποι τις ὕλαις, λιγνὺν δὲ ἐπὶ ταῖς ὑπεροπτηθείσαις τε καὶ καυθείσαις· ἔστι γὰρ ἡ μὲν λιγνὺς ἀναθυμίασις γεώδης, ὁ δὲ καπνὸς συμμιγὴς ἐξ ὑδατώδους τε καὶ γεώδους οὐσίας· οὐ μὴν οὐδ᾽ ὅπως χρὴ προνοεῖσθαι τῶν βραδυπεπτησόντων καὶ ὀξῶδες ἐρυγγανόντων ἀκούειν χρὴ ποθεῖν ἐν τῷδε· τῆς γὰρ ὑγιεινῆς πρα-
580K γματείας ἐστὶν ὁ περὶ τούτων λόγος. ἐνταῦθα | δὲ μόνον τῶν τοιούτων ἀπεψιῶν μνημονεύσομεν αἷς πυρετὸς ἕπεται. ἐπεὶ τοίνυν ἐφήμερος μὲν οὐχ ἕπεται πυρετὸς ταῖς ὀξυρεγμιώδεσιν ἀπεψίαις, ὁ δὲ ἐπὶ σήψει χυμῶν ἕπεταί ποτε, δεόντως εἰς τὸν ὑπὲρ ἐκείνων λόγον ἀνεβαλλόμην περὶ τῶν τοιούτων ἀπεψιῶν ἐρεῖν· οὐδὲ τότε καθ᾽ ὃν ἐν τοῖς Ὑγιεινοῖς τρόπον. ἐν ἐκείνοις μὲν γὰρ εἴπομεν ὁποῖ᾽ ἄττα χρὴ πρᾶξαι τὸν ἀπεπτήσαντα περὶ τὸν τῆς ὑγείας καιρόν· ἐνταῦθα δὲ ὁποῖα κατὰ τοὺς ταῖς ἀπεψίαις ἀκολουθήσαντας πυρετούς.

6. Σχεδὸν εἴρηταί μοι πάντα περὶ τῶν ἐφημέρων πυρετῶν· οἱ γὰρ ἐπὶ βουβῶσι πυρέξαντες οὐδὲ πυνθάνονται τῶν ἰατρῶν ὅ τι χρὴ ποιεῖν· ἀλλὰ τοῦ θ᾽

Now, in the teaching on the ephemeral fevers, having first added the following observation, I shall move on. A smoky exhalation would not, at any time, occur due to bradypepsia or a heartburn-producing apepsia because the cold vapors that flow out from such humors are altogether few and there are not many smoky and sooty exhalations. It is not possible to see smoke or soot in those things that are external, other than smoke due to materials that are "semiburned," as one might say, and soot due to things that have been overheated and burned. For soot is an earthy vapor while smoke is mixed from watery and earthy substance. It is not the case that you should give prior consideration to those things that have been digested slowly; moreover, you should not wish to hear about acidic eructations in this work. The discussion about these matters is for the treatise on health.[31] Here I shall mention only those apepsias which a fever follows. Therefore, since an ephemeral fever does not follow the heartburn-producing apepsias, but sometimes does follow those due to putrefaction of humors, I defer, as I should, the discussion about those to when I shall speak about such apepsias; nor will I speak about them in the same way as I did in my *On the Preservation of Health*. For in that work I did say something about what must be done for someone with apepsia while healthy. Here, however, [I am concerned with] what kinds of things [you should do] in fevers that follow apepsias.

580K

6. I have said almost everything [there is to say] about the ephemeral fevers, for those who are febrile due to swollen glands do not learn from doctors what they must

---

[31] See *De sanitate tuenda*, VI.422K.

ἕλκους ἐφ' ᾧπερ ἂν ὁ βουβὼν αὐτοῖς εἴη γεγεννη-
μένος, αὐτοῦ τε τοῦ βουβῶνος προνοησάμενοι, λούον-
ται κατὰ τὴν παρακμὴν τοῦ γενομένου παροξυσμοῦ.
κἂν φθέγξηταί τις τηνικαῦτα διάτριτον, ἅπαντες
καταγελῶσι καὶ σχολαστικὸν ἀποκαλοῦσι κατανο-
οῦντες, οἶμαι, φύσει μὴ δεῖν ὑπερβάλλειν τὸ μηδόλως
ἐσόμενον. ἀξιοῦσί τε τοιούτους εἶναι τοὺς ἰατροὺς ἐν
581K ἅπασι τοῖς ἄλλοις | ὅσα διαφεύγει τὰς αἰσθήσεις, οἷοί
περ αὐτοὶ περὶ τὰ φαινόμενα. τὸ δ' ἐν ἅπασιν εἶναι
τοιούτους οὐδὲν ἄλλο ἐστὶν ἢ τὸ γινώσκειν ὁπηνίκα
μὲν ἔσοιτο διὰ τρίτης ὁ παροξυσμός, ὁπηνίκα δ' οὐκ
ἔσοιτο. καὶ μέν γε καὶ ψυχθέντες, ἢ ἐγκαυθέντες, ἢ
κοπιάσαντες, ἤ τι τοιοῦτον ἕτερον παθόντες, ἐπειδὰν
πυρέξαντες τύχωσιν, εἶτα τῆς ἐν ταῖς παρακμαῖς
εὐφορίας αἴσθησιν ἔχωσιν ὑγιεινῆς, οὐδὲ τότ' ἀνέχονται τῶν τὴν διάτριτον φθεγγομένων. ὃ γὰρ εἴωθα
λέγειν πολλάκις, ὡς ἅπαντος σοφιστοῦ τῶν ἰδιωτῶν
ἕκαστος, ὃς ἂν ἔχῃ κατὰ φύσιν, ἀληθέστερα δοξάζει,
τοῦτο κἀπὶ τῆς θειοτάτης αὐτῶν διατρίτου θεάσασθαί
σοι πάρεστι, καταγελωμένης ὑπὸ πάντων τῶν κατὰ
φύσιν ἐχόντων.

ἔναγχος γοῦν τις ἰδιώτης ἐν δείπνῳ πολυτελεῖ
πλείω προσενεγκάμενος καὶ πιὼν ἧκε μὲν οἴκαδε,
σφαλλόμενός τε καὶ χειραγωγούμενος. ἐπιπολασάν-
των δὲ αὐτῷ τῶν βαρυνόντων τὴν γαστέρα, πάντ'
ἐξεμέσας αὐτά, διὰ νυκτὸς μὲν ἐπύρεξεν, ἐκοιμήθη δ'
ἐπὶ πολὺ τῆς ἐπιούσης ἡμέρας· εἶτα διαναστὰς καὶ
βραχέα περιπατήσας ἐλούσατο καταγελάσας τοῦ τὴν

do. But when they have taken care of the wound or ulcer due to which the swollen gland has arisen in them, and of the swollen gland itself, they bathe during the abatement of the paroxysm that occurs. And if, under these circumstances, someone speaks of a "three-day period," everyone laughs scornfully and calls him a pedant considering, I suppose, there is no need to go beyond what will not at all exist naturally. They think it right that the business of such doctors lies in all the other things which escape the senses, while they themselves [are concerned] with things that are apparent. In all things like these it is nothing else than to know at what time during the third day there will be a paroxysm and at what time there will not be. And even those who are cooled, or suffering heatstroke, or fatigued, or are affected in some other such way, whenever they happen to be febrile, if they have a sense of healthy well-being in the abatements, do not at that time tolerate those who speak of the "three day period." For what I have been accustomed to say on many occasions is that any layman of normal intelligence holds views that are truer than those of any sophist. This is there for you to see in regard to this divine three-day period of theirs, which is mocked by all normal people.

Anyway, recently a certain layman, after eating and drinking too much at a sumptuous dinner, came home stumbling and being led by the hand. As the food which remained undigested sat heavily in his stomach, he vomited it all, and during the night developed a fever, although he slept most of the following day. Then, when he got up and walked around a little, he bathed, laughing derisively at the

581K

582K διάτριτον αὐτῷ ἀναμένειν | συμβουλεύσαντος· ὅσον
μὲν γὰρ ἐπ᾿ ἐκείνῳ καὶ τὴν ἑξῆς ἡμέραν ἀσιτεῖν ἐχρῆν
αὐτόν· ὅσον δ᾿ ἐπὶ τοῖς ἀληθέσι καὶ οἷς αὐτὸς ἔπραξε
λουσάμενος καὶ μετρίως διαιτηθεὶς καὶ κοιμηθεὶς
ἀμέμπτως, ἕωθεν ἀναστάντα τῶν συνηθῶν ἔχεσθαι ἐν
αὐτῇ τῇ διατρίτῳ, καθ᾿ ἣν ἀσιτεῖν αὐτὸν ἐχρῆν πειθόμενον τοῖς διατριταρίοις ἰατροῖς, οὕτω γὰρ αὐτούς
τις ἐπισκώπτων ὠνόμαζε χαριέντως, καὶ δεδειπνη-
κότος δὲ ἤδη κατὰ τὸ σύνηθες τοῦ ἀνθρώπου, παρὼν ὁ
εὐτράπελος ἐκεῖνος ἐγελωτοποίει, ἀναμιμνήσκων τὸν
ἐμημεκότα τρίτης ἑσπέρας ὡς ἐχρῆν αὐτὸν ἄρτι δυοῖν
ἡμερῶν ἄσιτόν τε καὶ ξηρὸν καὶ ἄσης μεστὸν κατα-
κεῖσθαι, εἰς τὰς ὥρας ἀποβλέποντα κατὰ τὴν τῶν
διατριταρίων πρόσταξιν. ἔστι γὰρ, οἶμαι, καὶ τοῖς
ἰδιώταις εὔδηλα τὰ παρὰ τὴν ἐνάργειαν ἀποτετολμη-
μένα. κἀκ τούτου δεόντως ἰατροὺς μὲν ἑτέρους εἶναί
φασι, λογιάτρους δ᾿ ἑτέρους. πῶς γὰρ οὐ δίκαιοι
τούτου τοῦ προσρήματός εἰσι τυγχάνειν οἱ μὴ γινώσκοντες ἃ μηδεὶς ἰδιώτης ἀγνοεῖ; πῶς δ᾿ οὐ κατα-
γελᾶσθαι καὶ σκώπτεσθαι δικαιότατοι τυγχάνουσιν
ὄντες οἱ πρὸς τῷ τὰ τοιαῦτα ἀγνοεῖν Ἱπποκράτει
583K ἑαυτοὺς | προκρίνοντες; ἀλλὰ τὴν μὲν ἐκείνων ἀναι-
σθησίαν οὐδ᾿ ἂν ὁ Ἑρμῆς ἅμα ταῖς Μούσαις ἰάσαιτο.

7. Συντετελεσμένου δὲ ἡμῖν τοῦ περὶ τῶν ἐφημέρων

---

32 This term, *logiatros*, is also found in *De libris propriis*,
XIX.15K and is simply transliterated by Linacre. It is taken to

person who advised him to endure the three-day fast. As 582K
far as the latter was concerned, he ought to have fasted the
following day. As far as what he really did is concerned, after he had bathed, eaten moderately, and slept soundly, he
got up early and refrained from his customary activities
throughout the three day period during which he ought to
have fasted, if he had been persuaded by the "diatritarian"
doctors (for this is what someone who scoffed at them ele-
gantly named them). When he had already had his main
meal in the customary way, a man well known for his wit
who was present caused laughter by reminding the man
who had vomited during the third evening that he ought
to have been without food now for two days, and should
be dry and full of nausea in his disposition, watching the
hours according to the direction of the "diatritarians." For
it is, I think, quite clear, even to laymen, that things had
been done recklessly, contrary to the evident situation.
And from this they say that some doctors are as they ought
to be while others are "theoretical" doctors only.[32] For how
is it not right for them to acquire this title, if they don't
know those things of which no layman is ignorant? And
how is it not even more right that they happen to be de-
rided and scorned, in addition to being ignorant of such
things, when they place themselves above Hippocrates?
But Hermes and the Muses combined would not cure the 583K
stupidity of those men!

7. Since I have brought to completion the discussion of

mean those who theorize about medicine only and don't dirty
their hands in active practice. P. N. Singer (1997) uses the term
"word doctor," but this is perhaps misleading. "Merely theorists"
might capture the pejorative intent.

πυρετῶν λόγου κατὰ τὰς πικροχόλους φύσεις, ἐφ᾽ ὧν
τὸ μὲν θερμὸν στοιχεῖον τοῦ ψυχροῦ, τὸ δὲ ξηρὸν τοῦ
ὑγροῦ πλεονεκτεῖ, μεταβαίνωμεν ἤδη πρὸς ἑτέραν
κρᾶσιν τῶν σωμάτων οὐδὲν ἧττον τῆσδε κακόχυμον,
ἐν ᾗ τὸ μὲν θερμὸν τοῦ ψυχροῦ, τὸ δ᾽ ὑγρὸν τοῦ ξηροῦ
κρατεῖ. σηπεδονώδεσι γάρ τοι νοσήμασιν ἡ φύσις
αὕτη ἁπασῶν μάλιστα τῶν δυσκρασιῶν ἁλίσκεται,
διότι καὶ καθ᾽ ὃν ὑγιαίνει χρόνον ἐγγὺς σηπεδόνος
ἐστίν, ὡς ἔκ τε τῶν ἱδρώτων ἔνεστι τεκμαίρεσθαι δυσ-
ωδῶν ὑπαρχόντων, οὐχ ἥκιστα δὲ καὶ τῶν οὔρων καὶ
τῶν διαχωρημάτων, ἔτι τε καὶ τῆς ἐκπνοῆς. οἷα γὰρ ἡ
τῶν τράγων κρᾶσίς ἐστι, τοιαύτη καὶ ἡ τῶν τοιῶνδε
φύσεων. ἕτοιμον οὖν αὐτὴν νοσεῖν ἐξ αἰτίου παντὸς
ἄλλα τε νοσήματα πολλὰ καὶ πυρετούς. οὔτε γὰρ ὁπόταν στεγνωθῇ τὸ δέρμα, δυνατὸν τοῖς οὕτω κεκρα-
μένοις μὴ πυρέττειν, οὔθ᾽ ὅταν ἀπεπτῶσιν ἰσχυρῶς, ἢ
584K ἐγκαυθῶσιν, ἤ τι τοιοῦτον ἕτερον | πάθωσιν. αἱ μὲν
οὖν στεγνώσεις μάλιστα τῶν κακοχύμων ἅπτονται
φύσεων. ὥστε οὐδὲν ἧττον αἱ πικρόχολοι κράσεις τῶν
σηπεδονωδῶν ἁλώσονται πυρετοῖς ἐπὶ στεγνώσει. ὑπὸ
δὲ τῶν κνισωδῶν ἀπεψιῶν ἑτοιμότερον αἱ ὑγραὶ καὶ
θερμαὶ βλαβήσονται φύσεις, ἃς ἀρτίως ὠνόμασα
σηπεδονώδεις, ἐπειδὴ ῥᾷστα σήπονται κατὰ ταύτας οἱ
χυμοί.

περὶ δὲ τῶν ἐμφράξεων τί δεῖ καὶ λέγειν; ἃς διὰ
τοῦτο μάλιστα πυρετοὺς ἀνάπτειν ἔφαμεν, ὅτι σήψεις
ἐργάζονται κατὰ τὰ μὴ διαπνεόμενα σώματα. ἀλλὰ
περὶ μὲν τῶν τοιούτων πυρετῶν αὖθις ἐροῦμεν. οἱ δ᾽

ephemeral fevers in picrocholic natures in which the hot element prevails over the cold and the dry over the moist, let me move on now to another *krasis* of bodies which is no less *kakochymous* than this, in which hot prevails over cold and moist over dry. For certainly, in diseases which incline to putrefaction, this nature particularly succumbs to all the *dyskrasias* because during the time it is healthy it is close to putrefaction, as can be judged from the sweat being malodorous, and no less also the urine and feces, as well as the expired breath. For the *krasis* of billygoats is of the same kind as the *krasis* of such natures. There is a readiness, then, for this [nature] to suffer disease from any cause in respect of many other diseases and fevers. Whenever the skin is blocked up, it is impossible for those who have been "mixed" in such a way (i.e. have such a *krasis*) not to become febrile, just as it is not when they are strongly apeptic, or are suffering heatstroke, or are affected in some other such way. Therefore, the stoppages of 584K the pores particularly attack the *kakochymous* natures. As a result, the picrocholic *krasias,* no less than those that are putrefying, are seized by fevers due to such stoppages. The moist and hot natures will be more readily harmed by the steaming apepsias, which I just now termed "liable to putrefaction," since the humors putrefy very easily in these cases.

What else must be said about the blockages? On this point, I said they particularly kindle fevers because they bring about putrefactions in bodies that do not transpire. But I shall speak again about such fevers. All ephemeral fevers most readily subsist in the *krasias* spoken of due to

ἐφήμεροι πάντες ἑτοιμότατα μὲν ἐν ταῖς εἰρημέναις
συνίστανται κράσεσιν, ἐπὶ πᾶσι τοῖς αἰτίοις ὅσα
μικρὸν ἔμπροσθεν εἴπομεν ἐπὶ τῶν πικροχόλων φύ-
σεων· ἴασιν δὲ παραπλησίαν αὐτοῖς ἔχοντες ἐν ὀλίγοις πάνυ διαλλάττουσιν. ὑγρότεραι γὰρ αἱ φύσεις
ὑπάρχουσι, φέρουσιν ἐκείνων μᾶλλον ἀσιτίαν τε καὶ
δίψος. ὥσθ' ἧττον οὗτοι βλαβήσονται λιμαγχούμενοι
πρὸς τῶν διατριταρίων ἰατρῶν, ἐὰν μόνον ἀκωλύτως
585K   διαπνέωνται. διὰ ταῦτα δὲ | καὶ τῶν βαλανείων, ὡς
μανούντων μὲν τὸ δέρμα δέονται μᾶλλον, ἢ οὐχ ἧττόν
γε τῶν πικροχόλων· ὡς ὑγραινόντων δ' οὐ δέονται.
ὅσα δ' ἐπὶ τῶν τοιούτων φύσεων εἴρηται, ταῦτα κἀπὶ
τῶν ἐπικτήτων ἕξεων εἰρῆσθαι χρὴ νομίζειν. ἔνιοι
γὰρ ὑπάρχοντες εὔχυμοι φύσει πικρόχολοι γίγνονται,
συνελθόντων εἰς ἕνα χρόνον ἐνίοτε πλειόνων αἰτιῶν
ξηραινόντων τε καὶ θερμαινόντων τὸ σῶμα.

    φέρε γὰρ εἰς χωρίον ἀφῖχθαι τὸν ἄνθρωπον ὥρᾳ
θέρους θερμὸν καὶ ξηρόν· εἶναι δὲ καὶ τὴν ἐν τῷ τότε
χρόνῳ κατάστασιν θερμὴν καὶ ξηράν· ἐσθίειν τ' αὐτὸν
ἐδέσματα θερμὰ καὶ ξηρά· καὶ φροντίζειν καὶ λυπεῖ-
σθαι καὶ ἀγρυπνεῖν καὶ θυμοῦσθαι καὶ πονεῖν πάμ-
πολλα· δύναιτο δ' ἂν ὁ αὐτὸς ἄνθρωπος ἀφροδισίοις
τε συνεχέσι χρήσασθαι κατ' ἐκεῖνον τὸν καιρόν, ἐν
ἡλίῳ τε διατρίβειν τὰ πλείω καὶ πίνειν φαρμάκων
ξηραινόντων τε καὶ θερμαινόντων, ὥσπερ ἔνιοι μὲν
τῆς θηριακῆς, ἔνιοι δὲ τῆς ἀμβροσίας, ἔνιοι δὲ τῆς
ἀθανασίας, ἴσμεν γὰρ δήπου τὰ καλούμενα πρὸς τῶν
νεωτέρων ἰατρῶν φάρμακα τοῖς τοιούτοις ὀνόμασιν.

all the causes which I stated a little earlier in the case of the picrocholic natures, since they have a cure very similar to those, and differ only in a very few aspects. Moister natures tolerate fasting and thirst more than those natures. As a result, those who are reduced by hunger will be harmed less by the "diatritarian" doctors, just as long as they transpire without hindrance. Because of these things they also require more baths, as these make the skin porous, or at least not fewer baths than the picrocholic natures, as they do not require those things that are moistening. What was said in the case of such natures, you should think would also be said in the case of acquired states, for some who are *euchymous* in nature become picrocholic when, sometimes, a larger number of the causes that dry and heat the body come together at one time.

585K

So, for example, assume a person has come, in summer, to a place that is hot and dry, and at that time the weather conditions are hot and dry. Suppose he eats hot and dry foods, is anxious and distressed, is unable to sleep, and is angry and hard-pressed in every respect. Suppose the same man is able to indulge in frequent sexual activity at that time, to spend a long time in the sun, and to drink drying and heating medications, just as some drink theriac, some ambrosia and some athanasia, for we know, I am sure, the medications called such names by the younger doctors.[33] Someone to whom all the aforementioned

33 On the last two, see Galen, *De antidotis*, XIV.149K.

586K ὅστις οὖν ἐν ἅπασι τοῖς εἰρημένοις ἐγένετο | θέρους, ἀναγκαῖον αὐτῷ θερμὸν καὶ ξηρόν, καὶ διὰ τοῦτο πικρόχολον εἶναι τὸ σῶμα, κἂν ἔμπροσθεν εὐχυμότατον ἦν. καὶ τοίνυν καὶ πυρέξει ῥᾳδίως οὗτος ἐπὶ πᾶσι τοῖς ἔμπροσθεν εἰρημένοις αἰτίοις. οὕτω δὲ καὶ ὅστις εὔχυμος φύσει, χωρίον εὔκρατον οἰκῶν, εἰς ἕτερον χωρίον ὑγρὸν καὶ θερμὸν ἐν ἦρι μετέλθοι, καταστάσεως οὔσης θερμῆς καὶ ὑγρᾶς ἅπαντά τε τὰ διαιτήματα θερμὰ καὶ ὑγρὰ ποιήσαιτο, καὶ οὗτος ὁμοίως τῷ φύσει σηπεδονῶδει τοῖς τ᾽ ἄλλοις ἁλώσεται νοσήμασιν οἷσπερ κἀκεῖνοι· καὶ πυρέξει τούς τ᾽ ἄλλους πυρετοὺς ὁμοίως ἐκείνοις ἐπὶ τοῖς αὐτοῖς αἰτίοις, οὐχ ἥκιστα δὲ καὶ τὸν ἐφήμερον, ὑπὲρ οὗ νῦν ὁ λόγος.

8. Ὀκτὼ δ᾽ οὐσῶν τῶν πασῶν δυσκρασιῶν, ὡς ἐδείκνυμεν, ἑτοιμοτάτη μὲν εἰς πυρετοὺς ἡ θερμὴ καὶ ξηρά· καὶ ἢν μή τις αὐτήν, ὡς εἴρηται, διαιτήσῃ, τάχιστα μεταπίπτουσιν ἀπ᾽ αὐτῶν εἰς τοὺς ὀξεῖς πυρετοὺς οἱ ἐφήμεροι. ἐγγὺς δ᾽ αὐτῇ πυρετῶν γ᾽ ἕνεκα, καίτοι πρὸς ἄλλα γε ὑπάρχουσα χείρων, ἡ θερμὴ καὶ ὑγρά. τρίτη δ᾽ ἐπὶ ταύταις ἐστὶν ἡ ἁπλῆ δυσκρασία,
587K καθ᾽ ἣν ἡ μὲν ἑτέρα τῶν ἀντιθέσεων ἡ | κατὰ τὸ ξηρὸν καὶ ὑγρὸν ἄμεμπτος· ἐν δὲ τῇ λοιπῇ πλεονεκτεῖ τὸ θερμόν. ἑτοιμοτέρα γὰρ ἤδε τῶν ὑπολοίπων καὶ βλάπτεσθαι πρὸς ἁπάντων τῶν εἰρημένων αἰτίων καὶ μεταπίπτειν ἐκ τῶν ἐφημέρων εἰς τοὺς πολυημέρους· οὐδὲν γὰρ χεῖρον οὕτως αὐτοὺς ὀνομάσαι. ταύτῃ δ᾽ ἐφεξῆς ἐστιν ἡ κατὰ μὲν τὸ θερμὸν καὶ ψυχρὸν εὔκρατος, ἐπικρατοῦν δὲ τὸ ξηρὸν στοιχεῖον ἔχουσα τοῦ

things occur in summer, will of necessity have heat and 586K
dryness in him, and because of this, his body is picrocholic,
even if it was very *euchymous* before. Accordingly also,
this man will easily become febrile due to all the causes
previously mentioned. In this way, too, someone who is
euchymous in nature and dwells in a *eukratic* place, should
he move in the spring to another place that is moist and
hot, and when the weather conditions are hot and moist,
makes all his foods hot and moist, he too, like a nature that
is liable to putrefaction, will be seized by the other diseases, just as those men are. And he will become febrile in
respect of the other fevers in like manner to those due to
these same causes, and not least the ephemeral [fever],
which is what the discussion is now about.

8. Although there are, as I showed, eight *dyskrasias*
in all, the hot and dry *dyskrasia* is the most prone to fevers, and unless someone treats it with a regimen, as I said,
the ephemeral fevers arising from these *dyskrasias* very
quickly change to the acute fevers. Close to this *dyskrasia*,
as far as fevers go, is the hot and moist [*dyskrasia*], although in other respects it is in fact worse. Third in addition to these is the simple *dyskrasia* in which the other
opposition, that of dry and moist, is blameless. In the re- 587K
maining opposition, the hot predominates; for this is also
more prone than those remaining to be harmed by all
the causes mentioned, and to undergo a change from the
ephemeral to the polyhemeral (chronic), for it is no bad
thing to name the latter thus. Next in succession to this
dyskrasia is that where the hot and cold are *eukratic* but
there is a predominance of the dry element over the moist.

431

ὑγροῦ. μεθ᾽ ἣν ἡ μέση πασῶν ἡ εὔκρατός ἐστι κατ᾽
ἀμφοτέρας τὰς ἀντιθέσεις.

αἱ δ᾽ ὑπόλοιποι κράσεις αἱ τέτταρες οὔτε ῥᾳδίως
ἁλίσκονται πυρετοῖς ἐπὶ τοῖς εἰρημένοις αἰτίοις, οὔθ᾽
ἁλοῦσαί τινι τῶν ἐφημέρων, ἐὰν ἀσιτήσωσι, μεταπίπτουσιν εἰς τοὺς πολυημέρους. ἀλλ᾽ ἔνιαί γε αὐτῶν,
ἐπειδὰν τύχωσι πλῆθος ἠθροικυῖαι χυμῶν ἄνευ δια-
φθορᾶς, ἀθροίζουσι δὲ αἱ τοιαῦται συνεχῶς, ὑπὸ ἀσιτίας ὀνίνανται, μάλιστα μὲν ἡ ὑγρὰ καὶ ψυχρά, δευτέρα δὲ ἡ ὑγρά, καὶ τρίτη μετὰ ταύτας ἡ ψυχρά,
τετάρτη δ᾽ ἡ ξηρὰ καὶ ψυχρά. ἐπὶ γὰρ τοῖς ἐφημέροις
πυρετοῖς παυσαμένοις, ὅταν ἀναγκασθῇ τις οὐδὲν[22]
588K δέον ὑπερβάλλειν τὴν δαιμονιωτάτην | διάτριτον, ἕτε-
ρος μὲν οὐκ ἂν ἐκ τῆς ἀσιτίας ἀναφθείη πυρετὸς ἐν
ψυχραῖς κράσεσι· κακοῦται δ᾽ ἡ ξηρὰ καὶ ψυχρὰ
μάλισθ᾽ ἕξις, ὥστε πολλοῦ χρόνου χρῄζειν ἵν᾽ εἰς τὸ
κατὰ φύσιν ἐπανέλθωσιν οἱ οὕτω δύσκρατοι. συνού-
σης μέντοι τινὸς αὐτοῖς κατὰ τύχην ἐπικτήτου κακίας
χυμῶν, ἐγχωρεῖ μεταπεσεῖν ἐκ τῶν ἐφημέρων πυρετῶν
εἰς τοὺς πολυημέρους, ὅταν ἄσιτοι καὶ ἄλουτοι φυλα-
χθῶσι τὰς τρεῖς ἡμέρας. ἡ μέν τοι ξηρὰ μέν, ἀλλ᾽
εὔκρατος κατὰ τὴν ἑτέραν ἀντίθεσιν φύσιν σώματος,
ἐν ταῖς μακροτέραις ἀσιτίαις ἰσχνοῦται μὲν τῆς ξη-
ρᾶς καὶ ψυχρᾶς μᾶλλον, ἀνατρέφεται δὲ ῥᾷον ὑπὸ τῆς
προσηκούσης διαίτης.

9. Ἐξαλλαχθήσεται δὲ καὶ τοῖς τοιούτοις ἅπασιν ἡ
δίαιτα πολὺ δή τι πλέον ἢ τοῖς πρώτοις ἁπάντων
ῥηθεῖσι τῷ θερμῷ καὶ ξηρῷ καὶ ὑγρῷ. τὸν γοῦν ὑγρὸν

After this, and the middle of all, is the *eukrasia* in respect of both oppositions.

The remaining *krasias*, four in number, are not readily seized by fevers due to the causes mentioned; nor, should they be seized by one of the ephemeral fevers if people fast, do they change to the polyhemeral (chronic) fevers. But some of these *krasias* at least, whenever they happen to have collected together an abundance of humors without corruption and continue to gather them, are helped by fasting—particularly the moist and cold, and next the moist, and third after these the cold, and fourth the dry and cold. For in the case of the ephemeral fevers ceasing, whenever someone is compelled to go beyond the most miraculous three-day period more than is needed, another fever would not be kindled in the cold *krasias* from the fasting. The dry and cold state is particularly bad, so that those who are *dyskratic* in this way need a long time to return to normal. However, if perchance one of the bad humors acquired is present with them, there is the possibility of change from the ephemeral fevers to those that are polyhemeral (chronic), whenever they are kept fasting and not bathing for the three days. Indeed, the *dyskrasia* that is dry but otherwise *eukratic* in accord with the other natural opposition of the body is dried more in the prolonged fastings than the dry and cold *dyskrasia*, but is restored more easily by the apppropriate diet.

9. The diet will also be changed for all such cases much more, in fact, than for all those first spoken of—that is, in the hot, the dry, and the moist. At all events, we shall treat someone who is moist and cold in nature oppositely

588K

---

22 K; τίς εἰς οὐδέν B, *recte fort.*

καὶ ψυχρὸν φύσει τῷ θερμῷ καὶ ξηρῷ κατ' ἀμφοτέρας
τὰς δυσκρασίας ἐναντίως διακείμενον ἐναντίως διαι-
τήσομεν. ἄμεινον δ' οὐχ ἁπλῶς ἐναντίως εἰπεῖν, ἀλλὰ
κατὰ τὴν ἐκ τῆς κράσεως μόνης ἔνδειξιν, ὡς τήν γ' ἐκ
589K τοῦ πεπυρεχέναι τὸν | αὐτὸν πυρετὸν οὐχ ἑτέραν, ἀλλὰ
τὴν αὐτὴν ἕξουσιν ἔνδειξιν πάντῃ. λέλεκται γὰρ ἤδη
πολλάκις ὡς τὸ μὲν κατὰ φύσιν ἀεὶ χρῄζει τῶν ὁμοίων,[23] τὸ δὲ παρὰ φύσιν τῶν ἐναντίων· εἴ γε τὸ[24] μὲν
φυλάττεσθαι σκοπός, τὸ[25] δ' ἀναιρεῖσθαι. πυρετὸς μὲν
οὖν ἅπας τῶν παρὰ φύσιν ἐστὶ καὶ διὰ τοῦτο χρῄζει
διαίτης ὑγρᾶς καὶ ψυχρᾶς. αἱ κράσεις δὲ ποτὲ μὲν τῶν
ὁμοίων, ἔστι δ' ὅτε τῶν ἐναντίων χρῄζουσιν· αἱ μὲν
γὰρ ἄμεμπτοι τῶν ὁμοίων ἀεί, φυλάττειν γὰρ αὐτὰς
προσήκει· αἱ δύσκρατοι δ', ὡς κἀν τοῖς Ὑγιεινοῖς
ἐλέχθη, ποτὲ μὲν τῶν ὁμοίων, ἔστι δ' ὅτε τῶν ἐναντίων. καὶ τοῦτ' εὐλόγως πεπόνθασιν, οὔτε γὰρ ἀκριβῶς κατὰ φύσιν ὑπάρχουσι, πῶς γὰρ ἂν ἦσαν μεμ-
πτέοι; οὔτε πάντῃ παρὰ φύσιν ἔχουσι, διὰ παντὸς γὰρ
ἂν ἐνόσουν οἱ οὕτως κεκραμένοι· τῶν μὲν οὖν ὁμοίων
δέονται κατ' ἄλλα τέ τινα περὶ ὧν αὖθις εἰρήσεται καὶ
μέντοι καὶ κατὰ τὰ ἕλκη πάντα· δέδεικται γὰρ ἤδη
τοῦτο πολλάκις διὰ τῶν ἔμπροσθεν. τῶν ἐναντίων δὲ
κατὰ τὴν ὑγιεινὴν δίαιταν, ὡς ἐν τοῖς Ὑγιεινοῖς ἐπ-
590K εδείκνυμεν· οὐδὲ τούτων διαπαντός, | ἀλλ' ὅταν ἐπαν-
ορθοῦσθαι τὴν δυσκρασίαν αὐτῶν ἐθελήσωμεν. εἶναι

---

[23] K; τοῦ ὁμοίου B   [24] K; τῷ B
[25] K; τοῖς (sic) B

when he is affected oppositely in relation to both *dys-krasias* by heat and dryness. It is better not to say simply "oppositely," but "in relation to the indication from the *krasis* alone," as this is no different from what has kindled the same fever, but will have the same indication in every 589K way. I have already said, and often, that what is in accord with nature always needs like things, whereas what is contrary to nature always needs opposite things, if the objective in the first case is preservation and in the second removal. Every fever is contrary to nature and, because of this, needs a diet that is moist and cold. However, the *krasias* sometimes need like things and sometimes opposite things. Those that are without fault always need like things because it is appropriate to preserve them, whereas those that are *dyskratic*, as I said in *On the Preservation of Health*, sometimes need like things and sometimes need opposites.[34] And this is reasonable for those who have been affected, for they are not exactly in accord with nature. If they were, how would they be at fault? But they are not altogether contrary to nature because, if they were, those "mixed" in such a way would be continuously diseased. They require then similars, both in relation to certain other things—about which I shall speak again—and of course in relation to all wounds and ulcers. This has already been shown often by what has gone before. However, they do require opposite things in relation to a healthy regimen, as I showed in *On the Preservation of Health*, although not opposites constantly, but whenever 590K we wish their *dyskrasia* to be corrected.[35] Our aim ought

[34] Galen considers the *dyskrasias* in Book 6 of that work (VI.381–452K).   [35] See particularly *De sanitate tuenda*, VI.29K and VI.361–62K.

γὰρ ἡμῖν σκοπὸν διττὸν ἐπὶ τῶν ὑγιαινόντων, ἤτοι φυλάττειν ὑγιαῖνον τὸ σῶμα κατὰ τὴν ἀρχαίαν κρᾶσιν, ἢ καὶ ταύτην αὐτὴν βελτίω ποιεῖν.

οὕτως οὖν, οἶμαι, κἂν τῷ διαιτᾶν τοὺς τὸν ἐφήμερον πυρέξαντας ἤτοι τῇ πυρετώδει διαθέσει μόνῃ τὸ ἐναντίον, ἢ καὶ τῇ τοῦ κάμνοντος δυσκρασίᾳ παραλήψομαι. κατὰ μέντοι τὴν εὔκρατον φύσιν ἀναμφισβητήτως, ἀπὸ τοῦ πυρετοῦ μόνου τὴν τῶν ἐναντίων ἔνδειξιν λαμβάνοντες, ὑγρὰν καὶ ψυχρὰν εἰς τοσοῦτον ποιησόμεθα τὴν δίαιταν, εἰς ὅσον ἂν αὐχμωδέστερόν τε καὶ θερμότερον ἑαυτοῦ φαίνηται τὸ σῶμα γεγονός. ἀλλὰ τοῦτο μὲν εὔδηλόν τε καὶ εὐζήτητον, ἐπανέλθωμεν δ᾽ αὖθις ἐπὶ τὰς δυσκράτους φύσεις. ἐπειδὴ γάρ, ὡς ἐν τοῖς Ὑγιεινοῖς διῄρηται, τὰς τοιαύτας ἐγχωρεῖ μὲν καὶ φυλάττειν οἵας παρελάβομεν, ἐγχωρεῖ δὲ καὶ μετακοσμεῖν ἐπὶ τὸ βέλτιον, ἄμεινον οἶμαι τὴν ἐξάλλαξιν αὐτῶν τῆς διαίτης, ὅταν ἀμέμπτως ὑγιαίνωσιν, οὐχ ὅταν ἤτοι νοσῶσιν ἢ δυσαρεστῶνται, ποιεῖσθαι. χαίρουσι γὰρ αἱ φύσεις ἀεὶ τοῖς ἔθεσιν, ὡς |

591K   Ἱπποκράτης τε διὰ τῶν ἐναργεστάτων ἀπέδειξεν ἐν τῷ Περὶ διαίτης ὀξέων, ἅπασί τε σαφῶς φαίνεται τοῖς κατὰ φύσιν ἔχουσι. τοὺς γάρ τοι δυσαρεστουμένους, ὅτι μὲν ἔξω τοῦ χοροῦ τῶν κατὰ φύσιν ἐχόντων θετέον ἐστίν, οὐκ ἄν τις νοῦν ἔχων ζητήσειεν· εἴτε δ᾽ ἤδη παρὰ φύσιν, εἴτ᾽ οὐ φύσει μόνον ἔχουσι, τοῦτ᾽ ἄν τις σκέψαιτο. καὶ μὴν εἴπερ ἔνδειξίν τινα χρὴ κἀκ τῶν ἐθῶν λαμβάνειν, εἰ μὲν ἤδη φθάνοιμεν ἐκ πολλοῦ τὰς δυσκράτους φύσεις ἐναντίως²⁶ τῇ κρατούσῃ δυσκρα-

436

to be twofold in those who are healthy; to preserve the body in health in terms of the original *krasis* and to make this actual *krasis* better.

So then, I think, in treating those who are febrile with an ephemeral fever, I shall use either what is opposite to the febrile condition alone, or also what is opposite to the *dyskrasia* of the patient. And yet, in a nature that is indisputably *eukratic*, when we take the indication of the opposites from the fever alone, we shall make the diet moist and cold to the degree that the body has manifestly become drier and hotter than it should be. But as this is clear and readily ascertained, let me return once more to the *dyskratic* natures. For since, as defined in my work *On the Preservation of Health*, it is possible to preserve such *dyskratic* natures as we find them or to modify them for the better, I think it preferable to make the change of their regimen when they are faultlessly healthy and not when they are diseased or suffering distress. Natures always delight in those things that are customary, as Hippocrates   591K
showed with the greatest clarity in his *Regimen in Acute Diseases*,[36] and for all those who are normal this is clearly apparent. Certainly, in respect to those who are suffering, nobody in his right mind would doubt that they must be placed outside the group of those who are in accord with nature. What someone might consider is whether they are already abnormal or merely not normal. And even if it is necessary to take some indication from the customs, should we already be feeding the *dyskratic* natures over a long period beforehand in a manner opposite to the pre-

---

[36] See Hippocrates, *Regimen in Acute Diseases*, 31.

---

[26] B; ἐναντίον K

437

σίᾳ διαιτῶντες, οὕτω καὶ δυσαρεστούντων[27] πράξομεν· εἰ δὲ μὴ τῶν ὁμοίων τῇ κράσει, τὴν ἔνδειξιν ἕξομεν.

οὕτω δέ μοι δοκεῖ καὶ ὁ Ἱπποκράτης γινώσκειν, ἐπειδὰν λέγῃ, Αἱ ὑγραὶ πᾶσαι δίαιται τοῖσι πυρεταίνουσι ξυμφέρουσι, μάλιστα δὲ παιδίοισι καὶ τοῖσιν ἄλλοισι τοῖσιν οὕτως εἰθισμένοισι διαιτᾶσθαι. τοῖς μὲν γὰρ πυρεταίνουσιν ὑγραὶ δίαιται, καθόσον πυρεταίνουσιν ἐκ τῆς τῶν ἐναντίων ἐνδείξεως ὠφέλιμοι· τοῖς δὲ παιδίοις ἐκ τῆς τῶν ὁμοίων. ὡσαύτως δὲ καὶ τὴν ἀπὸ τῶν ἐθῶν ἔνδειξιν ἔλαβεν, ὡς καὶ τὴν ἀπὸ τῆς
592K ἡλικίας. ἥ τε γὰρ ἡλικία τῶν | κατὰ φύσιν οὖσα τῶν ὁμοίων ἑαυτῇ δεῖται διαιτημάτων, ἥ τε ἐκ τοῦ ἔθους ἔνδειξις, ὥσπερ καὶ ἡ ἐκ τῆς ἡλικίας ἐλήφθη· φυλάττειν γὰρ συμβουλεύει καὶ τοῦτο, καθάπερ καὶ τὴν οἰκείαν τῆς ἡλικίας κρᾶσιν. ὥστε καὶ ἡμᾶς ἀπὸ πάντων τῶν περὶ τὸν ἄρρωστον ἀεὶ χρὴ τὴν τοῦ συμφέροντος εὕρεσιν ποιουμένους ἐπισκοπεῖσθαι πηνίκα μὲν ἀλλήλαις αἱ ἐνδείξεις ὅμοιαι πᾶσαι, πηνίκα δ' ἐναντίαι γίνωνται. πασῶν μὲν γὰρ ὁμοίων οὐσῶν ἓν εἶδος ἀκριβὲς διαίτης συστησόμεθα· μαχομένων δ' ἀλλήλαις, τὰς ἰσχυροτέρας τε καὶ πλείους προκρινοῦμεν, ὥσθ' ἡμῖν ἐπικρατεῖσθαι μὲν ἐκ τούτων τὸ τῆς διαίτης εἶδος, οὐ μὴν παρῶφθαι οὐδὲ τὸ ἐναντίον. ἡ δ' ἐπικράτησις ἐνίοτε μὲν ἑκατέρων κεραννυμένων εἰς ἑνὸς βοηθήματος ὕλην, ἐνίοτε δ' ἐν μέρει παραλαμβανομένων γίνεται.

[27] K; δυσαρεστηθέντων B

438

vailing *dyskrasia*, so too will we do this when they are suffering malaise. If not, we shall have the indication of like things to the *krasis*.

It seems to me that Hippocrates knew this too when he said: "All the moist diets are suitable for those who are febrile, but especially for children and others who are accustomed to being fed in this way."[37] For to those who are febrile, insofar as they are febrile, moist diets are beneficial from the indication of the opposites, whereas for children, they are beneficial from the indication of similars. And just as he took the indication from the customs, so too did he from the age. For age, being one of those things that is normal, needs a regimen similar to itself, and the indication is taken from custom, just as it also is from the age; for he recommends to preserve this as well, just as he [recommends] the proper *krasis* for the age. So it behooves us always to consider all the things concerning the sick person when we are making the discovery of what is beneficial, and to know at what precise point of time the indications are all similar to each other and at what precise point of time they are opposite. For when they are all similar we shall contrive one exact kind of diet, while when they contend with each other we shall distinguish those that are stronger and more numerous, so that we choose from them the kind of diet that prevails, but certainly without the opposite being disregarded. Sometime the prevalence of each of the components of the *krasis* comes to the material of a single remedy and sometimes it is when they are taken individually.

592K

[37] See Hippocrates, *Aphorisms*, I.16.

κοινὸς μὲν οὗτος ὁ λόγος ἁπασῶν τῶν ἐμπεπλεγμέ-
νων ἀλλήλαις ἐναντίων ἐνδείξεων. ὅπως δ᾽ αὐτὸν χρὴ
μεταχειρίζεσθαι, ποτὲ μὲν αὐτῶν τῶν νοσημάτων ἐπι-
πλεκομένων, ἔστι δ᾽ ὅτε ἑνὸς μὲν ὑπάρχοντος τούτου, |
593K   τὴν δὲ ἔνδειξιν ἐναντίαν ποιουμένου τῇ φυσικῇ κράσει
τοῦ κάμνοντος, ἢ τῇ νῦν ἐπικτήτῳ, τῇ θ᾽ ἡλικίᾳ καὶ τῷ
ἔθει καὶ τῷ χωρίῳ καὶ τῇ ὥρᾳ, δι᾽ αὐτῶν τῶν κατὰ
μέρος ἐπιδείξω. νῦν οὖν ἡμῖν προκειμένου κατὰ τὸν
λόγον εὑρεῖν ἐπιτήδειον δίαιταν ἑκάστῃ τῶν εἰρημένων δυσκρασιῶν, ὁ κοινὸς τρόπος τῆς μεθόδου προ-
χειρισθεὶς ὡς ἐπὶ παραδείγματος ἐξεταζέσθω. φέρε
γὰρ ἁλῶναι πυρετῷ δύο ἀνθρώπους ἐν εὐκράτῳ μὲν
χωρίῳ γεγενημένους καὶ τεθραμμένους, οὐκ ὄντας δ᾽
ἐν αὐτῷ νῦν, ἀλλ᾽ ἐν μὲν θερμῷ καὶ ξηρῷ τὸν ἕτερον,
ἐν ὑγρῷ δὲ καὶ ψυχρῷ τὸν ἕτερον· εἶναι δὲ καὶ τῇ
κράσει διαφέροντας αὐτοὺς καὶ τῇ φύσει καὶ τοῖς
ἔθεσι καὶ ταῖς ἡλικίαις, εἰ δὲ βούλει καὶ κατὰ δια-
φερούσας ὥρας καὶ καταστάσεις ἀρρωστεῖν, ὥστε τὸν
μὲν θερμὸν καὶ ξηρὸν φύσει καὶ τῇ διαίτῃ πάσῃ
τοιαύτῃ συνειθίσθαι καὶ νεανίσκον εἶναι καὶ χειμῶνος
νοσεῖν ἐν ὑγρῷ καὶ ψυχρῷ χωρίῳ καὶ καταστάσει
τοιαύτῃ· τὸν δὲ ὑγρόν τε καὶ ψυχρὸν ἐν ἔθει τε διαίτης
ὁμοίῳ καὶ παῖδα τὴν ἡλικίαν· καὶ νοσεῖν δὲ θέρους ἐν
θερμῷ καὶ ξηρῷ χωρίῳ καὶ καταστάσει τοιαύτῃ.
594K   τίνες οὖν | ἐν²⁸ ἑκατέρῳ τούτων αἱ ἐνδείξεις; ἡ μὲν
ἀπὸ τοῦ πυρέττειν ἀμφοτέροις κοινή, τῶν ὑγραινόντων
τε καὶ ψυχόντων δεομένη, τῶν δ᾽ ἄλλων οὐκέτ᾽ οὐδεμία
κοινή. τῷ μὲν γὰρ προτέρῳ διά τε τὴν φύσιν καὶ τὴν

This argument is common when all the opposing indications are intermingled with each other. How we should manage the patient when sometimes the diseases themselves are combined and sometimes there is this particular one, producing the indication opposite to the natural 593K *krasis* of the patient, or to the *krasis* now acquired, or to the age, custom, place and season, I shall show, treating these things individually. Therefore, since what now lies before us in the discussion is to discover something suitable as a diet for each of the *dyskrasias* I have spoken about, let the general form of the proposed method be examined, as in the following example. Suppose there are two men seized with fever, men born and brought up in a *eukratic* region but not now in it, one being in a place that is hot and dry and the other in a place that is moist and cold. Suppose they also differ in *krasis*, nature, customs, and age, and if you wish, fall sick during different seasons and under different climatic conditions, so that one is a young man, hot and dry in nature and accustomed to every sort of diet, who is sick in winter in a moist and cold place, and with the same climatic conditions, whereas the other is a child in age, moist and cold in custom and with a diet that is similar, who is sick in summer in a hot, dry place and with the same climatic conditions.

What, then, are the indications in each of these? The 594K indicator from being febrile is common for both patients, which is the need for moistening and cooling agents. None of the other indicators is any longer common. For the first

28 K; ἐν *om*. B

441

ἡλικίαν ἐν ὁμοίοις τε αὐταῖς[29] διαιτήμασιν εἰθισμένῳ
ἐπιτήδεια τὰ ξηρότερα καὶ θερμότερα τῇ κράσει σιτία·
καὶ μέντοι καὶ διὰ τὴν ὥραν τε καὶ τὴν κατάστασιν
καὶ τὸ χωρίον· οὐ γὰρ τῶν ὁμοίων ἐστὶν ἡ ἀπὸ τούτων
ἔνδειξις, ἀλλὰ τῶν ἐναντίων. τῷ δ' ἑτέρῳ τῶν ὑποκειμένων διὰ μὲν τὴν ἡλικίαν καὶ τὴν φύσιν καὶ τὸ
ἔθος τῶν ὁμοίων αὐτοῖς· ὑγρῶν μὲν καὶ συμμέτρως
θερμῶν διὰ τὴν ἡλικίαν, ὑγρῶν δὲ καὶ ψυχρῶν διά τε
τὴν φύσιν καὶ τὸ ἔθος· διὰ δὲ τὸ χωρίον καὶ τὴν ὥραν
καὶ τὴν κατάστασιν τῶν ἐναντίων. ἅπερ ἐστὶ δήπου
καὶ αὐτὰ ψυχρότερα καὶ ὑγρότερα τοσούτῳ τῶν εὐκράτων, ὅσῳ πέρ ἐστι καὶ τὰ χωρία τῶν εὐκράτων
θερμότερα καὶ ξηρότερα· παρὰ μὲν γὰρ τῶν ἐθῶν ἡ
ἔνδειξις ἀεὶ τῶν ὁμοίων ἐστὶν ἄχρις ἂν νοσῶσιν οἱ
ἄνθρωποι. παρὰ δὲ τῶν ἡλικιῶν καὶ φύσεων καὶ χωρῶν ὡρῶν τε καὶ καταστάσεων εὐκράτων μὲν οὐσῶν
595K ἔνδειξις ἀεὶ | τῶν ὁμοίων διαιτημάτων καὶ βοηθημά-
των ἐστίν, ἄν θ' ὑγιαίνωσιν ἄν τε νοσῶσι. παρὰ δὲ
τῶν δυσκράτων ὡρῶν μὲν καὶ καταστάσεως καὶ χωρῶν ἔνδειξις τῶν ἐναντίων· ἡλικιῶν δὲ καὶ φύσεων οὐχ
ἁπλῶς, ἀλλὰ σὺν τῷ ἔθει. σκοπεῖσθαι δ' ἐπὶ τῶν
ἐναντίων ἐνδείξεων οὐδὲν οὕτω προσῆκεν ὡς τὸ μέγεθος ἑκάστου τῶν συνεισφερόντων, ὅπερ ἐδείκνυμεν ἐν
τοῖς Ὑγιεινοῖς ὑπομνήμασι διττόν· ἕτερον μὲν αὐτοῖς
τοῖς τῶν ἐνδεικνυμένων ἀξιώμασι μετρούμενον, ἕτερον
δὲ ταῖς οἰκείαις αὐτῶν ὡς ἂν εἴποι τις οὐσίαις, ἃς ἐκ

---

[29] B; ὁμοίοις ἑαυταῖς K

442

patient, because of his nature and age, is used to suitable foods that are drier and hotter in *krasis* with similar conditions to them. And of course, due to the season, climatic conditions and place, the indication is not from those things that are similar, but from those that are opposite. In the second of the patients proposed, due to his age, nature, and custom, [the need is] for the proposed things to be similar to them—that is, for foods that are moist and moderately hot due to his age, but for those that are moist and cold due to his nature and custom. But due to the place, the season and the climatic conditions, [the need is] for the opposites. These things are, of course, to be colder and moister than what is *eukratic* to the extent that the places are warmer and drier than what is *eukratic*. For the indication from the customs is always of similars whenever people are sick. The indications from age, nature, place, season and climatic conditions, when these are *eukratic*, are always of similars in terms of diets and remedies, whether people are healthy or sick. The indications from *dyskratic* seasons, climatic conditions and places, are of opposites. Indications from age and nature are not absolute, but in conjunction with the customs. In the case of the contrary indications, nothing is more appropriate to consider than the magnitude of each of the things we are administering at the same time. This I showed in the treatise *On the Preservation of Health* to be twofold:[38] one is measured by the actual worth of the things indicated, and one is measured by what one might speak of as their proper substances,

595K

---

[38] See Galen's *De sanitate tuenda*, VI.34K.

τοῦ παραβάλλειν τῷ συμμέτρῳ τε καὶ κατὰ φύσιν
ἔφαμεν χρῆναι λαμβάνειν. εἰ μὴ γὰρ οὕτω τις ἐξετά-
σειε τῶν ἐνδείξεων τὰς δυνάμεις, οὐχ οἷόν τε τὸ
μέτρον ἐξευρεῖν τῆς διαίτης. ἐνδεικνυμένου γὰρ ἀεὶ
τοῦ μὲν πυρετοῦ τὸ ψύχειν καὶ ὑγραίνειν, τῶν δ' ἄλλων
ἑκάστου, καθότι προείρηται, συμφωνούσης μὲν αὐτῶν
τῆς ἐνδείξεως οὐδὲν ἂν εἴη πρᾶγμα, διαφωνούσης δὲ
κατὰ τὰ πλείω καὶ μείζω τυποῦσθαι χρὴ τὴν δίαιταν.

596K   ἐπὶ γοῦν τῶν ὑποτεθέντων ἀρτίως ὅτι | μὲν πυρέτ-
τουσιν ἔνδειξις κοινὴ τῶν ὑγραινόντων καὶ ψυχόντων
ἐστίν, οὐ μὴν ἀπό γε τῶν ἄλλων. ἀλλ' ᾧ μὲν ἥ θ'
ἡλικία καὶ ἡ κρᾶσις ἅμα τῷ ἔθει καὶ ἡ χώρα καὶ ἡ
ὥρα καὶ ἡ κατάστασις τῶν ὑγραινόντων καὶ ψυχόντων
ἔχει τὴν ἔνδειξιν, οὐ μόνον ἀναμφισβήτητόν ἐστι τὸ
χρῆναι τοῦτον ὑγραίνειν καὶ ψύχειν, ἀλλὰ καὶ τὸ μὴ
μικρῶς μηδ' ὡς ἔτυχεν, ἰσχυρῶς δὲ πάνυ καὶ μεγάλως,
ἁπάντων γε τῶν σκοπῶν ἐνδεικνυμένων ταὐτόν. ᾧ δ' ἡ
τῶν ἄλλων ἁπάντων ἔνδειξις οὐχ ὑγρῶν καὶ ψυχρῶν
ὁμοίως τῷ πυρετῷ τῶν διαιτημάτων ἐστίν, ἀλλὰ τῶν
ἐναντίων αὐτοῖς τῶν θερμῶν καὶ ξηρῶν, ἀμφισβήτη-
σις γίγνεται καὶ μάχη τῶν ἐνδείξεων πρὸς ἀλλήλας,
ἡνίκα μάλιστά ἐστι χρεία τοῦ παραβάλλειν τόν τε
ἀριθμὸν καὶ τὰ μεγέθη τῶν ἐνδείξεων. ἡ γοῦν ἀπὸ τοῦ
πυρέττειν ἔνδειξις ἀριθμῷ μὲν λείπεται τῶν ἄλλων,
ὅταν ἅπασαι τὸν ἐναντίον ὑφηγῶνται τρόπον πράττειν, ὑπερέχει δὲ μεγέθει τῷ κατὰ τὴν ἀξίαν, ἢ τὴν
δύναμιν, ἢ ὅπως ἄν τις ὀνομάζειν ἐθέλοι καὶ μάλισθ'
597K ὅταν ἐπὶ πολὺ τοῦ κατὰ φύσιν ἐξεστήκοι.[30] | δύο γὰρ

444

which I said must be taken from a comparison with what is moderate and normal. If one does not examine the potencies of the things indicated in this way, it is not possible to discover the measure of the diet. For since the fever always indicates cooling and moistening, whereas for each of the other factors, just as I previously stated, when their indication is in agreement, there would be no problem, but when the indication does not agree, it is necessary to fashion the diet in terms of more and greater.

Anyway, in the features just now proposed, because there is an indication for those who are febrile of moistening and cooling agents, there is no indication for other agents. But in someone for whom age and *krasis* along with custom, place, season and climatic conditions have the indication for moistening and cooling agents, not only is this need to moisten and cool indisputable, and not slightly or contingently, but very strongly and vigorously, because all the indicators indicate the same thing. In someone in whom the indication of all the other things is not of a diet of foods that are moist and cold, as for a fever, but of the opposites to these—that is, of hot and dry [foods]—conflict and contention of the indications with each other arise, particularly when there is a need to collate the number and magnitude of the indications. At any rate, the indication from being febrile is numerically lacking compared to the others, whenever they all direct us to act in the opposite manner, although it is overriding in terms of value, or potency, or however someone might wish to term it, and particularly whenever it departs markedly from normal.

596K

---

30 B; ἐξέστηκε K

αὐτῷ τόθ᾽ ὑπάρξει μεγέθη, τό τ᾽ ἐκ τοῦ γένους τῶν πυρετῶν, τὸ κοινὸν ἅπασι, τό τ᾽ ἴδιον οἰκεῖον αὐτοῦ. καὶ γὰρ ὅτι πυρετός ἐστιν, ἐπιστρέφει μᾶλλον ἡμᾶς πρὸς ἑαυτὸν ἁπάντων τῶν ἄλλων· καὶ ὅτι τὸ μέγεθος αὐτῷ τὸ οἰκεῖον ἀξιόλογον ὑπάρχει καὶ διὰ τοῦτο κατακρύπτει τε καὶ συσκιάζει τὰς ἐναντίας ἐνδείξεις. εἰ δὲ μικρὸς ὑπάρχει, δύνανθ᾽ αἱ ἄλλαι πᾶσαι συντεθεῖσαι πρὸς αὐτὸν ἀντεξετάζεσθαι. καὶ εἴπερ ἴσαι φανεῖεν, οὔθ᾽ ἡ τοῦ πυρετοῦ κρατήσειεν ἂν ἔνδειξις ὡς ὑγραίνειν καὶ ψύχειν, οὔθ᾽ ἡ τῶν ἄλλων ὡς θερμαίνειν καὶ ξηραίνειν, ἀλλὰ τὸ μέσον εἶδος τῆς διαίτης ἐπὶ τῶν οὕτως ἐχόντων σωμάτων ἐκλεξόμεθα. κρατούσης δὲ τῆς ἑτέρας ἐνδείξεως εἰς τοσοῦτον ἀποχωρήσομεν τῆς εὐκράτου καὶ μέσης τῆς διαίτης, εἰς ὅσον ὑπερέχει τῶν ἐναντίων τὰ κρατοῦντα.

γεγυμνασμένον τε οὖν εἰς ταῦτα πάντα καὶ φύσει συνετὸν εἶναι χρὴ τὸν ἰατρόν, ἵν᾽ ἑκάστης ἐνδείξεως ἀκριβῶς ἐκλογισάμενος τὴν δύναμιν, ἀλλήλαις τε παραβάλλων ἁπάσας, ἕν τι κεφάλαιον ἀθροίσῃ τῶν οἰκείων τῷ | κάμνοντι διαιτημάτων. αἱ μὲν οὖν ἀπὸ τῶν ἐφημέρων πυρετῶν ἐνδείξεις ἐκ τούτων τῶν σκοπῶν λαμβάνονται, καὶ σπάνιον εἴ ποτε ἀπὸ τοῦ μέτρου τῆς τοῦ κάμνοντος δυνάμεως. αἱ δὲ τῶν πολυημέρων ἐξ ἀνάγκης μὲν καὶ τὴν δύναμιν προσλαμβάνουσιν· οὐ μικρὸς γὰρ ἐπ᾽ ἐκείνων ὁ ἀπὸ ταύτης σκοπός· οὐδὲν ἧττον δὲ καὶ τὴν γεννῶσαν αἰτίαν τοὺς πυρετούς, πλὴν τῶν ἑκτικῶν· οὐδεμία γὰρ ἀνάπτει τούτους αἰτία, καθάπερ οὐδὲ τὸ δεδεγμένον ἑκτικῶς εἰς ἑαυτὸ σῶμα

For two aspects of magnitude will exist in it at that time: 597K
one will be from the class of fevers, which is common to all
fevers, and the other will be that which is specific and peculiar to it. In that it is a fever, it turns us more toward itself
than all the other factors do, and because the magnitude
specific to it is significant, it thereby conceals and obscures
the opposite indications. If, however, it is slight, all the
other indications can be put together with it and compared. And, indeed, if they seem to be equal—that is, if
neither the indication of the fever prevails so as to moisten
and cool, nor that from the other things to heat and dry—
we shall choose an intermediate kind of diet in the case of
bodies that are like this. When, however, the other indications prevail, we shall depart from the *eukratic* and intermediate diet to the extent that the prevailing factors over-
ride their opposites.

Therefore, it is necessary for the doctor to be practiced
in all these things and naturally wise so that, after determining precisely the "strength" of each indication, and
comparing them all with each other, he may collate the
specific diets for the patient under one heading. Thus 598K
the indications from the ephemeral fevers are taken from
these indicators, and seldom, if ever, from the measure of
the capacity of the patient. But the indications from the
polyhemeral (chronic) fevers necessarily take into account
the capacity as well, for the indicator from this in those
fevers is of no small significance, and no less also is the
generating cause, except for the hectic fevers, for no cause
kindles these, just as a body (whether it be stone or wood)

447

GALEN

τὴν τοῦ πυρὸς θερμασίαν, εἴτε λίθος, εἴτε ξύλον ὑπάρχει. κοινὸν γὰρ αὐτοῖς τὸ μηκέτι γίνεσθαι πρὸς τοὺς
ἐφημέρους ἐστί, διαλλάττοντας δὲ ἐν τῷ τοὺς μὲν
ἐφημέρους κατὰ σχέσιν ἔχειν τὴν θέρμην, τοὺς δ᾽
ἑκτικοὺς ἐν ἕξει δυσλύτῳ. δύναιτο δ᾽ ἄν τις φάναι καὶ
τὴν ἀπὸ τῆς δυνάμεως ἔνδειξιν ἀεὶ παρεῖναι κατά τε
τἆλλα σύμπαντα νοσήματα καὶ τοὺς ἐφημέρους πυρετούς, ἡμᾶς δ᾽ ἐν τῷ νῦν λόγῳ μηδὲν αὐτῆς δεηθῆναι,
διότι πρώτως μὲν οὐκ ἐνδείκνυται, κατὰ συμβεβηκὸς
δέ, περὶ οὗ συμβεβηκότος ἐν τοῖς ἑξῆς διοριῶ.

which has easily received into itself the heat of the fire does not. What is common to these in relation to the ephemeral is that the cause no longer exists. Where they differ is in the ephemeral fevers having heat as a transient state, whereas the hectic fevers have it as an enduring state that is difficult to resolve. Someone could also say that the indication from the capacity is always present in all other diseases and in ephemeral fevers; but we have no need of it in the present discussion because it is not primarily but contingently indicated. I shall draw the distinction about contingency in what follows.

ΒΙΒΛΙΟΝ Ι

1. Τῶν ἐφημέρων πυρετῶν, ὦ Εὐγενιανέ, τὴν τῆς θεραπείας μέθοδον ἐν τῷ πρὸ τούτου γράμματι διῆλθον ὀγδόῳ τῆς ὅλης ὄντι πραγματείας. ἔστι δὲ τοῦ γένους τοῦδε τῶν πυρετῶν εἷς ὁ παροξυσμός, ἡμέρᾳ μιᾷ περιγραφόμενος ὡς τὸ πολὺ κατά γε τὴν ἑαυτοῦ φύσιν. ὅσα γὰρ ἢ διὰ τὴν τῶν ἰατρῶν ἀμαθίαν ἢ διὰ τὴν τῶν καμνόντων ἀκολασίαν ἢ διὰ τὴν τῶν ὑπηρετούντων πλημμέλειαν ἁμαρτανόμενα περὶ τοὺς τοιούτους πυρετοὺς εἰς ἕτερόν τι γένος αὐτοὺς μεθίστησιν,

οὐ κατὰ τὴν οἰκείαν αὐτῶν ἀποβαίνουσι | φύσιν. οὐδὲν γὰρ δήπου θαυμαστὸν ἐκ τῶν ἁμαρτανομένων οὐ μηκύνεσθαι μόνον ἢ μεταπίπτειν εἰς ἕτερόν τι γένος ἡντινοῦν τῶν νόσων, ἀλλὰ καὶ γεννᾶσθαι νῦν ἔμπροσθεν οὐκ οὖσαν. ὅπου γὰρ οὐδὲ τοῖς ὑγιαίνουσιν ἁμαρτάνειν ἀσφαλές, σχολῇ γ᾽ ἀκίνδυνον ἄν ποτε τοῖς κάμνουσι γένοιτο. λοιπῶν δ᾽ ὑπαρχόντων δυοῖν γενῶν, ὡς ἐν τῷ Περὶ τῆς διαφορᾶς τῶν πυρετῶν ὑπομνήμασιν ἐπιδέδεικται, τὸ μὲν ἕτερον ἐπὶ χυμοῖς σηπομένοις ἐδείκνυτο γίνεσθαι, τὸ δ᾽ ἕτερον αὐτὰ τὰ στερεὰ τοῦ ζῴου μόρια κατειληφὸς ἑκτικὸν ὀνομάζεται. μεταπίπτουσι δ᾽ εἰς ἑκάτερα γένη πολλάκις οἱ

BOOK IX

1. I have gone over the method of treatment of the ephemeral fevers, Eugenianus, in the book prior to this one,
which is the eighth of the whole treatise. In this class of fevers there is a single paroxysm limited to one day for the
most part by virtue of its own nature. All mistakes that are
made with regard to fevers of this kind, whether due to the
ignorance of doctors, the intemperance of patients, or the
fault of those in attendance, change them to some other
class and take them away from their own specific nature.
There is, of course, nothing remarkable [about the fact]
that, from these mistakes, not only do they last longer or
change into some other class of diseases altogether, but
they now also generate a nature that did not exist before.
For where it is not safe to make mistakes in those who are
healthy, it would hardly be without danger if this were to
occur at any time in those who are ill. Of the remaining two
classes [of fevers] that exist, one was shown to occur due to
putrefying humors, and the other, called "hectic," to involve the actual solid parts of the organism, as has been
shown in the treatise *On the Differentiae of Fevers*.[1] The
ephemeral fevers often change to either one of the [other]

[1] See Galen, *De differentiis febrium*, VII.304K ff., and 718K in
the present work.

GALEN

ἐφήμεροι πυρετοί, καθάπερ ἔμπροσθεν ἐδείκνυμεν, εἰ
μή τις αὐτοὺς προσηκόντως μεταχειρίζοιτο.

ὥσπερ δ᾽ ἐν τοῖς ἄλλοις ἅπασι πράγμασιν ὅσα
κοινωνεῖ ταῖς φύσεσιν ἑνὶ περιλαμβανόμενα γένει
καὶ διὰ τοῦτ᾽ εἰς ἄλληλα μεταπίπτειν δυνάμενα, τρίτη
τις ἰδέα κατὰ τὴν μετάπτωσιν αὐτῶν συνίσταται,
μηδετέρου τῶν ἄκρων ἀκριβοῦσα τὸν τύπον, οὕτω
κἂν ταῖς τῶν ἐφημέρων πυρετῶν μεταπτώσεσι φαν-
τασθείη τις ἂν ἕτερόν τι γένος ἐν τῇ μεταπτώσει
συνίστασθαι, μηδετέρῳ τῶν ἄκρων ταὐτόν. καί μοι |
601K δοκεῖ περὶ τοῦδε πρῶτον διοριστέον εἶναι, ἵν᾽ ὥσπερ
καὶ τἄλλα καὶ τοῦτο γνωρίζοιτό τε συνιστάμενον
αὐτίκα καὶ κωλύοιτο κατὰ τὴν πρώτην γένεσιν, ὡς
μηδὲ δυσίατον ἢ ἀνίατον αὐξηθὲν γενέσθαι. μέσον
δ᾽ αὐτὸ συνιστάμενον οὔτε γνωρίσαι δυνατὸν οὔτε
ἰάσασθαι τοῖς ἀγνοοῦσι τὴν ἰδέαν τῶν ἄκρων ὧν
ὑπάρχει μέσον. ἀλλά σοί γ᾽ ἐπισταμένῳ τά τε τῶν
ἐφημέρων πυρετῶν γνωρίσματα καὶ τὰ τῶν ὑπολοί-
πων δύο γενῶν, ἐπειδὰν ἤτοι νοθεύηται κατά τι ἢ
παρεμφαίνηται τὰ τῶν ἄλλων, οὐδὲν ἂν εἴη χαλεπὸν
ὑποπτεύειν ἤδη τὴν μετάπτωσιν.

ἐπ᾽ ἐγκαύσεσι μὲν οὖν, ἢ ἀπεψίαις, ἢ κόποις, ἢ
ἀγρυπνίαις, ἢ πάθεσι ψυχικοῖς, ἢ βουβῶσιν τοῖς διά
τινα αἰτίαν προφανῆ γεγενημένοις, μέχρι τρίτης ἡμέρας οὐκ ἄν ποτε προέλθοι πυρετοῦ παροξυσμός, ἄνευ
τοῦ πλημμεληθῆναί τι περὶ τὸν κάμνοντα. τοῖς δ᾽ ἐπὶ
στεγνώσει πυρέξασιν ἐγχωρεῖ καὶ ταύτης ἐξωτέρω
προβῆναι τὸν παροξυσμόν. ἐπειδὴ γὰρ ἡ στέγνωσις

452

two classes, as I showed before, unless someone manages them properly.

As in all the other instances, with things that form a community by virtue of their natures and are included in a single class, and because of this are able to change into each other, there exists a third form in relation to their change which does not conform precisely to the pattern of either of the extremes. In the same way, even in the changes of the ephemeral fevers, someone might envisage that another class exists within the change, which is not the same as either of the extremes. It seems to me that we 601K must make a distinction about this first so that, just as you may recognize the other things, you may also recognize this as soon as it exists and may prevent it at its inception, so that it does not increase to the point of becoming difficult or impossible to cure. It is not possible for those who do not know the forms of the extremes of which something is the intermediate to recognize and cure the intermediate form itself when it does exist. But to you, who do know both the signs of the ephemeral fevers and those of the remaining two classes, whenever they either depart from normal to some extent or the features of others appear alongside them, it would not be difficult to already suspect the change.

Therefore, with respect to heatstroke, apepsia, fatigue, insomnia, mental affections or buboes that have occurred due to some clearly apparent cause, the paroxysm of the fever would never come until the third day unless something was done wrongly concerning the patient. In those with fever due to obstruction of transpiration it is also possible for the paroxysm to advance beyond this. I stated with good

453

ἤτοι φραχθέντων ἢ μυσάντων γίγνεται τῶν πόρων,
ἢ καὶ τοῦ σώματος αὐτοῦ πιληθέντος ἀμετρότερον,
εὐλόγως ἔφαμεν ἐπί τε κρύει καὶ λουτροῖς ἤτοι τῇ |
602K θίξει ψύχουσιν, ἢ τῇ δυνάμει καὶ φαρμάκοις τοῖς
τοιούτοις, ἔτι δὲ τοῖς ξηραίνουσι τὸ δέρμα, καθάπερ
ἡλίῳ διακαεῖ καὶ τοῖς πιλοῦσιν, ὡς τῇ σκληρᾷ τρίψει
μετὰ τῶν ψυχόντων κατέχεσθαι μὲν ἐντὸς τοῦ σώματος τὰς οἷον καπνώδεις ἢ λιγνυώδεις ἀπορροάς,
ἀνάπτεσθαι δ' ἐπ' αὐτοῖς τὸν πυρετόν.

ἀλλ' ἐπὶ τούτοις μέν, ὡς εἴρηται, πίλησίν τέ τινα
γίγνεσθαι τῶν σωμάτων αὐτῶν καὶ μύσιν τῶν πόρων,
ἔμφραξιν δὲ διὰ γλίσχρους ἢ παχεῖς ἢ πολλοὺς
χυμούς. καὶ τοίνυν καὶ τοὺς πυρετοὺς ὅσοι μὲν ἐπὶ
μύσει τῶν πόρων ἢ πιλήσει προσφάτῳ τῶν σωμάτων
ἐγένοντο, παύεσθαί τε μετὰ τὸν πρῶτον παροξυσμόν,
ἐάν τις ὀρθῶς αὐτοῖς προσφέρηται καὶ ὄντως τούτους
ἐφημέρους γίγνεσθαι καὶ προσαγορεύεσθαι· ὅσοι δ'
ἐπ' ἐμφράξει τὴν γένεσιν ἔσχον, εἰ μὲν ὀλίγη τις αὕτη
συσταίη, λύεσθαι καὶ τούτους ὁμοίως γε τοῖς ἄλλοις
ἐφημέροις, ἐὰν ἰάσηταί τις αὐτοὺς προσηκόντως· εἰ δ'
ἰσχυρῶς οἱ χυμοὶ σφηνωθεῖεν, ὡς μοχλείας δεῖσθαι
σφοδροτέρας, ἐκτείνεσθαι μὲν ἀνάγκη τοὺς τοιούτους
πυρετοὺς ὑπὲρ τὴν μίαν ἡμέραν, ἐκπεπτωκέναι δὲ
603K δοκεῖν ἤδη τοῦ γένους τῶν ἐφημέρων. | πῶς γὰρ ἂν
ἐφήμερος ἔτι δύναιτο λέγεσθαι πυρετὸς ὁ μέχρι τρίτης ἡμέρας ἐκτεινόμενος; ἐν γὰρ τῇ δευτέρᾳ τῶν
ἡμερῶν προσήκει παύεσθαι τὸν τοιοῦτον πυρετόν,
ὅσον ἐπὶ τῇ προσηγορίᾳ. λέλεκται γὰρ ἔμπροσθεν ὡς

reason that, when restriction of transpiration occurs due to either closure or blockage of the pores, or when there is rather disproportionate thickening of the body itself, fever is kindled due to the following—chilling and baths, those things that cool by contact or potency, medications of 602K this sort, and further, those things that dry the skin like burning sun, and those that cause condensation, like hard rubbing with cooling agents that retain the smoky, as it were, and sooty outflows within the body.

But through the agency of these things, as I said, there occurs a certain contraction of bodies themselves and a closure of the pores, and blockage by viscid, thick or abundant humors. And therefore, those fevers which have occurred due to closure of the pores or recent contraction of bodies stop after the first paroxysm, if someone manages them correctly; they are truly ephemeral fevers and are termed such. However, those fevers that have their genesis in blockage, if this is slight, also resolve in the same way as the other ephemeral fevers, if someone treats them appropriately. But if the humors are severely impacted so as to need quite strong dislodgement, such fevers inevitably extend beyond one day and seem to have already fallen outside the class of ephemeral fevers. How could someone 603K still speak of an ephemeral fever which extends as long as the third day? For it is right that such a fever ceases on the second day, as its name implies. I have said before that a

τεττάρων καὶ εἴκοσιν ὡρῶν ὁ χρόνος ἐστὶ τῆς οὕτω
λεγομένης ἡμέρας, ὡς συναριθμεῖσθαι καὶ τὴν νύκτα
τῷ ὀνόματι τῆς ἡμέρας. καὶ μὴν εἰ μήτε σῆψις εἴη
χυμῶν μήτ' αὐτὸ τὸ σῶμα τὴν πυρεκτικὴν θερμασίαν
ἑκτικῶς ἀναδέξαιτο, τῶν δύο γενῶν ἐκπεπτωκὸς ἂν
εἴη. ἢ τοίνυν ἄλλο γένος ἐπὶ τοῖς εἰρημένοις τρισὶν
ἀναγκαῖόν ἐστι λέγειν ἐν τῇ πρώτῃ διαιρέσει τῶν
πυρετῶν, ἢ καταφρονοῦντα τῆς προσηγορίας ἐφημέρους ὀνομάζειν αὐτούς. οὐ γὰρ δὴ τῆς γ' οὐσίας ἦν
αὐτῆς τῶν τοιούτων πυρετῶν ὁ ἐφήμερος ὄνομα· προσηγορίας δ' οἰκείας ἀποροῦντες ἕνεκα σαφοῦς καὶ συντόμου διδασκαλίας ἀπὸ τοῦ πολλάκις αὐτοῖς ἑπομέ-
νου τοὔνομα ἐθέμεθα. φύσις μὲν οὖν αὐτῶν ἡ αὐτὴ
τοῖς ἐφημέροις ἐστίν, ὄνομα δὲ οὐ ταὐτόν.

2. Ἐφ' ὧν γὰρ ὁ παροξυσμὸς εἷς ἀπ' ἀρχῆς ἄχρι
παντὸς διαμένων εἰς πολλὰς ἡμέρας ἐκτείνεται συν-
604K όχους | ὀνομάζουσι τοὺς τοιούτους πυρετούς, οὐχ Ἑλ-
ληνικῷ μὲν ὀνόματι χρώμενοι, σολοικίζειν δ' ἑλόμενοι
μᾶλλον ἢ καταλιπεῖν ἀνώνυμον αὐτῶν τὴν ἰδέαν. ἀλλ'
ὥσπερ ἰδέα μία τῶν τοιούτων ἐστὶ πυρετῶν, ἀφ' ἧς
ὀνομάζουσι συνόχους αὐτούς, οὕτως ἡ φύσις οὐκέθ'
ἁπλῆ καὶ μία. τινὲς μὲν γὰρ αὐτῶν ἐναργὲς ἔχουσι τὸ
τῆς σήψεως σημεῖον· ἔνιοι δ' οὐδ' ὅλως ἔχουσιν, οὓς
ἐκ τοῦ γένους τῶν ἐφημέρων πυρετῶν ἔφαμεν εἶναι.
ἐπειδὴ γὰρ ἡ τοῦ παροξυσμοῦ παῦλα διαπνεομένων
εἴωθε γίγνεσθαι τῶν ζεσάντων χυμῶν, οὐ διαπνεῖται
δ' ἐπὶ ταῖς ἰσχυραῖς στεγνώσεσιν, ἀναγκαῖόν ἐστι
πολυήμερον γίγνεσθαι τὸν παροξυσμόν. ὅταν οὖν

time of twenty-four hours is spoken of as a day because the night is inclusively reckoned in the term "day." Further, if there were neither putrefaction of humors nor the body itself were to receive the feverish heat readily, the fever would fall outside the two classes. Accordingly, in the primary division of fevers, it is either necessary to speak of another class in addition to the three mentioned or, being disdainful of the term, to call them "ephemeral." For surely the name "ephemeral" would not be the term for the actual essence of such fevers! Since I was lacking a specific name, I applied a name from what often follows them for the purpose of clear and concise teaching. So their nature is the same as that of ephemeral [fevers], but the name is not the same.

2. [Doctors] term "continuous" ("nonintermittent")[2] those [fevers] in which there is a single paroxysm persisting throughout from the beginning and extending for many days. They don't use a Greek term, preferring to   604K
commit a solecism rather than leave their kind without a specific name. But although there is one kind of such fevers, from which they name them continuous, even so, their nature is no longer simple and single. Some of them have a clear sign of putrefaction, whereas some do not have this at all; these I said are from the class of ephemeral fevers. For since cessation of the paroxysm customarily occurs while the seething humors are dispersing, but they are not dispersed due to the severe impactions, a paroxysm occurring over many days (polyhemeral) is inevitable. There-

2 On "continuous," S has: "obsolete term for a continual febrile illness without intermittency as with malaria. Many cases were typhoid fever, but included many types of febrile illness."

μηδεμία συνῇ τῷ τοιούτῳ πυρετῷ πρόφασις ἀρχὴν
ἑτέρου παροξυσμοῦ γεννῶσα, πρὸς τῷ πολυήμερος
ὑπάρχειν οὐδ᾽ εἰσβολὴν ἐπισημασίας οὐδεμιᾶς ποιεῖ-
ται· μένει δ᾽ ἀπ᾽ ἀρχῆς ἄχρι τέλους εἷς ὁ πυρετός, οὔτ᾽
ἀμφημερινὴν οὔτε διὰ τρίτης οὔτε διὰ τετάρτης ἡμέρας ἔχων ἐπισημασίαν. ὅταν δ᾽ ἀκούσῃς μου λέγοντος ἐντὸς τοῦ σώματος στέγεσθαι τὸ πυρετῶδες θερμὸν ἐπὶ τῶν ἰσχυρῶς ἐστεγνωμένων σωμάτων, οὐχ
605K οὕτω χρὴ νομίζειν ἀκριβῶς γεγονέναι | τὴν πύκνωσιν,
ὡς μηδενὶ τῶν ἱκανῶς λεπτυνθέντων χυμῶν ἐπιτρέπειν
εἰς τοὐκτὸς διαρρεῖν. οὐδὲ γὰρ οὐδ᾽ ἐπὶ τῶν ὀστῶν
λεγόμενον ἀληθὲς ἂν εἴη τοῦτο, μήτοι γε δὴ τῶν
σαρκῶν τε καὶ τοῦ δέρματος. ἀλλ᾽ ὅταν ἡ μὲν ἐκ τῶν
ζεόντων χυμῶν ἀναθυμίασις ᾖ παμπόλλη, διαρρέῃ δ᾽
ἐκτὸς ὀλίγη, τὴν ὑπολειπομένην ἀνάγκη διαφυλάττειν
τὸν πυρετόν, οὐ μόνον τῷ μένειν αὐτήν, ἀλλὰ καὶ τῷ
τὰ πλησιάζοντα συνεκκαίειν.

3. Ἐπεὶ τοίνυν ἐν δυοῖν τούτοιν ἐστὶ τὸ φυλάττεσθαι τὸν πυρετόν, τῷ τε μὴ διαπνεῖσθαι πᾶσαν τὴν
ἀτμίδα καὶ τῷ συνεκθερμαίνειν τὰ ψαύοντα, τρεῖς
ἀνάγκη διαφορὰς ἐν τοῖς συνόχοις γίγνεσθαι πυρετοῖς, ἤτοι διαμένοντος ἴσου τοῦ μεγέθους, ὃ κατὰ τὸν
πρῶτον παροξυσμὸν ἔσχεν ἀκμάσας, ἢ προστιθέντος,
ἢ ἀφαιροῦντος ἀεί τι σμικρόν, καὶ ταῦτα ποιοῦντος ἢ
ὁμαλῶς, ἢ ἀνωμάλως. ἂν μὲν οὖν ἴσον ᾖ τῷ διαπνε-
ομένῳ τὸ ἀναπτόμενον, οὔτ᾽ αὔξησιν οὔτε μείωσιν ἕξει
τὸ πυρετῶδες θερμόν, ἀλλ᾽ οὕτως ἑαυτῷ διαμενεῖ
606K παραπλήσιον ὡς εἰ μήτε προσετίθετο μηδὲν | αὐτῷ

fore, whenever there is no cause associated with such a fever to generate the beginning of another paroxysm, in
addition to being of many days' duration (polyhemeral), it
does not create an attack of any accession; rather, the one
fever remains from beginning to end with an accession that
is neither quotidian, tertian, nor quartan. Whenever you
hear me say the feverish heat is held within the body in
bodies with severe obstruction of the pores, you must not
think, in this way, that thickening has occurred precisely in   605K
such a manner that you can rely on none of the humors that
are excessively thinned to flow to the outside. For it would
not be true to say this in the case of bones, and certainly
not in the case of flesh and skin. But whenever the exhalation from the seething humors is very great, and little flows
through to the outside, it is inevitable that what is left [of
the exhalation] maintains the fever, not only by virtue of it
remaining, but also by burning up the adjacent parts.

3. Therefore, since maintaining the fever lies in these
two things—in all the vapor not being dispersed, and in the
vapor making the things that contact it very hot like itself—there are, of necessity, three differentiae in the continuous fevers. These are: when the magnitude remains
equal to that which the fever had at the peak of the first
paroxysm; when it is always increasing and decreasing to a
small degree; and when it is doing these things either regu-
larly or irregularly. If what is being kindled is equal to what
is being dispersed, the feverish heat will have neither increase nor decrease but will remain of constant magnitude, as if nothing is either added to or taken away from   606K

459

μήτ᾽ ἀφῃρεῖτο. διαφέρει γὰρ οὐδὲν εἴς γε τὴν τῆς
ἰσότητος διαμονὴν ἢ διαφυλάττεσθαι τὴν αὐτὴν οὐσίαν διαπαντός, ἢ τῷ ἐκκενουμένῳ τὸ προστιθέμενον
ἴσον ὑπάρχειν· εἰ δέ γε θάτερον αὐτῶν ἐπικρατήσειεν,
εἰ μὲν τὸ κενούμενον, ἐλαττοῦσθαι τὸν πυρετὸν ἀναγκαῖόν ἐστιν· εἰ δὲ τὸ προστιθέμενον, αὐξάνεσθαι πάντως. εὔδηλον δὲ καὶ ὡς τὸ μᾶλλόν τε καὶ τὸ ἧττον
ἕτερον ἑτέρου τὸν πυρετὸν αὐξάνεσθαί τε καὶ μει-
οῦσθαι διὰ τὸ τῆς πλεονεξίας ἄνισον ἀναγκαῖόν ἐστι
γίγνεσθαι. ποτὲ μὲν γὰρ πλέονι μέτρῳ, ποτὲ δ᾽ ἧττονι
πλεονεκτούσης ἤτοι τῆς γενέσεως τῶν πυρετῶν ἢ τῆς
διαπνοῆς, οὔτε τοῖς αὐξανομένοις ἅπασιν ἴσην τὴν
αὔξησιν οὔτε τοῖς μειουμένοις τὴν καθαίρεσιν οἷόν τε
γίγνεσθαι· παμπόλλην δ᾽ εἶναι τὴν ἐν τῷ μᾶλλόν τε
καὶ ἧττον διαφοράν.

ἐν μὲν δὴ τοῦτο τὸ γένος ἢ εἶδος ἢ ὅ τί περ ἂν
ὀνομάζειν ἐθέλῃς ἐστι συνόχων πυρετῶν, ἐκ τῆς αὐτῆς
ὑπάρχον φύσεως τοῖς ἐφημέροις. ἕτερον δ᾽ ὅταν ἐν
ἅπασι τοῖς ἀγγείοις καὶ μάλιστα τοῖς μεγίστοις ὁμοτίμως οἱ χυμοὶ διασήπωνται. τῶν τοιούτων πυρετῶν
οὐδέτερον ἐν ἰσχνῷ καὶ ψυχρῷ σώματι φιλεῖ συν-
607K ίστασθαι, | τὰ πολλὰ δ᾽ ἐν πολυαίμοις τε καὶ πολυσάρ-
κοις ἡ γένεσις αὐτῶν. αἵ τε γὰρ ἐμφράξεις οὐχ ἱκαναὶ
γεννῆσαι πυρετὸν ἄνευ σήψεως, εἰ μὴ τὸ διαπνεόμενον
εἴη πολὺ καὶ θερμόν, ἥ τε σηπεδὼν τῶν χυμῶν εἰς
ἁπάσας τὰς μεταξὺ βουβώνων τε καὶ μασχαλῶν φλέβας ὁμοτίμως τε καὶ διὰ ταχέων ἐκταθῆναι κατὰ τὰς
ψυχρὰς ἕξεις οὐ δύναται. δεῖται γὰρ ἀεὶ τὰ σηπόμενα
θερμά τ᾽ εἶναι καὶ ὑγρὰ καὶ δυσδιάπνευστα.

460

it. It makes no difference to the continuation of equality whether the same substance is being preserved throughout or whether what is added is equal to what is evacuated. If, however, one or other of these does prevail, and if it is what is evacuated, the fever inevitably diminishes, whereas if it is what is added, in all instances the fever increases. But it is also clear that one fever is increased or decreased more than another due to the inevitable inequality of the preponderance. Either the genesis of the fevers or their dispersal predominates, sometimes by a greater measure and sometimes by a lesser; the increase in all those that do increase cannot be equal, and neither can the reduction in those that diminish. The difference in terms of more and less is very marked.

Indeed, there is this one class or kind (or whatever else you may wish to term it) of the continuous fevers which is of the same nature as the ephemeral. However, there is another [class] whenever the humors putrefy to an equal degree in all the vessels, but particularly in the largest. Neither of these fevers is wont to exist in a thin, cold body, their genesis in most instances being in those that are 607K blood-filled and very fleshy. For the blockages are not sufficient to generate a fever without putrefaction, unless what is being dispersed is large in amount and hot; in cold states putrefying humors can't be spread to all the veins between the axillae and the groins in equal degree and quickly. Those things that putrefy always need to be hot, moist and slow to disperse.

διὰ τοῦτο οὖν οὔτ᾽ ἐν ψυχραῖς ἡλικίαις οὔτ᾽ ἐν
κράσεσι σωμάτων ψυχραῖς, εἴτ᾽ οὖν ἐξ ἀρχῆς εἴτε καὶ
νῦν εἴη γεγονότα ψυχρά, σύνοχοι γεννῶνται πυρετοί,
καθάπερ οὐδ᾽ ἐν ἰσχνοῖς ἢ ἀραιοῖς, ἀλλὰ καὶ τῶν
ἡλικιῶν ταῖς θερμαῖς καὶ τῶν φύσεων καὶ τῶν ἐπικτήτων κράσεων ταῖς ὁμοίαις ἐγγίγνονται καὶ μᾶλλον
ὅταν εὔσαρκοί τε καὶ πολύαιμοι καὶ πυκνοὶ τὰς ἕξεις
οἱ κάμνοντες ὦσιν, ἢ περιττώμασι θερμοῖς πεπλη-
ρωμένοι. ταῦτ᾽ ἄρα τοῖς οὕτω πυρέττουσιν ὁ σφυγμὸς
μέγιστός ἐστι, ὁμαλὸς καὶ σφοδρός· καὶ τὴν σύστασιν τῆς ἀρτηρίας οὔτε σκληροτέραν οὔτε μαλακωτέ-
608K ραν τοῦ κατὰ φύσιν ἔχων, τάχους δ᾽ εἰς | τοσοῦτον
ἥκει καὶ πυκνότητος εἰς ὅσον ἂν ὁ πυρετὸς αὐτὸς ἥκοι
μεγέθους. κοινὰ μὲν οὖν ἀμφοῖν ταῦτα· πρόσεστι δ᾽
ἐξαίρετα θατέρῳ σημεῖα τὰ τῆς σήψεως, ἐν οὔροις τε
καὶ σφυγμοῖς καὶ τῷ τῆς θερμασίας ἀηδεῖ.

4. Καὶ δὴ παραδείγματος ἕνεκα ἀναμνήσω σε δυοῖν
νεανίσκοιν οὓς ἐθεάσω μεθ᾽ ἡμῶν. ἦν δὲ ὁ μὲν ἕτερος
αὐτῶν ἐλεύθερος καὶ γυμναστικός, ὁ δ᾽ ἕτερος δοῦλος
οὐκ ἀγύμναστος μέν, οὐ μὴν τά γε κατὰ παλαίστραν
δεινός, ἀλλ᾽ ὅσα δούλῳ πρέπει, τὰ ἐφήμερα γυμνάσιά
τε ἅμα καὶ ἔργα μεταχειρίζεσθαι. ὁ μὲν οὖν ἐλεύθερος
τὸν χωρὶς σήψεως ἐπύρεξε σύνοχον, ὁ δὲ δοῦλος
τὸν μετὰ σήψεως. ὁποίαν δ᾽ ἑκατέρῳ τὴν ἴασιν ἐποιησάμεθα καιρὸς ἂν εἴη λέγειν, ἐπειδὴ μάλιστα μὲν
χρὴ γυμνάζεσθαι τοὺς μανθάνοντας ὁτιοῦν ἐπὶ παρα-
δειγμάτων· οὐ γὰρ ἀρκοῦσιν αἱ καθόλου μέθοδοι πρὸς
τὴν ἀκριβῆ γνῶσιν.

Because of this, then, neither in cold age groups nor in the cold *krasias* of bodies, whether they have been cold from the beginning or have become cold now, are continuous fevers generated, just as they are not in those bodies that are thin and rarefied. But in the hot age groups and natures, and in the like kinds of acquired *krasias*, they do occur, and particularly whenever the patients are well-fleshed, blood-filled and thickset in habitus or are filled with hot superfluities. So it is, then, that the pulse in those who become febrile in this way is very great, even and strong, and the consistency of the artery is neither harder nor softer than normal, while the speed and frequency   608K
have come to a point commensurate with the magnitude of the fever itself. These features are common to both, where marked signs of the putrefaction are present in the urine and the pulse, but the unpleasant quality of the heat is present in the one only.

4. And now, by way of an example, I shall remind you of two young men whom you saw with me. One of them was a free man and skilled in exercise; the other was a slave who, while not untrained, was not proficient in matters pertaining to the wrestling school but in those appropriate for a slave—the day-to-day exercises and tasks to be carried out. The free man became febrile continuously without putrefaction; the slave with putrefaction. It is an opportune time to speak of the kind of cure I fashioned for each of them since it is particularly necessary for those learning to be practiced in various specific examples, because the methods in general are not enough for precise knowledge.

ἀμείνω δὲ τῶν παραδειγμάτων ἐστὶν ὧν αὐτόπται
γεγόναμεν· ὡς εἴ γε πάντες οἱ διδάσκειν ἢ γράφειν
609K ὁτιοῦν ἐπιχειροῦντες ἔργοις ἐπεδείκνυντο | πρότερον
αὐτά, παντάπασιν ἂν ὀλίγ' ἄττα ψευδῶς ἦν λεγόμενα.
νυνὶ δ' οἱ πλεῖστοι διδάσκειν ἄλλους ἐπιχειροῦσιν ἃ
μήτ' αὐτοί ποτ' ἔπραξαν μήτ' ἄλλοις ἐπεδείξαντο.
τοὺς μὲν οὖν πολλοὺς τῶν ἰατρῶν οὐδὲν θαυμαστὸν
ἀμελήσαντας ἤθους χρηστοῦ δοξοσοφίαν μᾶλλον ἢ
ἀλήθειαν σπουδάσαι. τὸ δ' ἡμέτερον οὐχ ὧδ' ἔχει. οὐ
γὰρ δὴ χθὲς ἢ πρώην, ἀλλ' εὐθὺς ἐκ μειρακίου φιλοσοφίας ἐρασθέντες ἐπ' ἐκείνην ἤξαμεν[1] πρῶτον. εἶθ'
ὕστερον τοῦ πατρὸς ὀνείρασιν ἐναργέσι προτραπέν-
τος ἐπὶ τὴν τῆς ἰατρικῆς ἄσκησιν ἀφικόμεθα καὶ δι'
ὅλου τοῦ βίου τὰς ἐπιστήμας ἑκατέρας ἔργοις μᾶλλον
ἢ λόγοις ἐσπουδάσαμεν. οὐδὲν οὖν θαυμαστὸν ἐν ᾧ
προσαγορεύουσιν ἄλλοι, περιθέοντες ὅλην τὴν πόλιν
ἐν κύκλῳ καὶ συνδειπνοῦσι καὶ παραπέμπουσι τοὺς
πλουτοῦντάς τε καὶ δυναμένους, ἐν τούτῳ τῷ χρόνῳ
παντὶ φιλοπονοῦντας ἡμᾶς ἐκμαθεῖν μὲν πρῶτον ὅσα
καλῶς εὕρηνται τοῖς παλαιοῖς, ἔπειτα διὰ τῶν ἔργων
αὐτὰ κρῖναί τε ἅμα καὶ ἀσκῆσαι.

τὸν τοίνυν γυμναστικὸν νεανίσκον ἀρξάμενον πυρέττειν ὥρας πρώτης τῆς νυκτὸς ἐθεασάμεθα κατὰ τὴν
610K ἐπιοῦσαν ἡμέραν | ὥρας που τρίτης. εὑρόντες δὲ
πυρετὸν ἱκανῶς μὲν θερμόν, ἀλλὰ καὶ τούς τε σφυ-

---

[1] B (q.v. Cobet, *Mnem.* 12, 1884, 446); ἤξαμεν K

Better, however, than examples are those things we have witnessed with our own eyes in that, if all those who attempted to teach or write about anything whatsoever, demonstrated them first through actions, there would un- 609K doubtedly be fewer false statements. At the present time, the vast majority try to teach others things which they themselves did not ever do or demonstrate to others. It is not surprising, then, that many doctors, being neglectful of good custom, are more eager for the pretense of wisdom than for truth. This is not the case with me. For not just yesterday or the day before, but right from when I was a young lad, gripped by a love of philosophy, did I eagerly turn to that [discipline]. Then later, urged on by the clear dreams of my father, I came to the practice of medicine, and throughout my whole life was zealous in both branches of knowledge, more through actions than words.[3] Therefore, there is nothing surprising [in the fact that], while others run around the whole city in circles making their greetings, dining with and cultivating the rich and powerful, I, having throughout that time acquired a love of labor, am diligent [in my endeavors] to first learn those things discovered well by the ancients, and then, through what I do, evaluate and, at the same time, practice them.

Accordingly, in the case of the athletic young man (i.e. the free man) who started to be feverish during the first hour of the night, I saw him about the third hour on the fol- lowing day. I found the fever to be excessively hot, but also 610K

[3] On the redirection of Galen's career from philosophy to medicine as a result of his father's dreams, see *De ordine librorum suorum ad Eugenianum*, XIX.59K.

465

γμοὺς ὁμαλοὺς καὶ μεγίστους καὶ ταχεῖς καὶ πυκνοὺς
καὶ σφοδροὺς καὶ τὴν τῆς θερμασίας ποιότητα τὸ
διαβρωτικὸν τῆς ἁφῆς οὐκ ἔχουσαν, ἔτι δὲ καὶ τὰ
οὖρα τῇ τε συστάσει καὶ τῇ χροιᾷ τῶν κατὰ φύσιν οὐ
πάνυ λειπόμενα, πυθόμενοί τε τοῦ τῶν γυμνασίων
ἔθους ἠμεληκότα τὸν ἄνθρωπον ἡμέραις ὡς τριάκοντα, τῇ δὲ προτεραίᾳ μόνῃ γεγυμνάσθαι σφοδρότερον
μέν, ἀλλ' οὐκ ἐπὶ πολύ, προσενηνέχθαι τε τὰ συνήθη
σιτία καὶ ταῦτα πεπέφθαι μέν, ἀλλὰ βραδέως καὶ
μόγις, ὡς ἂν ἐπιγενομένου κατὰ τὴν ἑσπέραν τοῦ
πυρετοῦ, φαινομένου δὲ ἐρυθροῦ καὶ μεστοῦ τοῦ ἀνθρώπου καὶ μέντοι καὶ πληρώσεως αὐτῷ τινα αἴσθησιν εἶναι λέγοντος, ἐν τούτῳ τε φθεγξαμένου περὶ
φλεβοτομίας τῶν παρόντων τινός, ἔδοξεν ἡμῖν ἀναβάλλεσθαι τὴν περὶ τοῦ βοηθήματος σκέψιν εἰς ἕτερον καιρόν, ἅμα μὲν ἵνα ἀκριβέστερον διαγνῶμεν ἐκ
ποίου γένους ἐστὶν ὁ πυρετός, ἅμα δ' ἐξ ἀνάγκης διὰ
τὴν προγεγενημένην βραδυπεψίαν. ἐπεὶ δὲ καὶ κατὰ
τὴν ἑσπέραν ὁμοίως ἀκμάζειν ὁ πυρετὸς ἐφαίνετο
611K μηδὲν ἀφαιρῶν | αἰσθητῶς, ὕποπτος ἦν ἤδη σύνοχος
ὑπάρχειν ἐπ' ἐμφράξει τε καὶ πολυαιμίᾳ καὶ τῇ διὰ τὸ
πλῆθος τῶν σαρκῶν στεγνώσει. διαφυλαχθέντος δὲ
τοῦ μεγέθους ἴσου δι' ὅλης τῆς νυκτός, ἐπὶ τῆς ὑστε-
ραίας ἐδόκει τοῖς ἐπισκοπουμένοις αὐτὸν ἰατροῖς ἅπασι φλεβοτομητέος εἶναι.

στάσεως δ' ἐγγινομένης περὶ τοῦ καιροῦ καὶ κρα-
τησάντων τῶν εἰς τὴν ἐπιοῦσαν ἀναβάλλεσθαι συμ-
βουλευόντων, ὁ πυρετὸς ἐναργῶς ἐφάνη δι' ὅλης τῆς

that the pulse was regular, very large, rapid, frequent and strong, and that the quality of the heat was not corrosive to the touch, and further, that the urine was not far removed from normal in both consistency and color. I learned that the man had neglected his customary exercises for about thirty days but had exercised quite vigorously only the day before, although still not to excess, and had taken his customary foods and digested these, albeit slowly and with difficulty. [I also learned that] when the fever came on in the evening, the man appeared red and full, and further, he said there was a certain sensation of fullness in him. When one of those present spoke about phlebotomy, I decided to put off consideration of this remedy to another time, partly so that we might make a more accurate diagnosis of the class of fever, and partly of necessity, because of the brady-pepsia that had previously occurred. However, since the fever seemed to reach a peak similarly during the evening with no perceptible relief, there was already a suspicion 611K that this was a continuous fever due to a blockage and excess of blood due to a stoppage of the pores resulting from the amount of flesh. When the fever maintained an equal magnitude throughout the whole night, it seemed to all the doctors observing him on the following day that phlebotomy was required.

When discord arose about the timing [of the phlebotomy], and the view of those who recommended deferring it to the following day prevailed, the fever clearly showed

GALEN

ἡμέρας ἐπακμάζων ἑαυτῷ. κἄπειτα τῆς ἐπιούσης
νυκτὸς τῆς τρίτης ἄλλος μὲν οὐκ ἐγένετο παροξυσμὸς
ὡς πρὸς τὸ πρῶτον ἐξ ἀναλογίας, ἀφόρητον δ᾽ ἦν τὸ
καῦμα τῷ κάμνοντι, καὶ τάσις ὅλου τοῦ σώματος ὡς
πεπληρωμένου καὶ σφυγμὸς τῆς κεφαλῆς, ἀγρυπνία
τε διὰ ταῦτα δεινὴ καὶ μεταρριπτοῦντος ἑαυτὸν ἄλλοτ᾽
εἰς ἄλλο σχῆμα τοῦ νεανίσκου. καὶ τοίνυν ὡς οὐκέτ᾽
ἔφερεν, ὥρας που τῆς νυκτὸς ὀγδόης, ἐκπέμψας οἰκέ-
την πρός με δεῖται παρ᾽ αὐτὸν ἀφικέσθαι διὰ ταχέων.
ὑπακούω δὴ καὶ ἀπέρχομαι καὶ καταλαμβάνω θερμό-
τατόν τε τὸν πυρετὸν καὶ τοὺς σφυγμοὺς οἵους ἔμπροσθεν εἶπον. ἐπεὶ δ᾽ οὔτ᾽ ἐν τούτοις οὔτε ἐν τοῖς
612K οὔροις οὔτ᾽ ἐν αὐτῇ τῇ τῆς | θερμότητος ποιότητι
σημεῖόν τι σηπεδόνος ἐφαίνετο χυμῶν, ἐδόκει κάλλιον
εἶναι τεμεῖν τὴν φλέβα, πρὶν ἄρξασθαι τὴν σῆψιν.

ἀφαιρῶ τοίνυν αὐτοῦ τοσοῦτον ἐξεπίτηδες ὡς λει-
ποθυμίαν ἐπιγενέσθαι, μέγιστόν τι βοήθημα τοῦτο
πυρετῶν συνόχων ἐν ἰσχυρᾷ δυνάμει καὶ τῷ λόγῳ καὶ
τῇ πείρᾳ δεδιδαγμένος. πρῶτον μὲν γὰρ εἰς ἐναντίαν
κατάστασιν ἀφικνεῖται τάχιστα ψυχόμενον ἐν τῇ λει-
ποθυμίᾳ τὸ σῶμα. τούτου δ᾽ οὔτε τοῖς κάμνουσιν οὔτ᾽
αὐτῇ τῇ διοικούσῃ τὰ ζῷα φύσει δύναιτ᾽ ἄν τις εὑρεῖν
ἥδιον ἢ χρηστότερον. ἔπειτα δ᾽ ἐξ ἀνάγκης ἐν τοῖς
τοιούτοις σώμασιν ἕπεται διαχώρησις γαστρός, ἔστι
δὲ ὅτε καὶ χολῆς ἔμετος, ἐφ᾽ οἷς αὐτίκα νοτίδες ἀπὸ
παντὸς τοῦ σώματος, ἢ ἱδρῶτες. ἅπερ οὖν κἀκείνῳ
πάνθ᾽ ἑξῆς γενόμενα παραχρῆμα τὸν πυρετὸν ἔσβε-
σαν, ὥστε τινὰς τῶν παρόντων εἰπεῖν, Ἔσφαξας ἄν-

an increase in strength through the whole day. And when, during the following, third night, another paroxysm in proportion to the first did not occur, the burning was intolerable to the patient, there was a tension of the whole body as if it had been filled, and a pulse in the head. Because of these things, the insomnia was terrible and the young man was throwing his body from one side to the other. Therefore, as he could no longer bear it, somewhere around the eighth hour of the night, he sent out a household slave to me with a request to come to him as quickly as possible. I listened of course and set out, finding on arrival that the fever was very hot and the pulse was of the kind I previously stated. Since neither in the pulse, nor in the urine, nor in the actual quality of the heat, was there any apparent sign of putrefaction of the humors, it seemed to be better to cut the vein before putrefaction began.

612K

Accordingly, I deliberately took from him an amount sufficient to bring about fainting, since I had been taught that both on theoretical grounds and from experience this is a most important remedy for continuous fevers in someone of strong capacity. For first, a body being cooled in the fainting very quickly comes to the opposite state. Nobody would be able to discover anything more pleasant or useful than this, either for patients or for the actual nature governing animals. Since, of necessity, in such bodies gastric excretion follows, while there is sometimes also vomiting of bile, the result is that moisture is immediately removed from the whole body, or there are sweats. Therefore, when these things all happened in sequence in that man, I quenched the fever right away so that some of those pres-

θρωπε τὸν πυρετόν· ἐπὶ τούτου μὲν δὴ πάντες ἐγελά-
σαμεν. ὅπως δὲ πληρώσαιμι τὴν διήγησιν, οὐδὲν ἂν
εἴη χεῖρον ὀλίγα προσθεῖναι.

μετὰ δύο γὰρ τῆς φλεβοτομίας ὥρας βραχύ τι
προσδοὺς τροφῆς τῷ κάμνοντι καὶ κελεύσας ἡσυχά-
613K ζειν ἀπηλλαττόμην. ἀφικόμενος | δὲ πέμπτης ὥρας
τῆς ἡμέρας οὕτω βαθέως ὑπνοῦντα κατέλαβον ὡς
ἁπτομένου μου μηδ' ὅλως αἰσθάνεσθαι. λεγόντων δὲ
καὶ τῶν ὑπηρετουμένων αὐτῷ βαθὺν οὕτως εἶναι τὸν
ὕπνον ὡς μηδ' ὅταν ἀπομάττωσιν αὐτοῦ τὰς νοτίδας
ἐξεγείρεσθαι, συνεβούλευον οὕτω πράττειν, εἶναι γὰρ
ἀκριβῶς ἤδη τὸν ἄνθρωπον ἀπύρετον. ἀφικόμενος δ'
αὖθις ὥρας δεκάτης εὗρον ἔτι καὶ τότε κοιμώμενον
αὐτόν. ἐξελθὼν δὲ πάλιν ἐπ' ἄλλους ἀρρώστους
ἐπανῆλθον ὥρας πρώτης νυκτὸς οὐκέτι μετὰ σιωπῆς,
ἀλλ' ἐξεπίτηδες μέγα φθεγγόμενος, ὅπως ὁ κάμνων
διεγερθείη τοῦ ὕπνου. καὶ τοίνυν οὕτω γενομένου
πτισάνης χυλῷ μόνῳ θρέψας αὐτὸν ἀπηλλαττόμην.
ἐπιμετρήσας δὲ τὴν ὑστεραίαν ἐπὶ τὸ λουτρὸν ἀπ-
έλυσα τῇ μετ' αὐτήν. τὰ μὲν δὴ κατὰ τοῦτον οὕτως
ἐπράχθη.

τὰ δὲ κατὰ τὸν ἕτερον, ἤδη σοι δίειμι. δι' ὅλης
ἡμέρας ἐκεῖνος ὁ ἄνθρωπος καμὼν πολλά, κἄπειτα
λουσάμενος ὀλίγα τε προσενεγκάμενος ὑπήρξατο πυρέττειν ἐν τῇ νυκτὶ συνάψας αὐτῇ καὶ τὴν ἑπομένην
ἡμέραν. ἐθεασάμεθα δ' ἡμεῖς αὐτὸν μετὰ τὴν δευτέραν
614K νύκτα τὰ μὲν οὖν ἄλλα πάντα τῷ προειρημένῳ | παρα-
πλησίως διακείμενον, ἐναργῆ δὲ τὰ τῆς σηπεδόνος

ent said, "You, sir, have slain the fever!," upon which we all laughed. It would be a good idea to add a little more so that I might complete the story.

Two hours after the phlebotomy, when I gave the patient a little food and directed him to rest, I departed. When I arrived at the fifth hour of the day, I found the man 613K so deeply asleep that when I touched him he did not feel it at all. And when those attending him said that his sleep was so deep he did not wake up when they wiped away his moisture, I advised them to do this, for the man was already completely afebrile. When I came again during the tenth hour, I found him at that time still asleep. Having left once more to [attend to] other patients, I returned during the first hour of the night and no longer remained silent but deliberately spoke loudly so the patient was wakened from sleep. And accordingly, when this happened, I nourished him with the juice of ptisan only and departed. On the day after that I ordered an increase in bathing and left. This was how things were done for the man in question.

I shall now set out for you what was done for the other patient. That man, having suffered greatly throughout the whole day and then bathed and eaten a little, developed a fever continuing in the night itself and for the following day. I saw him after the second night, and while in all other respects he was in a similar state to the man previously 614K

471

τῶν χυμῶν ἔχοντα γνωρίσματα. φλέβα τοίνυν αὐτῷ
παραχρῆμα διελόντες ἄχρι λειποθυμίας ἐκενώσαμεν.
ἐφ' ᾗ διαλιπόντες αὔταρκες ἐθρέψαμεν, μελικράτῳ μὲν
πρῶτον, μετὰ δὲ ὥραν ἐκείνου μίαν πτισάνης χυλῷ.
καὶ πάντ' ἐπέπρακτο ταῦτα πέμπτης ὥρας ἐντός. ὁμοίου δὲ αὐτῷ διαμένοντος τοῦ πυρετοῦ σύνοχον εἶναι
προσεδοκήσαμεν ἐπὶ σήψει. καὶ τοίνυν καὶ οὕτως
ἀπέβη. θεασάμενοι γὰρ αὐτὸν ὥρας που δευτέρας
νυκτὸς ἐν ἴσῳ μεγέθει, τὸν διὰ τρίτης παροξυσμὸν
ἤτοι γ' ἐσόμενον ἢ οὐκ ἐσόμενον, ἀκριβῶς ἠβουλήθη-
μεν παραφυλάξαι νυκτὸς ὥραν ἑβδόμην τὴν ὕποπτον
ἔχοντα.

ὄρθρου δὴ βαθέος ἐπὶ τὸν ἄνθρωπον ἐλθόντες εὕρομεν ὅπερ ἠλπίσαμεν. οὔτε γὰρ ὁ διὰ τρίτης ἐγεγόνει
παροξυσμὸς ἐφαίνετό τε βραχύ τι μικρότερος ὁ πυρετὸς οὗ κατελείπομεν ἐπὶ τῆς ἑσπέρας. ὡς δὲ καὶ τῆς
μεσημβρίας ἰδὼν αὐτόν, ἣν ἤδη βεβαιότατος ἀφαιρεῖν τι βραχὺ καὶ σύνοχον εἶναι παρακμαστικόν,
ἄμεινον ἐδόκει θρέψαι καὶ τόθ' ὁμοίως τὸν ἄνθρωπον.
διελθούσης δὲ καὶ τῆς τετάρτης νυκτὸς ἐναργῶς ἐλάτ-
615K των ἑαυτοῦ κατὰ τὴν τετάρτην ἡμέραν | ἦν, ἐν ᾗ πάλιν
ὁμοίως αὐτὸν θρέψαντες ἠκολουθήσαμεν ἀφαιροῦντες
τοῦ μεγέθους, δι' ὅλης τε τῆς ἡμέρας ἐκείνης καὶ τῆς
ἐπιούσης νυκτὸς τῆς πέμπτης, ὥστ' ἐναργῶς τῇ πέμπτῃ τῶν ἡμερῶν ἐλάττονα φαίνεσθαι τοῦ πρόσθεν.
ἀνάλογον δὲ τῇ μειώσει τοῦ πυρετοῦ καὶ ἡ τῶν οὔρων
πέψις προὐχώρει. καὶ ἦν δῆλον ὡς κατὰ τὴν ἑβδόμην
ἡμέραν παύσαιτο, καὶ οὕτως ἐγένετο. παρακμαστικὸς

mentioned, he clearly had the signs of putrefaction of humors. Therefore, having immediately opened a vein in him, I drained [blood] to the point of fainting. After leaving a sufficient interval from this, I nourished him, first with melikraton and one hour later, with the juice of ptisan. All these things had been done within five hours. When the same fever persisted in him, I considered it to be continuous due to putrefaction. And so it proved to be. When I saw at some point during the second hour of the night that it was of the same magnitude, I wanted to observe precisely whether the paroxysm during the third night would follow or not, the suspicion being that it would occur in the seventh hour of the night.

On attending the man very early in the morning, I found what I expected. A paroxysm had not occurred during the third night, and the fever seemed to be a little less than when I left him in the evening. And so, when I saw him at midday, I was already very confident that I had got rid of the fever to some extent and the continuous fever was past its peak, so it seemed better to nourish the man in similar fashion at that time too. Also, when the fourth night went past, the fever was clearly less than it was on the fourth day on which, after I had nourished him again in 615K like manner, I followed by reducing the magnitude [of the fever] throughout the whole of that day and the following fifth night, so that, on the fifth day, it appeared clearly less than before. The concoction of the urine was improving in proportion to the reduction of the fever. It was clear that [the fever] would cease on the seventh day, and so it happened. This continuous fever due to the putrefaction of

οὖν ἀκριβῶς οὗτος ἡμῖν ὁ σύνοχος ἐπὶ σήψει χυμῶν
ὤφθη.

πολλοὺς δ᾽ ἄλλους συνόχους ἐθεασάμεθα τοὺς μὲν
ἐπακμαστικούς, τοὺς δὲ ὁμοτόνους, ἢ ἀκμαστικούς, ἢ
ὅπως ἄν τις ἐθέλοι καλεῖν, ἐνίους δ᾽ ὥσπερ καὶ οὗτος ὁ
προειρημένος παρακμαστικούς· καὶ τινὰς μὲν ἐν τῇ
πρώτῃ τῶν ἡμερῶν τὰ τῆς σήψεως ἔχοντας γνωρίσματα, τινὰς δ᾽ ἐν τῇ τρίτῃ σχόντας, ἢ τὸ πλεῖστόν
γε τῇ τετάρτῃ, ποτὲ μὲν ἀμαθίᾳ τῶν ἰατρῶν μὴ τε-
μνόντων τὴν φλέβα, ποτὲ δ᾽ αὐτῶν τῶν καμνόντων
δειλίᾳ. τῶν γὰρ ἐπ᾽ ἐμφράξει μόνῃ πυρεξάντων οὐδεὶς
μετέπεσεν εἰς τὸν ἐπὶ σηπεδόνι πυρετὸν ἀποχέαι τι
616K φθασάντων αἵματος. ἰσχυρᾶς μὲν | οὖν ὑπαρχούσης
τῆς δυνάμεως καὶ τῆς ἡλικίας συγχωρούσης, ἄχρι
λειποθυμίας ἄμεινον ἄγειν. εἰ δέ τι τούτων ἐνδέοι κατά
τι, βέλτιον ἀφελεῖν μὲν ὅσον ἂν ἱκανὸν εἶναι φαίνηται, τήν γε πρώτην· ἐπαφαιρεῖν δ᾽ αὖθις ὅσον ἐνδέοι.
μὴ φλεβοτομηθέντες γὰρ οἱ οὕτως πυρέττοντες εἰς
ἔσχατον ἥκουσι κινδύνου, πλὴν εἴ ποτε ῥώμη δυνάμεως, ἢ αἱμορραγία λάβρος, ἢ ἱδρὼς πολὺς ἐξαρπά-
σειεν αὐτοὺς ὀλέθρου προφανοῦς. ἀλλ᾽ ὅμως ἔνιοι τῶν
ἰατρῶν ὁρῶντες τὴν φύσιν ἐναργέστατα δι᾽ αἱμορραγίας ἐκσῴζουσαν οὐκ ὀλίγους τῶν οὕτω κινδυνευόντων, ἀποδιδράσκουσι τὴν φλεβοτομίαν, οὔτ᾽ ἐμπειρίαν οὐδεμίαν οὔτε λόγον ἀληθῆ προστησάμενοι.
τούτους μὲν οὖν ἐάσωμεν αὐτάρκως γε πρὸς αὐτοὺς
διειλεγμένοι καθ᾽ ἓν ὅλον γράμμα τὸ Περὶ φλεβοτο-

the humors was, then, seen by me to be entirely past its peak.

And I saw many other continuous fevers, some coming to a peak, others being of equal strength, either at their peak, or however someone might wish to term it, and some, like the man previously described, that had passed the peak. Also, I saw some who have the signs of putrefaction on the first day, some who have them on the third day, or the majority who have them on the fourth day, sometimes due to the ignorance of the doctors in not opening a vein, and sometimes due to the cowardice of the patients themselves. For none of those who were febrile due to blockage alone, and who beforehand shed some blood, underwent a change in the fever due to putrefaction. Therefore, when the capacity is strong and the age permits, it is better to take [the bloodletting] to the point of fainting. If one of these [requirements] is lacking in some respect, it is better to take off as much as seems sufficient, at least on the first day, and to again take as much as might be necessary. For if those who are febrile like this are not phlebotomized, they come to the ultimate point of danger, unless at some time the strength of their capacity, or the vigorous flow of blood, or considerable sweating were to snatch them away from obvious destruction. Nevertheless, there are some doctors who, although they see that Nature very clearly saves quite a number of patients through bleeding in this way, shy away from phlebotomy without putting forward any experience or true logic. So allow that I have sufficiently refuted them in one entire book, *On*

616K

μίας πρὸς Ἐρασίστρατον, αὐτοὶ δ' αὖθις ἐπὶ τὸ προκείμενον ἴωμεν.

5. Ἦν δ' ἐν τῷ νῦν λόγῳ προκείμενον ὑπὲρ ἐκείνων
διελθεῖν τῶν πυρετῶν[2] ὅσοι στεγνώσεσιν ἕπονται ταῖς
δι' ἔμφραξιν. οὕς, εἰ μὲν ἄνευ σηπεδόνος εἶεν, ἐκ τοῦ
617K τῶν | ἐφημέρων ἔφασκον ὑπάρχειν γένους, εἰ δέ τις
αὐτοῖς ἢ ἐξ ἀρχῆς εὐθέως ἢ ἐξ ὑστέρου σῆψις ἐπι-
γίγνοιτο, κατὰ μὲν ἁπάσας τὰς φλέβας καὶ μάλιστα
τὰς μεγάλας συνισταμένης αὐτῆς τοὺς συνόχους γεννᾶσθαι πυρετούς, ἐν ἑνὶ δὲ μορίῳ τοῦ ζῴου τοὺς κατὰ
περίοδόν τινα παροξυνομένους. ἀλλὰ περὶ μὲν τούτων
αὖθις. ὃ δὲ χρὴ καὶ δὶς οἶμαι καὶ τρὶς καὶ πολλάκις
αὖθίς τε καὶ αὖθις εἰπεῖν, εἰ μέλλοιεν οἱ οὕτως νοσοῦν-
τες ὀρθῶς θεραπευθήσεσθαι, τὸ τῆς φλεβοτομίας
παραληπτέον ἐστίν. ὅσοις γὰρ ἐπὶ πλήθει χυμῶν
δυσδιάπνευστον τὸ σῶμα γενόμενον ἤθροισε τοσαύτην θερμασίαν ὡς ἤδη πυρέττειν, ἀφαιρεῖν αἵματος
χρὴ τοσοῦτον ἂν ὅσον ἡ δύναμις φέρει, γινώσκοντας
ὡς εἰ μὴ παραληφθείη τὸ βοήθημα τοῦτο, πνιγήσε-
σθαι τοὺς οὕτω διακειμένους τὴν φύσιν, ἢ συγκοπή-
σεσθαι[3] γε πάντως αὐτούς, εἰ μήποθ', ὡς εἴρηται,
ῥώμη τῆς φύσεως ἢ ἱδρὼς πάμπολυς αὐτοὺς ἢ λάβρος
αἱμορραγία ῥύσηται τοῦ θανάτου. δῆλον μὲν οὖν ἐστί,
κἂν ἐγὼ μὴ λέγω, τοῖς γε τὸ Περὶ πλήθους γράμμα
618K καλῶς ἀναλεγομένοις, οὐδὲν δ' ἧττον | εἰρήσεται καὶ

2 B (*et alia MSS, apud* Boulogne; cf. febribus KLat); ἰατρῶν K
3 B; συγκόπτεσθαι K

Phlebotomy Against Erasistratus; and let me proceed again to what lies before me.[4]

5. What is proposed in the present discussion is to go over in detail those fevers that follow stoppages of the pores due to blockage. I said these [fevers] are from the 617K class of the "ephemeral" if they are without putrefaction, whereas if some putrefaction supervenes in them, either right from the start or later, when this exists in all the veins, and particularly in the large ones, it generates continuous fevers. When these are provoked in one [particular] part of the organism they have a certain time course. But more about these matters later. What I think must be said two or three times and often repeated is that you must make use of phlebotomy, if those who are ill in this way are going to be treated correctly. In those in whom the body, which has become slow to disperse due to the abundance of humors, has built up such a heat as to be already febrile, it is necessary to remove as much blood as their capacity tolerates, recognizing that, if this remedy is not put to use, those who are in such a condition in terms of their nature will choke or suffer syncope unless, as I said, the strength of their nature, or substantial sweating, or a violent loss of blood saves them from death. This is clear, even if I don't say so, at least to those who read through the book *On Plethora*[5] properly, although I shall speak of it no less now, even 618K

---

[4] *De venae sectione adversus Erasistratum*, XI.147–86K. There is also the *De venae sectione adversus Erasistrateos Romae degentes*, XI.187–249K. Both have been translated into English by P. Brain (1986).

[5] *De plenitudine*, VII.513–83K.

πρὸς ἡμῶν ὅτι νῦν ὁ λόγος ἐστὶν οὐ περὶ τοῦ πρὸς τὴν
δύναμιν πλήθους. οὔτε γὰρ ἐμφράττει τοῦτο τὰ στό-
ματα τῶν ἀγγείων, ὡς κωλύειν τὴν ἀνάψυξιν, οὔτε
τείνει τοὺς χιτῶνας αὐτῶν οὔτ᾽ ἔρευθος οὔτ᾽ ὄγκον
ἐργάζεται, πολὺ δὲ δὴ μᾶλλον οὐδὲ τὰς τῆς σαρκὸς ἢ
τοῦ δέρματος ἐπέχει διαπνοάς.

ἀλλὰ περὶ τοῦ μηκέτι δυναμένου χωρεῖσθαι πρὸς
τῶν ἀγγείων, ὃ καὶ διὰ τοῦτ᾽ αὐτὸ ῥηγνύναι τε καὶ
ἀναστομοῦν εἴωθεν αὐτὰ καὶ τἆλλα συμπτώματα τὰ
νῦν εἰρημένα φέρει τὴν ἔμφραξιν, τὴν τάσιν, τὸν
ὄγκον, τὸ ἔρευθος, ὁ σύμπας μοι λόγος ἐπεράνθη τε
καὶ νῦν οὐχ ἧττον περανθήσεται. τὸ γάρ τοι πλῆθος
τοῦτο χρηστὸν ὂν συναύξεσθαι πέφυκε τῇ ῥώμῃ τῆς
δυνάμεως· ὡς εἴ γε βαρύνειέ ποτ᾽ αὐτήν οὔτ᾽ αὐξηθή-
σεται τοῦ λοιποῦ καὶ παύσεται χρηστὸν ὑπάρχον. εἰ
γὰρ ἅπαξ ἀπολέσειε τὸν ἐκ τῆς φύσεως κόσμον,
ἀδύνατον αὐτῷ μὴ σαπῆναι, διότι μηδ᾽ ἄλλο μηδὲν
ὑγρὸν καὶ θερμὸν σῶμα τουτωνὶ τῶν ἐκτὸς ὁρῶμεν
ἄσηπτον διαμένον. ταῦτά τοι χρὴ σπεύδειν ἀφαιρεῖν
τοῦ αἵματος, ὅπως ἡ τῶν ἀγγείων φύσις ἐκ τοῦ δια-
619K πνεῖσθαί τε καὶ ῥιπίζεσθαι | τὴν φυσικὴν εὐκρασίαν
φυλάττουσα κατὰ τὸν ἐξ ἀρχῆς τρόπον ἐπικρατῇ τῶν
χυμῶν· ὡς εἴ γε δύσκρατος γενομένη κάμοι, κίνδυνος
αὐτοῖς σαπῆναι. ἡ δυσκρασία δ᾽ ἐν αὐτῇ γίνεται διὰ
τὴν τοῦ πυρετοῦ θέρμην. ἣν ὅταν τις ἀμαθὴς ἰατρὸς
αὐξήσῃ τῷ μὴ κενῶσαι τὸ τῆς στεγνώσεως αἴτιον,
ἄπορος ἡ λοιπὴ πᾶσα βοήθεια γίγνεται. τοῦ μὲν γὰρ
πλήθους κενοῦσθαι δεομένου, τῆς θέρμης δ᾽ ἐμψύχε-

though the present discussion is not about abundance in respect of the capacity. For this does not block up the openings of the vessels so as to prevent cooling, nor does it stretch their walls, nor does it produce redness or swelling, and much more certainly it does not impede the transpirations of the flesh or the skin.

But my whole argument was advanced (and will be advanced no less now) about what can no longer be contained by the vessels, and what, because of this, is itself accustomed to break and create an opening in these, and bring the other symptoms now mentioned, i.e. blockage, tension, swelling and redness. For this abundance, as long as it is useful, increases naturally with the strength of the capacity so that, if it should weigh this down at some time, it will not be further increased and will cease to be useful. If it loses its natural good order once and for all, it is impossible for it not to putrefy, since we see none of the other moist and hot bodies in the external world remaining unputrefied. Therefore, we must hasten to remove blood so that, by transpiration and ventilation, the nature of the vessels, preserving the natural *eukrasia*, prevails over the humors in the same way from the beginning because, if it suffers by becoming *dyskratic*, there is a danger of putrefaction to the vessels. The *dyskrasia* occurs in this due to the heat of the fever. Whenever some ignorant doctor increases this by not evacuating the cause of the stoppage of the pores, every remedy henceforth becomes ineffectual. For when the abundance needs to be evacuated and the

619K

σθαι, μάχη καὶ στάσις εἰς ἄλληλα τοῖς ἰωμένοις
βοηθήμασι γίγνεται, τῶν μὲν ἐμψυχόντων, εἰ καὶ τὴν
εὐκρασίαν ἐκπορίζοιτο τῇ φύσει, κατεχόντων γοῦν τὸ
πλῆθος ἐν τῷ σώματι, τῶν δὲ τῇ μανώσει τοῦ δέρματος ἐκκενούντων αὐτὸ θερμαινόντων ἁπάντων, ὥστ' εἰ
μὲν τοῦτο κενοῦν ἐθέλοις, αὐξήσεις τὸν πυρετόν, εἰ δ'
ἐκεῖνον ἐμψύχειν, καθέξεις τοῦτο. ταύτην τὴν ἀπορίαν
τῶν ἰαμάτων οἱ μὴ κενώσαντες εὐθὺς ἐξ ἀρχῆς ἐργάζονται, πεισθέντες Ἐρασιστράτῳ τῷ μηδὲ τὰς αὐτο-
μάτους αἱμορραγίας μιμήσασθαι δυνηθέντι.

κάλλιστον μὲν οὖν εὐθέως ὡς ἐθεάσω διαπαντὸς
ἡμᾶς ποιοῦντας, οὐχ ἡμερῶν ἀριθμῷ ἐπὶ παντὸς πράγματος προσέχειν τὸν νοῦν, ἀλλὰ τῇ ῥώμῃ μόνῃ τῆς
620K δυνάμεως | ἐπὶ τῶν τοιούτων πυρετῶν· εἰ γὰρ αὐτὴ
διασῴζοιτο, μὴ μόνον ἑκταίους ἢ ἑβδομαίους, ἀλλὰ
καὶ κατὰ τὰς ἑξῆς ἡμέρας φλεβοτομεῖν. εἰ δ' ἀναγ-
κασθείης ποτὲ θεραπεύειν ἄρρωστον, ᾧ μὴ μόνον
ἔμπροσθεν παρελείφθη τὸ τῆς φλεβοτομίας βοήθημα, ἄλλα καὶ νῦν ἤτοι διὰ τὴν ἀμαθίαν τῶν ἰατρῶν, ἢ
τὴν τοῦ κάμνοντος, ἢ τὴν τῶν οἰκείων αὐτῶν δειλίαν,
ἐπὶ προήκοντι τῷ χρόνῳ κωλυθείης ἀφαιρεῖν αἵματος,
ἐπὶ τὴν τοῦ ψυχροῦ δόσιν ἔρχεσθαι διορισάμενος
ἀκριβῶς ὁπόση τις ἐξ αὐτοῦ γενήσεται βλάβη. μι-
κρᾶς μὲν γὰρ ἢ οὐδ' ὅλως ἐσομένης διδόναι πίνειν
ἀκραιφνὲς ψυχρὸν ὅσον ἂν ὁ κάμνων βούληται· ἔτι
καὶ μᾶλλον θαρρῶν, εἰ ψυχροπότης εἴη. μεγάλης δὲ
τῆς βλάβης προσδοκωμένης ἀφίστασθαι μὲν τῆς
δόσεως, τοῖς δὲ ἄλλοις χρῆσθαι βοηθήμασιν, ἃ τάς τ'

heat needs to be cooled, natural struggle and discord arise among the curing remedies. This is because those remedies that are cooling, if they bring about a *eukrasia* in nature, retain the abundance in the body, while all those that are heating evacuate [the abundance] by the rarefaction of the skin. The result is that, if you wish to evacuate this, you will increase the fever, whereas if you wish to cool the fever, you will retain this. Those who do not carry out evacuation right from the start bring about this ineffectiveness of the cures, being persuaded by Erasistratus, who was not able to imitate the spontaneous hemorrhages.

Therefore, just as you saw me doing throughout, it is best not to direct your attention immediately to the number of days in regard to the whole matter, but only to the strength of the capacity in such fevers, for if this is to be preserved, carry out phlebotomy, not just on the sixth or seventh day but also on the days following. If, however, you are compelled at some time to treat a sick person in whom the remedy of phlebotomy has not only been neglected previously but in whom now also, due to the ignorance of the doctors or the timidity of the patient or of the members of his household, you are prevented from withdrawing blood by the progression of time, then resort to giving a cooling remedy, after determining precisely how great a harm will arise from this. If the harm will be small or nonexistent, give pure cold water to drink, as much as the patient wishes, and be confident about giving still more if he is a cold water drinker. However, when severe harm is anticipated, refrain from giving [cold water] and use instead other remedies which remove the obstructions, evacuate

620K

ἐμφράξεις ἐκφράττει καὶ τὸ πλῆθος κενοῖ καὶ πραΰνει
τὸ ζέον τῶν πυρετῶν. αἱ δ' ἐξ ἀκαίρου ψυχρᾶς πόσεως
ἢ ἀμέτρου βλάβαι κατὰ τάδε γίγνονται, τοὺς γλίσχρους καὶ παχεῖς καὶ πολλοὺς χυμούς, εἴτ' ἔμφρα-
621K ξιν, εἴτε σῆψιν, εἴτε φλεγμονήν, εἴτε | ἐρυσιπελατώδη
διάθεσιν, ἢ σκιρρώδη τύχοιεν, ἢ οἰδηματώδη πεποιη-
μένοι, κωλύει λεπτύνεσθαί τε καὶ διαφορεῖσθαι.

ὅταν οὖν ἐκ τούτων μὲν ὁ πυρετὸς ἀνάπτηται,
μηδὲν δὲ εἰς τὴν κένωσιν αὐτῶν ἡ τοῦ ψυχροῦ δόσις
ὠφελῇ, παραχρῆμα μὲν οὐκ ὀλίγην φέρει τὴν ῥαστώ-
νην ἐπὶ τὸ σβέσαι τὸν ἤδη γεγονότα πυρετόν, ἅτε δὲ
διαμενούσης τῆς αἰτίας αὖθις ἕτερον ἀναγκαῖον ἀνάπτεσθαι καὶ πολλάκις γε χαλεπώτερον τοῦ πρόσθεν,
ὅταν ἐκ τοῦ ψυχροῦ πυκνωθῇ τὸ σῶμα. τουτὶ μὲν δή
σοι βλάβης εἶδος ἓν οὐκ εὐκαταφρόνητον. ἕτερον δὲ
τοιόνδε· πολλὰ τῶν ἀσθενεστέρων τοῦ κάμνοντος μορίων εἴτε διὰ φυσικὴν δυσκρασίαν εἴτε δι' ἐπίκτητον
βλάβην ὑπὸ τοῦ ψυχροῦ πλήττεται. τῷ μὲν γὰρ στόμαχος ἔπαθεν οὕτως ἰσχυρῶς ὡς μόγις καταπίνειν, τῷ
δ' ἡ γαστὴρ ὡς μόγις πέττειν, ἄλλῳ δὲ τὸ στόμα τῆς
γαστρός, ἢ τὸ ἧπαρ, ἢ τὸ κῶλον, ἢ ὁ πνεύμων, ἢ αἱ
φρένες, ἢ καὶ νὴ Δία νεφροὶ καὶ κύστις ἤ τι τοιοῦτον
ἕτερον ὑπὸ τοῦ ψυχροῦ πληγὲν ἄρρωστον εἰς τὴν
οἰκείαν ἐνέργειαν ἐγένετο. τινὲς δὲ αὐτῶν ἐξ ἀκαίρου
622K τε καὶ ἀμέτρου πόσεως οὐκ εἰς μακρὰν οὐδ' εἰς | ὕστε-
ρον, ἀλλ' αὐτίκα δυσπνοίαις καὶ σπασμοῖς καὶ τρό-
μοις ἁλίσκονται καὶ συλλήβδην εἰπεῖν κακοῦνται πᾶν
τὸ νευρῶδες γένος.

the abundance, and ameliorate the seething of those who are febrile. The harms which arise from an untimely or excessive cold drink prevent the thinning and dispersal of viscid, thick and abundant humors when these have created obstruction, putrefaction, inflammation, or an erysipelitic condition, whether it happens to be scirrhous or edematous.   621K

Therefore, whenever the fever is kindled from these things, while the giving of cold water is of no benefit to the evacuation of these humors, it does immediately bring significant relief by quenching the fever that has already occurred. But insofar as the cause still remains, another fever is inevitably kindled, and this is often harder to deal with than the previous one when the body is thickened from the cold. Certainly this is one kind of harm you should not take lightly. And another is this: many of the weaker parts of the patient are struck, either by a natural *dyskrasia* or by an acquired injury due to cold. For the esophagus is strongly affected by this in such a way that swallowing is difficult, and the stomach in such a way that digestion is difficult. Another is when the opening of the stomach, or the liver, colon, lungs or diaphragm, or also, by Zeus, the kidneys or bladder or some other such thing, is struck by cold, and becomes weak in terms of its specific function. Some people, from an untimely and excessive drink, are seized after   622K a short interval or even straight afterward by dyspnea, spasms and tremors and, in short, become distressed with respect to whole nervous class (nervous system).

οἶδ' ὅτι φοβερὸν ἄν σε πρὸς τὴν τοῦ ψυχροῦ δόσιν
εἰργασάμην ἐξ ὧν εἶπον, εἰ μή με πολλάκις ἐθεάσω
μὲν χρησάμενον, ἀεὶ δ' ὠφελήσαντα χωρὶς τοῦ βλάψαι τι τὸν κάμνοντα σαφές. ὅσοις μὲν γὰρ ἐν κυρίῳ
μορίῳ φλεγμονώδης ὄγκος ἢ οἰδηματώδης ἢ σκιρ-
ρώδης ἐστίν, οὐ χρὴ τούτοις διδόναι τὸ ψυχρόν· οὐ
μὴν οὐδ' ὅσοις ἔμφραξις ἢ σῆψις χυμῶν ἄπεπτος. εἰ
δ' ἐναργῆ βλέποις τὰ τῆς πέψεως σημεῖα, χωρὶς τῶν
εἰρημένων ὄγκων ἐκεῖνο μόνον ἔτι διάσκεψαι, μή τι
μόριον οὕτω ψυχρὸν εἴη τὴν κρᾶσιν ὡς εἰς αὐτὸ
κατασκῆψαι τὴν βλάβην. ἐρυσίπελας δὲ τὸ γοῦν
ἀκριβὲς οὐκ ἂν ἄλλως ἰάσαιο. συμμιγὲς δ' εἴπερ εἴη
φλεγμονῇ, τὰ τῆς πέψεως ἀναμεῖναι χρὴ γνωρίσματα.
ταῦτα μὲν οὖν ἐπιπλέον ἢ κατὰ τὴν ἐνστῶσαν ὑπό-
θεσιν, ὅθεν ἴσως καὶ αὖθις ἀναγκαῖον ἔσται ποτὲ
δόσεως ψυχροῦ μνημονεῦσαι.

νυνὶ γὰρ ὁ μὲν ἕτερος τῶν συνόχων ἐμφράξει μόνῃ
623K τὴν γένεσιν εἶχεν, ὁ δ' ἕτερος | ἅμα σήψει καθ' ὅλας
τὰς φλέβας. ὅταν οὖν ποτ' ἐπ' αὐτῶν ἴδῃς τὰ τῆς
πέψεως τῶν χυμῶν σημεῖα, περὶ ὧν αὐτάρκως ἐν
τοῖς Περὶ κρίσεων εἴρηται, θαρρῶν διδόναι τὸ ψυχρόν. ἡ γάρ τοι φύσις τῶν στερεῶν τοῦ ζῴου μορίων
ῥωσθεῖσα τοῖς προλελεπτυσμένοις ἐπιτίθεται χυμοῖς·
ὥσθ' ὅσοι μὲν χρηστοὶ καὶ τρέφειν ἱκανοί, τούτους
μὲν ἕλκειν εἰς ἑαυτά,[4] τοὺς δ' ἀχρήστους ἐκβάλλειν
ἤτοι διὰ τῆς γαστρὸς ἢ διὰ τοῦ δέρματος. εἰ δὲ

⁴ K; αὐτά B

484

I know that I might have made you apprehensive of the administration of cold water from what I said, if you hadn't seen me use it often, always benefiting the patient without there being apparent harm. You must not give cold water to those in whom there is an inflammatory, edematous or scirrhous swelling in an important part, nor to those in whom there is a blockage or putrefaction of humors which are unconcocted. If you see the clear signs of concoction without the aforementioned swellings, still examine that alone, lest some part is cold in terms of *krasis* in such a way that harm befalls it. Anyway, you would not cure pure erysipelas otherwise. And if it is mixed with inflammation, you must await the signs of concoction. These things have a broader relevance than just in relation to the present hypothesis, on which account it will perhaps be necessary to make mention again of the administration of cold water at some point.

For the present, in one case it pertains to the genesis of continuous fevers from blockage alone; in another case it 623K pertains to that with putrefaction in all the veins at the same time. Therefore, whenever at any time you see the signs of concoction of the humors in the veins (enough was said about these in the writings *On Crises*),[6] give cold water with confidence. For once strengthened, the nature of the solid parts of the organism adds to the humors that have already been thinned, so that they draw to themselves those things that are useful and sufficient to nourish, but cast out those things that are useless, either through the stomach or through the skin. If the patient is a drinker of

[6] *De crisibus*, IX.550–768K; see particularly Book 2, chapter 6 (IX.662–66K).

ψυχροπότης ὁ κάμνων εἴη, πάνυ θαρρῶν δίδου τὸ
ψυχρόν, αὐτῇ τῇ πείρᾳ δεδιδαγμένος ἀνέχεσθαι πάντα τὰ σπλάγχνα τῆς ὁμιλίας αὐτοῦ.

πάντως γὰρ εἴ τι ψυχρὸν οὕτως ὑπῆρχεν ὡς πλήτ-
τεσθαι πρὸς αὐτοῦ κατὰ τὸν τῆς ὑγείας χρόνον, ἐναργῆ τὴν βλάβην ἐνεδείξατο ἄν· μηδενὸς δὲ μηδὲν
μορίου βλαβέντος οὐδ᾽ ἂν ἐν τῷ πυρέττειν βλαβείη τι.
ὅπου γὰρ καὶ ἀήθεις ἔνιοι ψυχροῦ, διὰ καυσώδη
πυρετὸν ἀναγκασθέντες πιεῖν οὐδὲν ἐβλάβησαν, οὔ-
που γε⁵ τῶν ἐθάδων ἄν τις βλαβείη, πρόβλημα μέγιστον ἔχων ἐξ ἐπιμέτρου τὸ πλῆθος τῆς θερμασίας.
624K αὕτη γὰρ ἐν ταῖς εὐρυχωρίαις | τῶν ἀγγείων πολλὴ
περιεχομένη, τῆς ἀερώδους οὐσίας ἐν αὐτοῖς καὶ
προσέτι τῶν χυμῶν ἁπάντων ἐκπεπυρωμένων, ὅσον
ὑπὸ τοῦ ψυχροῦ πάσχει ψυχομένη, τοσοῦτον εἰς αὐτὸ
δρᾷ θερμαίνουσα. καὶ διὰ τοῦτο τοῖς ὀλίγον αἷμα καὶ
σάρκας ἔχουσιν ἡ πόσις τοῦ ψυχροῦ σφαλερωτέρα·
ταχὺ γὰρ ἐπὶ τὰ στερεὰ τοῦ ζῴου μόρια διικνεῖται,
μηδενὶ προσεντύγχανον⁶ ὑφ᾽ οὗ θραυσθήσεται. διὰ
τοῦτο δὲ καὶ οἱ ἑκτικοὶ τῶν πυρετῶν οὔτ᾽ ἀκραιφνοῦς
ὁμοίως οὔτε πολλοῦ χρῄζουσι τοῦ ψυχροῦ, λεπτοῖς
καὶ ὀλιγαίμοις τοὐπίπαν ἐγγινόμενοι σώμασιν. ἀλλὰ
περὶ μὲν ἐκείνων αὖθις εἰρήσεται·

ἰάματα δὲ συνόχων πυρετῶν δύο ταῦτ᾽ ἐστὶ μέγιστα, φλεβοτομία καὶ ψυχρόν. ἀλλ᾽ ἐκείνη μὲν ἐν παντὶ

---

⁵ K; ἢ πού γε B   ⁶ B; προσεντυγχάνων K

---

⁷ It is not entirely clear what Galen means here. What he

cold water, you may be particularly confident about giving it to him, for I have been taught by experience itself that all the organs tolerate association with it.

All in all, if something is so cold as to be smitten by this at a time of health, it would clearly have revealed the damage. Providing no part is damaged in any way, nothing would be harmed at all during the course of the fever. Also, when some who are unaccustomed to cold are forced to drink cold water due to a burning fever are not harmed at all, how much less would those who are accustomed [to drinking it] be harmed. This is because the abundance of heat acts as a significant defense into the bargain. For since   624K the major part of this heat is contained in the lumina of the vessels, the airy substance in them, and in addition, all the humors have been burned up, to the extent that while it is being cooled it suffers due to the cold drink, so much when it is heating does it do this.[7] And because of this, for those who have little blood and flesh, a cold drink is more dangerous, as it penetrates quickly to the solid parts of the organism, meeting nothing by which it will be broken up. Also because of this, the hectic fevers require cold that is neither pure, similar in degree, nor considerable in amount, since they are in thin and, in general, relatively bloodless bodies. But more will be said about those in due course.

There are these two major cures of continuous fevers: phlebotomy and cold [water]. The former is appropriate at

seems to be saying is that the heat of the fever is, for the most part, contained in the lumina of the blood vessels—that is, in the *pneuma* and overheated humors—so that whatever harm might be inflicted by the cooling drink is countered by what the heat itself does.

καιρῷ, φερούσης γε τῆς δυνάμεως· ἡ δὲ τοῦ ψυχροῦ
πόσις, ὅταν μὲν τοῖς σφυγμοῖς καὶ τοῖς οὔροις τὰ τῆς
πέψεως ἐναργῆ σημεῖα βλέπῃς,[7] μέγιστος δὲ πυρετὸς
ᾖ.[8] προσεπιβλέπειν δὲ τοῖς τῆς φλεβοτομίας σκοποῖς
τά τε προηγούμενα καὶ τὰ πάντως ἐπακολουθήσοντα.
προηγησαμένης γὰρ ἀπεψίας σιτίων, τοσοῦτον χρό-
625K νον ἀναβάλλεσθαι κέλευε τὴν φλεβοτομίαν ὅσος | ἂν
ἱκανὸς εἶναί σοι δόξῃ πρός τε τὴν πέψιν αὐτῶν καὶ τὴν
τῶν περιττωμάτων ὑποχώρησιν. ἑπομένης δέ τινος ἐξ
ἀνάγκης κενώσεως, ἀπολιπεῖν αὐτῇ τοῦ περιττοῦ τοσοῦτον ὅσον μέλλει κενώσειν. ὥστε εἴτε καταμήνια
τύχοι κινηθέντα[9] κατὰ τὸν τῆς φλεβοτομίας καιρόν,
εἴθ᾽[10] αἱμορροῒς ἀναστομωθεῖσα θεασάμενος τοῦ φε-
ρομένου τὴν ὁρμήν, εἰ μὲν ἱκανὸν αὐτὸ φαίνοιτο μόνον
ἐκκενῶσαι τὸ δέον, ἐπιτρέπειν τῇ φύσει τὸ σύμπαν· εἰ
δὲ μή, τοσοῦτον ἀφαιρεῖν αὐτοῦ, ὡς ἐξ ἀμφοῖν συν-
τεθέντων ἀνυσθῆναι τὸ προσῆκον.

ὥσπερ δ᾽ ἐπὶ τούτων τὴν φυσικὴν κένωσιν αἰσθητὴν οὖσαν οὐ μικρὸν χρὴ τίθεσθαι σκοπόν, οὕτως
ἑτέρωθι τὴν φύσιν ἢ τὴν ἡλικίαν ἢ τὴν κατάστασιν.
ἔνιοι μὲν γὰρ εὐδιαφόρητοι φύσει, παῖδες δ᾽ ἀεὶ διὰ
τὴν ἡλικίαν· ἡ κατάστασις δ᾽ ὅταν ᾖ θερμή τε ἄγαν
καὶ ξηρά. καλῶ δὲ δηλονότι κατάστασιν τὴν τοῦ
περιέχοντος ἡμᾶς ἀέρος κρᾶσιν, ἐν ᾗ καὶ χώρα καὶ
ὥρα περιείληπται· καὶ γὰρ καὶ τούτων ἑκάτερον ἔχει
τὴν ἔνδειξιν ἐκ τῆς τοῦ περιέχοντος κράσεως. εὐδια-
626K φόρητοι | δ᾽ εἰσὶν οἱ ὑγροὶ τὴν φύσιν ἅπαντες καὶ
μᾶλλον ἅμα θερμότητι, καὶ οἱ ἀραιοὶ τὴν ἕξιν, ἔτι δ᾽

any time, at least when the capacity will bear it; a cold drink
is appropriate whenever you detect clear signs of concoction in the pulse and urine, and the fever is very severe.
Among the indicators of phlebotomy, look as well at what
preceded, and especially at what will follow. When failure
of digestion of foods preceded, direct the phlebotomy to
be delayed for such a time as seems to you sufficient for the   625K
digestion of these foods and the excretion of the superfluities. Since some evacuation necessarily follows, leave
behind for it as much of the superfluity as it is going to
evacuate. So that, if either the menstrual flow happens to
be set in motion at the time of the phlebotomy, or piles are
opened up, after observing the rush of what is being
brought forth, if it seems to be sufficient alone to evacuate
what is required, entrust the whole matter to Nature. If
not, remove an amount of blood such that what is appropriate is achieved by the joint action of both.

Just as in these matters, natural evacuation, since it is
perceptible, should be placed as a major indicator, so too it
must otherwise be placed in the nature, age, and climatic
conditions. For some have a natural propensity to disperse
(children always have this due to their age), and climatic
conditions when they are hot and very dry [favor disper-
sion]. Obviously, I call climatic conditions the *krasis* of the
air surrounding us in which both the place and season are
included, for each of these also has the indication of the
krasis of the ambient air. Those with a natural propensity   626K
to disperse are all those who are moist in nature, and particularly if this is associated with heat, and those who are
thin in habitus, and further, those in whom the opening

---

7 K; βλέπηται σημεῖα B, *recte fort.*   8 B; εἴη K
9 K; κενωθέντα B   10 B; εἴτ᾽ K

οἷς τὸ στόμα τῆς γαστρὸς ἢ πικρόχολον, ἢ ἄρρωστον,
ἢ πέρα τοῦ δέοντος αἰσθητικόν. οὗτοι μὲν οὖν οἱ
ἀντιπράττοντες τῇ φλεβοτομίᾳ σκοποί. συνενδεικνύ-
μενοι δ' αὐτὴν τῇ πρώτως χρῃζούσῃ διαθέσει, σκληρὰ μὲν ἡ ἕξις καὶ πυκνὴ καὶ τὸ σύμπαν φάναι δυσδια-
φόρητος· ὑγρὸν δὲ ἢ ψυχρὸν τὸ περιέχον.

εὐλόγως δήπου τὰ μὲν συνενδείκνυται τὴν φλεβοτομίαν, τὰ δὲ ἀντενδείκνυται· κένωσις μὲν γὰρ αὐτῆς ὁ
σκοπός. τῶν δ' εἰρημένων τὰ μὲν ἐπέχει τὰς διαπνοάς,
τὰ δὲ προτρέπει· καὶ τὰ μὲν εὐκένωτά ἐστι, τὰ δὲ οὔ.
κωλύει μὲν οὖν διαπνεῖσθαι τὰ σώματα στέγνωσίς τε
τῶν πόρων καὶ τοῦ περιέχοντος ἡμᾶς ἀέρος ὑγρότης
τε καὶ ψύξις, προτρέπει δὲ τὰ τούτων ἐναντία. τῶν
κενουμένων δ' αὐτῶν τὰ μὲν ὑγρὰ καὶ θερμὰ καὶ λεπτὰ
διαφορεῖται τάχιστα, τὰ δὲ παχέα καὶ γλίσχρα καὶ
ψυχρὰ δυσδιαφόρητά ἐστι. ταυτὶ μὲν οὖν εἴς γε τὰ
627K παρόντα περὶ φλεβοτομίας | τε καὶ ψυχρᾶς πόσεως
ἐγνῶσθαι κάλλιον. ἅπερ ἀμφότερα πολλοὺς ἐθεάσω
δεδιότας ἀγυμνάστους ἰατρούς· ὧν τοὺς μὲν αἱμοφό-
βους, τοὺς δὲ ψυχροφόβους ὀνομάζομεν, ὥσπερ ὑδρο-
φόβους τοὺς λυττῶντας· εἰσὶ γὰρ ἀμέλει καὶ τούτων
ἔνιοι διὰ τὴν περὶ τὰ δόγματα σπουδὴν οὐκ ἐν βρα-
χείᾳ λύττῃ.

καταλιπόντες οὖν αὐτοὺς ἐχώμεθα τῶν ἐξ ἀρχῆς
ἡμῖν προκειμένων ἀναμνησθέντες ὡς ἐκ τῶν ἐφημέρων πυρετῶν ὁ λόγος εἰς τοὺς συνόχους ἀφίκετο τῇ
κοινωνίᾳ τῶν συμβαινόντων αὐτοῖς. ἐπειδὴ γὰρ ἡ τῆς
ἐμφράξεως διάθεσις, ὅταν μὲν αὐτή τε σμικρὰ τύχοι

(cardiac orifice) of the stomach is either full of bitter bile (picrocholic) or weak, or overly sensitive. These are the indicators of things that act in opposition to phlebotomy. Those things jointly indicating this (i.e. phlebotomy) primarily with the needful condition are a state that is hard and thick, and, in summary, that is difficult to disperse, and ambient air that is moist and cold.

It is reasonable, of course, that some things jointly indicate phlebotomy, while others contraindicate it, since its objective is evacuation. Some of the things mentioned hold back transpiration but others urge it on; and some things are easily evacuated while some are not. Stoppage of the pores prevents bodies transpiring as do moistness and coldness of the ambient air, whereas the opposites of these urge it on. Of the evacuations themselves, those that are moist, hot and thin are dispersed very quickly, while those that are thick, viscid and cold are difficult to disperse. It is better to know these things, at least in relation 627K to the present issues of phlebotomy and cold drinks. Both of these are things you saw many unpracticed doctors afraid of—those doctors whom I call "hemophobes" and those I call "psychrophobes," just as I call those bitten by a rabid dog "hydrophobes." For some of them, due no doubt to their zeal for their own dogmas, are to no small degree mad.

Leaving aside these doctors, let me concern myself with those things I put forward at the beginning, when I mentioned that the argument proceeds from the ephemeral to the continuous fevers due to the commonality of the things occurring together in them. For the condition of blockage, whenever it happens to be slight and is well at-

GALEN

καὶ καλῶς παιδαγωγηθῇ, τὸν ἐφήμερον ἐργάζεται
πυρετόν. ὅταν δ' ἤτοι διὰ μέγεθος ἢ διὰ ἀμαθίαν
ἰατρῶν ἐκπέσοι τοῦ συνήθους χρόνου τῶν ἐφημέρων,
ἤτοι σύνοχον ἤ τινα τῶν περιοδιζόντων, εἰκότως ὁ
περὶ τοῦ τοῖς ἐφημέροις ὁμογενοῦς συνόχου λόγος
ἡμᾶς ἐξεδέξατο· διὰ δὲ τούτων εὐθέως καὶ ὁ τοῦ μετὰ
σήψεως χυμῶν. ἔδοξέ τε βέλτιον εἶναι σαφηνείας
628K ἕνεκα καὶ | γυμνασίας ἑνὸς ἀρρώστου μνημονεῦσαι
καθ' ἑκάτερον εἶδος τῶν πυρετῶν.

6. Ἐπεὶ τοίνυν πέπρακται τοῦθ' ἡμῖν, ἐναργῶς τε
δέδεικται δι' αὐτῶν, ὅπερ ἐν τῷ πρὸ τούτου γράμματι
κατὰ τὴν τελευτὴν ἐλέχθη, τὸ χρῆναι τὰς ἐνδείξεις ἐπὶ
τῶν ἄλλων πυρετῶν, ὅσοι μηκέτ' εἰσὶν ἐφήμεροι, μὴ
μόνον ἀπὸ τῶν ἐν ἐκείνῳ τῷ λόγῳ διδαχθέντων σκοπῶν, ἀλλὰ καὶ τῆς τῶν πυρετῶν ἐργαζομένης αἰτίας
καὶ τῆς τοῦ κάμνοντος δυνάμεως λαμβάνεσθαι, πάλιν
ἐπὶ τὴν καθόλου γυμνασίαν ἀνέλθωμεν, εὖ εἰδότες ὡς
οὐχ οἷόν τ' ἐστὶ τέχνης οὐδεμιᾶς ἐπιστήμην κτήσασθαι χωρὶς τοῦ μέθοδον μέν τινα διὰ τῶν καθόλου
λεγομένων θεωρημάτων, ἄσκησιν δὲ διὰ τῶν ἐν μέρει
λαμβάνειν παραδειγμάτων. οὔτε γὰρ οἷόν τε χωρὶς
τοῦ γυμνάσασθαι πολυειδῶς ἐν τοῖς κατὰ μέρος ἐπὶ
τῶν καμνόντων, ἃ χρὴ πράττειν· οὔτ' αὐτὴν τὴν
γυμνασίαν ἐγχωρεῖ γίγνεσθαι προσηκόντως ἄνευ τῆς
τοῦ καθόλου γνώσεως· ἐν ἐκείνοις μὲν γὰρ ἡ μέθοδος,
ἡ δ' ἄσκησις ἐν τοῖς κατὰ μέρος. ὥσπερ οὖν ὅσοι
βαδίσαι τινὰ ὁδὸν ἐφίενται τοῖς σκέλεσιν ἀμφοτέροις
629K ἐν | μέρει χρῶνται, θατέρῳ δ' εἴ τις σκάζων μόνῳ

492

tended to, brings about an ephemeral fever, whereas it falls outside the characteristic time of the ephemeral fevers whenever, due to its magnitude or the ignorance of the doctors, it is either continuous or is one of the periodic fevers. Therefore, my discussion about the continuous fevers being cognate with the ephemeral fevers reasonably took my attention; and via these, the discussion of the fever with putrefaction of humors immediately took my attention also. And it seemed better, for the sake of clarity and 628K practice, to mention a specific patient, sick due to each kind of fever.

6. Accordingly, having done these things, I have shown clearly by means of them what was said at the end in the book prior to this one, which is that in the other fevers, those that are no longer ephemeral, there is a need to take the indications not only from the indicators taught in that discussion, but also from the cause creating the fever, and from the capacity of the patient. Let me return once more to [the matter of] practice in general, knowing full well that it is impossible to acquire the knowledge of any craft without a method based on principles stated generally and practice based on examples taken individually. For in the individual patient, it is impossible to know what you must do without being practiced diversely in particular patients. Nor is it possible for the practice itself to be appropriate without knowledge of the general, for the method lies in the principles while the practice lies in individual instances. Therefore, just as those who proceed to walk along a certain path using both legs, use them in turn, 629K whereas someone who is limping and uses one leg only,

493

χρῷτο, παμπόλλῳ τε χρόνῳ καὶ μετὰ τοῦ σφάλλεσθαι
πολλάκις ἀνύσει τὴν πορείαν, οὕτως ὅστις ἐπὶ τέλος
ἡστινοσοῦν ἀφικέσθαι τέχνης ἐθέλει, χρηστέον αὐτῷ
τοῖς δύο τούτοις οἷόν περ σκέλεσιν ἢ ὀργάνοις ἢ ὅπως
ἄν τις ὀνομάζειν βούληται· μεθόδῳ μὲν ἐν τοῖς καθόλου θεωρήμασιν, ἀσκήσει δ᾽ ἐν τοῖς κατὰ μέρος.

7. Ἀνελθόντες οὖν αὖθις ἐπὶ τὴν μέθοδον ὑπὲρ τῆς
τοῦ κάμνοντος ἐπισκεψώμεθα δυνάμεως, εἴτε συντελεῖ
τι πρὸς τὴν θεραπείαν, εἴτ᾽ οὐδὲν ὅλως, ὡς ἔδοξέ τισιν.
ἔοικε δὲ τοῖς ἄχρηστον εἰς θεραπείας εὕρεσιν εἶναι
φαμένοις τὴν δύναμιν οὐκ οἰκεία τοῖς ἔργοις τῆς
τέχνης ἡ σκέψις, ἀλλὰ λογικωτέρα μᾶλλον, ὡς σοφισταῖς γεγονέναι. λέγουσι μὲν γὰρ ἐκ τῆς διαθέσεως
ἣν θεραπεύομεν εἶναι τὴν ἔνδειξιν ὧν χρὴ πράττειν,
οὐκ ἐκ τῆς δυνάμεως. ἀμέλει καὶ ὥρας καὶ χώρας καὶ
καταστάσεις καὶ ἡλικίας καὶ ἔθη καὶ κράσεις σωμάτων ἀχρήστους εἶναί φασι κατὰ τὸν αὐτὸν λόγον. εἶθ᾽
ὅταν ἀναγκάζονται τὴν θεραπείαν ἐξαλλάττειν, διὰ
ταῦτα τῶν ἐν ἀρχῇ ῥηθέντων ἑαυτοῖς ἐπιλανθανό-
630K μενοι, τὰ μὲν αὐτῶν ἀντενδείκνυσθαί | φασιν ἐνίοτε
τοῖς θεραπευτικοῖς σκοποῖς, τὰ δ᾽ εἰς ὕλας βοηθημάτων εἶναι χρήσιμα, καθάπερ καὶ τὰ μόρια τοῦ ζῴου.
καὶ γὰρ ἐπ᾽ ἐκείνων τὴν αὐτὴν στρέφονται στροφήν,
ἐν ἀρχῇ μὲν τῶν λόγων ἄχρηστα πρὸς τὴν τῆς θεραπείας εὕρεσιν ὑπάρχειν αὐτὰ φάσκοντες, ὕστερον δ᾽
ὅτ᾽ ἂν ἐξελέγχωνται, πρὸς μὲν τὴν ἔνδειξιν τῆς θεραπείας ἄχρηστα λέγοντες, εἰς δὲ τὴν τῆς ὕλης ἐξάλλα-

manages to walk but takes a very long time and makes many mistakes, so someone who wishes to achieve the aim of any craft whatsoever must use these two limbs, as it were, or organs, or whatever one might wish to call them— method based on general principles and practice based on individual instances.

7. Therefore, since I am returning again to the method, let me consider the capacity of the patient and whether it has some bearing on the treatment, or none at all, as it seems to some. To those who say the capacity is useless for the discovery of treatment, it looks as though it is not a specific consideration for the actions of the craft, but rather something more theoretical, which is for "experts." They say the indication of the things we must do is from the condition which we are treating and not from the capacity. Of course, according to the same argument, they also say that seasons, places, climatic conditions, age, customs and the *krasis* of the body are useless. Then, when they are forced to change the treatment because of these factors, they forget what they themselves said at the beginning, and say 630K that sometimes these factors do give contrary indications to the therapeutic objectives, or are useful in regard to the materials of remedies, just as the parts of the organism are. Also, in the case of latter, they make the same volte-face, when they say at the start of their discussions that these are useless for the discovery of treatment, and yet later, when they are refuted in saying they are useless for the indication of treatment, they concede that they are necessary

ξιν, ἢ εὕρεσιν, ἢ διαφοράν, ἢ ποιότητα, καὶ γὰρ καὶ
ὀνομάζουσιν οὐχ ὡσαύτως ἅπαντες, ἀναγκαῖα συγ-
χωροῦντες ὑπάρχειν.

ὅπερ οὖν εἶπον ἀρτίως, ἀναλήψομαι καὶ νῦν, ὡς ἐκ
τοῦ χρησίμου τῆς τέχνης μεταβαίνοντες εἰς λογικὴν
ἀφικνοῦνται σκέψιν. ἔστι μὲν γὰρ τὸ χρήσιμον ἐξευρεῖν βοηθήματα δι᾽ ὧν ὁ κάμνων θεραπευθήσεται·
καταλιπόντες δ᾽ ἐκεῖνοι τοῦτο περὶ ὀνομάτων ἐρίζου-
σιν, ἐνδείξεις τε καὶ ἀντενδείξεις λέγοντες καὶ κοινό-
τητας καὶ σκοποὺς καὶ ὕλας βοηθημάτων, ὅσα τ᾽
ἄλλα τοιαῦτα τῇ διαφορᾷ τῶν ὀνομάτων ἀπολογεῖ-
σθαι νομίζοντες, ὑπὲρ ὧν ἀπεφήναντο ψευδῶς. ἐγὼ δ᾽
οὐ κωλύω μὲν αὐτοὺς ὀνόμασιν οἷς ἂν ἐθέλωσι χρῆ-
631K σθαι· μεμνῆσθαι δὲ ἀξιῶ τῶν ἐξ ἀρχῆς | προτεθέντων,
ἅπερ ἐστὶ βοηθημάτων εὑρέσεις· εἰς ἃς ὅ τι περ ἂν
φαίνηται συντελοῦν ὁπωσοῦν ἀναγκαῖον αὐτὸ φατέον
εἰς τὴν θεραπείαν ὑπάρχειν. οὕτως οὖν καὶ ἡ δύναμις
εἴτ᾽ ἀναγκαία σκοπεῖσθαι τοῖς ἰατροῖς ἐπὶ τῶν ἀρρω-
στούντων εἴτ᾽ οὐκ ἀναγκαία σκοπῶμεν. ἐγὼ μὲν γάρ
φημι πολλάκις ἀναγκαιοτάτην ὑπάρχειν αὐτήν, ὡς
πάντα σχεδόν τι τὰ περὶ τὸν κάμνοντα δι᾽ ἐκείνην
πράττεσθαι μόνην, ἔστι δ᾽ ὅτε μετρίως ἀναγκαίαν, ὡς
λαμβάνεσθαι μέν τι καὶ ἐξ αὐτῆς εἰς τὰ ποιητέα, μὴ
μέντοι πρὸς αὐτήν γε τὸ πᾶν κῦρος ἀναφέρεσθαι τῶν
πρακτέων. αὖθις δ᾽ ἄν σοι δείξαιμι τὴν δύναμιν οὕτω
βραχὺ συντελοῦσαν εἰς τὴν τῶν βοηθημάτων εὕρεσιν,
ὡς διὰ σμικρότητα καὶ λανθάνειν ἐνίοτε καὶ παραπέμ-
πεσθαι καὶ σιωπᾶσθαι, καθάπερ ὅλως οὐκ οὖσαν ἐκ

for the change of material (or discovery, or difference, or quality, for they do not all name this in the same way).

Therefore, I shall take up again what I said just now—that those who go beyond what is useful for the craft arrive at logical disputation. What is useful is to discover remedies with which the patient will be cured. When those men forsake this, they wrangle over names, speaking about indications and contraindications, about "communities" and indicators, and about materials of remedies and other such things, thinking they are speaking in defense of the differentiation of terms, which they speak about incorrectly. I am not preventing them using whatever names they might wish. I do, however, think it worthwhile to mention those things proposed from the beginning, which are the discoveries of remedies. In regard to these, whatever might, in some way or another, seem to accomplish something, we must say is necessary to treatment. Therefore, in like manner, the capacity is either something that it is necessary for doctors to consider in those who are sick, or it is not necessary. I say the capacity is frequently very necessary, insofar as almost everything that is done for the patient is done for the sake of capacity alone, and sometimes moderately necessary, in that something is taken from the capacity regarding what is to be done, although not everything important for what is done is referred to it. Contrariwise, I could show you the capacity contributing little in this way to the discovery of remedies because, due to its being small, it is sometimes neglected and overlooked, as if it were altogether not among the useful indicators. But what applies

631K

τῶν χρησίμων σκοπῶν. ἀλλ' ὅπερ ἐπὶ τῆς δυνάμεώς
ἐστι, τοῦτο καὶ ἐπὶ τῶν ἄλλων ἁπάντων ὑπάρχει τῶν
ἐνδεικνυμένων. ἄλλοτε γὰρ αὐτῶν ἄλλο τὸ μὲν ἧττον
δύναται, τὸ δὲ πλέον· ἐνίοτε δὲ οὕτως ἰσχυρόν ἐστιν
ὡς μόνον ἐνδείκνυσθαι δοκεῖν, ἢ οὕτως ἀσθενὲς ὡς
παραλείπεσθαι. |

632K   8. Ταῦτ' οὖν ἐπιδεικνύντι μοι πρόσεχε τὸν νοῦν,
ἐκεῖνο διὰ μνήμης ἔχων ὡς οὔτ' ἐν ἄλλοις τισὶ τὸ
μεθόδῳ θεραπεύειν ἐστὶν οὔτ' ἄλλος τις πρὸ ἡμῶν
διωρίσατο πάνθ' ἑξῆς αὐτά, καίτοι γε ὑφ' Ἱπποκράτους εὑρημένης τῆς ὁδοῦ. ταυτὶ γὰρ ἃ νῦν ἐγὼ
μέλλω διέρχεσθαι τὴν θεραπείαν ἐκδεικνύμενα πρῶτος ἁπάντων ἐκεῖνος ἔγραψεν· ἀλλ' ὡς ἂν πρῶτος
εὑρίσκων οὔτε τὴν προσήκουσαν ἅπασιν ἐπέθηκε
τάξιν οὔτε τὴν ἀξίαν ἑκάστου τῶν σκοπῶν ἀκριβῶς
ἀφωρίσατο, παρέλιπέ τέ τινας ἐν αὐτοῖς διορισμούς,
ἀσαφῶς τε τὰ πλεῖστα διὰ παλαιὰν βραχυλογίαν
ἑρμήνευσε. καὶ δὴ καὶ περὶ τῶν ἐπιπεπλεγμένων διαθέσεων ὀλίγιστα παντάπασιν ἐδίδαξε. συνελόντι δὲ
φάναι τὴν ἐπὶ τὰς ἰάσεις ὁδὸν ἅπασαν μέν μοι δοκεῖ
τέμνεσθαι, δεομένην μέντοι γ' ἐπιμελείας εἰς τὸ τέλει-
ον, ὥσπερ καὶ νῦν ὁρῶμεν ἐνίας τῶν ἐπὶ τῆς γῆς ὁδῶν
τῶν παλαιῶν ἢ πηλῶδές τι μόριον ἑαυτῶν, ἢ λίθων, ἢ
ἀκανθῶν πλήρες, ἢ λυπηρῶς ὄρθιον, ἢ κάταντες σφα-
λερῶς, ἢ θηρίων πλῆρες, ἢ διὰ μέγεθος ποταμῶν
633K δύσβατον, ἢ μακρόν, ἢ | τραχὺ κεκτημένας.

    ἀμέλει ταῦτ' ἐχούσας ἁπάσας τὰς ἐπὶ τῆς Ἰταλίας
ὁδοὺς ὁ Τραϊανὸς ἐκεῖνος ἐπηνωρθώσατο, τὰ μὲν ὑγρὰ

in the case of the capacity also applies in the case of all the other things that act as indicators; for at one time one of them carries less weight and at another time more. Sometimes one is so strong that it seems to be the sole indicator; at other times one is so weak as to be left aside.

8. Heed me then, as I demonstrate these things, keep- 632K ing this in mind: treatment lies in method and not in anything else, and that nobody else before me distinguished all these things in order. And yet the path was, in fact, discovered by Hippocrates. It was Hippocrates who first of all wrote these very things that I now intend to go through as indicating the treatment. But, because he was the first to discover them, he neither established the proper order for all of them, nor determined the worth of each of the indicators precisely. And he left out some distinctions between them, and explained the majority without clarity due to the ancient [predilection for] brevity of speech. Moreover, all in all, he taught very little about combined conditions. In short, it seems to me the whole road to the cures was cut short and certainly requires attention to the end, just as we now see in the land some of the roads of the ancients which have some part that is muddy, or rock-strewn, or full of thorns, or laboriously steep, or with a dangerous incline, or full of wild animals, or impassable due to the great size of the rivers, or long, or rough.

The famous Trajan was, of course, the man who re- 633K paired all the roads in Italy that were like this, paving over

καὶ πηλώδη μέρη λίθοις στρωννύς, ἢ ὑψηλοῖς ἐξαίρων
χώμασιν, ἐκκαθαίρων δὲ τά τε ἀκανθώδη καὶ τραχέα
καὶ γεφύρας ἐπιβάλλων τοῖς δυσπόροις τῶν ποταμῶν·
ἔνθα δ᾽ ἐπιμήκης οὐ προσηκόντως ὁδὸς ἦν, ἐνταῦθα
σύντομον ἑτέραν τεμνόμενος· ὥσπερ καὶ εἰ δι᾽ ὕψος
λόφου χαλεπή, διὰ τῶν εὐπορωτέρων χωρίων ἐκτρέ-
πων· καὶ εἰ θηριώδης ἢ ἔρημος, ἐξιστάμενος μὲν
ἐκείνης, ἐφιστάμενος δὲ εἰς τὰς λεωφόρους, ἐπανορ-
θούμενος δὲ καὶ τὰς τραχείας. οὔκουν χρὴ θαυμάζειν
εἰ μαρτυροῦντες Ἱπποκράτει τὴν εὕρεσιν τῆς θεραπευτικῆς μεθόδου γράφειν ἐπεχειρήσαμεν αὐτοὶ τήνδε
τὴν πραγματείαν. οὐ γὰρ ὡς οὐδ᾽ ὅλως εὑρεθείσης
αὐτῆς, ἀλλ᾽ ὡς δεομένης ὧν ὀλίγον ἔμπροσθεν εἶπον,
ἐπὶ τήνδε τὴν συγγραφὴν ἧκον οὐδένα τῶν πρὸ ἐμοῦ
συμπληρώσαντα τὴν μέθοδον εὑρών. ἔνιοι μὲν γὰρ
οὐδ᾽ ἔγνωσαν ὅλως αὐτήν, ἔνιοι δὲ γνόντες οὐκ ἠδυνή-
θησαν προσθεῖναι τὸ λεῖπον· εἰσὶ δ᾽ οἳ καὶ κατα-
κρύψαι καὶ συσκιάσαι προείλοντο καὶ ἀφανῆ ποιῆσαι |
634K   παντάπασιν· οἵτινες δ᾽ εἰσὶν οὗτοι προϊόντος ῥηθήσε-
ται τοῦ λόγου. νυνὶ δ᾽ ὅπερ ὑπεσχόμην ἤδη ποιήσω
πάντας ἑξῆς ἐκθήσομαι τοὺς θεραπευτικοὺς σκοπούς.

9. Ἀρχὴ δ᾽ αὐτῶν εἰς σαφήνειαν χρήσιμος ἀνάμνη-
σις ὧν ἐν τῇ πρὸ τούτου γράμματι διῆλθον, ὑπὲρ τῶν
ἐφημέρων πυρετῶν διαλεγόμενος. ἐφαίνετο γὰρ ἐπ᾽
ἐκείνων ἔνδειξις ἡ πρώτη μὲν καὶ ὡς ἂν εἴποι τις κυ-
ριωτάτη τὴν διάθεσιν τοῦ νοσοῦντος ἐκκόπτειν, ὥσπερ
ἐπὶ τῶν ὑγιαινόντων ἐδείχθη φυλάττειν. ἀλλ᾽ ὥσπερ
τηρεῖται διὰ τῶν ὁμοίων, οὕτως ἀναιρεῖται διὰ τῶν

the wet and muddy parts with stones, raising them up with high banks, removing things that were thorny and sharp, and throwing bridges over rivers that were difficult to cross. Where a road was inappropriately long, there he cut another one that was short, just as also, if a road was difficult due to a high ridge, he redirected it through more readily traversable places. If it was infested with wild animals or desolate, he altered it, establishing highways and repairing the rough roads. You shouldn't be surprised, then, that although I bear witness to Hippocrates' discovery of the method of medicine, I myself turned my hand to writing this particular treatise. I came to this book, as I said a little earlier, not because the method itself was entirely undiscovered, but because it was lacking in some respects, since I found that none of my predecessors had completed the method. Indeed, some did not know it at all, while those who did know it were unable to add what was lacking. There are some who deliberately choose to hide, obscure and conceal everything; who these people are will be   634K
mentioned as my discussion proceeds. Now I shall do as I already promised—I shall set out in order all the therapeutic indicators.

9. A beginning for these, useful for clarity, is recollection of those things I went over in detail in the book prior to this one when I discussed the ephemeral fevers. For in those, the primary, and one might say most important, indication appears to be the eradication of the condition of the person who is sick, as it was shown to be the preservation of the condition in those who are healthy. But just as it [the capacity] is preserved through similars, in the same

ἐναντίων. συνενδείκνυσθαι δ' ἐλέγομεν εἰς τὴν τῶν
ἰαμάτων εὕρεσιν αὐτήν τε τοῦ σώματος τὴν κρᾶσιν
ἅμα τοῖς ἔθεσι καὶ τὴν χώραν καὶ τὴν ὥραν καὶ τὴν
κατάστασιν. ἐφ' ὧν δὲ νοσημάτων ἐξαίρετός τις ἐν ἑνὶ
μορίῳ τοῦ ζῴου διάθεσις ἐγένετο, καθάπερ ἐπὶ τῶν
ἐγκαύσεων ἐν τῇ κεφαλῇ, λαμβάνεσθαί τινα κἀκ τοῦ
μέρους τούτου ἔνδειξιν. οὐ μὴν ἐν ἐκείνῳ γε τῷ γράμματι περὶ τῆς δυνάμεως ἢ τῆς τὴν διάθεσιν ἐργαζομένης αἰτίας εἴρηταί τι, διὰ τὸ τὴν μὲν δύναμιν οὐδεμιᾶς
635K ἐξαιρέτου δεῖσθαι προνοίας ἐν τοῖς | ἐφημέροις πυρε-
τοῖς, ὡς ἂν ἐρρωμένην τοὐπίπαν· διάθεσίν τε οὐδεμίαν
εἶναι τὸν ἀκριβῶς ἐφήμερον πυρετὸν ἤτοι γεννῶσαν ἢ
αὐξάνουσαν, ἑκατέρως γὰρ ἐγχωρεῖ λέγειν. ἀλλὰ νῦν
γε περὶ πρώτης διελθὼν τῆς δυνάμεως ἑξῆς ἐπὶ τὰς
γεννώσας αἰτίας τὸν πυρετὸν ἀφίξομαι τῷ λόγῳ.

10. Διοικοῦσι τὸ ζῷον, ὡς ἐν τοῖς Περὶ τῶν Ἱπποκράτους καὶ Πλάτωνος δογμάτων ἐπεδείκνυτο, τρεῖς
ἑτερογενεῖς ἀλλήλων δυνάμεις, ὥσπερ ἐκ πηγῆς τι-
νος[11] ἰδίας ἑκάστη παντὶ τῷ σώματι διανεμόμεναι.
καλεῖ δ' αὐτὰς ὁ Πλάτων ψυχάς, ἰδίαν ἑκάστης εὑρίσκων τὴν οὐσίαν. ἔστι δ' ἡ μέν τις αὐτῶν εἰς τὸ
τρέφεσθαι τὸ ζῷον ἀναγκαία καὶ κοινὴ πρὸς τὰ φυτά,
τὴν μὲν οἷον πηγὴν ἔχουσα τὸ ἧπαρ, ὀχετοὺς δ' ἐξ
αὐτῆς εἰς ὅλον τὸ σῶμα διασπειρομένους τὰς φλέβας·
ἣν εἴτ' ἐπιθυμητικήν, εἴτε φυσικήν, εἴτε θρεπτικὴν
ὀνομάζοις, οὐδὲν διοίσει, καθάπερ οὐδὲ εἰ ψυχὴν ἢ

───

11 K; τινος om. B

502

way it is destroyed through opposites. I said that the *krasis* of the body jointly indicated the actual discovery of the cures along with the customs, the place, the season, and the climatic conditions. In those who are diseased, what stands out is some condition that has occurred in one part of the organism—for example, in those with heatstroke in the head—and an indication is taken from this part. Nothing was in fact said in that book about capacity or about the cause producing the condition because there is no pressing need to give forethought to the capacity in the ephemeral 635K fevers, inasmuch as it would be strong in every respect. It is possible to say, in another way, that there is no condition that generates an ephemeral fever exactly or that exacerbates it. But now, when I have gone over the primary capacity, I shall next come to the discussion of the causes generating the fever.

10. There are three capacities, different in class from each other, which govern the organism, as was shown in my treatise *On the Doctrines of Hippocrates and Plato*.[8] Each is distributed, as if from some specific source, to the whole body. Plato calls these "souls," identifying the specific substance of each. One of them, essential to nourishing the organism and in common with plants, is the liver which is as it were the fount, and the veins are the conduits distributing from this to the whole body. Whether you call this "appetitive," or "physical" or "nourishing" will make no difference, just as it makes no difference if you call it "soul" or

[8] See *De placitis Hippocrates et Platonis*, V.181–805K, and in particular V.506 and 532K.

δύναμιν. ἑτέρα δ' οὐ μόνον ὡς φυτοῖς ἡμῖν ἢ ζῶσιν,
ἀλλὰ καὶ ὡς ζῴοις ὑπάρχουσα ψυχή, κατὰ τὴν καρδίαν ἵδρυται πηγή τις οὖσα καὶ ἥδε τῆς ἐμφύτου
636K θερμασίας· ὀχετοὶ δὲ καὶ ταύτης | τῆς πηγῆς αἱ ἀρτη-
ρίαι, καλουμένης καὶ αὐτῆς ὀνόμασι πολλοῖς· καὶ γὰρ
δύναμις ζωτικὴ καὶ δύναμις θυμοειδὴς καὶ ψυχὴ ζωτικὴ καὶ ψυχὴ θυμοειδὴς ὀνομάζεται. τρίτη δ' ἐν ἐγ-
κεφάλῳ καθίδρυται ψυχὴ λογική, τῶν κατὰ προαίρεσιν ἐνεργειῶν ἅμα ταῖς αἰσθήσεσιν ἐξηγουμένη,
χρῆται δὲ μορίοις καὶ ἥδε καθάπερ ὀχετοῖς τισι τοῖς
νεύροις, αἴσθησίν τε καὶ κίνησιν ἐπιπέμπουσα δι'
αὐτῶν τῷ ζῴῳ παντί. τὸ μὲν δὴ ταύτας φυλάττειν τὰς
δυνάμεις οὐδὲν ἄλλο ἐστὶν ἢ τὸ φυλάττειν τὴν ζωήν·
ἐδείχθη γὰρ ὅπως ἀπολομένης ἡστινοσοῦν ἐξ αὐτῶν
μιᾶς ἀναγκαῖόν ἐστι καὶ τὰς λοιπὰς συναπόλλυσθαι.
καὶ διὰ τοῦτο κατὰ τὴν ὑγιεινὴν πραγματείαν ὁ σκοπὸς ἡμῖν οὗτος ἦν. ἐπὶ δὲ τῆς θεραπευτικῆς ὁ μὲν
πρῶτος σκοπὸς ἀνάλογον τῷ κατὰ τὴν ὑγιεινὴν ἡ τοῦ
νοσήματος ἀναίρεσίς ἐστιν· οἷον γὰρ ἐπ' ἐκείνης τὸ
φυλάξαι τὴν ὑγείαν ὑπάρχει, τοιοῦτον ἐπὶ ταύτης
ἐκκόψαι τὴν νόσον.

ἐν δ' ἀμφοτέραις αὐταῖς κοινὸν ἡ φυλακὴ τῆς ζωῆς·
ἐπὶ μὲν τῆς ὑγιεινῆς πραγματείας ἑπομένη τῇ φυλακῇ
τῆς ὑγείας· ὃ γὰρ ἂν ὑγείας ἕνεκα πράττηται, τοῦτ'
637K εὐθέως ἐστὶ καὶ τῆς ζωῆς φυλακτικόν· | ἐπὶ δὲ τῆς
θεραπευτικῆς οὐκέτι, διὰ τὸ τὰ λυτικὰ τῶν παρὰ
φύσιν ἐν ἡμῖν διαθέσεων οὐκ ἐξ ἅπαντος φυλάττειν
τὴν ζωήν· ἔνια γὰρ ἐξ αὐτῶν ἐστι τοιαῦτα ταῖς δυνά-

"capacity." Another, which is not only in plants and animals but also in us as living creatures, is a soul seated in the heart, and this is also a source of the innate heat. The arter- 636K ies are the conduits of this fount which is called by many names, for it is termed vital capacity or passionate capacity or vital soul or passionate soul. The third, which is the rational soul, is seated in the brain and governs the functions of volition along with those of sensation. This uses parts including certain nerves as conduits, sending via these sensation and movement to the whole organism. Certainly, to preserve these capacities is nothing other than to preserve life. For it was shown that, if there is destruction of any one of these whatsoever, of necessity those remaining are destroyed with it. And on this account, this was my objective in the work *On the Preservation of Health*.[9] In therapeutics, the primary objective is the elimination of disease, which is analogous to that relating to health, for in the latter it is the preservation of health but in the former it is the eradication of disease.

The preservation of life is the one thing common to both. In the work on health, life follows the preservation of health because whatever is done for the sake of health is also immediately preserving of life. In the work on treat- 637K ment this is no longer [the case] because those things that dispel abnormal conditions in us do not, in every instance, preserve life, for some of these agents are such in their po-

[9] *De sanitate tuenda*, VI.1–452K.

μεσιν, ὥστε ἀμετρότερον ἢ ἀκαιρότερον αὐτοῖς χρησαμένων ἀπόλλυσθαι τὴν ζωήν. ἐπὶ γοῦν τῆς φλεβοτομίας, ἣν ὀλίγον ἔμπροσθεν ἄχρι λειποθυμίας ἐλέγομεν χρῆναι ποιεῖσθαι, χάριν τοῦ σβέσαι τὴν φλόγα
τῶν ἐπὶ στεγνώσει συνόχων, οὐ σμικρόν τι τὸ βλάβος
εἰκὸς ἀκολουθήσειν, εἰ μὴ κατὰ τὸν προσήκοντα καιρὸν ἢ τὸ δέον ἀποτελεσθείη μέτρον.

δύο οὖν ἀνθρώπους ἀπολλυμένους εἶδον ἐν αὐταῖς
τῶν ἰατρῶν ταῖς χερσί, λειποθυμήσαντας μέν, ἀνακο-
μισθέντας δ' οὐκέτι. πολλοὶ δὲ εἰ καὶ μὴ παραχρῆμα
διεφθάρησαν, ἀλλ' ἐξ ὑστέρου γε διὰ τὸν τῆς δυνάμεως κάματον· οὓς εἴ τις ἐκένωσεν ἄνευ τοῦ καταλῦσαι
τὴν δύναμιν, οὐκ ἂν ἀπώλοντο. καὶ μέν γε καὶ εἰς
νόσον ἔνιοι μακρὰν ἐξέπεσον, ἐπὶ κενώσεσιν ἀμέτροις
ἐκλυθείσης τῆς δυνάμεως. ἄλλοι δ' εἰς τὸν ἐφεξῆς
βίον ἅπαντα τὴν κρᾶσιν ὅλην τοῦ σώματος ἔσχον
ψυχροτέραν, οὐ δυνηθέντες οὐκέτ' ἀνακαλέσασθαι τὴν
638K ἐκ τῆς ἀμέτρου κενώσεως | βλάβην· ἐξ ἧς ψυχρότητος
οἱ μὲν ἄχροοί τε καὶ καχέκται καὶ ῥᾳδίως ἐπὶ παντὶ
βλαπτόμενοι διετέλεσαν, ἄλλοι δ' ἐξ αὐτοῦ τούτου
νοσήμασιν ἑάλωσαν ὀλεθρίοις, ὑδέροις καὶ ὀρθοπνοί-
αις καὶ ἥπατος ἀτονίαις καὶ γαστρός, ἀποπληξίαις τε
καὶ παραπληξίαις. εἰς τοσοῦτον οὖν ἀναγκαίας οὔσης
τῆς δυνάμεως, ἁπάντων δ' αὐτὴν τῶν κενωτικῶν βοηθημάτων ὅταν ἀμετρότερον αὐτοῖς τις χρήσοιτο βλα-
πτόντων, ἐναντιούμεναι δηλονότι πρὸς ἀλλήλαις αἱ
ἐνδείξεις γίγνονται κατὰ τὰ τοιαῦτα τῶν σωμάτων ἐν
οἷς ἡ μὲν διάθεσις ἵνα λυθῇ δεῖται κενώσεως ἀξιολό-

tencies that, when used rather excessively or in too untimely a way, they destroy life. Indeed, in the case of phlebotomy, which I said a little earlier needs to be carried out to the point of fainting for the sake of quenching the fire of the continuous fevers due to stoppage of the pores, no little harm is likely to follow, if it is not done at the appropriate time and with due moderation.

I saw two men perish at the very hands of doctors, these men, when they fainted, were not restored to life again. There are many who, if they are not destroyed immediately, are destroyed later due to exhaustion of the capacity, but who would not have perished if someone had carried out evacuation without dispersing the capacity. And, in truth, there were also some who suffered prolonged illness when the capacity was dispersed by excessive evacuations. Others, who had the whole *krasis* of the body colder for the rest of their lives, were no longer able to have the harm done by excessive evacuation repaired. From the coldness 638K they were pallid and cachectic and continued to be easily harmed by anything, while others were, from the same coldness, seized by fatal diseases—dropsies, orthopneas, weaknesses of the liver and stomach, apoplexies, and hemiplegias. Since the capacity is so essential, and since all evacuating remedies harm it when someone uses them excessively, obviously the indications are in opposition to each other in those bodies in which the condition, in order that it be resolved, requires a major evacuation, while the

γου, φέρειν δ᾿ αὐτὴν ἡ δύναμις οὐ δύναται. πολλάκις δ᾿ οὐδὲν ὑπὸ τῆς κενώσεως ἡ δύναμις βλαπτομένη παρορᾶται τοὐντεῦθεν ὑπὸ τῶν ἀσκέπτων τε ἅμα καὶ προπετῶν ἰατρῶν, ὡς οὐδέποτε οὐδὲν ἐνδεικνυμένη. σφάλλονται δέ, ὡς εἴρηται, διὰ τὸ μὴ γινώσκειν ὡς ἄλλου μέν τινος ἕνεκεν ἡ δύναμις σκοπὸς τῶν ποιητέων ἐστίν, εἰς δὲ τὴν λύσιν τῆς διαθέσεως οὐδὲν ἐνδείκνυται.

ὥσπερ οὖν ἐπὶ τῶν ὑγιαινόντων ἕπεται διὰ παντὸς τοῖς ἕνεκα τῆς ὑγείας πραττομένοις καὶ οὐδὲν ἔστιν 639K εὑρεῖν τὸ φυλακτικὸν ὑγείας | ἀναιρετικὸν τῆς δυνάμεως, οὕτως ἐπὶ τῶν νοσούντων ἐνίοτε μὲν ἕπεται τοῖς ὡς πρὸς τὴν λύσιν αὐτοῦ πραττομένοις, ἐνίοτε δ᾿ ἀναιρεῖται πρὸς αὐτῶν. εἰ μὲν γὰρ ὀλίγον τῆς δυνάμεως εἴη τὸ πλῆθος ἰσχυρότερον, ἐπὶ ταῖς συμμέτροις κενώσεσιν οὐ μόνον οὐδεμία βλάβη τῆς δυνάμεως, ἀλλὰ καὶ ὠφέλεια μεγίστη γίνεται τοῦ βαρύνοντος αὐτὴν ἀρθέντος. εἰ δ᾿ ἵνα μὲν ἡ διάθεσις ἰαθῇ, πολλὴ χρεία τῆς κενώσεως εἴη, καταλύοιτο δὲ ὑπὸ τῆς τοσαύτης ἡ δύναμις, οὐ μόνον αὖ πάλιν οὐδὲν ὀνήσεται πρὸς αὐτῆς ὁ ἄνθρωπος, ἀλλὰ καὶ κινδυνεύσει τὰ μέγιστα. κατὰ μὲν οὖν τοὺς ἐπὶ στεγνώσει συνόχους ἰσχυρᾶς οὔσης ὡς τὰ πολλὰ τῆς δυνάμεως ἀκίνδυνος ἡ κένωσις, ἐν ἑτέροις δὲ νοσήμασιν ἐσχάτως ἐστὶ κινδυνώδης, οἷον ἐπὶ τῆς διαφθορᾶς εἰ τύχοι τῶν χυμῶν· ἐν καιρῷ γὰρ ἂν εἴη μνημονεύειν αὐτῆς ὑπὲρ τοῦ καὶ τοὺς τῆς φλεβοτομίας ἅπαντας σκοποὺς ἐν τούτῳ τῷ βιβλίῳ διορίζειν.

508

capacity is unable to bear this. Often the capacity suffers no harm from the evacuation and is henceforth disregarded by rash and unreflecting doctors as never indicative of anything at all. However, they fall into error, as I said, because they don't know that the capacity, on account of some other factor, is the indicator of what must be done, but indicates nothing regarding the resolution of the condition.

Therefore, just as the capacity continually responds to those things done for the sake of health in those who are healthy, and there is nothing to find that is preserving of health and destructive of capacity, in a similar way, in those who are diseased, the capacity sometimes follows those things that are done for the resolution of the disease and is sometimes destroyed by them. For if the abundance is a little stronger than the capacity, not only does no harm befall the capacity from moderate evacuations, but also a very considerable benefit arises from the removal of what is burdening it. If, in order for the condition to be cured, there is great need of evacuation, while the capacity is dissipated by such an evacuation, not only will the person again derive no benefit from it, but he will also be very greatly endangered. Therefore, in the continuous fevers due to stoppage of the pores, since the capacity is generally strong, evacuation is without danger, whereas in other diseases the danger is extreme, if for example it should happen in the corruption of the humors. It is an appropriate time to mention this for the purpose of defining the objectives of phlebotomy in this book.

639K

τὸ τοίνυν διεφθαρμένον ἀλλότριόν ἐστι τῇ φύσει,
τὸ τοιοῦτον δ' ἐνδείκνυται τὴν ἄρσιν. ἐὰν οὖν ἡ δια-
φθορά ποτε μετὰ δυνάμεως ἀσθενοῦς συμπέσῃ, μάχε-
640K ται τῆς διαθέσεως τὸ βοήθημα | τῇ φυλακῇ τῆς δυνά-
μεως. ἡ μὲν γὰρ διάθεσις ἐνδείκνυται τὴν κένωσιν
ἤτοι διὰ φλεβοτομίας ἢ καθάρσεως· οὐδετέραν δ' αὖ ἡ
ἄρρωστος οἴσει δύναμις. ἐπεὶ τοίνυν ἐν ἁπάσαις ταῖς
τοιαύταις μάχαις ἄπορος ἡ βοήθεια γίνεται, ποτὲ μὲν
ὅλως ἀνίατος ἡ διάθεσις ἔσται, ποτὲ δὲ ἐν χρόνῳ καὶ
μόγις ἰατροῦ μεγάλου τυχοῦσα δύναιτ' ἂν ἰαθῆναι.
χρὴ γὰρ δηλονότι κατὰ τὰς τοιαύτας ἐναντιώσεις τῶν
ἐνδείξεων κατὰ βραχὺ μὲν ἐκκενοῦν τὸ μοχθηρόν,
κατὰ βραχὺ δ' ἀντ' αὐτοῦ τὸ χρηστὸν ἐντιθέναι.
καλεῖται δ' ἐπίκρασις ὑπὸ τῶν ἰατρῶν ἡ τοιαύτη
θεραπεία τῆς κακοχυμίας.

11. Μέμνησο δέ μοι πρὸς τὰ μέλλοντα καὶ τοὺς τῆς
φλεβοτομίας σκοπούς. ἐπειδὴ γὰρ αἵματός ἐστι κένωσις ἡ φλεβοτομία καὶ χρηστὸν τῇ φύσει τὸ αἷμα, χρὴ
δήπου καλῶς αὐτὴν γίγνεσθαι τὸ ἄχρηστον τῇ φύσει
κενοῦσαν. ἄχρηστον δὲ γίγνεται τῇ φύσει τὸ αἷμα
διττῶς· ἢ τὸ μὴ φυλάττον ἀκριβῶς τὴν ἑαυτοῦ ποιότητα, μηδὲ τρέφειν ἔτι δυνάμενον ὡς πρόσθεν χρηστὸν ἢ τὸ πλῆθος τοσοῦτον γενόμενον, ὡς ἤτοι βαρύ-
νειν τὴν δύναμιν, ἢ τείνειν, ἢ ῥήσσειν, ἢ ἐμφράττειν |
641K τάς τε ἀρτηρίας καὶ τὰς φλέβας. ἐν τούτοις μὲν ἡ
φλεβοτομία χρήσιμος, ὡς ἕν τι καὶ αὐτὴ τῶν κενωτικῶν βοηθημάτων· ἐν ἑτέροις δ' ὡς ἀντισπαστικὸν ἢ
παροχευτικόν, ὅταν ὁρμὴν χυμῶν σφοδροτέραν ἤτοι

Accordingly, what is corrupted is alien to nature, and to this extent indicates removal. If at some time the corruption should happen along with a weakened capacity, the remedy of the condition conflicts with the preservation of 640K the capacity. For the condition indicates evacuation, either by phlebotomy or purging, neither of which, on the other hand, a weak capacity will bear. Since, therefore, in all such conflicts, the remedy is ineffectual, sometimes the condition will be altogether incurable, while at other times it can be cured over a period of time and with difficulty, when it happens upon a good doctor. It is necessary, obviously, in such oppositions of the indications, to evacuate what is distressing gradually and to introduce what is useful gradually in response to this. Such a treatment of *kakochymia* is called by doctors *epikrasis* ("tempering").[10]

11. Direct your attention for me to what comes next, including indicators of phlebotomy. Since phlebotomy is the evacuation of blood, and blood is useful in nature, it is of course necessary for phlebotomy to occur properly, evacuating what is useless in nature. Blood is useless in nature in two ways: when it does not exactly preserve its own quality and is no longer able to provide useful nourishment as before, or when the amount of blood has become such that it either weighs down the capacity, or stretches, breaks, or obstructs both the arteries and veins. In these [circum- 641K stances], phlebotomy is useful as it is also one of the evacuating remedies. In other [circumstances], [it is useful] as a revulsive or diversionary [remedy] whenever we either draw a stronger impulse of the humors to its opposite by

[10] A term normally applied to wine. In the Kühn index there is no other reference to its use by Galen.

γε εἰς τοὐναντίον ἀντισπάσωμεν δι' αὐτῆς, ἢ παροχε-
τεύσωμεν εἰς τὰ πλάγια. δεῖται δ' ἀεὶ δυνάμεως ἰσχυρᾶς ἐξαρκούσης τῷ ποσῷ τῆς κατ' αὐτὴν κενώσεως. ἐκ
τούτων οὖν τῶν σκοπῶν ἐνίοτε μὲν ἅπαξ ἀφαιροῦμεν
αἵματος ἐνίοτε δὲ δὶς ἢ τρὶς ἢ πλεονάκις ὡς κἂν τοῖς
ἑξῆς ἔσται δῆλον. ἐν δὲ τῷ παρόντι τοῦθ' ἡμῖν δέδεικται σαφῶς ὅτι τῶν κενωτικῶν βοηθημάτων ἐνίοτε
μὲν ἀλύπως ἡ δύναμις ἀνέχεται, μήτε ὀνιναμένη πρὸς
αὐτῶν μηδὲν ὅ τι καὶ ἄξιον λόγου μήτε βλαπτομένη.
πολλάκις δ' ἤτοι μᾶλλον ἢ ἧττον ὠφελεῖταί τε καὶ
βλάπτεται καὶ ὡς ὁπόταν μὲν ἀξιόλογον γένηται τὸ
βλάβος ἐκ τῶν κενωτικῶν βοηθημάτων, ἐναργῶς φαίνεται τηνικαῦτα ἡ δύναμις ἐνδεικνυμένη τι καὶ αὐτὴ
χρήσιμον, ὥσπερ γε καὶ ὁπόταν ὠφελεῖται σαφῶς.
ὁπόταν δὲ μήτε ὄφελος αὐτῇ τι μήτε βλάβος ἐκ τῶν
τὴν διάθεσιν ἀνασκευαζόντων γένηται, παρορᾶται
μὲν τηνικαῦτα ὡς ἄχρηστος. |

642K   12. Μία δὲ κἀκ ταύτης[12] ἐστὶ διαπαντὸς ἡ τοῦ
συμφέροντος ἔνδειξις, ὥσπερ καὶ τῶν ἄλλων ἁπάντων. κατὰ γὰρ τὴν ἑαυτοῦ φύσιν ἐνδεικνύμενον ἕκαστον ἓν ἐνδείξεται διαπαντός, ὡς ἂν καὶ μίαν ἔχον τὴν
φύσιν. καὶ τοῦτο τὸ ἓν εἰ μὲν ἁπλοῦν εἴη, τὸ τὴν
ἔνδειξιν ποιούμενον ἁπλοῦν ἔσται καὶ αὐτό· συνθέτου
δ' ὑπάρχοντος ἐκείνου καὶ τοῦτο ἔσται σύνθετον. ἔνια
μὲν οὖν ἐνδείκνυται τὴν ἑαυτῶν[13] φυλακήν, ἔνια δὲ τὴν
ἀναίρεσιν· ὁποτέρως δ' ἂν τοῦτο ποιῇ, τὸ μὲν ἁπλοῦν
ἁπλῆν καὶ τὴν ἔνδειξιν ἔχει, τὸ δ' οὐχ ἁπλοῦν οὐχ
ἁπλῆν. διὰ τοῦτο ἐν τοῖς ἔμπροσθεν ἀεὶ τὴν μὲν

means of phlebotomy, or divert it to the sides. However, it always requires a capacity of sufficient strength in proportion to the amount of the evacuation relating to it. From these indicators, then, we sometimes remove the blood all at once and sometimes in two or three sittings, or more often, as will be clear from what follows. For the present, I have shown clearly that with the evacuating remedies, the capacity sometimes remains unaffected, neither being benefited by them nor significantly harmed. Often, however, it is either benefited or harmed more or less and as whenever the harm from the evacuating remedies becomes significant, the capacity, under these circumstances, manifestly indicates something and this is useful, so also is this clearly the case whenever it is benefited. Whenever there is neither benefit nor harm to it from the things that clear away the condition, under these circumstances, it is disregarded as useless.

12. There is always one indication of benefit from the condition itself, just as there also is from all the other factors. For each thing indicating one thing by virtue of its own nature will always indicate it, as it would also have the one nature. And if this one thing should be simple, then that which produces the indication will also be simple itself, whereas if it is compound, then this too will be compounded of that. Thus some things indicate their own preservation and some their removal. However, in whichever way it might do this, what is simple also has an indication that is simple, and what is not simple does not. Because of this, in what has gone before I always said that

642K

---

12 κἀκ ταύτης K; καὶ ἀπ᾽ αὐτῆς B, *recte fort.*
13 B; ἑαυτοῦ K

ἁπλῆν δυσκρασίαν ἁπλῆν καὶ τὴν ἔνδειξιν ἔφαμεν
ποιεῖσθαι, καθάπερ εἰ τύχοι ἐπὶ ψύξεως τὴν θερμό-
τητα, τὴν δ᾽ οὐχ ἁπλῆν οὐδὲ τὴν ἔνδειξιν ἔχειν ἁπλῆν.
τὴν γὰρ θερμὴν καὶ ξηρὰν δυσκρασίαν ἐνδείκνυσθαι
τὴν ἴασιν ἑαυτῆς διὰ τῶν ὑγραινόντων τε καὶ τῶν
ψυχόντων.

13. Ἕξει τοίνυν καὶ ἡ δύναμις ἀεὶ τὴν ἔνδειξιν
μίαν, ἥτις ἐστὶν ἑαυτῆς φυλακή. ἐκ τίνων οὖν φυλαχθήσεται; κατὰ μὲν πρῶτον λόγον ἐκ τοῦ τῆς
643K οὐσίας αὐτῆς τὸ κενούμενον | ἢ ἀλλοιούμενον ἐπανορ-
θοῦσθαι, κατὰ δεύτερον δ᾽ ἐκ τοῦ κωλύειν ἡμᾶς τὰ
κενοῦν ἢ ἀλλοιοῦν τὴν οὐσίαν αὐτῆς δυνάμενα. περι-
λαμβανέσθω δ᾽ ἐν τοῖς ἀλλοιοῦσι βραχυλογίας ἕνεκα
καὶ τὰ τὴν συνέχειαν ἑαυτῆς[14] διασπῶντα. ταῦτ᾽ οὖν
ἀεὶ τῆς δυνάμεως ἐνδεικνυμένης, ἐπὶ μὲν τῶν ὑγιαινόντων, ὡς εἴρηται, συμφωνία τῶν πραττομένων
ἁπάντων ἐστὶν εἰς τὴν διαμονὴν αὐτῆς, ἐπὶ δὲ τῶν
νοσούντων οὐκ ἀεί. κἂν τούτῳ χρὴ σκοπεῖσθαι τὸ
ἐπικρατοῦν ἤτοι κατὰ τὸ μέγεθος ἢ καὶ κατὰ τὴν
ἀξίαν, ὡς καὶ πρόσθεν ἐλέγετο. μέγιστον μὲν οὖν
ἀξίωμα τὸ τῶν δυνάμεών ἐστιν· αὐτὸ γὰρ τὸ ζῆν ἡμῖν
ἐκ τῆς τούτων ὑπάρχει φυλακῆς, ἐπειδὴ καὶ τὴν ζωὴν
ἀναγκαῖον ἤτοι αὐτὰς τὰς δυνάμεις ἢ τὰς ἐνεργείας
αὐτῶν ὑπάρχειν· οὐδὲν δ᾽ ἂν εἴη τῆς ζωῆς πρότερον ἐν
τοῖς ζῴων σώμασιν, ὅτι μηδ᾽ ὑγιαίνειν οἷόν τε χωρὶς
τοῦ ζῆν. ἡ δύναμις οὖν ἁπάντων πρῶτον φυλακτέα
τοῖς ἀνθρώποις ἐστὶν ἐκ τοῦ μήτε κενοῦσθαι τὴν
οὐσίαν αὐτῆς μήτ᾽ ἀλλοιοῦσθαι τοσαύτην ἀλλοίωσιν,

a simple *dyskrasia* creates a simple indication, just as, should it so happen, in the case of a cold *dyskrasia*, the indication is for heat. However, the *dyskrasia* that is not simple does not have a simple indication, for the hot and dry *dyscrasia* indicates the cure of itself through those things that are moistening and cooling.

13. Therefore, the capacity will always have the one indication, which is its own preservation. By what means will it be preserved? The first is by the restoration of what has been evacuated or changed in its own substance, while the second is by our prevention of the things that are able to evacuate or change its substance. For the sake of brevity, let us also include in the things that cause change, those that break up its continuity. Thus, since the capacity always indicates these things, in those who are healthy, there is, as I said, a concord of all the things done toward its continuation, whereas in those who are diseased this is not always so. And even in this it is necessary to consider what prevails, in terms of either magnitude or worth, as I also said before. The capacities have greatest worth; our very life depends on the preservation of these capacities in that either the capacities themselves or their functions are essential to life. However, there is nothing prior to life in the bodies of living things because it is not possible to be healthy without living. Therefore, the capacity is the first of all things to be preserved in people against the evacuation of its substance or a change such that its substance is

643K

---

[14] K; αὐτῆς B, *recte fort.*

ὅση πλησιάσει φθορᾷ. καὶ ταύτης τῆς φυλακῆς ἐστι
μόριον, ὅταν τὴν εὐκρασίαν ἤτοι φυλάττομεν[15] οὖσαν |
644K ἢ οὐκ οὖσαν ἐργαζόμεθα.

μετὰ δὲ τὴν δύναμιν ἡ ἀπὸ τῶν διαθέσεών ἐστιν
ἔνδειξις, ἃς θεραπεύειν πρόκειται, κἄπειτα ἑξῆς τἆλλα. πᾶν οὖν ὅπερ ἂν μάχηται τῇ ἀπὸ τῆς δυνάμεως
ἐνδείξει δεύτερόν ἐστι καὶ ὕστερον ὧν ἐκείνη κελεύει.
τὰ δ᾽ εἰς ταὐτὸ συμβαίνοντα τοῖς ἐκείνῃ συμφέρουσιν
ἀμφοτέρας ἔχει τὰς ψήφους, καὶ τὴν τῶν πρωτείων ὡς
ἂν εἴποι τις καὶ τὴν τῶν δευτερείων. ὥστε ταῦτα μὲν
ἀναμφισβητήτως ποιητέα, τὰ δ᾽ ἐναντιούμενα ποτὲ
μὲν οὐδ᾽ ὅλως ἐστὶ ποιητέα, ἐνίοτε δ᾽ ἐπ᾽ ὀλίγον. εἰ
γὰρ ἡ διάθεσις ἣν θεραπεύειν προαιρούμεθα χρῄζει
κενώσεως, ἄρρωστος δ᾽ ἡ δύναμις ἐσχάτως ὑπάρχει
τῶν τὴν διάθεσιν ἰωμένων, οὐδὲν τῷ κάμνοντι προσ-
άξωμεν ἐν ἐκείνῳ τῷ χρόνῳ παντί, καθ᾽ ὅσον ἀνα-
τρέφομεν τὴν δύναμιν. ὅταν δ᾽ εἰς τοσοῦτον ἰσχύος
ἥκειν αὐτὴν ὑπολάβωμεν ὡς ἤτοι μηδὲν ἢ ὀλίγον ὑπὸ
τῆς κενώσεως βλάπτεσθαι, τηνικαῦτα καὶ πρὸς τὴν
τῆς διαθέσεως ἴασιν ἀφιξόμεθα. πρῶτος μὲν οὖν
ἁπάντων σκοπὸς ὁ τῆς δυνάμεώς ἐστιν· οὐ μὴν ὡς
θεραπευτικός γε πρῶτος, ὅς γε τὴν ἀρχὴν οὐδὲ τῶν
θεραπευτικῶν ἐστιν, ἀλλ᾽ ὡς ἂν εἴποι τις τῶν ζω-
645K τικῶν. | οὐ γὰρ ἵν᾽ ὑγιαίνωμεν, οὐδ᾽ ἵν᾽ ἀπαλλαγῶμεν
νόσων, ἐξ αὐτῆς τι πρώτως λαμβάνομεν, ἀλλ᾽ ἵνα
ζῶμεν. πρόσκειται δ᾽ ἐν τῷ λόγῳ τὸ πρώτως, ὅτι γε καὶ
κατὰ συμβεβηκός ποτε τῆς δυνάμεως ἡ ῥῶσις εἰς τὴν

altered so that it comes near to destruction. And part of this preservation is when we either preserve a *eukrasia* that exists or bring it about if it doesn't exist.   644K

After the capacity, the indication comes from the conditions which it is proposed to treat, and then the other things in order. Everything which might contend with the indication from the capacity is secondary and subsequent to those things which [indication] bids. Things that coincidentally agree with that indication have a vote on both counts —that of things of the first rank, as one might say, and that of things of the second rank. As a result, those things are indisputably to be done, while their opposites are sometimes not to be done at all and at other times to be done gradually. If the condition which we choose to treat requires evacuation, but the capacity is extremely weak in comparison to those things curing the condition, we shall introduce nothing to the patient in that whole time during which we are building up the capacity. When we suppose that it has come to such a degree of strength as either not to be harmed at all or only a little by evacuation, we shall, under these circumstances, approach the cure of the condition. First of all, then, is the indicator of the capacity, but it is not in fact first as a therapeutic indicator. Actually, to begin with, it is not among the therapeutic indicators, but is, one might say, among the vital indicators. We do not   645K
take something primarily from the capacity so that we are healthy or freed from diseases but so that we may live. And what is primarily proposed in the discussion is that the strength of the capacity is sometimes contingently useful

---

15 K; φυλάττωμεν B, *recte fort.*

GALEN

τῆς διαθέσεως ἴασιν ὑπάρχει χρήσιμος. ἀλλὰ τοῦτο
μὲν ἐν τοῖς ἑξῆς ἐπιδειξόμεθα.

14. Ἐπὶ δὲ τὸ προκείμενον αὖθις ἐπανέλθωμεν, εἰς
ὀλίγα κεφάλαια συνόψεως ἕνεκα τὸ πᾶν ἀθροίσαντες.
οἱ θεραπευτικοὶ σκοποὶ παρά τε τοῦ νοσήματός εἰσι
καὶ τῆς τοῦ σώματος κράσεως καὶ τρίτου τοῦ περιέχοντος ἡμᾶς ἀέρος. εἰ δ' ἔμπροσθεν ἕτεροί τινες εἶεν,
ἢ καὶ αὖθις οὐχ οὗτοι μόνον, ἀλλὰ καὶ ἄλλοι ῥηθεῖεν
ἐκ τῆς τούτων τομῆς ἅπαντες φανοῦνται γιγνόμενοι. ἡ
μὲν γὰρ κρᾶσις τοῦ σώματος ἡ κατ' ἐκεῖνον τὸν
καιρόν, ἡνίκα νοσεῖν ἤρξατο, πρὸς τὴν εὕρεσιν τῶν
ἰαμάτων ὑπάρχουσα χρήσιμος, ἐκ φύσεως καὶ ἡλικίας καὶ ἐθῶν καὶ τοῦ περιέχοντος ἡμᾶς ἀέρος γεν-
νᾶται. τὸ περιέχον δ' αὐτὸ τοιοῦτον εἰργάσθη διά τε
τὴν χώραν καὶ τὴν ὥραν καὶ τὴν κατάστασιν. ἐν
τοίνυν ἑκάτερον αὐτῶν ὑπάρχον ἐκ πολλῶν γίνεται
646K τοιοῦτον. ἡ δέ γε διάθεσις οὐχ ὡς | εἰς αἰτίας τέμνε-
σθαι πέφυκεν, ἀλλ' ὡς εἰς διαφοράς· ἤτοι γὰρ ἐν τοῖς
ὁμοιομερέσιν ἐστίν, ἢ ἐν τοῖς ὀργανικοῖς συνίσταται
μορίοις. ἐν μὲν οὖν τοῖς ὁμοιομερέσιν ἤτοι γ' ἁπλῆ τις
ἢ σύνθετος ὑπάρχουσα δυσκρασία, ἐν δὲ τοῖς ὀργανικοῖς περὶ τὴν τῶν ἁπλῶν μορίων γιγνομένη θέσιν, ἢ
διάπλασιν, ἢ ἀριθμόν, ἢ μέγεθος, ὅσαι τε καθ' ἕκαστον τούτων ἐδείχθησαν οὖσαι διαφοραί. ταῦτ' οὖν
ἐστι τρία τὰ πρῶτα τὰ τὴν θεραπείαν ἐνδεικνύμενα, τὸ
περιέχον ἅμα ταῖς διτταῖς τοῦ σώματος ἤτοι διαθέσεσιν ἢ κατασκευαῖς ἥδιόν σοι φάναι, μιᾷ μὲν τῇ
κατὰ φύσιν, ἑτέρᾳ δὲ τῇ παρὰ φύσιν, ἥτις τὸ νόσημά

518

to the cure of the condition. But I shall demonstrate this in what follows.

14. Let me return again to what was proposed, after gathering everything together under a few headings for purposes of recapitulation. The therapeutic indicators are from the disease, from the *krasis* of the body, and third, from the air surrounding us. If, however, there are others that are prior, or again not only these but also others spoken of all will obviously arise from the division of those [stated above]. For the *krasis* of the body at the time when it began to be diseased, which is useful for the discovery of the cures, is generated from nature, age, customs and the ambient air. This air is itself produced by the place, season and climatic conditions. Therefore, each of these things which is one is such that it arises from many [things]. However, the condition is not such as to be divisible into causes naturally but into differentiae, for it subsists either in the *homoiomeres* or in the organic parts. In the *homoiomeres*, any *dyskrasia* is either simple or compound, whereas in the organic parts, those differentiae that pertain to each of the simple parts were shown to involve position, conformation, number or magnitude. There are, then, these three primary things which indicate the treatment—the ambient air together with the two conditions of the body (or states if you prefer to say this), one being normality (accord with nature) and the other abnormality (contrariety to na-

646K

ἐστιν. ὅτι δ᾽ οὐκ ἐνδέχεται αὐτὰ τὰ πρῶτα γένη τῶν
θεραπευτικῶν σκοπῶν πλείω τῶν εἰρημένων τριῶν
ὑπάρχειν ἐξ αὐτῆς τῆς τῶν πραγμάτων οὐσίας ἔνεστί
σοι μαθεῖν. τὸ μὲν γὰρ νόσημα καὶ ἡ ὑγεία διαθέσεις
εἰσὶ τοῦ σώματος, ἡ μὲν τὴν ἀναίρεσιν ἐνδεικνυμένη
τὴν ἑαυτῆς, ἡ δὲ τὴν τήρησιν· τὸ περιέχον δέ, οὗ
χωρὶς οὔτ᾽ ἀνελεῖν οἷόν τε τὴν νόσον οὔτε κτήσασθαι
τὴν ὑγείαν, τουτὶ μὲν οὖν τὸν ὧν οὐκ ἄνευ λόγον
ἐπέχει.

    τὸ νοσοῦν δ᾽ ἀλλοιοῦντες, εἰ μὴ γινώσκοιμεν εἰς ὅ
647K τι | μεθιστῶμεν οὐδ᾽ ὁπότε παυσόμεθα τῆς ἀλλοιώ-
σεως εἰσόμεθα. καθόλου οὖν ἐπὶ πάσης τῆς θεραπείας
ἐν τούτοις τοῖς τρισὶ γένεσιν οἱ σκοποὶ τοῖς πλείστοις
εἰσί. κατὰ μέρος δέ, ὡς ἔμπροσθεν εἴρηται, τινὲς μὲν
τῶν ἁπλῶν μορίων εἰσὶν ἴδιοι, τινὲς δὲ τῶν συνθέτων.
ἓν δή τι τῶν ἐγγινομένων τοῖς ἁπλοῖς ἐστιν ἡ κατὰ τὸ
θερμὸν δυσκρασία. ταύτης δ᾽ εἰς ὅλον ἐκταθείσης τὸ
σῶμα πυρετὸς τοὔνομα. καὶ τοίνυν ἅπαντος πυρετοῦ,
καθόσον ἐστὶ πυρετός, ὑγρότης καὶ ψύξις ἰάματα.
μόνου μὲν οὖν ὄντος αὐτοῦ σκοπὸς ἁπάντων τῶν
βοηθημάτων κοινὸς ὁ νῦν εἰρημένος, ὥστε οὐδὲν ἔτι
δεῖ πρὸς τὴν ἴασιν ἀλλ᾽ ἢ τὰς ὕλας ἐξευρεῖν ἐν αἷς ὁ
σκοπός, εὑρόντας δὲ τὰς ὕλας τὸν μὲν καιρὸν ἐξ αὐτῶν
τῶν τεσσάρων τοῦ πυρετοῦ λαμβάνειν καιρῶν, τὸ
ποσὸν δ᾽ ἐκ τοῦ παραβάλλειν τῷ κατὰ φύσιν τὴν
νόσον. εἰς ὅσον γὰρ ἀποκεχώρηκε τοῦ κατὰ φύσιν, εἰς
τοσοῦτον χρὴ ψύχειν τε καὶ ὑγραίνειν τὸν κάμνοντα.

    τὸ περιέχον δέ, εἰ μὲν ἐναντίαν ἔχοι τῷ νοσήματι

ture), which is disease. It is possible for you to learn from the actual substance of the matters that the primary classes of the therapeutic indicators themselves do not admit of more than the three mentioned. For disease and health are conditions of the body, the one indicating removal of itself and the other preservation. The ambient air, without which it is impossible to take away the disease or acquire health, is therefore the particular feature which provides the *sine qua non*.

When we change what is diseased, if we do not know what we are changing it to, we will not know when to stop the change. In general, then, in every treatment, the indicators are at most in these three classes. Individually, as I said before, some are specific to the simple parts and some to the compound parts. Certainly, one of the things arising in the simple parts is the *dyskrasia* due to heat. When this extends to the whole body, it is termed a fever. Accordingly, for every fever, insofar as it is a fever, moisture and cold are cures. When it is the fever alone, the common indicator of all the remedies is that which was just now spoken of, so that nothing more is needed for the cure except to discover the materials in which the indicator is, and having discovered the materials, to take the appropriate time from the four times of fever themselves[11] and the amount from how much the disease deviates from normal. For it is necessary to cool and moisten the patient to an extent dependent on how far he has departed from normal (an accord with nature).

The ambient air, if it is opposite in *krasis* to the disease,

647K

[11] Presumably the ephemeral, the continuous, the intermittent (including its variants), and the hectic; see Figure 9.

τὴν κρᾶσιν, ἔν τι τῶν βοηθημάτων ἐστίν· εἰ δὲ ὁμοίαν,
ἔν τι τῶν νοσωδῶν αἰτίων. διότι δὲ τὰς μὲν ἄλλας ὕλας
648K ἐκλέξασθαί | τε καὶ φυγεῖν δυνάμεθα, τὴν δ᾽ ἐκ τοῦ
περιέχοντος οὐχ οἷόν τ᾽ ἐστὶ φυγεῖν, ἀλλ᾽ ἀναγκαῖον
χρῆσθαι τῇ παρούσῃ καταστάσει, διὰ τοῦτο καὶ ταύτῃ προσέχειν ἀναγκαζόμεθα, πολλάκις μὲν ὁμολογού-
σῃ τῇ κατὰ τὸ νόσημα ἐνδείξει, ὡς εἴπομεν, πολλάκις
δ᾽ ἐναντιουμένῃ τε καὶ ἀντιπραττούσῃ. λόγον μὲν οὖν
ἕξει τὸ περιέχον, ὡς εἴπομεν, ἐνίοτε μὲν ὕλης ἰωμένης,
ἐνίοτε δὲ ὡς αἰτίου νοσάζοντος. εἰ μὲν γὰρ ὑγραίνεσθαι καὶ ψύχεσθαι δεομένης τῆς νόσου ταῦτ᾽ ἐργάζοιτο, λόγον ὕλης ἔχει θεραπευτικῆς· εἰ δὲ θερμαίνοι
καὶ ξηραίνοι, νοσῶδες αἴτιον γίγνεται. συμφωνοῦντος
μὲν οὖν αὐτοῦ ταῖς θεραπευτικαῖς ἐνδείξεσιν ἀσμενι-
στέον, ἐναντιουμένου δὲ κωλυτέον τὴν βλάβην, ἐπιτεί-
νοντάς τε καὶ παραυξάνοντας τῶν ἄλλων βοηθημάτων
τήν θ᾽ ὑγρότητα καὶ τὴν ψύξιν.

ἐπιτηδευτέον δὲ καθ᾽ ὅσον ἐγχωρεῖ αὐτὸ τὸ περιέχον οἷον χρὴ κατασκευάζειν, ὑγροτέρους τε καὶ ψυ-
χροτέρους οἴκους αἱρούμενόν τε καὶ κατασκευάζοντα·
καθάπερ εἰ καὶ ψυχρὸν καὶ ὑγρὸν εἴη τὸ εἶδος τῆς
νόσου, τὸ θερμαίνειν καὶ ξηραίνειν ἐνδεικνυμένης,
αἱρεῖσθαί τε καὶ κατασκευάζειν οἴκους θερμαίνοντάς |
649K τε καὶ ξηραίνοντας. αἱρεῖσθαι μὲν οὖν λέγω τοὺς ἐξ
ἑαυτῶν τοιούτους, κατασκευάζειν δὲ ἐπιτεχνώμενον, ἐν
μὲν τῷ θέρει καταγείους τε καὶ πρὸς ἄρκτον ἐστραμ-
μένους αὐτούς, ῥαίνοντα συνεχῶς τὸ ἔδαφος αὐτῶν, ἐξ
εὐρίπου τέ τινος αὔραν εἰσπνεῖν ἐπιτεχνώμενον, ἄνθη

is one of the remedies. However, if it is the same, it is one of the causes of the disease. Because we are able to choose or avoid the other materials, whereas it is not possible to 648K avoid the ambient air, but necessary to use the existing climatic conditions, we are, for that reason, compelled to give attention to this which, as I said, often agrees with the indication from the disease, but is often contrary and counteracting. Therefore, the ambient air will, as I said, sometimes have the ground of curative material but sometimes that of disease causation. For if, when the disease needs to be moistened and cooled, it does this, it has the ground of therapeutic material, whereas if the ambient air is hot and dry, it is a cause of disease. So when this is in accord with the therapeutic indications, you must gladly accept it, whereas when it is opposing, you must prevent the harm, increasing in intensity and augmenting the moistness and coldness of the other remedies.

As far as possible you must make a suitable choice about the kind of ambient air that you should provide, choosing and preparing moister and colder houses. Similarly, if the kind of disease is cold and moist, indicating heating and drying, choose and prepare houses that are 649K heating and drying. I mean, choose those that are like this of themselves, and prepare what can be devised. In summer, [prepare houses] that are below ground and turned toward the north, continually sprinkle their floors with water, contrive to blow in air from a ventilator,[12] and strew

---

[12] "Ventilator" is a figurative development from the word for a narrow strait of seawater which is subject to marked changes in current.

τε καὶ βλαστήματα δυνάμεως ὑγρᾶς τε καὶ ψυχρᾶς
ἐπὶ τοῦ ἐδάφους καταρρίπτοντα, χειμῶνος δὲ τἀναντία
τούτων ἐπιτεχνώμενον, ὅταν γε τὸ νόσημα κελεύῃ
ποιεῖν, ὡς εἴρηται· μὴ κελεύοντος δέ, τὴν ἀμετρίαν
μόνην ἑκατέρας τῆς ὥρας κωλυτέον. ὑπὲρ ἧς ἐξ ἀνάγκης μοί τι καὶ μετὰ ταῦτα λεχθήσεται. νυνὶ γὰρ οὐ τὰς
ὕλας τῶν βοηθημάτων ἔγνωκα κρίνειν, ἀλλὰ τὰς μεθόδους ἐκδιδάσκειν.

15. Ἀνάγωμεν οὖν αὖθις ἐπὶ ταύτας τὸν λόγον,
ἐπιδεικνύντες ὅπως ὀλίγοι τε καὶ πολλοὶ γίγνονται
σκοποὶ θεραπευτικοὶ κατὰ διαφόρους τομάς. τρεῖς τε
γὰρ οἷόν τε ποιῆσαι τοὺς πάντας, ὡς ἀρτίως ἐποιήσαμεν, εἰς ὀλίγα κεφάλαια τὸ πᾶν ἀνάγοντες· ἕκαστόν
τε πάλιν αὐτῶν τέμνοντες πολλοὺς ἐξ αὐτῶν ποιή-
650K σασθαι, καθάπερ καὶ τοῦτο ἤδη δέδεικται. | καὶ μέντοι
καὶ αὐτῶν τῶν τμημάτων ἕκαστον οἷόν τε τέμνειν
αὖθις, ἄχρις ἂν ἐπί τι τῶν ἀτμήτων ἀφικώμεθα. καὶ τὰ
πρῶτα δὲ πάντων εἰρημένα τρία δυνατὸν εἰς ἐλάττω
συναγαγεῖν, ὡς εἶναι δύο τὰ πάντα κεφάλαια τῶν τὴν
θεραπείαν ἐνδεικνυμένων σκοπῶν τήν τ᾽ ἐνεστῶσαν
διάθεσιν, ἣν νόσον ὀνομάζομεν, ἥν τ᾽ ἔμπροσθεν εἶχεν, ἣν ὑγείαν τε καὶ τὸ κατὰ φύσιν ἐκαλοῦμεν. οὐ
γὰρ δὴ ἄλλο γέ τι τὸ ὑγιάζειν ἐστὶ παρὰ τὸ τὴν νῦν
οὖσαν ἐν τῷ σώματι διάθεσιν εἰς τὸ κατὰ φύσιν ἄγειν·
ἐναντίων δ᾽ οὐσῶν τῶν διαθέσεων ἀναγκαῖόν ἐστι δι᾽
ἐναντίαν γίγνεσθαι τὴν εἰς ἀλλήλας αὐτῶν μετάπτωσιν. καὶ οὕτως αὖ πάλιν εἰς μόνος ὁ γενικώτατος
ἔσται σκοπὸς τῶν ἰαμάτων ἡ ἐναντίωσις.

flowers and seeds of moist and cold potency all over the floor, but contrive the opposite of these things in winter, at least whenever the disease directs [you] to do [this], as I said. When it does not, you must prevent only what is excessive during each season. I shall, of necessity, say something about this subsequently. Right now my intention is not to evaluate the materials of the remedies, but to teach the methods thoroughly.

15. So let me once more bring the discussion back to these [methods], demonstrating how there are both few and many therapeutic indicators, depending on the different divisions. For it is possible to make three in all, as I did just now, subsuming the whole under few headings and to divide each of these again to make many from them, as has also been demonstrated already. And yet it is possible 650K to divide each of these divisions again until we come to something that is indivisible. It is also possible to bring together to a lesser number the first three of all those mentioned, so that there are two headings in all of the indicators which indicate treatment: the existing condition which we call disease, and what was present prior, which we call health and accord with nature. For producing health is, in fact, nothing other than to bring the condition contrary to nature, which is now present in the body, to an accord with nature. When there are opposite conditions, it is necessary for there to be the change of one to the other brought about by what is opposite. And in this way again, the one and only absolutely general indicator of cures will be opposition.

ἐντεῦθεν οὖν πάλιν ἀρξάμενοι τοὺς κατὰ μέρος
ἐξεύρωμεν. ἐπειδὴ τὸ νοσοῦν ἐναντίον ἐστὶν ὑγιαί-
νοντι, δι' ἐναντίων ἡ ὁδὸς ἐπ' αὐτῶν. πῶς δ' εὑρεθήσε-
ται τὸ ποσὸν ἐν τοῖς ἐναντίοις; οὐ γὰρ ἀρκεῖ δήπου
προσαγαγεῖν τὰ ψύχοντα ταῖς θερμαῖς διαθέσεσιν
ἄνευ τοῦ προσήκοντος μέτρου· κίνδυνος γὰρ ἐλλείπον-
651K τας μὲν ἀπολιπεῖν | τι καὶ τῆς νόσου λείψανον, ὑπερ-
βάλλοντας δὲ γένος ἐναντίον ἀπεργάσασθαι νοσήματος. ὁ γὰρ ἐπὶ πλέον ἢ χρὴ ψύχων τὸ θερμὸν νόσημα
τὴν ὑγιεινὴν ὑπερβὰς συμμετρίαν, ἕτερον ἐργάσεται
νόσημα ψυχρόν. ὅπως οὖν τοῦτο μὴ γίγνοιτο, τὴν
φύσιν ἐπίστασθαι χρὴ τοῦ θεραπευομένου σώματος,
ἵν' ὁπόσον ἀφέστηκεν αὐτοῦ ἡ νόσος τῆς συμμετρίας
γνόντες ἐπιθῶμεν τῷ ψύχοντι βοηθήματι τὸ μέτρον.
ἀλλ' οὐχ οἷόν τε τὴν ἀπόστασιν αὐτῶν ἐξευρεῖν, εἰ μὴ
γνοίημεν εἰς ὅσον ἦν ὑγρόν, ἢ ξηρόν, ἢ θερμόν, ἢ
ψυχρὸν ὅθ' ὑγίαινεν, εἰς ὅσον δ' ἐξέστηκεν ἐν τῷ
νοσεῖν. τὸ μὲν γὰρ ἐπὶ πλεῖστον ἀποκεχωρηκός[16] τοῦ
κατὰ φύσιν ἰσχυρῶν δεῖται τῶν ἐναντίων, τὸ δ' οὐκ ἐπὶ
πλεῖστον ἀσθενῶν.

ἐπὶ πλεῖστον μὲν οὖν ἀφέστηκε τὸ ψυχρὸν νόσημα
τοῦ φύσει θερμοῦ σώματος, ὥσπερ γε καὶ τὸ θερμὸν
τοῦ ψυχροῦ· ἔλαττον δὲ ἀφέστηκε τὸ μὲν θερμὸν τοῦ
θερμοῦ, τὸ δὲ ψυχρὸν τοῦ ψυχροῦ, ὥστ' ἐξ ἀμφοῖν ἡ
ἔνδειξις. ἀλλ' ἐπεὶ τῶν μὲν ἄλλων ἁπάντων βοηθημάτων τῆς ὕλης ἡ ἔκλεξις ἐφ' ἡμῖν ἐστι, τῆς δὲ τοῦ
περιέχοντος οὐκ ἐφ' ἡμῖν, εὔλογον δήπου κἀκ τούτου |
652K τὴν ἔνδειξιν λαμβάνειν. ἀλλ' ἐπεὶ διαγινώσκειν προσ-

Therefore, here again making a start from this point, let us discover these individually. Since being diseased is the opposite to being healthy, the path in the case of these is through opposites. How will the amount be discovered in the opposites? It is not enough, obviously, to apply cooling agents to hot conditions without an appropriate measure because there is a danger that you may leave out and omit   651K
some remnant of the disease which remains, or go too far and bring about the opposite class of disease. If, when cooling a hot disease, you overstep a healthy moderation more than is necessary, this will bring about a different cold disease. Therefore, in order that this does not happen, we must know the nature of the body being treated, so that by knowing how much its disease departs from a balance, we may apply the cold remedy in due measure. But it is not possible to discover the departure [from balance] of these bodies if we do not know to what degree it was moist, dry, hot or cold when healthy, and to what degree it has changed in being diseased. For if it has departed from what is normal to a very great extent, it needs opposites that are strong, whereas if not to a very great extent, then weak ones.

Therefore, the cold disease departs to a very great degree from the body hot in nature just as the hot also does from the cold, whereas the hot departs less from the body hot in nature, and the cold from the cold, so that the indication is from both. But since we have a choice of the material of all other remedies but not of the ambient air, it is reasonable, presumably, to take the indication from this. However, because it is appropriate to diagnose the *kra-*   652K

---

16 B (cf. recessit KLat); προκεχωρηκός K

ἥκει τὰς κράσεις τῶν σωμάτων ἀκριβῶς, ἔστι δὲ πάνυ
χαλεπὸν τὸ τοιοῦτον, πειρατέον μὲν δήπου καὶ ἐκ τῶν
ἐν τοῖς Περὶ κράσεων εἰρημένων[17] ἐξευρίσκειν αὐτάς·
ἐξ ἐπιμέτρου δὲ πρὸς ἐπιστημονικωτέραν διάγνωσιν,
ἀπό τε τῆς ἡλικίας καὶ τοῦ προηγουμένου βίου παντός, ὃς ἐξ ἐπιτηδευμάτων τε καὶ διαίτης συνέστηκε καὶ
τῆς ἐν τῷ τοιῷδε χωρίῳ καὶ ὥρᾳ τῇδε καὶ καταστάσει
τῇδε διατριβῆς. εἰ μὲν γὰρ ἰσχνὸς φύσει καὶ θερμὸς
ἁπτομένῳ καὶ χολώδης καὶ θυμικὸς καὶ δασύς, θερμὸς
οὗτός ἐστι καὶ ξηρὸς τὴν κρᾶσιν, οὐ μὴν ἐξ ἀνάγκης
γε καὶ νῦν, ὁπότε νοσεῖν ὑπήρξατο, τὴν αὐτὴν ἀκριβῶς ἔχει διάθεσιν. ἐγχωρεῖ γὰρ αὐτῷ βίον ἀργὸν ἐν
ὑγροῖς καὶ ψυχροῖς ἐδέσμασί τε καὶ ποτοῖς γεγονέναι·
καὶ πρὸς τούτοις ἔτι καὶ βαλανείοις ἐπὶ τροφαῖς καὶ
ὕπνοις πολλοῖς ἐν ὑγρῷ καὶ ψυχρῷ χωρίῳ χειμῶνος,
ὑγρᾶς καὶ ψυχρᾶς ἐπικρατούσης καταστάσεως, ὥσπερ γε τἀναντία τούτων ἐγχωρεῖ προηγήσασθαι,
τουτέστι τὰ θερμαίνοντα καὶ ξηραίνοντα καθ᾽ ἕκαστον τῶν εἰρημένων γενῶν. ἐξ αὐτῶν οὖν τῶν προηγη-
653K σαμένων | ἀναπληρώσομεν ὅσον ἂν ἐνδεὲς ἔχωμεν εἰς
τὴν τῆς ἐνεστώσης διάγνωσιν κράσεως, ἐξ ἐκείνης δὲ
εὑρεθείσης τὴν τῶν ποιητέων ἔνδειξιν ἕξομεν.

16. Ἐλέγετο γοῦν ἔμπροσθεν ἐπὶ τῆς τοῦ ψυχροῦ
δόσεως ἔν τι καὶ τοῦτο ἐπισκεπτέον εἶναι, μή τι τῶν
μορίων φύσει ψυχρὸν ὂν ἐκ τῆς πόσεως αὐτοῦ βλαβείη μεγάλως. ἆρ᾽ οὖν ἐγχωρεῖ τὴν κρᾶσιν ἑκάστου
τῶν μορίων ἐπὶ τοῦ κάμνοντος ἀνθρώπου, χωρὶς ὧν
εἴπομεν σημείων ἐν ταῖς περὶ τούτων πραγματείαις

sias of bodies precisely—and such a thing is particularly difficult—we must also attempt, presumably, to discover them from what was said in the work *On Krasias (Mixtures)*[13] and, into the bargain, from a more scientific diagnosis from the age and the whole preceding life which is made up of activities and regimen, and from the time spent in the particular place, and the particular season and climatic conditions. For if a particular person is thin in nature and hot to the touch, and is bilious, irascible and hirsute, he is hot and dry in *krasis*. It is not, however, inevitable that now, when he begins to be ill, he has this exact same condition. It is possible for him to have led an inactive life with moist and cold foods and drinks; and besides these things, with bathing after food as well, and a lot of sleep in a moist and cold place during winter when the prevailing climatic conditions are moist and cold, just as it is possible to propose the opposites of these—that is to say, things that are heating and drying in each of the previously mentioned classes. Therefore, from these preceding factors, we shall fill out the deficiencies in the diagnosis of the existing *krasis*, while from the discovery of the *krasis*, we shall have the indication of what is to be done. 653K

16. Anyway, I said before, in the case of a cold drink, there is this one thing you must consider: whether one of the parts, being cold by nature, is greatly harmed by the drink. Is it possible, then, to discover the *krasis* of each of the parts in the sick person apart from the signs I spoke of in the treatises *On Krasias (Mixtures)*? There are, in point

[13] *De temperamentis libri III*, I.509–694K; there is an English translation by P. N. Singer (1997).

[17] B; εἰρημένοις K

ἐξευρεῖν; ἐκεῖνοι μὲν οὖν οὔτε ζητοῦσιν οὔτε ἴσασιν,
ἀλλὰ καὶ κατὰ τὴν παροιμίαν Ἑνὶ καλάποδι[18] πάντας
ὑποδοῦσιν.[19] ἡμεῖς δὴ ζητήσαντες εὕρομεν ὅτι διὰ
πολλῶν καὶ δυσκόλων γνωσθῆναι γνωρισμάτων ἑκάστου τῶν μορίων ἡ κρᾶσις εὑρίσκεται, στοχαστικῶς
μᾶλλον ἢ ἐπιστημονικῶς. ἐὰν τοίνυν φανῇ τι γνωρισμάτων πολλῶν τε καὶ μακρῶν ἀκριβέστερον ἕν,
οὐκ ἂν ἀπορρίψαντες ἐκεῖνα τοῦτο ἀντὶ πάντων ἑλοί-
μεθα; πᾶς, οἶμαι, νοῦν ἔχων ὁμολογήσει. καὶ μὴν οὐδ᾽
ἐγγύς ἐστι τῇ πίστει τὰ γνωρίσματα τῶν κράσεων
654K ἅπαντα συνελθόντα πρὸς τὴν ἐκ τοῦ | ἔθους ἔνδειξιν.
εἰ γὰρ ὑγιαίνων τις ἔπινεν ἀεὶ ψυχρόν, οὔθ᾽ ἧπαρ οὔτε
κύστιν οὔτε κοιλίαν οὔτ᾽ ἄλλο τι τῶν τοιούτων βλαπτόμενος, εὔδηλον ὡς ἰσχυρά τ᾽ ἐστὶν ἅπαντα ταῦτα
καὶ οὐδ᾽ ἂν νῦν ὑπὸ τοῦ ψυχροῦ βλαβείη. μὴ τοίνυν
μικρόν τινα καὶ φαῦλον σκοπὸν εἰς βοηθημάτων
εὕρεσιν ὑπολαμβάνωμεν εἶναι τὸ ἔθος, ὥσπερ οὐδ᾽ εἰς
ὑγείας φυλακήν. οὐδεὶς γὰρ ἀνθρώπων οὕτως ἐστὶν
ἔμπληκτος ὡς μεγάλως βλαπτόμενος ὑπὸ τῆς τοῦ
ψυχροῦ πόσεως εἰς πολυχρόνιον ἔθος ἀγαγεῖν τὴν
χρῆσιν· ἐν γὰρ τῷ μεταξὺ βλαβεὶς ἐναργῶς ὑπ᾽ αὐτοῦ
καὶ νοσήσας ἀποστήσεται πάντως. ὥσπερ οὖν ἄλλα
πολλὰ παρὰ τὰς κοινὰς ἐννοίας ἐνίοις ἐρρέθη τῶν
ἰατρῶν, οὕτω καὶ τὸ μηδεμίαν εὕρεσιν ἢ διαίτης ὑγιεινῆς ἢ θεραπείας νοσούντων ἐξ ἔθους γίγνεσθαι.
φαίνεται γὰρ οὐχ ὥσπερ ἄλλ᾽ ἄττα σμικρὰν καὶ
φαύλην ἔχον τὴν δύναμιν, ἀλλὰ μεγίστην τε καὶ

530

of fact, those who neither seek nor know, but as the proverb says, "They bind everyone on one [bootmaker's] last." Certainly, when we do seek, we find that the *krasis* of each of the parts will be discovered through signs that are many and difficult to recognize, and by guesswork more than scientifically. Therefore, if one of the many and long-lasting signs seems more accurate, would we not reject those, and choose this one before all the others? Anyone in his right mind will, I believe, agree. And yet all the signs of the *krasias* do not come near the indication from custom in terms of reliability. For if someone, when he is healthy, always drinks cold water, and neither the liver, bladder, or stomach, nor any other such [organ] is harmed, it is clear that these [organs] are all strong and would not now be injured by the cold drink. So let us not suppose custom is some minor and trivial indicator in the discovery of remedies, just as it is not in the preservation of health. For there is no man so stupid as to make the use of cold drinks a longstanding custom, if he is greatly harmed by them. If he is clearly harmed by this in between times, when he is sick he will avoid it altogether. Therefore, just like many other things said by some doctors that run counter to common concepts, [there is the statement] that no discovery, either of a regimen for health or of a treatment for those who are diseased, arises from custom. It is obvious that unlike other things, custom does not have some small and trivial

654K

---

18 B; καλόποδι K
19 B; ὑποδεοῦσιν K

κυριωτάτην, ὡς ἂν ἐνδεικνύμενον ἑκάστου τῶν σωμάτων τὴν φύσιν.

ἡμεῖς μὲν οὖν ἐξ αὐτοῦ μεθόδῳ προερχόμενοι, τὴν
διάθεσιν τῶν μορίων εὑρίσκεσθαί φαμεν, εἶτ᾽ ἐξ ἐκεί-
655K νης αὖθις ἐπὶ τὴν ἔνδειξιν ἔρχεσθαι τῶν | ποιητέων.
ὁ δ᾽ ἐμπειρικὸς ὑπερβαίνειν φησὶ τὴν ζήτησιν τῆς
φύσεως, ἐφ᾽ ὃ γὰρ ἂν ὑπ᾽ αὐτῆς ἤχθη, τοῦτο ἔχειν ἤδη
γνῶσιν τοῦ μὴ βλάπτεσθαι τὸ τοῦ κάμνοντος σῶμα
πόσει ψυχροῦ, τοῦτο δ᾽ εἶναι τὸ ἐξ ἀρχῆς ἡμῖν προκείμενον οὐ τῷ γνῶναι τὴν φύσιν. εἶτ᾽, οἶμαι, προσηκόντως ἡμᾶς μὲν ὡς μακρὰν ὁδὸν εἰκῆ περιερχο-
μένους ἐπισκώψειεν· τοῖς δ᾽ ὅλως ἄχρηστον εἶναι
λέγουσι τὸ ἔθος, ὡς μηδὲ τὸν κοινὸν νοῦν ἔχουσιν,
εἰκότως μέμψαιτο. τί ποτ᾽ οὖν φήσεις πρὸς αὐτοὺς
ἐροῦμεν ἡμεῖς οἱ μεθόδῳ τὴν τέχνην συνιστάμενοι; τί
δ᾽ ἄλλο γε ἢ ὅτι τἀληθῆ λέγουσιν ἐπὶ τῶν οὕτως
ἐναργῆ τὴν ὠφέλειαν ἢ βλάβην ἐχόντων ἐθῶν; ἴσασι
γὰρ ἅπαντες αὐτὰ καὶ πρὸ τῆς μεθόδου.

τῶν δ᾽ ἄλλων ἐθῶν ἔνια τῶν κατὰ βραχὺ μεγάλην
ἀθροιζόντων τὴν βλάβην οὐκέθ᾽ οἷόν τ᾽ ἐστὶ δι᾽ ἐμπειρίας ἐξευρεῖν· εἴρηται δὲ περὶ τῶν τοιούτων ἐπὶ
πλέον ἐν τοῖς Ὑγιεινοῖς. ὅθεν εἰ κἂν τῇ τοῦ ψυχροῦ
δόσει χωρὶς τῆς μεθόδου τὸ ἔθος ἱκανόν, οὐδ᾽ ἡμεῖς
δήπου τὴν ἐμπειρίαν ἐκφεύγομεν, ἀλλ᾽ ὅσα περ ἐκεῖ-
656K νοι πάντα ἐξ αὐτῆς | ὠφελούμενοι προστίθεμεν
ἔξωθεν, οἵ γε προστιθέναι δυνάμενοι τὴν ἐκ τῆς τῶν
πραγμάτων φύσεως ἔνδειξιν. ἐν δὲ τῷ παρόντι λόγῳ
τὴν ἀκολουθίαν τῆς μεθόδου φυλάττοντες ὁμολογοῦ-

force, but a large and very important force, as indicating the nature of each of the bodies.

Therefore I say, when we proceed from this by method, the condition of the parts is discovered, and from that in turn, we advance to the indication of the things to be done. 655K The Empirics say they pass over the search for the nature; they said they were led to this partly by already having the knowledge that the body of the patient is not harmed by a cold drink, and partly because this is established in us originally and does not come from a knowledge of nature. Then, I think, they will rightly mock us as needlessly taking the long way around, while they say that for them, custom is altogether useless, and as they do not have a general concept (i.e. a method), it is a reasonable criticism. What, you might ask, will we, who have established the craft by method, say to them? What else, in fact, than what they say is true, since there are clearly customs that have benefit and customs that are harmful? Everyone knows these things prior to method.

There are, however, some of the other customs which, when gradually gathered together, cause great harm, but which are impossible to discover by experience. I spoke about these in greater detail in the work *On the Preservation of Health*.[14] Wherefore, if in the giving of cold apart from method, the custom is sufficient, we do not, of course, shun experience. Rather, we incorporate from without all the things those men find beneficial—at least 656K they are able to apply the indication from the nature of the matters. In the present discussion, although we do defend

14 This is taken to be a general reference to Galen's *De sanitate tuenda*, VI.1–452K, but see particularly VI.361K.

μὲν ὁδῷ μακροτέρᾳ ταὐτὸν εὑρίσκειν, ὃ διὰ βραχυτέ-
ρας ἐνῆν ἐξευρεῖν ἀναλογισμοῦ χωρίς. ἑωρακότες
γάρ, οἶμαι, πολλάκις ἤδη πολλοὺς τῶν καυσουμένων
ἐν πυρετοῖς, ὅταν ἤδη μετρίως ὦσιν οἱ χυμοὶ πεπεμ-
μένοι, παραχρῆμα λύσαντας ἤδη τὸν πυρετὸν ἐπὶ τῇ
πόσει τοῦ ψυχροῦ, κἂν ἀήθεις ὦσιν αὐτοῦ, πολὺ δήπου
μᾶλλον ἐλπίσομεν ἐπὶ τῶν εἰθισμένων ἄνευ πάσης
βλάβης ἔσεσθαι τὴν ὠφέλειαν. ἔχεις οὖν ἤδη καὶ τὸν
περὶ τῶν ἐθῶν λόγον, εἰ καὶ κατὰ τὸ πάρεργον, ἀλλ'
ἱκανῶς διωρισμένον. ἓν γάρ τι καὶ τοῦτό ἐστι τῶν
ἐνδεικνυμένων τὰς ἰάσεις διὰ μέσης τῆς τοῦ νοσήματος φύσεως, ὥσπερ γε καὶ ἡ ἡλικία. καὶ γὰρ καὶ
αὕτη διὰ μέσης τῆς κράσεως ἔχει τὴν ἔνδειξιν, οἷον ἡ
τῶν παιδίων ὑγρὰν καὶ θερμὴν ἔχουσα τὴν κρᾶσιν
ἑτοίμως διαφορεῖται.

17. Πρόσχες γάρ μοι κἀνταῦθα τὸν νοῦν τῇ διαφορᾷ τῶν σωμάτων. ἡ μὲν τῶν λίθων οὐσία μόνιμός
657K ἐστι | διὰ ξηρότητα καὶ ψύξιν· οὕτως δὲ ἡ τοῦ χαλκοῦ
καὶ τοῦ σιδήρου καὶ ἡ τοῦ χρυσοῦ καὶ συνελόντι
φάναι πάντων τῶν γεωδῶν σωμάτων. ἡ δέ γε τῶν
ὑγρῶν καὶ θερμῶν εἰς τὸ περιέχον ἀεὶ σκίδναται
τάχιστα, ποτὲ μὲν εἰς αἰσθητοὺς ἀτμοὺς λυομένη,
ποτὲ δ' εἰς ἀδήλους μὲν τῇ ὄψει, τῷ λόγῳ δὲ γνωστὰς
ἀπορροάς. αὗται μὲν οὖν αἱ φύσεις τῶν σωμάτων οἷον
ἐκ διαμέτρου πρὸς ἀλλήλας ἀντίκεινται· τῶν δ' ἄλλων
ἁπασῶν αἱ μὲν[20] τῇ ξηρᾷ καὶ ψυχρᾷ φύσει πλησι-
έστερον ὑπάρχουσαι δυσδιαφόρητοι μέν εἰσιν, ἀλλ'
ἧττον σιδήρου καὶ λίθου καὶ ἀδάμαντος. αἱ[21] δὲ τῇ

following the method, we concede that we discover the same thing by a longer road, which it is possible to discover by a shorter road without reasoning. For, I believe, often when I saw many of those who were burning up in fevers, whenever the humors were already moderately concocted, I immediately resolved the fever with a cold drink, even if they were not used to this. Much more, of course, will I expect there to be benefit without any harm in those that are accustomed [to cold drinks]. Thus, you now also have the discussion about customs. Even if it is a digression, at least the matter has been defined sufficiently. For custom, too, is one of the things that indicate the cures by way of the nature of the disease, just as the age also is. For this is also an indication by way of the *krasis*—for example, the *krasis* of children, being moist and hot, is readily dissipated.

17. Direct your attention for me here to the variance of bodies as well. The substance of rocks is enduring because 657K of dryness and coldness. The same applies to the substance of copper, iron and gold and, in a word, of all earthy bodies. But the substance of those things that are moist and hot is always very quickly dispersed to the surroundings, sometimes being resolved into perceptible vapors and sometimes into things invisible to sight, but which are known by reason to be flowing away. Therefore, the actual natures of the bodies are, as it were, diametrically opposed to each other. Of all the others, those that are nearer to the dry and cold in nature are difficult to disperse, but less difficult than iron, stone and adamant. Those nearer to the hot and

---

20 B; ἐν μὲν . . . ἐν δὲ . . . K
21 αἱ B; ἐν K

θερμῇ καὶ ὑγρᾷ διαφοροῦνται μὲν ἑτοιμότερον, ἀλλ᾽ ἧττον καὶ αὗται τοῦ θερμανθέντος ὕδατος. χωρὶς μὲν γὰρ θερμότητος οὐχ οἷόν τ᾽ ἐστι διαφορεῖσθαί τι σῶμα, μενούσης αὐτοῦ τῆς οὐσίας ἐσφιγμένης τε καὶ πεπιλημένης· ὑπὸ θερμότητος δ᾽ οὐχ ἅπαν ἀλλ᾽ ὅσον ἐπιτήδειόν ἐστιν εἰς ἀτμοὺς λύεσθαι· τοιοῦτον δ᾽ ἐστὶ δήπου τὸ ὑγρόν. ἡ τοίνυν τῶν παιδίων οὐσία ῥᾷστα διαφορεῖται καὶ σκεδάννυται, διότι πασῶν μέν ἐστιν
658K ὑγροτάτη, ψυχροτάτη²² | δ᾽ οὐδεμιᾶς. ὥστε οὐ χρῄζει κενωτικοῦ βοηθήματος, ἐξ ἑαυτῆς ἔχουσα τὸ κενοῦσθαι σύμφυτον.

γίγνεται δὲ δήπου καὶ διὰ χώραν θερμὴν καὶ ξηράν, ὥραν τε θερινὴν καὶ κατάστασιν ἱκανῶς θερμὴν καὶ ξηρὰν ἡ διὰ τῆς ἐπιφανείας τοῦ σώματος ἀξιόλογος κένωσις, ὥστε²³ καὶ διὰ ταύτην²⁴ τοὺς δεομένους αἵματος ἀφαιρέσεως ἢ οὐδ᾽ ὅλως φλεβοτομήσομεν ἢ ὀλίγον ἀφαιρήσομεν. ἔσονται δὲ καὶ κατὰ τὴν τομὴν τήνδε πολλοὶ σκοποί, ἡ φύσις ἡ ἐξ ἀρχῆς, τὸ ἔθος, ἡ ἡλικία, τὸ χωρίον, ἡ ὥρα τοῦ ἔτους, ἡ κατάστασις, ἔτι τε πρὸς τούτοις ἡ νοσώδης διάθεσις ἣν θεραπεύομεν, ἑτέρῳ τε τρόπῳ πρὸ αὐτῆς ἡ τοῦ κάμνοντος δύναμις. οὐσῶν δὲ δηλονότι πολλῶν τῶν διαθέσεων ἰδίᾳ καθ᾽ ἑκάστην αὐτῶν ἡ ἔνδειξις· ἀπὸ μὲν τῆς θερμῆς ἡ τοῦ ψύχειν, ἀπὸ δὲ τῆς ψυχρᾶς ἡ τοῦ θερμαίνειν, ἑκάστης τε τῶν ἄλλων ὑπεναντίωσις, ὥστε ὀκτὼ δυσκρασιῶν οὐσῶν ὀκτὼ καὶ τοὺς κατὰ μέρος εἶναι σκοποὺς ἁπάντων τῶν ἐπιγιγνομένων νοσημά-

moist in nature are more readily dispersed, but are less easily dispersed than hot water. Without heat, it is not possible for any body to be dispersed because its substance remains bound together and compressed. Not everything is broken down to vapor by heat, but only what is suitable; the same applies, of course, with respect to moisture. Accordingly, the substance of children is very easily dispersed and dissipated because it is the most moist of all, but colder than none. As a result, it has no need of an evacuating remedy, having an innate evacuation of its own.

658K

Of course, a significant evacuation through the surface of the body also occurs due to a hot dry place, a summery season, and weather conditions that are excessively warm and dry. And consequently, in those who need to have blood withdrawn, we either do not carry out phlebotomy at all or we shall withdraw very little [blood]. There will also be, in relation to this cut (i.e., venesection), many indicators: the nature which exists from the beginning, custom, age, place, season of the year, climatic conditions, and in addition to these, the disease condition which we are treating, and taking priority in another way, the capacity of the patient. When there are clearly many conditions, the indication relates to each of them specifically: from heat, the indication is for cooling and from cold it is for heating, and from each of the others an opposite so that, since there are eight *dyskrasias*, there are also the eight indicators individually of all the diseases supervening in simple bodies,

---

22 K; ψυχροτέρα B, *recte fort.*

23 B; ὥσπερ K

24 K; ταῦτα B, *recte fort.*

των τοῖς ἁπλοῖς σώμασιν, ἐξ ἐπιμέτρου δὲ τὸν κοινὸν πρὸς τὰς τῶν ὀργανικῶν διαθέσεις, οὗ μηδὲν ὄνομα ἦν τοῖς ἔμπροσθεν,

659K ἀλλ᾽ ἡμεῖς | ἀναγκασθέντες ἑρμηνεύειν αὐτὸ συνεχείας λύσιν ἐκαλέσαμεν, ἐπεδείξαμέν τε τὸν κοινὸν κἀνταῦθα τῆς θεραπείας σκοπὸν εἶναι τὴν ἔνωσιν. ὅσοι δ᾽ ἐξ αὐτοῦ τούτου πάλιν τεμνόμενοι κατὰ μέρος συνίστανται, διὰ τῶν τεττάρων ὑπομνημάτων ἐδείχθησαν, τοῦ τρίτου τῶν ἐν τῇδε τῇ πραγματείᾳ καὶ τετάρτου καὶ πέμπτου καὶ ἕκτου. νυνὶ μὲν οὖν ἔτι τὸν περὶ τῶν δυσκρασιῶν ἐν τοῖς ἑξῆς ὑπομνήμασι ποιήσομαι λόγον. συντελέσας δ᾽ αὐτὸν ἐπὶ τὰ τῶν ὀργανικῶν μορίων ἀφίξομαι νοσήματα καὶ δείξω κατὰ τὴν αὐτὴν μέθοδον ἐν αὐτοῖς τούς τε πρώτους ἁπάντων σκοπούς, οὓς γενικωτάτους ὀνομάζειν οὐδὲν κωλύει, καὶ τοὺς ἐκ τῆς τούτων τομῆς γενομένους· εἶτ᾽ αὖθις τοὺς ἐξ ἐκείνων, ἄχρι περ ἂν ἐπί τι τῶν μηκέτι ἐγχωρούντων τομὴν ἀφικώμεθα· ταύτην γὰρ ἡμᾶς ὁ Πλάτων ἐδίδαξε μέθοδον ἁπάσης τέχνης συστάσεως. ἀλλὰ περὶ μὲν τῶν ὀργανικῶν, ὡς ἔφην, αὖθις εἰρήσεται.

νυνὶ δ᾽ ἀναμνησθέντες ἐν τίνι μέρει τῆς προκειμένης πραγματείας ἐσμέν, ἐπιθῶμεν ἤδη τῇ παρούσῃ διεξόδῳ τὴν προσήκουσαν τελευτήν. ἐν δή τι γένος

660K ἐστὶ νοσήματος ἐγγινόμενον τοῖς | ὁμοιομερέσι καὶ ἁπλοῖς ὀνομαζομένοις τοῦ ζῴου μορίοις, ὃ καλεῖται δυσκρασία. ταύτης τῆς δυσκρασίας ἐνίας μὲν ἐδείξαμεν ἀλλοιουμένων τῶν σωμάτων κατὰ ψιλὰς ποιότητας, ἐνίας δὲ μεθ᾽ ὕλης τινὸς εἰς αὐτὰ κατασκηπτού-

and in addition, what is common to the conditions of organic bodies but was not named by my predecessors.

But when I was forced to explain it, I called it "dissolu-   659K
tion of continuity," and showed the common indicator of
treatment here to be union. Those indicators that arise
from this itself (i.e. dissolution of continuity) individually
were shown in the four treatises and books three, four, five
and six of this work.[15] Therefore, I shall now further pursue the discussion of the *dyskrasias* in the books that follow. When I have completed this, I shall come to the diseases of the organic parts, and I shall demonstrate by the
same method the primary indicators of all in these, which
nothing prevents me calling the most generic, and those
arising from the division of these, and then again, from the
division of those, until I reach the point where no further
division is possible. For Plato taught us that this was the
method for every existing craft.[16] But I shall speak again
about the organic [parts], as I said.

Now, bearing in mind what stage I am at in the treatise
before me, let me put a suitable end to the present passage. There is one class of disease that happens to the   660K
homoiomeres and what are termed the simple parts of the
organism. It is called a *dyskrasia*. I showed that some instances of this *dyskrasia* are brought about when bodies
undergo change in their "bare" qualities, while some instances are brought about by an inflow of some material

[15] The four treatises are presumably those on the differentiae
and causes of diseases and symptoms; see VI.831–80 and VII.1–
272, and I. Johnston (2006).

[16] See Plato, *Phaedrus*, 266b and 273d.

σης ἀποτελεῖσθαι, γίγνεσθαι δ᾽ ἑκατέρας αὖ πάλιν
ἤτοι περὶ ἓν ἢ πλείω μόρια τῶν ζῴων ἢ καὶ σύμπαν τὸ
σῶμα. διελθόντες οὖν ἐν τῷ τῆσδε τῆς πραγματείας
ἑβδόμῳ γράμματι τὰς περὶ τὰ μόρια συνισταμένας
δυσκρασίας ἀλλοιουμένων μόνων τῶν ποιοτήτων,
ἑξῆς ἐν δυοῖν τούτοιν ἐπὶ τὰς εἰς ὅλον ἐκτεταμένας τὸ
ζῷον ἀφικόμενοι περὶ πρώτων ἔγνωμεν διαλεχθῆναι
τῶν πυρετῶν. ἐπεὶ δὲ καὶ τούτων αὐτῶν ἔνιοι μὲν ἔτι
γίνονται, τῆς ἐργαζομένης αὐτοὺς αἰτίας ἐν τῷ σώματι περιεχομένης, ἔνιοι δ᾽ ἐγένοντο μέν, οἴχεται δ᾽
αὐτῶν τὸ ποιῆσαν αἴτιον, ἄμεινον ἐδόκει τὸν λόγον ἐπὶ
τούτους πρώτους ἄγειν, ἁπλῆν τὴν πρώτην ἔνδειξιν
ἔχοντας, εἶθ᾽ ἑξῆς αὐτῶν ἅψασθαι τῷ λόγῳ κἀκείνων
τῶν πυρετῶν ὧν ἐν τῷ σώματι τὸ ποιοῦν ἐστιν αἴτιον.

into these [qualities], and again, that each in turn may occur involving one or many parts of the organism or the whole body. Therefore, since I went over the *dyskrasias* involving the parts when there is a change of the qualities alone in the seventh book of this treatise, and in the next two books, those *dyskrasias* which extend to the whole organism, I realize I have come to the point where the primary fevers are to be discussed. Because some of these same fevers are still occurring, when the cause bringing them about is contained in the body, while some have occurred and their effecting cause has gone, it seemed better to direct the discussion to the latter first, since they have a simple, primary indication, and then in turn, to touch on in the argument those fevers of which the effecting cause remains in the body.